METHODOLOGY IN SYSTEMS MODELLING AND SIMULATION

METHODOLOGY IN SYSTEMS MODELLING AND SIMULATION

Proceedings of the Symposium on
Modelling and Simulation Methodology,
held at the Weizmann Institute of Science,
Rehovot, Israel, August 13-18, 1978

edited by

BERNARD P. ZEIGLER (editor in chief)
Weizmann Institute of Science
Israel

MAURICE S. ELZAS
Wageningen Agricultural University
The Netherlands

GEORGE J. KLIR
State University of New York
U.S.A.

TUNCER I. ÖREN
University of Ottawa
Canada

1979

NORTH-HOLLAND PUBLISHING COMPANY – AMSTERDAM · NEW YORK · OXFORD

ISBN: 0 444 85340 5

Publishers:

NORTH-HOLLAND PUBLISHING COMPANY
AMSTERDAM • NEW YORK • OXFORD

Sole Distributors for the U.S.A. and Canada:

ELSEVIER NORTH-HOLLAND INC.
52 VANDERBILT AVENUE, NEW YORK, N.Y. 10017

Library of Congress Cataloging in Publication Data

Symposium on Modelling and Simulation Methodology,
    Weizmann Institute of Science, 1978.
    Methodology in systems modelling and simulation.

    Includes index.
    1. Digital computer simulation--Congresses.
2. Mathematical models--Congresses. I. Zeigler,
Bernard P., 1940-   II. Rehovot, Israel. Weizmann
Institute of Science.
QA76.9.C65S94 1978   001.4'24   79-15205
ISBN 0-444-85340-5

PRINTED IN THE NETHERLANDS

FOREWORD

The first meeting ever to devote itself entirely to modelling and simulation
methodology was held at the Weizmann Institute of Science in Rehovot, Israel during
August 1978*. The present volume emerged from this Symposium but it is much more
than a record of its proceedings. The book contains papers which were selected
from those delivered to the meeting, and subjected to a rigorous refereeing process
by the editors. A paper was finally accepted only if it presented important ideas
relevant to the main theme in a clear and readable manner. Other papers presented
at the Symposium are listed herein, qualified by index descriptions and may be
obtained from the authors.

Beyond this, the editors' goals were to document the progress of methodology to
date and to facilitate its development in the next five to ten years. Accordingly,
the volume contains a number of integrative features. There are, in the order of
appearance:

- An index of major subject categories referencing the authors who deal with each
  category. This is intended to provide a top down overview of the book's themes
  and their treatment. (It should be noted that no effort has been made to har-
  monize divergent views which are sometimes apparent in the various articles.
  Indeed, such disagreements should be taken as pointers in the direction of research needed to clarify the points involved.) For further details the reader
  may refer to the subject index where major categories are further refined in an
  order preserving manner.

- A definition of modelling and simulation methodology given in analogy to design
  methodology.

- An exploration of future directions open to development in methodology by leading
  experts.

- Reviews of selected works relevant to methodology for the reader who wishes to
  acquire a broader perspective.

The volume is divided into six sections. Section 1 provides four related frameworks for modelling and simulation methodology. Klir's formulation, the most fundamental, views the activities of modelling from a broader systems research standpoint. Ören's and Elzas' frameworks derive their impetus from contemporary simulation practice but go well beyond the latter's conceptual constraints. Indeed,
their "bottom up" approach goes some distance toward meeting Klir's "top down"
formulation. Finally, Zeigler closes this section with a set of principles for
organizing the model/data base component of advanced simulation methodologies.

Section 2 deals with methodological approaches and issues in several multi-
disciplinary contexts. Jones, Innis and Dickhoven discuss computer assisted integrative modelling in the energy, ecology and socio-economic domains respectively.
Roth et al. report on practitioner reactions to proposals for improving federal
management of the modelling processes.

Section 3 presents some state-of-the-art simulation languages and sytems. Such

---

*Symposium on Modelling and Simulation Methodology. August 13-18, 1978, sponsored
by the Weizmann Institute of Science, the Israel Academy of Sciences and Humanities,
the European Research Office of the U.S. Army, the Society for Computer Simulation,
the International Society for Mathematics and Computers in Simulation, the Society
for General Systems Research and IFIPS group on modelling and simulation.

software tools serve as a basis for model description and construction in advanced methodologies (which will make several other forms of model manipulation available as well). The richness of the discrete, continuous, and combined concepts is nicely brought out in the variegated approaches of Davies, Cellier, Hogeweg, Sampson and Carver.

Section 4 deals with a most fundamental aspect of modelling methodology - how should one search through the candidate model space. Milstein, Vansteenkiste and Broekstra take approaches ordered according to increasing breadth of model space considered. Halfon voices concern about the limitation that data quality may place on progress toward acceptable models.

Section 5 discusses modelling in the context of design. Wong, Riddle and Mendelbaum take approaches to computer system and software design which employ formalisms ranging from detailed simulation languages to high level automaton abstractions. Bandler surveys approaches to the design of components produced in large numbers where yield may be maximized by proper choice of nominal values.

Section 6 presents some of the emerging attempts to establish a theoretical foundation for modelling methodology. Kleijnen, O'Neill, Aggarwal and Melamed adopt increasingly structural approaches to model simplification. Averbukch asserts that general systems theory provides the framework for integrating reductionist and holistic world views.

I wish to thank my fellow editors for their help in all aspects of the Symposium organization and book preparation. In particular, Ören is to be credited with the bi-level index, while Elzas and Klir wrote many of the book reviews. My thanks also go to the typing staff at the Virginia Polytechnical and State University and to Sandra Padova at the Weizmann Institute for their forbearance in dealing with many details required to produce this volume.

Bernard P. Zeigler

Rehovot

TABLE OF CONTENTS

- o -

MAJOR CATEGORY INDEX

Acceptability of Models (Wrt Real System, Other Models) :
  KLIR, ÖREN, ZEIGLER, ROTH, BROEKSTRA, HALFON, O'NEILL, AGGARWAL, MELAMED

Bases, Model/Data :
  ZEIGLER, JONES, DICKHOVEN, SAMPSON

Experimental Frame :
  ÖREN, ELZAS, ZEIGLER, JONES, CELLIER, RIDDLE, AGGARWAL

Experimentation (Computer Assisted Simulation, Statistical) :
  ÖREN, ELZAS, DAVIES, SAMPSON, KLEIJNEN

Man-Machine Interface :
  ÖREN, INNIS, DAVIES, SAMPSON

Methodology, Modelling :
  KLIR, ZEIGLER, JONES, INNIS, HOGEWEG, VANSTEENKISTE, HALFON, AVERBUKCH

Methodology, Modelling/Design :
  JONES, WONG, RIDDLE, MENDELBAUM, BANDLER

Methodology, Parameter Identification/Sensitivity :
  ZEIGLER, MILSTEIN, VANSTEENKISTE, HALFON, O'NEILL

Methodology, Structure Identification :
  KLIR, VANSTEENKISTE, BROEKSTRA, O'NEILL

Model Description Language :
  ÖREN, DAVIES, CELLIER, HOGEWEG, SAMPSON, MILSTEIN, WONG, RIDDLE, MENDELBAUM

Model Manipulation :
  ÖREN, ZEIGLER, BROEKSTRA, WONG, RIDDLE, O'NEILL, AGGARWAL, MELAMED

Morphism :
  ZEIGLER, AGGARWAL, MELAMED

Recommendation for Improving Modelling Process :
  ÖREN, DICKHOVEN, ROTH

Simulation Language (Types, Properties) :
  ÖREN, ELZAS, ZEIGLER, DICKHOVEN, ROTH, DAVIES, CELLIER, HOGEWEG, CARVER

Software Tools :
  KLIR, ÖREN, ELZAS, ZEIGLER, DICKHOVEN, DAVIES, CELLIER, SAMPSON, VANSTEENKISTE, WONG, RIDDLE, MILSTEIN

WHAT IS MODELLING AND SIMULATION METHODOLOGY?

In the good old days, people fashioned flint axes, clay pots and straw huts. Later they built cotton gins, steam engines and steel bridges. It took them a few thousand years to realize that you didn't always have to start in directly to work the raw materials into the desired form. Sometimes, you could make a plan - figure out which materials to use, decide on tools before-hand - and save yourself a lot of wasted materials and labor, not to mention occasional catastrophes.

People gave the name design to this exciting idea and as time went on, there emerged classes of artisans specializing in design - architects, engineers, inventors. Of course, these designers were quite restricted in the range of things they could design (it was, and is, unusual, but not impossible to find a dress designer who could design electric circuits). Although they were aware of good designs and bad ones, and good designers and bad ones, it hardly occured to anyone that design itself could be designed.

But then, suddenly, someone invented the computer, and then within a few years the FORTRAN language, and then all hell broke loose. Now, every activity known to man could be improved with the computer's help, all you had to do was write an appropriate program. Alas, things didn't go as smoothly as expected. People weren't very good at writing reliable programs. After some painful experiences, the nature of the problem was preceived. People were writing programs much the same way that stone-age man was chipping out flints axes. Software had to be designed just like anything else in the twentieth century if it were to work. Aside from the paucity of programming experience (a mere decade), program design introduced one major new element. Until now, design was context dependent, as indicated above. But the computer was a general purpose device, and an approach to design was desired (perhaps naively) which would be applicable to any program, independent of its area of application. Suggestions as to proper program design - structured programming, top down refinement, and the rest - brought home the realization that various methodologies (complexes of methods) of design were being proposed.

This somewhat apocryphal history of design methodology helps to put the less familiar notion of modelling and simulation methodology into perspective. First, let us note how modelling and design were interlinked activities right from the start. Design involves manipulation of elements representing real possibilities, i.e., it relies on credible models of reality for its efficacy. (Imagine what architecture would be like if distances on paper could not be reliably correlated to real counterparts, or bridge design without Hooke's Law.) Thus the awareness of design stimulated the investigation of nature and the construction of models - compact, manipulable representations of real phenomena. With time, modelling emerged as an independent activity, but like design, it remained largely bound to discipline. Unlike design, the possibility of a transdisciplinary methodology for modelling (the scientific method) was recognized quite early on by philosophers. However, the philosophical account, abstract and in dispute as it is, is not intended, nor does it provide the tools, to become operational.

Again, the advent to the digital computer radically changed the situation by introducing an unprecedented medium for the symbolic representation of models and the numeric computation of their behavior. Concurrently, the ideal of isolating phenomena was losing ground to the more comprehensive systems view. Models grew in size and complexity, well beyond solution by analytic means, and often beyond computational resources and supporting data as well. If large programs were unreliable, large models were doubly so.

A model, in the computer context, can be viewed as a set of instructions for generating behavior (time sequences of values of variables). As a model of a real system, the model's behavior must be comparable with some behavior of interest in the real system. A computer, under the control of a program which implements to model, may be employed to generate the model's behavior. The process of behavior generation is called simulation. The program implementing a model is called a simulation program and is often written in a specialized simulation language.

It is tempting to identify the model with its program implementation. While ultimately mistaken, this identification suggests an analogy - constructing a model is (like) designing a program. So, we can regard modelling and simulation methodology as a special case of program design methodology. Once the positive features of the analogy have been set forth, exposing its negative aspects will help point out distinguishing characteristics of modelling and simulation methodology.

Figure 1 depicts a version of current views of design methodology. The process starts with specification of the performance desired from the system once it is built, and a description of the available technology. The initial output of the design process is a blueprint (wiring diagram, flow chart, etc.), bearing the claim: Faithfully follow my directions and you will build a system within the available technology and satisfying the given performance specifications. After the blueprint is implemented in the technology, the result is tested to verify that the performance specifications have been achieved. We shall not enter into the various elaborations of the skeleton in Figure 1 (see Selected Readings Relevant to Modelling and Simulation Methodology, this volume). We do wish to point out some observations that carry over to modelling methodology as well.

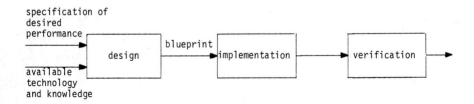

Figure 1. The Design Process

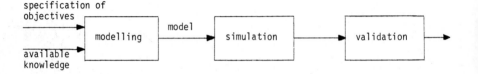

Figure 2. The Modelling Process

1. Work in design methodology is an attempt to rationalize the procedures naturally, but perhaps unconsciously, employed by good designers.
2. This work attempts to decompose the design process and identify component subprocesses.
3. The motivation underlying the work, is to improve the design process, and hence indirectly the design product. The methodologist may hope that this improvement will arise spontaneously as the designer's attention is focused on the existence and necessity of the subprocesses (e.g., the prior specification of desired performance). Or he may go beyond this by forcefully advocating the adoption of certain paradigms over others (e.g., top down versus bottom up design).
4. Recognition of the subprocesses may be followed by the awareness that new tools are required to facilitate their proper execution.

Clearly work in methodology involves a host of commitments and suppositions. Basically, the methodologist is proposing a model of the design process - both in the sense of how it is done and how it should be done. In this he is presupposing a model of human problem solving and psychology. Above all, he is committed to the principle that adherence to a methodology will enhance creativity rather than stifle it. These commitments and suppositions are open to question. If methodology is not to rest on dogma, then it must be exposed to test and modification (as is any other design or model).

The above remarks apply directly to modelling viewed as programming. Indeed, one may expect that simulation program writing will benefit from software engineering developments. But, a model is more than a program and it is time to see where the design and modelling methodologies are similar and where they diverge.

Figure 2 emphasizes the analogy between design and modelling. The following correspondences are evident.

DESIGN		MODELLING
specification of desired performance	———	specification of objectives
available technology and knowledge	———	available knowledge
blueprint	———	model
implementation	———	simulation
verification	———	validation

Modelling, like design, takes in a specification of something desired (objectives vs. performance) and is constrained by something available (knowledge vs. technology). It puts out something abstract (a model vs. a blueprint) which must be interpreted in concrete form (simulation vs. implementation) and checked against the specifications (validation vs. verification).

Hidden in these formal correspondences are differences in detail and in emphasis which result in essential differences in the problems methodologists must consider.

1. Partial versus Complete Specification of Behavior

Both the "objectives" of modelling, and the "desired performance" of design, relate to behavior but the extent which this behavior is known differs. In modelling, there is a real system, some of whose behavior has been observed. The objectives in modelling relate to reproducing this known behavior and predicting behavior not yet seen. In design, it is the real system which is to be built and a complete specification of its behavior is the ideal. More plainly, in designing a system, one wants to dictate what it will do in all circumstances - there should be no surprises. In modelling, one expects to be surprised most of the time!.

Thus the "objectives" of modelling embody an emphasis on knowledge acquisition while the "desired performance" of design embodies knowledge application. (Thus as indicated earlier, design presupposes modelling.)

2. Definite versus Indefinite Life Span of Process

The design of a process takes place in a limited time frame. Modelling of a system, especially a large scale or multifaceted one, does not. That is to say, as part of the specifications of design is a statement of delivery date. Ideally the system should be completely verified by that date. No such date is meaningful for modelling of a multifaceted system - while models of some of its aspects may eventually be validated, no date can be fixed beforehand. Thus modelling methodology must acknowledge living with uncertainty. It must provide means for maintaining, testing, and modifying competing models in partial states of validation over indefinite time spans.

3. Functioning versus Representational Product

A fundamental difference in emphasis is in the status of the final product. In design, this product is a functioning real system. The blueprint serves only in an intermediate role. In modelling, it is the model which is the final product. This means that a model has the freedom to remain at a high level of abstraction that a functioning system has not. This difference in stringency of demand is illustrated by the following juxtaposition:

a) build a speech transduction machine,
b) model human speech transduction.

To build a speech transducer, we have to produce a system that will "listen to" real human speech and make intelligible speech responses. In modelling speech transduction, we may be content with a model which takes an abstracted representation of speech as input and produces an abstracted representation of speech as output. The abstractions employed will depend on objectives: spectral distributions, if we are interested in physiology; phoneme sequences, if we are interested in syntactical processing. Clearly, Turing's test for complete indistinguishibility of machine and human behavior is applicable as test of verification in case a) (design) but would not be meaningful in case b) (modelling).

4. Multiplicity versus Unicity of Product

In modelling, it may be advantageous to develop a multiplicity of models even when operating within the same objectives. Such models offer different views of the same system and each may provide its own insight. For example, the same linear system may be modelled by a transfer function in the time domain or by an equivalent frequency operator in the frequency domain. The first representation may offer more insight into transient response, while the second is more convenient for steady state response.

In contrast, the design process must yield a unique final product. It is true that a number of alternative blueprints may be examined before final decision, but they will always be regarded as competitors. The aspect of complementary, or mutual reinforcement of alternative representation, so important in modelling is necessarily absent in design.

An expanded view of modelling and simulation methodology implied by the preceding contrasts is depicted in Figure 3. Note the addition of model and data bases which together constitute the knowledge available at any time. When new objectives are formulated, relevant models in the model base should be retrievable for use in the modelling process. Such models may be simplified, decomposed, modified and coupled together with new components to form the model intended to meet the given objectives. Similarly relevant data should be retrievable from the data base for use in calibrating the model. Experimentation may be undertaken when the available data for calibration is insufficient, or may be initiated in the subsequent validation phase. Data so acquired may be added to the data base. Likewise, the model, together with a confidence estimate of its validity may be added to the model base. The model and data bases should be organized in such a way as to facilitate the modelling process just described.

This expanded view evidently embodies the characteristic features of the modelling and simulation process enumerated above:

1.   continued and indefinite span of process,
2.   partial data availability and partial model validity,
3.   multiple complementary and competitive representations in the model base,
4.   models at various levels of abstraction, simplification and aggregation.

In sum, work in modelling and simulation methodology attempts to rationalize and structure the modelling process. It hopes to improve the process through advocacy of preferred methods of approach and the development of new tools to facilitate their execution. Such advanced computer assisted methodologies are especially needed in dealing with multifaceted systems, whether in the large scale (ecological, socio-economical, etc.) or in the small scale (brain, cellular, etc). Progress in such domains hinges on adequate tools to greatly amplify the limited human capacity for managing complexity.

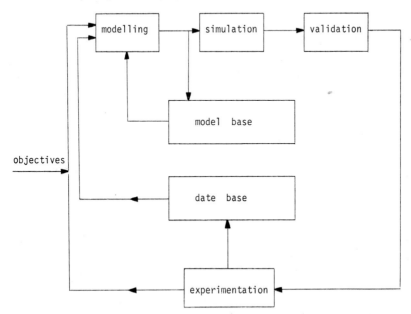

Figure 4. Expanded View of Modelling Process

SECTION I
CONCEPTUAL BASES

METHODOLOGY IN SYSTEMS MODELLING AND SIMULATION
B.P. Zeigler, M.S. Elzas, G.J. Klir, T.I. Ören (eds.)
© North-Holland Publishing Company, 1979

GENERAL SYSTEMS PROBLEM
SOLVING METHODOLOGY

George J. Klir
Department of Systems Science
School of Advanced Technology
State University of New York
Binghamton, New York, U.S.A.

A package of computer-aided methodological tools for systems
modelling is described. The package is viewed as one component
of a larger package of methodological tools for general systems
problem solving. An example from the area of performance
evaluation of aircraft pilots in a flight simulator is used to
illustrate the described methodological tools.

1. Introduction

Various approaches to general systems research have been suggested during
the last two decades or so. One of the approaches, to which this author has become committed since the late 1960's, is oriented toward general systems problem
solving. This approach aims at the characterization of systems problems, their
classification, and the development of mathematical, computational or heuristic
methodological tools for solving the individual classes of systems problems.

A conceptual framework based on the approach has developed during the last
decade. It was recently described in terms of an interactive general systems
problem solver (GSPS) by Cavallo and Klir (1978a). GSPS consists of a taxonomy of
systems and systems problems according to which particular systems and systems
problems are classified into system types and problem types. A package of methodological tools is then associated with each problem type. Problems are basically of two kinds: (i) Given a particular initial system, determine a terminal
system of a specified type, say type t, such that specified requirements (objectives, constraints) are satisfied; a solution of this kind of problem is a particular system of type t. (ii) Given two particular systems, determine some
property, specified by given requirements, of the relationship between the two
systems.

The utility of GSPS is dependent on interaction with users who are specialists in various disciplines of science, engineering, humanities or the arts. The
interaction represents the communication channel between, on the one hand, the
real-world systems which the specialists represent and, on the other, the expanded
set of concepts represented by the linguistic structure of GSPS. The communication is accomplished by a set of query procedures through which the user is
assisted in the identification and formulation of his problem. Furthermore, GSPS
provides the user with lists of minor modifications for some problems which cannot be solved by GSPS, in order to make them solvable in slightly modified forms

or only partially. In case of problems which cannot be solved by the methodological tools available, GSPS may provide the user with useful information such as
references to program libraries, papers or books relevant to his problem.

The kernel of the GSPS conceptual framework is a <u>hierarchy of epistemological</u>
<u>levels of systems</u> which forms a basis for the most fundamental classification of
systems and systems problems (Klir, 1978c).

At the lowest level of the hierarchy, denoted as level 0, a system is defined
by a set of variables, a set of potential states declared for each variable, and
some operational way of describing the meaning of both the variables and their
states in terms of some associated real world attributes and their manifestations.
The name "<u>source system</u>" was chosen for systems defined at this level to indicate
that it is appropriate to view such systems as sources of experimental data.

The set of variables and the set of corresponding attributes are usually
partitioned into two subsets, referred to as <u>basic and supporting variables or</u>
<u>attributes</u>. Aggregate states of all supporting variables form a support set,
usually referred to as a <u>parameter set</u>, within which changes in states of the individual basic variables occur. The parameter set is usually ordered, either
totally (linearly) or partially, and is frequently associated with a metric. The
most frequent instances of supporting variables are time, spacial location and
various sorts of populations of individuals of the same kind (social groups, manufactured products of the same kind, etc.).

Any useful relation in the set of states of each individual variable which
can be justified as a homomorphic image of a relation manifested by the corresponding attribute is usually included in the source system. However, no knowledge regarding a way in which the variables are related is available at the source
system level. Such knowledge is associated with higher epistemological levels;
they are distinguished from each other by the level of knowledge regarding the
variables of the corresponding source system. A higher level system entails all
knowledge of the corresponding systems at any lower level and, at the same time,
it contains some additional knowledge which is not available at the lower levels.

When a source system is supplemented by data regarding the variables, which
are either results of some observation or measurement procedures (as in the problem of empirical investigation) or are defined as desirable data (as in the problem of system design), the new system is called a <u>data system</u>. Data systems are
associated with level 1 in the epistemological hierarchy.

At level 2, systems are defined by an overall relation which is parameter-
invariant and through which the corresponding data can be generated (in a deterministic or stochastic fashion), given appropriate boundary conditions; systems
at this level are called <u>generative systems</u>. At level 3, the overall relation is
implemented by a set of parameter-invariant relations, coupled together in some
way; such systems are called <u>structure systems</u>.

At higher levels, the structure system is allowed to change in time, space or some other parameter set within which the variables are defined. At level 4, the changes are described by a single procedure; such systems are called meta-systems. At level 5, the procedure is allowed to change in the parameter set according to a higher level procedure or metaprocedure; such systems are called meta-metasystems (or metasystems of second order). In a similar fashion, meta-systems of higher orders can be defined.

Regardless of the level at which the system is defined, the basic variables may be classified into input and output variables. Under such classification, states of input variables are viewed as conditions which affect all other events in the system. That is to say, statements describing various events associated with the system are conditional: "If the aggregate state of input variables is s, then ...." Input variables are thus not subject of inquiry but are viewed as being determined by some agent which is not part of the system under consideration. Such an agent is referred to as the environment. It is important that the notion of input variables be not confused with the notion of independent variables.

Systems whose variables are classified into input and output variables are called directed systems; those for which no such classification is given are called neutral systems. This dichotomy of systems holds for each of the epistemological levels.

The types of systems defined at the individual epistemological levels are summarized in Table 1, including symbols for both neutral and directed systems at each level. A somewhat similar hierarchy has been recognized by Zeigler (1974, 1976a,b).

Table 1

Hierarchy of Epistemological Levels of Systems

Level	Name	Neutral System	Directed System
0	Source system	S_{0N}	S_{0D}
1	Data system	S_{1N}	S_{1D}
2	Generative system	S_{2N}	S_{2D}
3	Structure system	S_{3N}	S_{3D}
4	Metasystem	S_{4N}	S_{4D}
5	Meta-metasystem	S_{5N}	S_{5D}

. . . .

Two disjoint classes of important systems problems follow naturally from the outlined hierarchy of epistemological levels of systems:

1. Problems associated with transitions from higher to lower levels or
those in which the level does not change; they are basically problems of deductive
nature in which no new information is produced; they are frequently referred to
as problems of <u>systems analysis</u>.

2. Problems involving transitions from lower to higher levels; these are
problems in which some new information is produced, either by designing a higher
level system which conforms to the given system at a lower level or, in the case
of indigenous systems, by some kind of inductive inference. These two alternatives are referred to as <u>systems design</u> and <u>systems modelling</u>, respectively.

The classification of systems by the <u>epistemological criteria</u> is refined by
additional criteria referred to as <u>methodological criteria</u>. They apply to the
variables or relations involved in the system specifications. For instance,
variables are classified into nominal, ordinal or metric scale variables, well
defined or fuzzy variables, etc.; the parameter set (time, space) is unordered,
partially ordered, linearly ordered, etc.; relations are deterministic (functional) or probabilistic, memoryless or memory-dependent, etc. The epistemological
and methodological classification criteria, which are complementary to each other,
form a basis for defining system types, problem types and, eventually, the whole
framework for GSPS.

GSPS is a basis within which software packages for various problems of systems analysis, design and modelling can be integrated. A particular effort has
been devoted by the author during the last two years to developing a package of
such methodological tools for systems modelling which are applicable to nominal
or ordinal scale variables, well defined or fuzzy variables, totally or partially
ordered parameter sets, and which allow for both deterministic and probabilistic
representations of systems. This methodological package is described in the
paper in more detail. Individual methodological tools are illustrated by a
comprehensive example described in Section 4.

2. Relevant Concepts

Let $A = \{a_i | i \epsilon I_n\}$ be a set of basic attributes chosen by the investigator to
represent an object of interest for some specific purpose; $I_n = \{1,2,\ldots,n\}$ is
an index set determined by the number n of basic attributes. Let A_i denote the
set of potential appearances of basic attribute a_i and let R_{A_i} denote a set of
useful relations (e.g., ordering of some kind) which are recognized in set A_i.

Let $B = \{b_j | j \epsilon I_m\}$ be a set of supporting attributes (time, space, etc.)
chosen by the investigator ($I_m = \{1,2,\ldots,m\}$); let B_j stand for the set of potential appearances of supporting attribute b_j and let R_{B_j} denote a set of useful
relations which are recognized in set B_j. Then, the pair

$$OBJ = (\{(a_i, A_i, R_{A_i}) | i \epsilon I_n\}; \{(b_j, B_j, R_{B_j}) | j \epsilon I_m\})$$

will be referred to as <u>neutral object system</u>.

Let v_i, V_i and R_{V_i} $(i \epsilon I_n)$ denote basic variables, sets of states of these variables and a set of relations defined in V_i, respectively, and let supporting variables, their sets of states and sets of relations in these sets be denoted by w_j, W_j and R_{W_j} $(j \epsilon I_m)$, respectively. Then, the pair

$$IMG = (\{(v_i, V_i, R_{V_i}) | i \epsilon I_n\}; \ \{(w_j, W_j, R_{W_j}) | j \epsilon I_m\})$$

will be referred to as <u>neutral image system</u>.

In order to get a meaningful basis for data gathering and data interpretation, a correspondence between the entities of the object and image systems must be introduced. This is accomplished by: (i) a one-to-one mapping $f_a : A \rightarrow V$; (ii) a one-to-one mapping $f_b : B \rightarrow W$; (iii) a family of mappings $\{g_i : A_i \rightarrow V_k | i, k \epsilon I_n \ ; \ v_k = f_a(a_i)\}$ such that R_{V_k} is a homomorphic image of R_{A_i} for each of the functions g_i; (iv) a family of mappings $\{h_j : B_j \rightarrow W_\ell | j, \ell \epsilon I_m \ ; \ w_\ell = f_b(b_j)\}$ such that R_{W_ℓ} is a homomorphic image of R_{B_j} for each of the functions h_j. Let the mappings (i) - (iv) be called the <u>interface</u> between the object system and image system, and let the symbol INT be used to denote the interface. The collection of an object system, an image system and an interface between them forms a <u>neutral source system</u>. Hence,

$$S_{ON} = (OBJ, IMG; INT)$$

In some experimental arrangements, there exist dependencies among the basic attributes and variables which are solely due to observation or measurement procedures (e.g., one attribute represents an arithmetic average of others). In such cases, the definition of the source system should be augmented by a description of all these observation-bound dependencies.

<u>Directed source systems</u> can be obtained from a given neutral source system by introducing desirable classifications of attributes/variables into input and output attributes/variables.

Let $V = \underset{i \epsilon I_n}{X} V_i$ and $W = \underset{j \epsilon I_m}{X} W_j$ denote the sets of aggregate states of the basic and supporting variables, respectively. Each source system implicitly contains all pairs included in $W \times V$. A meaningful restriction of $W \times V$ to one particular function, say $\delta : W \rightarrow V$, which in the modelling problem is determined by data gathering, constitutes <u>well defined data</u> regarding the variables.

The restriction to one meaningful function by data gathering is frequently not feasible. In such cases, δ may be viewed as a fuzzy relation defined on $W \times V$ or, alternatively, as a set of fuzzy relations, each defined on $W \times V_i (i \epsilon I_n)$; data of this sort are usually referred to as <u>fuzzy data</u> (Klir, 1975).

When a neutral source system is augmented with data (well defined or fuzzy), we obtain a <u>neutral data system</u>. Hence,

$$S_{1N} = (S_{ON}; \delta)$$

Modifications to various <u>directed structure systems</u> is obvious.

Let $v_{i,w}$ denote the state of variable v_i at parameter instance w and let a set of variables $s_k (k \epsilon I_q)$, referred to as <u>sampling variables</u>, be introduced by the equation

$$s_{k,w} = v_{i, \lambda_r(w)},$$

where $s_{k,w}$ stands for states of sampling variable s_k at parameter instance w and λ_r denotes a parameter-invariant <u>translation rule</u> which for any given parameter instance w determines one or several other instances of the parameter.

Let Λ denote the set of all translation rules under consideration, let X denote the set of all basic variables (including possible <u>internal variables</u>, introduced hypothetically), and let the relation

$$M \subset \Lambda \times X$$

specify which translation rules are applied to which variables. Set M is called a <u>mask</u> (Klir, 1969, 1975); sets $M_i \subset M$ such that $(x,y) \epsilon M_i$ iff $x = v_i (i \epsilon I_n)$ are called submasks of the overall mask M, each associated with one basic variable.

Let $s_{k,w} \epsilon S_k$. Then, clearly, $S_k = V_i$ if sampling variable is defined in terms of basic variable v_i.

Given a source system and a mask, a set of sampling variables s_k is uniquely defined by assigning values of the identifier k to the individual elements of the mask.

For example, when the parameter set is linearly ordered (as in the case of time) and represented by the set of positive integers, each translation rule can be described by a simple equation

$$\lambda_r(w) = w + a,$$

where a is an integer. Sampling variables are in this case defined by the equation

$$s_{k,w} = v_{i,w+a}$$

and the mask can conveniently be represented by a matrix in which the rows and columns are labelled with the i's and a's, respectively, and the entries are identifiers k of the sampling variables or blanks. For example, given four variables v_1, v_2, v_3, v_4, a mask can be defined by the following matrix:

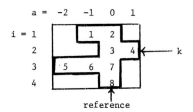

The column labelled a=0 in the mask is called <u>reference</u>. When the mask is applied
to data to read a sample, the reference column coincides with the value of the
parameter at which it is applied (e.g., present time). When the parameter set is
partially ordered, $\lambda_r(w)$ may stand, e.g., for parameter instances which are pre-
decessors (or successors) of w with a particular distance from w.

Let S denote the set of all potential aggregate states of all sampling variables defined for a particular source system by a mask and let B denote the set
of those aggregate states which have actually been observed. Then

$$B \subseteq S,$$

where the equality is reached only when the variables are completely independent
of each other (unconstrained). Set B is referred to as a <u>basic behavior</u>.

To employ the basic behavior for generating data, some order of the parameter set must be chosen. If the parameter set has no natural order, it may be
artificially ordered in some convenient way. Once a generative order in the
parameter set is decided, such relation

$$R \subset B \times B,$$

can be defined whose elements are pairs of successive states of the sampling
variables with respect to this generative order. Relation R is called a <u>state-
transition relation</u>(or ST-relation, in abbreviation). The ST-relation may be
supplemented by probabilities $p(s'|s)$ of a next state s', given a present state
s ($s,s' \in B$; $(s,s') \in R$).

Another way of obtaining a generative relation is to modify the basic behavior by partitioning the involved sampling variables into those which are
generated and those which generate. When the parameter space is ordered, the
partition is unique: the generated variables are defined in terms of translation
rules $\lambda_{r_i} \in M_i$ for each $i \in I_n$ such that $\lambda_{r_i}(w) \geq \lambda_r(w)$ for all λ_r such that $(\lambda_r, v_i) \in M_i$.
Let s_g denote that portion of state $s \in B$ which is associated with the generated

variables and let $\bar{s}_g$ denote the rest of s (s without s_g). Let $s_g \epsilon S_g$ and $\bar{s}_g \epsilon \bar{s}_g$. Then we can define a relation

$$GB \subset \bar{S}_g \times S_g$$

such that $(\bar{s}_g, s_g) \epsilon GB$ iff $\bar{s}_g, s_g$ are portions of state s and $s \epsilon B$. This relation is called the <u>generative behavior</u>; it may be supplemented by probabilities $p(s_g | \bar{s}_g)$.

Systems represented by a behavior or ST-relation are called <u>generative</u> <u>systems</u>. It is well known that the behavior and ST-relation are just two differ- ent forms of the same thing; one can be derived from the other if the mask is properly adjusted (Klir, 1975; Broekstra, 1976a).

Each generative system (say system S_3, either neutral or directed) is defined by a triple

$$S_3 = (S_0; M; G),$$

where S_0 is a source system (neutral or directed, respectively), M is a mask and G is a relation (basic or generative behavior or ST-relation, with or without probabilities) through which desirable data can be generated for appropriate boundary conditions.

At epistemological level 3, each individual system (called a <u>structure sys-</u> <u>tem</u>) consists of a set of generative systems (or, sometime, lower level systems), referred to as <u>subsystems</u> of the larger system and <u>couplings</u> among the subsystems-- variables shared by several subsystems. Such variables indicate a form of data exchange among the individual subsystems as well as the way in which generative relations associated with the individual subsystems are joined or composed.

The structure system can formally be defined in a number of alternative ways (Klir, 1969, 1978c). An alternative is chosen here in which the structure system is defined as a set of subsystems, each characterized in a convenient manner, which are subject to conditions of pair-wise distinguishability and mutual com- patibility. It is assumed here for the sake of simplicity, that all subsystems are defined at epistemological level 2 in terms of their basic behaviors.

Let $\alpha_i (i \epsilon I_r)$ denote identifiers (names, labels) of subsystems of a structure system, let β_i denote all basic variables of subsystem α_i, and let $\gamma_i = (S_{0,i}; M_i; B_i)$ denote the generative system which represents subsystem α_i; $S_{0,i}$, M_i, B_i stand for the source system, mask and basic behavior of the gener- ative system γ_i, respectively. Let

$$C_K = \bigcap_{k \epsilon K} \beta_k \text{ for some } K \subset I_r$$

and let $pr_{C_K} B_i$ denote the projection of B_i which disregards all sampling variables defined by mask M_i except those which are defined in terms of variables in set C_K.

Each structure system, say system S_3 (neutral or directed), can be now defined by a set of triples

$$S_3 = \{(\alpha_i; \beta_i; \gamma_i = (S_{0i}, M_i, B_i)) \mid i \epsilon I_r\}$$

which satisfy the following conditions:

(i) $\alpha_i \neq \alpha_j$ for all $i, j \epsilon I_r$ and $i \neq j$;

(ii) β_i consists of exactly those basic variables which are included in S_{0i};

(iii) all source systems $S_{0i} (i \epsilon I_r)$ are defined in terms of the same parameter set;

(iv) for every $K \subset I_r$, $pr_{C_K} \gamma_i = pr_{C_K} \gamma_j$ for all $i, j \epsilon K$.

Condition (i) guarantees that each subsystem has a unique identifier (name) which distinguishes it from any other subsystem included in the structure system. Condition (ii) guarantees that the set of basic variables β_i is the same set for which the generated system γ_i is defined. Conditions (iii) and (iv) guarantee that all subsystems of a given structure system are compatible.

3. A Package of Methodological Tools for Systems Modelling

Given a purpose and constraints of some empirical investigation, systems modelling basically attempts to identify a suitable system on a relevant object. The purpose of investigation can be viewed as a set of questions which the investigator (or his client) wants to answer. Constraints associated with an investigation consist of financial and time limitations, available measuring instruments, limited manpower, and various other restrictions imposed upon the investigation. The object of investigation is loosely defined as a part of the world which can be identified as a single entity for an appreciable period of time and which is desirable for a particular investigation.

The process of systems modelling begins with the selection of an object; this selection is guided purely by the purpose of investigation. Since objects can almost never be studied in all of their complexities, the next step in systems modelling consists of a set of decisions through which a source system is defined on the object. These decisions are made on the basis of both the purpose and constraints of the investigation.

The definition of the source system consists of three natural steps: (i) selection of an object system; (ii) selection of an image system; (iii) specification of the interface between the object and image systems.

No formal procedure is possible for executing step (i). It is likely that the better the investigator understands the questions which motivate the investigation and the more knowledge and experience he has in the general area within which the investigation is subsumed, the more meaningful source system would result from his considerations. Too many attributes seem usually relevant, at least potentially, to the purpose of investigation but, due to the various constraints, the investigator can afford to select only a small subset of them. Which of the subsets to select constitutes a basic dilemma for the investigator. Computer-aided modelling attempts to somewhat reduce this dilemma by providing the investigator with such methodological tools which allow him to be less restrictive in his selection of attributes. First, they allow him to handle considerably more attributes than he could handle otherwise and, secondly, they allow him to explore, within acceptable time, different subsets of the desirable attributes and compare the obtained results. Steps (ii) and (iii) in defining the source system are strongly related to the chosen observation or measurement procedures.

Once a source system is defined, the next step in empirical investigation is data gathering. Its objective is to collect information about the actual states of basic variables within the chosen parameter set. The result of data gathering is a data system.

The procedure of empirical investigation is schematically summarized in the first part of the flow diagram in Figure 1. It is followed by further investigation which is theoretical in nature. It basically involves appropriate processing of the data through which attempts are made to model the given data system by suitable systems at higher epistemological levels. The major objectives of the modelling consist of: (a) representing the data in a parsimonious fashion--descriptive characteristics of the model; (b) identifying useful structural properties of the data such as reconstructability of the overall data set from data sets based on subsets of the basic variables or subsets of the parameter set-- explanatory characteristics of the model; (c) extending the data beyond the limits of the parameter set chosen in the empirical investigation--predictive or retrodictive characteristics of the model. All of the characteristics of systems are eventually utilized as a basis for decision making, e.g., decision making regarding actions to be taken to regulate or control the attributes involved in any desirable direction.

At each epistemological level, systems are defined by certain specific types of systems traits which uniquely identify a category of systems; they are referred to as primary traits of the individual categories of systems. It is a general property of the epistemological hierarchy that the set of primary traits characterizing a particular level is a subset of the set of primary traits associated with any higher level. At each level, the primary traits are exactly those through which systems are defined at that level.

Besides the primary traits, a system may be supplemented by some additional traits which are not identificatory for that system. For instance, a generative system may be supplemented by a structure system which conforms to it. Since

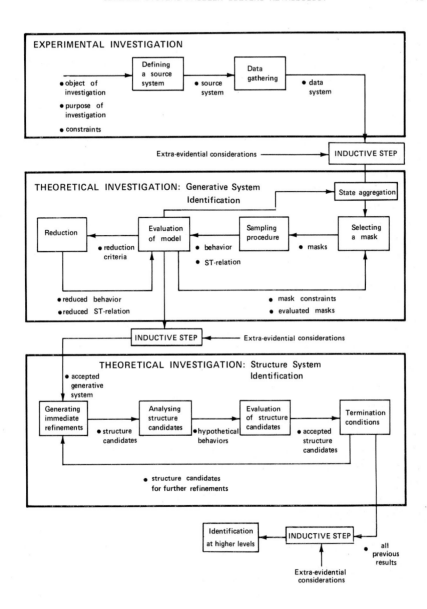

Figure 1   Summary of a package of methodological tools for systems
modelling.

there is a set of different structure systems which all conform to the same
generative system, the latter is invariant with respect to any change in the
structure system within this set. The additional traits which are associated with
a system but are not a part of its definition are called secondary traits of the
system.

The primary traits of a system are required to be completely known and invariant with respect to the chosen parameter set in order to identify the system.
No such requirements are imposed upon the secondary traits. They may be completely unknown or known only partially, and are not required to be parameter-
invariant.

If a primary trait of a system changes, then, by definition, the system
changes too. On the other hand, if a secondary trait changes, the identity of
the system is not affected. For example, while a source system is not affected
by the set of available data regarding its variables, a particular data system
changes when some data are excluded from it or added to it.

In the process of systems modelling, which is associated with transitions
from one epistemological level to higher levels, it is often useful to redefine
the system in the sense that some secondary traits are accepted as primary
traits; such changes in the definitions of systems correspond to inductive inferences.

In pure empirical investigation, the system is defined at level 0 as a source
system; all other traits such as data regarding variables of the source system,
behaviors representing the data, structures representing the behaviors, meta-
systems representing the structures, etc., are viewed as secondary traits. When
the investigator becomes confident that the collected set of data is sufficient
to fully characterize the variables of the source system, he may accept the data
set as a primary trait. This means that he redefines the source system, making
it now a data system. Such change represents an inductive step because it is
based on the assumption that the available set of data contains all information
regarding the investigated variables. The investigator may decide to make this
inductive step on the basis of both the data and some other aspects such as a
comparison with similar investigations performed previously, relation of the data
to the existing body of knowledge, investigator's intuition, and the like; these
latter aspects are usually referred to as extraevidential considerations.

When the source system is redefined as the data system, the next objective
is to find a suitable behavior to represent the data system. The behavior is a
secondary trait of the data system at this level of investigation. There are
different behaviors conforming to the data and each of them generates not only
the data array involved but other data arrays as well. Whether the best behavior
(for whatever adopted criterion of "goodness") with respect to the data system is
also the best one for describing all data the source system may produce depends

on how well justified the inductive step was. The transition from the data system to a justifiable generative system is called the <u>generative system identifica-
tion</u>.

When the investigator becomes confident that he determined the right behavior,
he may redefine the data system by including the behavior as a primary trait.
This again involves an inductive step because the given data set is extended to
all sets of data which conform to the behavior. Whether these are sets of data
producible by the source system depends on the justification of both the previous
inductive steps.

The system is now defined at level 2 (generative system) and further investigation is oriented to the <u>structure system identification</u>. Once the investigator accepts a particular structure system, he may then define the system
as the structure system (another inductive step) and focus, if desirable, on the
identification of a meaningful systems at higher levels.

A computer-aided procedure for the identification of generative systems,
which is described in details in a previous paper (Klir, 1975), is summarized in
the second part of the simplified flow chart in Figure 1. The procedure consists
of the following basic steps:

(a) If desirable, the resolution level of the data is appropriately reduced
by some aggregation of states of the variables.

(b) A mask is selected as a paradigm for representing the data.

(c) The data are sampled within the whole parameter set (in an appropriate
order) on the basis of the chosen mask. The total numbers of the individual
samples are determined and used for the calculation of probabilities representing
the basic behavior for the mask; other forms of the generative system can easily
be derived from the basic behavior when desirable (Klir, 1975; Broekstra, 1976a).

(d) The generative system obtained by the sampling procedure is evaluated
by the investigator. If it is satisfactory, the investigation at level 2
terminates and, if desirable, the structure system identification may be initiated.
If the results are too complex to provide the modeler with any insight, they can
be reduced in many different ways. This may provide the modeler with a spectrum
of simplified generative systems, all based on the same data and mask; the modeler
can choose reduction criteria from a set of available options.

The whole procedure can be repeated for different masks and the generative
systems obtained for the individual masks compared by some subjective or objective criteria. In case of objective criteria, the procedure can be repeated
automatically until an optimal mask is reached within given constraints. The
procedure can also be repeated for different resolution levels of the data.

The entropy function is used in the procedure for generative system identification as a reasonable measure of <u>generative uncertainty</u> through which
individual masks can be objectively evaluated and compared with respect to the

given data (Klir, 1975). For neutral systems, entropy H_M for mask M is given by
the formula

$$H_M = - \sum_{\bar{s}_g \in \bar{S}_g} p(\bar{s}_g) \sum_{s_g \in S_g} p(s_g|\bar{s}_g) \log_2 p(s_g|\bar{s}_g),$$

where the symbols have the same meaning as defined in Section 2. A slightly
modified formula must be used for directed systems (Klir, 1975). Normalized
reduction of uncertainty A_M in the generative process, due to mask M, can be then
evaluated by the formula

$$A_M = \frac{H_{max} - H_M}{H_{max}},$$

where H_{max} denotes the largest possible entropy for the mask and the source system under consideration. A quality Q_M of mask M is viewed as the reduction of
generative uncertainty per sampling variable of the mask, i.e.,

$$Q_M = \frac{A_M}{n},$$

where n stands for the number of sampling variables in mask M.

A software package has been developed in which these measures are employed
to evaluate and compare submasks of a given maximal acceptable mask. All admissible submasks and their behavior and/or ST-relations can be determined
automatically by the procedure; mask M is called admissible iff any mask M'≠M
whose behavior gives a better approximation in accounting for the data contains
more sampling variables. Alternatively, masks with the highest quality Q_M can
be determined, if this is desirable.

Once a generative system has been accepted, the next objective of systems
modeling focuses on the structure system identification. There are basically two
alternatives of this problem:

(a) The problem of identifying meaningful decompositions of the generative
system in which new coupling variables are introduced.

(b) The problem of identifying structure systems whose elements represent
projections of the overall behavior of the given generative system, each associated with a subset of variables of the generative system, and through which the
overall behavior can be reconstructed with an acceptable degree of approximation.

Regarding problem (a), a quotation from a previous paper (Cavallo and Klir,
1978b) seems appropriate:

"These situations arise most often in problems of design
and are also evidenced in some modelling approaches. If
there is no restriction on the use of internal variables,
the designer or modeller has much more freedom and,
consequently, such cases admit much more easily of the

derivation of acceptable structure systems. We can draw
a parallel here between the work of a designer--who has
available for use in construction or design a set or
store of available "elements" (whether these are actual
or conceptual)--and an investigator trying to develop a
"theory" regarding some object of investigation. In this
latter case the elements may be previously completed
theoretical or empirical studies which overlap the domain
of interest of the investigator. In this case, however,
the use of these "elements" imposes significant constraint
on the investigator and the approach becomes very similar
to standard analytic procedures. On the other hand, it
does not seem to be conceptually significant to allow the
theory-builder general freedom in the use of internal
variables as their use would then entail essentially no
external reference. For these reasons, studies relating
to the use of internal variables are quite tricky at the
general level since there is no semantics which naturally
accrues to them. We thus conceive that the use of internal
variables implicitly defines a new system including these
variables in the original configuration and we restrict our
attention to structure determination which is always based
on a fixed set of variables."

A method for solving problem (b) was proposed by Klir (1976) and further developed by Klir and Uyttenhove (1976a,b, 1977). This method, which has been
fully implemented on a computer, will be now briefly summarized. It is assumed
here that the systems under consideration are neutral; modifications to directed
systems are rather trivial. Alternative approaches to problem (b), based on some
concepts of information theory, has been developed by Broekstra (1976b, 1977,
1978) and Krippendorff (1978a,b).

Let V denote the set of variables involved in the given data system and the
generative system derived from it. A structure system involved in problem (b)
can uniquely be defined by a family of subsets of V and the following conventions:
(1) Each subset in the family, say subset E_a, identifies a subsystem of the
structure system; (2) the behavior of each subsystem E_a is a projection of the
overall behavior of the given generative system into the subspace associated with
the submask based only on variables in set E_a; (3) couplings between pairs of
subsystems are defined in terms of intersections between the corresponding subsets in the family.

Let C denote a family of subsets of V which, using the described conventions,
uniquely defines a structure system. Let $Q_C \subset V \times V$ denote a binary relation such
that $(x,y) \in Q_C$ if and only if $(x,y) \in E_a$ for at least one $E_a \in C$, let $Q_a \subset V \times V$ denote
a binary relation such that $(x,y) \in Q_a$ if and only if $(x,y) \in E_a$, and let $Tr(Q)$ denote the transitive closure of a binary relation Q. In order to be able to reconstruct relation R_V through structure system C, all variables in set V must be
included in C. Hence, we require that C satisfy the following condition:

$$(\alpha) \quad \bigcup_{E_a \in C} E_a = V.$$

Furthermore, to guarantee that every subsystem of the structure system contains some unique empirical information which is neither included in the other subsystems nor can be obtained hypothetically through a sequence of joins of relations[*] associated with the other elements according to the pattern of couplings, we require that C also satisfy another condition:

(β) there is no $E_a \in C$ such that $Q_C \subseteq Tr(Q_C - Q_a)$.

Structure systems which satisfy conditions (α) and (β), as well as the conventions described previously, will be called <u>structure candidates</u>. Given a relation R_V, each structure candidate derivable from it, say candidate C, represents a scheme for reconstruction of R_V from its projections included in C. Conditions (α) and (β) will be referred to as <u>axioms of structure candidates</u>.

Let C_i and C_j denote two structure candidates associated with the same set of variables. Then, we view C_i as a <u>structure refinement</u> of C_j if and only if for each $A \in C_i$ there exists $B \in C_j$ such that $A \subseteq B$. The relation "being a structure refinement of," which will be denoted by $\leq$, is a partial ordering. Moreover, all structure candidates for each specific set of variables together with this partial ordering form a lattice.

Let L_n denote the <u>lattice of structure candidates</u> based on a set V containing n variables. Then the greatest candidate in L_n, say candidate C_o, consists of one element which includes all variables of V; its relation, say relation R_o, is the same as the overall relation R_V. The least candidate in L_n, say candidate C_ℓ, contains n elements, each associated with only one variable. While candidate C_o is a basis for viewing the set of variables as a whole, candidate C_ℓ represents the view that the variables are pair wise independent; the remaining candidates stand for all the other views which are somewhere in-between the two extreme views. Given two structure candidates in L_n, say C_i and C_j, described by relations $Q_i, Q_j \subset V \times V$, respectively, it is not guaranteed that $Q_i \neq Q_j$; if $Q_i = Q_j$ we view the candidates C_i, C_j as equivalent.

If $C_i \leq C_j$, then candidate C_i is called a <u>successor</u> of C_j in the appropriate lattice. If $C_i \leq C_j$ and there is no C_k in the lattice such that $C_i \leq C_k \leq C_j$, then C_i

---

[*] A join $R*Q$ of two binary relations $R \subset A \times B$ and $Q \subset B \times C$ is a ternary relation defined as $R*Q = \{(a,b,c) | (a,b) \in R \text{ and } (b,c) \in Q\}$. If probabilities $p(a,b)$ and $p(b,c)$ are associated with elements (a,b) and (b,c) of relations R and Q, respectively, then probabilities $p(a,b,c)$ of elements (a,b,c) of the join $R*Q$ are calculated by the formula $p(a,b,c) = p(a,b) \cdot p(c|b)$.

is called an <u>immediate successor</u> of C_j in the lattice.

An algorithm for generating all immediate successors of a given structure candidate was developed by Klir and Uyttenhove (1976a). Although the algorithm is computationally surprisingly simple, its proof, also presented by Klir and Uyttenhove (1976b), is rather complicated. When combined with the approach to structure identification proposed by the author (Klir, 1976), the algorithm provides a feasible basis for a computer-aided structure identification procedure. The procedure consists of the following four steps:

1. All immediate sucessors of a given structure candidate in the appropriate lattice L_n are generated; either C_o or a structure candidate specified by the user is taken as the initial structure candidate.

2. For each candidate, say candidate C_k, generated in Step 1, total join of relations associated with its elements is determined according to the pattern of couplings among the elements. This is accompanied by the calculation of probabilities $p_k(s)$ of aggregate states of the overall relation, say relation R_k, and results in behavior

$$B_k = \{(s,p_k(s)) \,|\, s \epsilon R_k; \; 0 < p_k(s) \leq 1; \; \textstyle\sum_s p_k(s) = 1\}$$

3. Behavior B_k obtained in Step 2 for candidate C_k is compared with behavior B obtained directly from the data which is the same as the behavior of C_o. Although the comparison can be based on various measures expressing different objective criteria (Klir, 1976; Broekstra, 1978; Cavallo, 1978; Uyttenhove, 1978b), let us consider here a simple distance d_k between behaviors B_k and B, defined by

$$d_k = \textstyle\sum_s |p_k(s) - p_o(s)|$$

Clearly,

$$0 \leq d_k \leq 2$$

and $d_k = 0$ represents perfect structure candidates; $d_o = 0$ for any B.

4. If the smallest distance determined in Step 3 is larger than acceptable and/or the minimal distance increases from one iteration to the next more than acceptable, then the procedure terminates. Otherwise, the procedure is repeated for structure candidates with the smallest distance or, if desirable, for those whose distance is as close to the smallest distance as the user requires.

Further refinement of only structure candidates with the smallest distance or those whose distance is in the neighborhood of the smallest distance (Step 4) is justified by the following considerations. When $C_i \leq C_j$ and $C_i \neq C_j$, then relations of some elements of C_i which are directly derived from the data as

projections of the accepted overall relation R_v are derived in C_i only partially
from the data (i.e., they are reconstructed in C_i from smaller projection). This
does not hold the other way around: all information in C_i derived directly from
the data is also available in C_j. Hence, if C_i as a whole conforms perfectly to
R_v, then C_j conforms perfectly to R_v too. That is to say, if $d_i = 0$, then $d_j = 0$.
This means that if the aim of structure identification is to determine the most
refined perfect structure candidate, it is sufficient to consider at each level
of refinement only candidates whose distance is zero. If the candidates are not
perfect, it is likely that $d_i \geq d_j$, although this inequality is not theoretically
guaranteed. It is also likely that the structure candidate with the smallest
distance at a particular level of refinement is derivable from candidates with
the smallest distance at the previous refinement level or from those whose distance is "sufficiently close" to the smallest distance. In order to evaluate
this likelihood and to provide the modeller with some practical guidelines for
proper interpretation of values of the distance, increment in the distance between
iterations, distance discriminations among a set of evaluated structure candidates
at the same level of refinement, and the dependence of the distance on the size
of available data, extensive sets of experiments based on computer simulation
with full information were performed (Klir and Uyttenhove, 1977). In each of the
experiments, a data set of some size was generated by a computer program through
a specific structure system and analyzed then by the structure identification
procedure.

For instance, the plots in Figure 2 summarize a set of simulation experiments
for structure candidates with 5 variables. Plot 1 shows the dependence of the
average distance of the correct structure candidates (at the correct level of
refinement) on the number of observations in the analyzed data. Plot 2 shows the
average distance of the structure candidate with the smallest distance except the
distance of the correct candidate at the same (correct) level of refinement.
Plot 3 summarizes the average distance of all the structure candidates evaluated
at the correct level of refinement except the correct candidate.

The experiments, which are reported in more details elsewhere (Klir and
Uyttenhove, 1976b, 1977; Uyttenhove, 1978b), showed that the procedure identified
the correct path in the lattice of structure candidates in about 90% of experiments. The remaining 10% of experiments, in which the correct path was not
followed, were clearly experiments with small number of observations. In case of
data with more than 250 observations, the correct path was identified in 100% of
cases.

The experiments also showed that a correctly identified structure system has
a strong ability to recover aggregate states which are not included in the
available data, although the variables can assume them. For instance, in the
simulation experiments with 5 variables, all states except 10 out of a total of

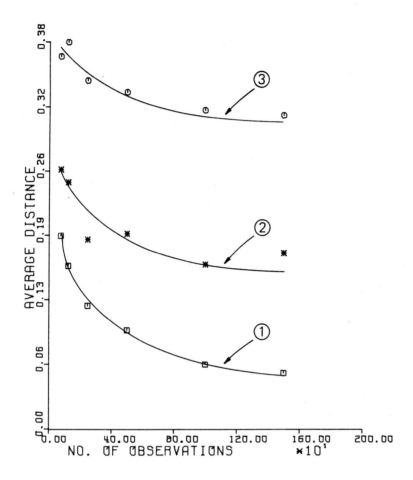

Figure 2   Characteristics based on simulation experiments for the structure
system identification.

255 missing states in the analyzed data were recovered through the identified
structure systems.   This indicates a powerful predictive mechanism.   It is based
on the fact that segments of an overall state may appear in other overall states
as well.   Consequently, it is more likely, when collecting data, to come across
all desirable segments of the overall state (segments corresponding to projection

of the identified structure candidate) before finding the state itself.

The state prediction is contingent upon the identification of a correct structure system even though it need not be the correct structure system at the highest level of refinement. Although the predictive power increases with an increase in the structure refinement, an overly refined system may (and usually does) generate spurious states. Consequently, the credibility of the predicted states is closely related to the credibility of the identified structure system through which the predictions are made. The latter can be evaluated in terms of experimental characteristics for the structure identification procedure such as those in Figure 2. If the distance of a structure candidate which generates some additional states is sufficiently small in terms of these characteristics, then the additional samples are likely to exist for the investigated variables and should be viewed as predictions following from the structure system modelling. The confidence of such predictions depends on the distance of the candidate and the data size evaluated in terms of the characteristics in Figure 2. The confidence, say c, in predictions based on structure candidate C_k can reasonably be defined as

$$
c = \begin{cases} \dfrac{d_{max}-d_k}{d_{max}} & \text{when } d_{max}-d_k \geq 0 \\ \\ 0 & \text{when } d_{max}-d_k < 0, \end{cases}
$$

where d_{max} stands for the largest acceptable distance and d_k denotes the distance of candidate C_k. It seems appropriate to choose the largest acceptable distance somewhere between plot 1 (a pessimistic choice) and the middle point between plots 1 and 2 (an optimistic choice) for each particular number of observations. Each choice of d_{max} leads to a prediction range of distances for each particular data set (number of observations).

The experience with both artificial data and real world data indicates a rather profound systems principle: information about the structure system is implicitly included in data and can be utilized, using appropriate structure identification procedure for developing models in areas where knowledge regarding the structure is not available. When repeatedly applied in specific application areas, it is hoped that the structure identification will eventually lead to the discovery of reconstructability patterns in these areas of inquiry. In areas with persistenly poor reconstructability at all levels of structure refinement, systems modelling at higher epistemological levels, allowing changes in the structure system within the parameter set (time, space, etc.), may lead to more satisfactory results.

The problem of <u>metasystem identification</u> has been explored by Uyttenhove
(1978a,b). While subsystems of a structure system are associated with subsets of
basic variables, subsystems of a metasystem are associated with subsets of the
parameter set. Uyttenhove developed and tested a procedure for metasystem identification which is based on the assumption that the parameter set is linearly
ordered and that the metasystem consists of a set of catenated subsystems in the
parameter set. The essence of the procedure can be described as follows:

(i) Let α be a given integer, β be a given rational number, $w_i \leq w_j$ for all
$w_j \in W$, and let k=1.

(ii) Determine the behavior corresponding to segment $[w_i, w_i + k\alpha]$ of the
parameter set and calculate its generative uncertainty H_o.

(iii) Replace k by k+1; if $w_i + k\alpha \notin W$, go to (vi).

(iv) Determine the behavior corresponding to segment $[w_i, w_i + k\alpha]$ of the
parameter set and calculate its generative uncertainty H_k.

(v) If $(|H_k - H_{k-1}|/H_k) < \beta$, go to (iii); otherwise, record $w_i + (k-1)\alpha$ as an
approximate point of catenation of two subsystems, replace w_i by $w_i + (k-1)\alpha$, make
k=1, and go to (ii).

(vi) END.

The procedure is based on the observation that the generative uncertainty of
a behavior which is invariant with respect to a parameter set quickly converges
to a constant value, after some initial erratic changes. A large increase in the
uncertainty, after its constant value is closely approached, provides thus an
incentive for viewing the whole system as broken down into catenated subsystems,
each associated with a subset of the parameter set.

The sensitivity and efficiency of the procedure considerably depend on the
chosen values of α and β. Uyttenhove has performed a number of simulation experiments on the basis of which he prepared some guidelines for the user regarding these values. These experiments and the resulting guidelines, more detailed
formulation of the procedure and some applications are included in his dissertation (Uyttenhove, 1978b).

The whole package of modelling tools which are described in this paper have
been completely implemented in the APL language and some initial studies of the
computational complexity of the individual tools were undertaken (Uyttenhove,
1978b) and applied to modelling problems in various areas (Klir, 1978a,b; Klir
and Uyttenhove, 1977; Uyttenhove, 1978b; Cavallo, 1978).

4. An Example

An example from the area of performance evaluation of aircraft pilots in a
flight simulator is used in this section to illustrate the whole package of
methodological tools described in this paper. The example is a modified version
of a previously described project (Uyttenhove, 1978b; Comstock and Uyttenhove,
1978).

The source system consists of four basic attributes/variables describing
certain flight characteristics of a jet aircraft: speed-S; altitude-A; TACAN
radial-R;[*)] TACAN DME direction-D[*)]. The supporting attribute/variable is time.

The data system describes, in terms of the four variables of the source
system, a typical (ideal, correct) tactical navigation landing approach by a jet
aircraft at some specific approach site. The full time period of the landing
approach was divided into 170 equal time intervals and for each of them an
aggregate state of the variables was given. Variables S and R distinguish 5
states each, while variables A and D distinguish 7 and 4 states, respectively;
states of each variable are labelled by positive integers. Data set is given in
terms of the following matrix, in which the columns are ordered according to the
natural order of the time set.

```
S   44445555555555555555555555555555555555555555555554444444333
A   77766666666666666666666666666666666666666666666666666655555
R   32222222222222222222222222222222222222222222222222222222333
D   44444444444444444444444444444444444443333333333333333333333
```

```
S   333333333333333322222222222222211111111111111111111111111111
A   5555555555554444444444444444444433333333333333333333333333
R   33344444444444444444444444444444444444444444444444444444444
D   33333333333333333222222222222222222222222222222222222222222
```

```
S   11111111111111122222333333333333333333333333333333
A   3222222111111112222223334444444444444444444444444
R   44444444444444444444455555555555555555555555555555
D   22222222221111111111111111112222222222222222222222
```

The maximal acceptable mask was specified by the client as the mask with
two successive columns. The mask

reference

was determined as the mask with the highest quality; $\bar{S}$ stands for the sampling
variable defined by the equation

$$\bar{S}_t = S_{t-1} ,$$

_____

[*)]These are technical terms which are defined in <u>Air Force Manual</u> 51-37, Air
Force Dept. of Instrument Flying, Washington, D.C., 1971.

where t denotes the defined (discrete) time. When the data is exhaustively
sampled by this mask, the following basic behavior is obtained:

S̄SARD	prob.	S̄SARD	prob.	S̄SARD	prob.
11141	.0296	22443	.0059	43533	.0059
11142	.0178	23251	.0059	44523	.0118
11242	.0355	32443	.0059	44623	.0237
11342	.1479	33351	.0178	44624	.0059
11442	.0118	33443	.0118	44724	.0118
12241	.0059	33451	.0237	45624	.0059
21442	.0059	33452	.1124	54623	.0059
22241	.0237	33533	.1006	55623	.0592
22442	.0651	33543	.0533	55624	.1893

When this behavior is taken as input to the structure identification procedure, no structure system (at any level of refinement) is perfect; the smallest
distance is 0.0287. This was not acceptable to the client but he was willing to
accept a metasystem description with up to 3 catenated subsystems, provided that
each subsystem would be represented by some perfect and sufficiently refined
structure system.

The data were analyzed for the maximal acceptable mask by the metasystem
identification for several values of α and $\beta = 0.1$ (a value suggested by results
of a number of simulation experiments; see details in Uyttenhove, 1978b) and the
determined points of catenation obtained for different values of α were averaged.
This led the following three subsets of the time set: 1-70, 71-125, 126-170.
Mask evaluation and structure identification applied to each of the subsystems
leads to the results which are summarized in Figure 3. These results, which
describe the desirable performance of the pilot have been successfully utilized
in developing effective ways of evaluating the actual performance of students
who learn these skills by operating a flight simulator (Comstock and Uyttenhove,
1978).

5. Conclusions

Systems modelling used to be viewed as basically an intuitive process, highly
resistent to any algorithmic description. Although intuition remains important
in systems modelling, it is increasingly recognized that many aspects of systems
modelling can be made algorithmic. General systems researchers has been trying
to identify basic principles of systems modelling, make them operational and
utilize the computer to assist the investigator (modeller) in the modelling
process.

The package of methodological tools for systems modelling, as described in
this paper, is a component of a larger set of methodological tools referred to as
the general systems problem solver. The described tools for systems modelling as

G.J. KLIR

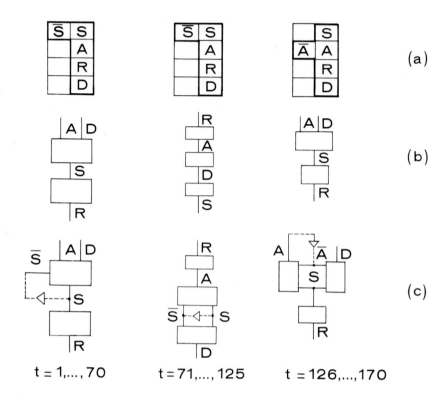

Figure 3   Metasystem identified in the example describing a typical
           landing approach by a jet aircraft: (a) masks with the highest
           qualities for each subsystem; (b) structure systems based on
           basic variables of the individual subsystems; (c) structure
           systems based on all sampling variables (the triangles indicate
           delays of one discrete time).

well as the whole general systems problem solver are characterized by the
following features:

     1. They are based on a symbiotic man-computer interactive mode of operation.
The computer generates hypotheses, analyze and evaluate them, and compare the
various alternatives; the human user provides information based on current
knowledge in the respective field of inquiry and interpretes the computer results
within the context of the investigated phenomena.

     2. The focus is on techniques which are applicable to variables of any
scale (quantitative as well as qualitative) and to both well defined and fuzzy
variables. In conformity with the nature of empirical data, the techniques have
been developed for discrete variables.

3. It is recognized that several models, each reflecting certain aspects of the data, may give the investigator much better insight than any of them could afford alone.

A number of research problems remain to be solved to extend and improve the described package of software tools. Among them are the following problems:

(i) the problem of optimally introducing internal (hypothetical, unobserved) variables into the investigated data system in the sense that the generative uncertainty for a given mask and a given number of states of the internal variables becomes maximally reduced.

(ii) The problem of developing a generalized structure identification procedure characterized by weakening axiom (β) of structure candidates into

$$E_a \not\subset E_b \text{ for any pair } E_a, E_b \in C,$$

as well as procedures for other desirable classes of structure candidates characterized in a recent paper by Cavallo and Klir (1978b).

(iii) Combining top-down and bottom-up procedures of structure identification in order to allow the user to explore the structure neighborhood of any given structure system.

(iv) Developing of more sophisticated methods for the identification of metasystems of various levels.

REFERENCES

BROEKSTRA, G. (1976a), Some comments on the application of informational measures to the processing of activity arrays. International Journal of General Systems, 3, No. 1, pp. 43-51.

BROEKSTRA, G. (1976b), Constraint analysis and structure identification. Annals of Systems Research, 5, pp. 67-80.

BROEKSTRA, G. (1977), Constraint analysis and structure identification II. Annals of Systems Research, Vol. 6, pp. 1-20.

BROEKSTRA, G. (1978), Structure modelling: A constraint (information) analytic approach. In: Applied General Systems Research, edited by G.J. Klir, Plenum Press, New York.

CAVALLO, R. (1978), The Role of Systems Methodology in Social Science Research. Martinus Nijhoff, Leiden and Boston.

CAVALLO, R. and G.J. KLIR (1978a), A conceptual foundation for systems problem solving. International Journal of Systems Science, Vol. 9, No. 2, pp. 219-236.

CAVALLO, R. and G.J. KLIR (1978b), The structure of reconstructable relations. Proc. Fourth European Meeting in Cybernetics and Systems Research, Linz, Austria.

COMSTOCK, F.L. and H.J.J. UYTTENHOVE (1978), Computer Implemented Grading of Flight Simulator Students. In: Proceedings AIAA Flight Simulation Technologies Conference, Arlington, Texas.

KLIR, G.J. (1969), An Approach to General Systems Theory. Van Nostrand, New York.

KLIR, G.J. (1975), On the representation of activity arrays. International Journal of General Systems, Vol. 2, No. 3, pp. 149-168.

KLIR, G.J. (1976), Identification of generative structures in empirical data. International Journal of General Systems, Vol. 3, No. 2, pp. 89-104.

KLIR, G.J. (1978a), The problem of choosing appropriate structure system in systems modelling. Proc. Intern. Conference on Systems Modelling in Developing Countries, Asian Inst. of Technology, Bangkok, pp. 3-20.

KLIR, G.J. (1978b), Computer-aided systems modelling. In: Systems Theory in Ecology, edited by E. Halfon, Academic Press, New York.

KLIR, G.J. (1978c), General systems concepts. In: Cybernetics: A Sourcebook, edited by R. Trappl, Hemisphere, Washington, D.C.

KLIR, G.J. and H.J.J. UYTTENHOVE (1976a), Procedure for generating hypothetical structure in the structure identification problem. Proc. Third EMCSR, Vienna, edited by R. Trappl, et al.

KLIR, G.J. and H.J.J. UYTTENHOVE (1976b), Computerized methodology for structure modelling, Annals of Systems Research, Vol. 5, pp. 29-66.

KLIR, G.J. and H.J.J. UYTTENHOVE (1977), On the problem of computer-aided structure identification: Some experimental observations and resulting guidelines. International Journal of Man-Machine Studies, Vol. 9, No. 6, pp. 593-628.

KRIPPENDORFF, K. (1978a), A spectral analysis of relations: Foundations. In: Recent Developments in Systems Methodology for Social Science Research, edited by R. Cavallo, Martinus Nijhoff, Leiden and Boston.

KRIPPENDORFF, K. (1978b), A spectral analysis of relations: Further developments. Proc. of the Fourth European Meeting on Cybernetics and Systems Research, Linz, Austria.

UYTTENHOVE, H.J.J. (1978a), Metasystem identification: A procedure for detection and structural composition in time dependent systems. In: Applied General Systems Research, edited by G.J. Klir, Plenum Press, New York, pp. 147-160.

UYTTENHOVE, H.J.J. (1978b), Computer-Aided Systems Modelling: An Assemblage of Methodological Tools for Systems Problem Solving. Ph.D. Dissertation, Systems Science Department, SUNY at Binghamton.

ZEIGLER, B.P. (1974), A conceptual basis for modelling and simulation. International Journal of General Systems, 1, No. 4, pp. 213-228.

ZEIGLER, B.P. (1976a), Theory of Modelling and Simulation. John Wiley, New York.

ZEIGLER, B.P. (1976b), The hierarchy of system specifications and the problem of structural inference. In: F. Suppe and P.P. Asquith (editors), PSA 1976, Philosophy of Science Assoc., East Lansing, Mich., pp. 227-239.

METHODOLOGY IN SYSTEMS MODELLING AND SIMULATION
B.P. Zeigler, M.S. Elzas, G.J. Klir, T.I. Ören (eds.)
© North-Holland Publishing Company, 1979

CONCEPTS FOR ADVANCED COMPUTER ASSISTED MODELLING

Tuncer I. Ören
Computer Science Department
University of Ottawa
Ottawa, Ontario K1N 6N5
Canada

ABSTRACT: After a brief rationale as to why advanced
computer-assistance is needed in modelling and model
manipulation, some terminological clarifications and
a classification of mathematical models are given.
The useability of models is discussed in three categories, i.e., communication, experimentation, and algorithmic model manipulation. Criteria for the assessment of the acceptability of models and relevant computer programs and data are systematized. The importance of the referability of models is stressed.

1. INTRODUCTION

1.1 Motivation

As Bell expresses it " A number of countries in the West, the United States among
them, are now passing from an industrial into a post-industrial phase of society.
... The strategic resource of the post industrial society becomes theoretical
knowledge, just as the strategic resource of an industrial society is money capital, and the strategic resource of a pre-industrial society is raw materials"
(Bell 1976). Post industrial technology is intellectual technology as opposed to
machine technology of industrial society and craft of pre-industrial society.
Similarly, methodology for pre-industrial society is common sense, trial and error
and experience, while methodology for industrial society is empiricism and experimentation. In post-industrial society, however, the methodology is based on
abstract theory, models, simulations, decision theory, and system analysis.

Within this framework, it is therefore quite natural that "The U.S. Government is
the largest sponsor and consumer of models in the world. Estimates have indicated that over one-half billion dollars is being spent annually on developing,
using, and maintaining models in the Federal Government" (Roth, Gass, Lemoine
1978). Ways to improve management of federally funded computerized models were
prepared by US GAO (1976).

However, some professionals at positions to influence science policy, and therefore responsible for relevant scientific and methodological developments are yet
to be convinced about these facts. For example, the President of the Weizmann
Institute of Science of Israel writes: "At the same time, you will be well aware
that the merits of the theoretical approach to modelling are still a matter of
some contreversy" (Sela 1978).

To speed up the availability of computer assistance in formulating and manipulating models to manage complex multifaceted systems, appropriate science policies
in post-industrial societies are urgently needed.

1.2 Some Terminological Clarifications

Given two objects A and B and an observer C, the object A is said to be a <u>model</u> of
the object B if the observer C can use it to answer questions that interest him
about this object B. This definition given by Minsky (Minsky 1965), has several
practical implications. The object B is most often called "real system," or
object under investigation. An implication of this definition is that for an
object under investigation there is not "<u>the</u> model" to represent it, but a set of
models, called <u>partial models</u> to represent its different aspect. In this paper,
<u>mathematical models</u> are considered. By changing the observer or his goals for
modelling or his level of understanding the "real-system," one may end up having
distinct partial models of the same system.

<u>Experimentation</u> is controlled observation and can be done either with the real-
system or with one of its models. <u>Simulation</u> is experimentation with models. In
<u>computerized simulation</u>, necessary computations to carry out the experiments are
performed by a computer.

In the 60's, the terms "simulation" and "modelling" were used interchangeably as
the following quotation taken from the first issue of <u>Simulation</u> shows: "So, with
the power vested in me by the SCi Board of Directors as Editor of the Journal, I
proclaim the name to be SIMULATION and simulation to mean 'the act of representing
some aspects of the real world by numbers or symbols which may be easily manipula-
ted to facilitate their study.'" (McLeod 1963).

In the early 70's, at least some professionals realized that the terms "modelling"
and "simulation" should not be used interchangeably: "What is modelling, and what
is simulation? For years your Editor considered the meanings to be practically
identical, but showed a strong preference for simulation because that's the name
of our game. However, a distinction seems to be evolving, and I'm in favor of it.
After all, it seems wasteful to have two perfectly good words meaning *exactly* the
same thing!" (McLeod 1970).

Computerized simulation overshadowed the usages of models for purposes other than
experimentation. However, in the late 70's at least, we have to distinguish
modelling, model manipulation and simulation. As it will be discussed in the
later sections of this article, simulation is one of the several possible usages
of models. Hence, the scope of this paper is intended to be broader than the
scope of articles restricted to simulation only. So far as models are concerned,
in addition to the modelling process, we can identify three relevant broad concepts, i.e., useability, acceptability and referability of models. This paper
is intended to explore conceivable, desirable, and feasible possibilities for
computer assistance in modelling as well as for enhancing the useability, acceptability, and referability of models.

As also pointed out by Rubinji, the three levels of usages of computers are
mechanization, automation, and cybernetics (Rubinji 1975). Therefore, it is
possible to conceive three levels of computer involvement in model-related activities:

<u>The first level</u> of computer involvement is mechanization (or computerization) of
different model-related activities. Mechanization (computerization) is the re-
placement of human efforts by machines (computers). In computerized simulation,
simulated behaviour is generated by a computer. Therefore, computerized simulation is an example to first level of computer-assistance in model-related activities. However, as they are also discussed in different sections of this paper,
several computerization possibilities still exist at totally or partially unex-
plored stages such as building and maintaining computerized model files (ören,
Zeigler 1979).

The second level of computer usage is automation of several model-related activities. Automation, which requires a higher level of mechanization, forms a series
of systems into a self-regulating process. As application at this level, one can
cite for example, automatic invoking of algorithmic consistency checks of models
with respect to some modelling formalisms as part of a modelling and model manipulation system (Ören, Zeigler 1979).

The third level of computer assistance invokes self-learning capability for the
software system in addition to self-regulating features. "Cybernetics, which
deals with systems capable of receiving, storing, and processing information and
utilizing it for purposes of communication and control, is the most appropriate
scientific discipline on which to base our efforts in developing new methodologies
to cope with complexity." (Ören, 1978a, p.182) Maybe then, instead of crisis
management, humanity can learn to design its own future.

Artificial Intelligence techniques are already applied to software engineering
problems (SIGPLAN/SIGART 1977). What is needed, is not only more advances in this
field but also imbedding artificial intelligence techniques in computer-assisted
modelling and model manipulation. The latter can drastically improve the way we
manage complexity. The rationale for large-scale system simulation software based
on cybernetics and general system theories was discussed in another article
(Ören 1978b).

2. MODELLING

2.1 Classification of Models

A classification of mathematical models is useful to explore systematically the
possibilities for computer-assisted modelling. In Figure 1, a classification of
mathematical models is given based on seven groups of classification criteria
which are variables, their functional relationships, formalism used to describe
the models, intended use of models, disposition of submodels, goals to be pursued
and organization of submodels.

Figure 1. A Classification of Mathematical Models

(m.: model, s.m.: system model, [ ] optional)

CLASSIFICATION BASED ON:	types of variables:
VARIABLES	
- INPUT VARIABLES, EXISTENCE OF	autonomous m.
	non-autonomous m.
- INPUT & OUTPUT VARIABLES, ""	closed s.m.
	open s.m.
- STATE VARIABLES, EXISTENCE OF	memoryless m.
	memory m.
RANGE SETS OF	discrete-state m.
	continuous-state m.
	mixed-state m.
- TIME: TIME SET	discrete-time m.
	continuous-time m.
	mixed-time m.

- TIME: TIME DEPENDENCE   time-varying m.
                                           time-invariant m.
                                               static m.
                                               dynamic m.
- TRAJECTORY OF DESCRIPTIVE VARIABLES discrete [-change] m
                                           continuous [-change] m.
                                           combined [-change] m.

FUNCTIONAL RELATIONSHIPS OF VARIABLES
  - DETERMINISM   deterministic m.
                                           stochastic m. (or probabilistic m.)
  - ANTICIPATION OF THE FUTURE   non-anticipatory m.
                                           behaviorally-anticipatory m.
  - LINEARITY   linear m.
                                           non-linear m.
  - STIFFNESS   stiff m.
                                           non-stiff m.

FORMALISM USED TO DESCRIBE THE MODELS   differential equation m.
                                               distributed-parameter m.
                                               lumped-parameter m.
                                           difference equation m.
                                           Markov chain m.
                                           finite-state automaton
                                               Mealy, Moore
                                               deterministic, stochastic
                                               cellular automaton
                                           process-interaction m.
                                           discrete-event m.
                                           activity-scanning m.

INTENDED USE   value-free m.
                                               descriptive m.
                                               explanatory m.
                                               predictive m.
                                           value-dependent m. (or normative m.)
                                               evaluative m.
                                               prescriptive m.

DISPOSITION OF SUBMODELS   monolithic m.
                                           modular m.
                                               coupled m.
                                                   m. with nested coupling
                                                   hierarchical m.
                                               cellular m.

GOALS TO BE PURSUED	allotelic s.m.
	   teleonomic s.m.
	   (or goal-determined s.m.)
	     adaptive s.m.
	       self-stabilizing s.m.
	       learning s.m.
	   teleological s.m.
	   (or goal-seeking s.m.)
	   teleozetic s.m.
	   (or goal-selecting s.m.)
	autotelic s.m.
	   teleogenic s.m.
	   (or goal-generating s.m.)

ORGANIZATION OF SUBMODELS	allopoietic s.m.
	   system has no rules for changing
	   its organization
	     mechanistic s.m.
	autopoietic s.m.
	   system has rules for changing
	   its organization
	     variable-structure m.
	       m. with time-varying coupling
	     self-organizing s.m.
	     self-reproducing s.m.
	   system has meta-rules to modify
	   its rules
	     self-learning s.m.
	     self-regulating s.m.
	     evolutionary s.m.

Classification Based on Variables:

From a system-theoretic point of view, one can identify three systemic elements in a model, i.e., inputs, states, and outputs. As depicted in Figure 2, each systemic element has to be represented by appropriate variables with well defined range sets, i.e., possible values, and appropriate functions to specify how the values are organized in time.

Figure 2. Components of Systemic Elements

Systemic Elements	Variables	Values (Range sets)	Functions
Input	✓	✓	✓
State	✓	✓	✓
Output	✓	✓	✓

In black box representation of models, input and output variables can also be named "input ports" and "output ports," respectively. As seen in Figure 3, depending on the existence of the state variables, memoryless and memory models are identified. A memoryless model does not have state variables and is also named instantaneous function. A model which has at least one state variable is a memory model.

Figure 3. Types of System Models Depending on the Existence of State, Input, and Output Variables

			existence of the descriptive variables			pictorial representation
			state v.	input v.	output v.	
memoryless model (instantaneous function)				✓	✓	 (box diagram)
memory model	autonomous model	without output (closed s. model)	✓			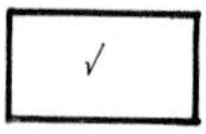 ✓
		with output	✓		✓	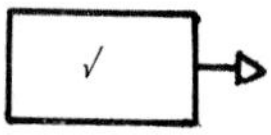 ✓
	non-autonomous model	without output	✓	✓		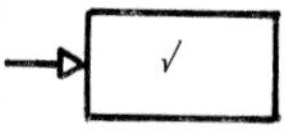 ✓
		with output	✓	✓	✓	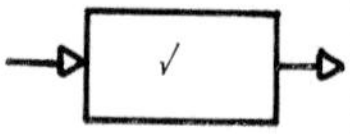 ✓

Depending on the existence of input variables in a memory model, autonomous and non-autonomous models are distinguished. An autonomous model is a memory model without input variable. A non-autonomous model is a memory model with at least one input variable. A closed system model does not have input or output variables. An open system model is a memory model which has either input variable or output variable or both.

The range of a variable is the set of all values that the variable can assume. Figure 4 depicts different possibilities for the ranges of model's descriptive variables.

Figure 4. Possibilities for Sets of Values (Ranges) of Descriptive
           Variables of Models

	Discrete Range	Continuous Range	Mixed Range
Input Variables	✓	✓	* ✓
State Variables	✓	✓	✓
Output Variables	✓	✓	✓
Type of Model:	Discrete-state Model	Continuous-state Model	
	Mixed-state Model		

* Some of the input variables have discrete-range and some others
  continuous range.

In a discrete-state model, descriptive variables of models, i.e., input, state,
and output variables, all assume discrete values. In a continuous-state model,
the ranges of the model's descriptive variables can be represented by one or
several regions of the real number axis. In a mixed-state model, both types of
variables are present. Normally, models should be allowed to be mixed-state
models. Several possibilities exist for mixed-state models. In the most general
case, input, state, and output variables may have some components with discrete
values and some other components with continuous values. In special cases, input
variables, for example, can be discrete valued, while state and output variables
are continuous valued.

Time set of a model is the time frame within which a model is studied. Time set
of a model can be discrete or continuous. All the descriptive variables of a
model are indexed by time. As it is seen in Figure 5, there are four possibilities
for the availability, at continuous or discrete time, of the descriptive variables
which can assume continuous or discrete values (Ören 1979).

Figure 5.   Availability of Continuous or Discrete-Valued Descriptive
            Variables at Continuous or Discrete Time

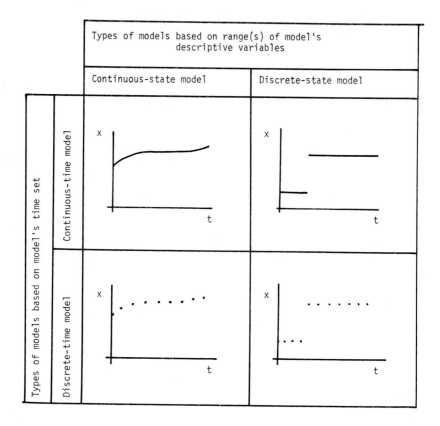

In a <u>continuous-time model</u>, time can assume all or part of the real numbers. In
<u>discrete-time models</u>, time is represented by integral numbers. Therefore in con-
tinuous-time models, descriptive variables are available at any time instant and
in discrete-time models descriptive variables are updated and available at discrete-valued time instances only. Continuous-time models can be continuous-state
models, discrete-state models, or mixed-state models.

<u>Time-trajectory of a variable</u> is a function which maps the set of time instants
where the variable is defined to the set of values (or range) of that variable.
In other words, a specific time-trajectory of a variable represents how the values
of the variable are organized in time. Depending on the nature of the trajectories of the descriptive variables of a model, one can identify continuous-change
models (or continuous models, in short), discrete models, and combined models.

A classification of modelling formalisms that are mostly used in simulation is given in Figure 6.

Figure 6. A Classification of Modelling Formalisms
(Mostly Used in Simulation)

Trajectory of model's descriptive variables	Time set of model	Modelling formalism	Range of variables		Groups
			Cont.	Discr.	
COMBINED [-CHANGE] MODEL	CONTINUOUS [-CHANGE] MODEL / DISCONTINUOUS [-CHANGE] MODEL — CONTINUOUS-TIME MODEL	Partial Differential Equation	✓		GROUP 1
		Ordinary Differential Equation	✓		GROUP 1
	DISCRETE [-CHANGE] MODEL — DISCRETE-TIME MODEL	Activity Scanning	✓	✓	GROUP 2 / GROUP 4 / GROUP 5
		Difference Equation	✓	✓	GROUP 2 / GROUP 4 / GROUP 5
		Finite-State Machine		✓	GROUP 2 / GROUP 4 / GROUP 5
		Markov Chain		✓	GROUP 2 / GROUP 4 / GROUP 5
	CONTINUOUS-TIME MODEL	Discrete Event	✓	✓	GROUP 3 / GROUP 4 / GROUP 5
		Process Interaction	✓	✓	GROUP 3 / GROUP 4 / GROUP 5

(Attributes within [ ] may be omitted)

Continuous-change models or continuous models, in short, are described by differential equations. They are continuous-time models. As it is apparent from Figure
6, all continuous-time models are not necessarily continuous-change models. For
example, discrete-event models are continuous-time models but are not necessarily
continuous-change models.

Discontinuous-change models (or discontinuous models or piecewise continuous
models) are also described by differential equations and may have basically two
types of discontinuities, i.e., derivative discontinuity and state-variable discontinuity. Derivative discontinuity occurs when the derivative of a state
variable assumes discontinuous values at some points. One should note that at a
point of derivative discontinuity, the state variables are still continuous.
State-variable discontinuity or jump discontinuity occurs when a state variable
assumes discontinuous values. In Figure 7, where x and t represent state variable
and time, there is a derivative discontinuity at point t_1 and a jump discontinuity
at point t_2. State variable discontinuity corresponds to reinitialization of a
state variable. Derivative discontinuity corresponds to a modification of a model.

Figure 7. Discontinuity

(Derivative discontinuity at point t_1 and
Jump discontinuity at point t_2)

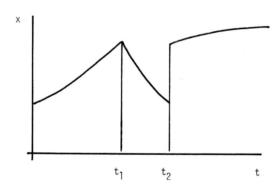

A discontinuous model is, therefore, a model described by differential equations
where at least one state variable and/or its derivative may assume discontinuous
values.

Discrete-change models (or discrete models, in short) are updated (change their
state) at discrete instances of time. Between two updating time their state
remain constant. Discrete models can be discrete-time, continuous time, or
mixed-time models.

As seen in Figure 6, some of the formalisms (Group 2 models) used to specify discrete-time models (which are necessarily discrete models) are activity scanning,
difference equation, finite state machine and Markov chain representations.

Activity scanning in simulation modelling is also named time-slicing.

In this approach, time is augmented by constant increments. After every time increment, a set of conditions are tested to decide whether relevant activities can be performed or not.

Finite-state machines can be Meally or Moore machine depending whether or not the output at a certain time depends on the current value of the input variable. Some other related modelling formalisms are stochastic finite-state automaton and cellular automaton.

Markov chains are autonomous memory models. The model has a set of state variables. The transition from one state to another is given by its probability of occurrence. Markov chains can be closed system models. However, one can define one or more output variables to be able to couple a Markov chain model to other types of models. In this case, like a Moore machine, for every state one defines corresponding output value(s).

Important formalisms for continuous-time discrete models (Group 3 models in Figure 6) are discrete-event and process representations. Mixing Group 2 and Group 3 formalisms in a model leads to mixed-time discrete models.

Combined continuous/discrete-change models (or combined-change models, or combined models, in short) are partly expressed by differential equations and partly by discrete-change models. There are several possibilities for combining modelling formalisms. Therefore there are several types of combined-change models. Accordingly, combined-change models can be either continuous-time or mixed-time models.

One such possibility is combining Group 1 modelling formalism(s) with Group 4 formalisms to have continuous-time combined-change models. Even in this case, there are two basically different modelling approaches. One possibility is to use events to model (and thereafter manipulate) discontinuities of piecewise continuous models. Another approach is to have in Group 3 models (i.e., discrete event and process interaction models) some variables which have to change continuously and which therefore require Group 1 modelling formalisms. For example, an ingot can be created at the occurrence of an event. The temperature change of the ingot can then be modelled by a differential equation. Another type of event may represent putting several ingots in an oven. The sudden change of the temperature of the oven can be modelled as a jump discontinuity (or state-variable discontinuity at the occurrence of this event.

Yet another possibility is to combine Group 1 modelling formalism(s) with Group 2 formalisms(s) to have mixed-time combined-change models. Modelling formalisms and specially developed appropriate algorithms to handle this type of models already exist (Wait 1969, Palusinski, Wait 1977a, 1977b, 1978). Most general type of combined-change models would correspond to the combination of Group 1, Group 2, and Group 3 modelling formalisms.

Based on the time-dependence of models, time-variant and time-invariant models can be distinguished. "In a time-varying model, time may enter explicitly as an argument of the rules of interaction which may thus appear to be different at different times" (Zeigler 1976, p.23). In a time-invariant model, the sequence of the state values that the model will assume is independent of the initial moment of time for a given initial state and input sequence. Time-invariant models are also named stationery models.

Classification Based on Functional Relationships:

Based on the functional relationships the following criteria can be used: nature of the determinism, nature of the anticipation of the future, linearity, and stiffness.

<u>Deterministic models</u> are also named <u>state-determined models</u> (Padulo, Arbib 1974,
p. 25). In a <u>deterministic model</u>, current state and current input, if any,
uniquely determine next value of state variable(s). A <u>non-deterministic model</u>
(or <u>stochastic model</u>, or <u>probabilistic model</u>) has at least one random variable and
therefore, at a given instant in time the next state of the model is not uniquely
determined.

Most (if not all of the) models commonly used are <u>non-anticipatory models</u>. In a
<u>non-anticipatory model</u>, at a given time, the future values of the input variable(s)
are not taken into consideration to compute next values of the state variable(s).
Non-anticipatory modelling is the essence of state-determined modelling which is
a very powerful and important paradigm in modelling different types of systems.
Modelling formalisms commonly used are non-anticipatory (or non-anticipative) in
nature. For example, "for a differential equation to uniquely describe a system,
it must be understood that the system is nonanticipative." (DeRusso, Roy, Close
1966, p. 48). However, non-anticipatory modelling has its limitations in modelling real systems. Consider for example a manufacturer of umbrellas. The number
of umbrellas to be manufactured for the next rainy season should depend, among
other things, on the anticipated characteristics of the next rainy season.

A methodologically consistent approach is to recourse to behaviorally-anticipatory
modelling. In a <u>behaviorally-anticipatory model</u>, a special module of the model
has the special task of anticipating the future, i.e., anticipating the characteristics of the input variable(s). Once the current image of the future is anticipated, the internal working of a behaviorally-anticipatory model is similar to
a non-anticipatory model. The only difference is that at a certain point in time,
next values of the state variable(s) of a behaviorally-anticipatory model depend
on the current value(s) of state and input variables, as well as on the currently
anticipated values of the future values of the input variable(s).

A <u>linear model</u> changes state and emits outputs according to linear transformations.
A <u>linear transformation</u> $L:V \rightarrow W$ satisfies the superposition principle, i.e.:

for all v_1, $v_2 \in V$ and all scalars c_1, c_2

$$L [c_1 v_1(t) + c_2 v_2(t)] = c_1 L [v_1(t)] + c_2 L [v_2(t)].$$

In other words, if the responses of the model to two different inputs $v_1(t)$ and
$v_2(t)$ are $y_1(t)$ and $y_2(t)$, respectively and if c_1 and c_2 are scalars, then the
response of the model to

$$v(t) = c_1 v_1(t) + c_2 v_2(t) \text{ is}$$
$$y(t) = c_1 y_1(t) + c_2 y_2(t)$$

for all values of v_1, $v_2 \in V$ and c_1 and c_2. (DeRusso, Roy, Close, 1966, p.3)

<u>Coupled linear/nonlinear models</u> have important application areas especially in
stiff system models (Palusinski, Wait 1977a, 1977b, 1978). In a <u>stiff system
model</u> time constants are widely separated. Special algorithms exist to detect and
handle the integration of stiff system models (Shampine, Gordon 1975, Enright,
Hull, Lindberg 1975, Palusinski, Wait 1978).

<u>Classification Based on Formalisms Used to Describe the Models</u>:

Based on the formalism used to formulate the models, it is possible to identify
differential equation models (partial, ordinary, and several types thereof),
activity scanning models, difference equation models, finite-state machines,
Markov chain models, discrete-event models, and process interaction models. A
categorization of these modelling formalisms based on the trajectories of the
descriptive variables of models and the time set of models is given in Figure 6.

Classification Based on Intended Use of the Models:

There are two basic types of decisions, i.e., value-free and value-dependent decisions. Therefore, there are two types of models, i.e., value-free models which
are used for value-free decisions and normative models which are used for value-
dependent decisions. Value-free models are further distinguished as descriptive
models, explanatory models, and predictive models. Normative models are basically
evaluative models and prescriptive models. Some explications on the respective
decision making techniques are given in relevant subsection of "experimentation"
section of Part 3 titled "useability of models" of this paper.

Classification Based on the Disposition of Submodels:

Based on the existence of submodels, one distinguishes monolithic models and
modular models. A monolithic model does not have submodels. A modular model consists of several submodels. Depending on the nature of the disposition of each
submodel vis-a-vis to other submodels, there are two types of modular models, i.e.,
coupled models and cellular models. In a coupled model, coupling specification
defines the input/output relationships of the submodels. (Ören 1971, 1975, 1978e,
Elmqvist 1978) In a model with nested-coupling, at least one of the submodels is
itself a coupled model.(Ören 1971)

A hierarchical model consists of several submodels. The system coupling represents the inhibitions about the input/output relationships of the submodels based
on their place in the hierarchy. (Mesarovic, Macko, Takahara 1970, Elliott,
Talavage 1978, Mahmoud 1977) Time scales of submodels belonging to different
levels of hierarchy may be the same or different depending on the time-lags of the
outputs or on the input filtering of the submodels which are at higher levels of
the hierarchy. Since a hierarchical system model is basically a coupled model
with special restrictions for the input/output relationships of the submodels, the
special types of models which are possible for coupled models are also possible
for hierarchical models. Especially two possibilities are important, i.e., hierarchical system models with nested coupling and hierarchical system models with
time-varying structure.

In a hierarchical system model with nested coupling, at least one submodel in the
hierarchy is already expressed as a coupled model. Depending on the need, the
submodels of this particular submodel may be coupled as a hierarchical model.
However, this is not a prerequisite for a hierarchical system model with nested
coupling.

Classification Based on Goals to be Pursued:

Based on the goals to be pursued, one can distinguish allotelic and autotelic
system models. (Locker, Coulter Jr. 1975, 1976) In an allotelic system model
the goals are set or formulated by somebody outside the system. An autotelic
system model generates its own goals. The two types of allotelic system models
are teleonomic system model and teleozetic system model. A teleonomic system
model (or a goal determined system model) acts according to fixed goals.

In an adaptive model, the structure of the model changes several times to improve
its performance in time with respect to a reference performance. Powerful new
concepts applicable to adaptive control systems, such as actively adaptive controller have already been developed: "... An actively-adaptive controller utilizes, in addition to the available real-time information, the knowledge that
future observations will be made, and regulates its adaptation (learning). This
is done by anticipating how future estimation will be beneficial to the control
objective. On the other hand, a passively adaptive controller, while utilizing
the available real-time measurements, does not account for the fact that future
observations will be made. Thus any learning in such a case will occur in an

accidental manner." (Tse, Bar-Shalom 1976)

A recent bibliography covers major publications on adaptive system models (Asher
et.al. 1976). A system model is called a self-stabilizing system model, "if and
only if, regardless of the initial state and regardless of the privilege selected
each time for the next move, at least one privilege will always be present and the
system is guaranteed to find itself in a legitimate state after a finite number of
moves" (Dijkstra 1974). The paradigm of self-stabilizing system models is important to have built-in robustness in complex systems. An important application
area is realization of robust computer operating system (Sammes 1975).

A teleological system model (or a goal-seeking system model) "may be characterized
by a class of possible input functions, a class of possible output functions, a
performance index which may (and usually, although not necessarily, does) provide
some measure of system effectiveness or efficiency, and a system goal. The system
goal may be either: 1) to optimize the performance index for a given input; or
2) to optimize the performance index based on only those outputs which conform to
certain prescribed standards or specifications." (Pollin, Sanders 1969)

Set theoretic definitions of several types of goal-seeking system models (type 1
goal-seeking system, type 2 goal-seeking system, functional goal-seeking system,
etc. ...) are given in the literature (Pollin 1969, Pollin, Sanders 1969). A
teleozetic system model (or a goal selecting system model) is able to choose among
a set of goals. An autotelic system model is responsible for its own goals and is
therefore a teleogenic or a goal generating system model. A teleogenic system
model is able to generate previously non-existing goals for its own activity.
A very simple mode of teleogenesis is combination of already existing goals
(Locker, Coulter Jr. 1975, 1976, Mago, Locker 1975).

Classification Based on Organization of Components:

Zeleny, like Maturana and Varela, distinguishes organization and structure of
systems as follows: "A network of interactions between the components, renewing
the system as a distinct unity constitutes the organization of the system. The
actual spacial arrangement of components and their relations, integrating the
system temporarily in a given physical milieu, constitutes its structure."
(Zeleny 1977)

Based on the organization of components of a system, there are two types of system
models: allopoietic and autopoietic system models. An allopoietic system model
has no rules to change its organization. This type of system model is also called
mechanistic system model. Mechanistic system models should be used with great
care if they are going to represent organic systems, i.e., biological or social
organizations. If the time span of the study of an organic system is short (relative to the speed of its organizational change) then one can use mechanistic
system models to represent them. This is like assuming that the earth is flat,
locally. However, beyond the region of fit, this assumption is no longer valid.
"Allopoietic systems are organizationally open, they produce something different
than themselves. Their boundaries are observer-dependent, their input and output
surfaces connect them mechanically with their environment. Their purpose, as an
interpretation of their input/output relation, lies solely in the domain of the
observer." (Zeleny 1977)

An autopoietic system model is able to change its organization and can do so by
using some rules. Organic systems can best be represented by autopoietic system
models, if they are studied during a long time span. Basically, there are two
types of autopoietic system models: 1) System has fixed rules for changing its
organization. 2) System has meta-rules to modify its rules.

Some types of autopoietic system models without meta-rules are variable-structure model, self-organizing system model, and self-reproducing system model. In a variable-structure model, the submodels of a model may be created, destroyed, or modified at the occurrence of some conditions and they may have time-varying input/output relationships (Ören 1975).

In a model with time-varying coupling, the input/output relationships of the submodels of a model (or their coupling) can be altered based on the occurrence of some conditions (Ören 1975). Another interesting concept in modelling complex system is hierarchical system model with variable structure. This is a combination of hierarchical system model and variable-structure model.

In a self-organizing system model, independent of the physical distribution of the submodels and the characteristics of their surrounding, every type of submodel and the rules according to which they interact with each other and with their surrounding, are specified. In a self-organizing system model, some submodels may have catalytic effects to foster or hinder some types of organizational changes. Use of self-organizing system models in simulation is relatively recent (Hogeweg, Hesper 1979).

As it is pointed out by Humphries, the term "self-reproducing" almost implies that an object reproduces itself and is therefore logically inconsistent. However, from another point of view, the prefix "self" means "... either (a) that whatever initiates the process is an integral part of the object, or (b) that once initiated, the process is internally directed, proceeding spontaneously with respect to its environment, or both. Or put in another way, the conditions that are both necessary and sufficient for initiating and/or maintaining the process are an integral part of the object, and not part of the environment." (Humphries 1973) Methodological refinements for self-reproducing system models can be very useful to study especially the anomalies in the reproduction of the units. In this way one can simulate, for example, different types of cancers.

Autopoietic systems with meta-rules can be represented by self-learning system models, self-regulating system models, or by evolutionary system models. In all of them self-reference is important (Varela 1975) and can be realized by having submodel(s) responsible for self-observation.

2.2 Computer-Assisted Modelling

On-line modelling has been a reality for some time. As the next level of achievement, what is needed is computer-assisted modelling (Ören 1977, 1978c). With the help of specially developed programs, computer assistance may be helpful in several aspects of the modelling process. However, one needs, first of all, the specification of language(s) for computerized modelling and model manipulation. This point of view embraces, but is not limited to, thinking in terms of simulation languages only (Ören 1978d). In this approach, one needs to have a computer language to specify models based on different model formalisms which are outlined in the classification of models part of this paper. A particular implementation of this modelling language may, of course, be limited to a subset of all the modelling methodologies that the language specification may allow.

Modelling and model manipulation languages(s) should have its (their) modelling formalisms based on adequate theories. Theory-based modelling is not only possible but also highly desirable, since it has several advantages such as uniformity of the descriptions and availability (or possibility of developing) algorithms to handle such models for different purposes including algorithmic tests of the completeness and correctness of the model specifications (Ören 1974, Klir, Uyttenhove 1976).

Several system theories, such as deductive system theories of Wymore or Mesarovic (Ören 1974), or inductive system theory of Klir (Klir 1969), or the theory of modelling and simulation Zeigler (Zeigler 1976, 1978) provide the basics for this purpose. Uniform modelling formalisms applicable to several classes of models can provide useful frameworks to define necessary man/machine dialogues for computer-assisted modelling (Ören 1974, 1978 a,b,d, Ören, Dulk 1978, Ören, Zeigler 1979, Zeigler 1976). Furthermore, uniform modelling formalisms are also important for computer-assistance in the use, assessments of acceptability of models, and enhancement of the referability of models.

Computer-assisted modelling can work for several types of system paradigms. "For example, if the modeller wants to develop a model of a micro-programmed computer, he may want to model it as a composition of two automata, i.e., the operational and control automata (Agerwala 1976). Since the operational automaton is a Moore machine (Hartmanis, Stearns 1966), the tutorial program can assist the modeller by providing him a framework within which the modeller can specify input, state, and output alphabets, as well as state transition and output functions. Further sophistication can also be conceived where such a model may automatically be checked for minimization and can then be minimized if required. Such a tutorial package can guide the user to correctly specify a system. This approach is more valuable than letting the modeller to define a system without any assistance and then algorithmically checking for any model inconsistencies (Ören 1978b, p.155).

Modular-system modelling has several advantages such as to increase the efficiency of the modelling process by allowing several teams of modellers to work on different modules simultaneously and to increase the possibilities to understand, modify and refer to parts or totality of models. Other advantages are the possibilities offered for hierarchical and adaptive system modelling as well as modelling of systems with variable-structures.

Parametric system models are generalization of similar system models where individual copies of models may be distinguished by specific values of model parameters (Ören 1971). Parametric system modelling ability can easily be provided in a computer-assisted modelling environment to provide flexibility in modelling. Non-procedural modelling is the expression of a model without taking into account the requirements of the computational operations such as the sequence of the operations. Computer-assistance is essential in translating user-oriented non-procedural models into machine-oriented procedural representations.

Some important software engineering concepts such as top-down programming and step-wise programming (Wirth 1973) have their counter parts directly relevant in the modelling process itself, i.e., top-down modelling and step-wise modelling (Hunt 1978). Top-down modelling reflects the natural sequence of human thinking in developing model(s) of a system. It involves identification of the components of a system as well as the attributes of the components; classification of the attributes as parameters, inputs, states, and outputs; specification of the functional relationships of different attributes for every component, and specification of the input/output relationships of the submodels.

Most of the existing simulation languages have completely different "world views." Programs written in most of the existing simulation languages start with the "initial values" of the state variables and the values of the parameters. Even though these initial values are essential in simulation, they do not have to be presented first in a model description. Step-wise model refinement ability can be integrated in a computer-assisted modelling system. One can then, interactively, modify the characteristics of the elements of a model and their functional relationships.

An important concept yet to be incorporated in modelling formalisms is model robustness.

"It is evident that the concept of robustness is an extension of the work already extant in areas such as system reliability and fault-tolerant computing. System robustness should go one step further, however, by recognizing that, whereas system reliability and fault-tolerant computing are concerned with assuring correct and perfect system functioning through techniques such as redundancy, system robustness implies that the system may not always function correctly and perfectly but it should degrade predictably and gracefully, rather than catastrophically. We cannot demand that a system function precisely as designed when subjected to an abnormal environment, but we can demand that it survive the abnormal environment rather than become self-destructive (Hunt 1978, p.147).

In computer-assisted modelling, the specifications of a model have to be transformed into a program by a translator. Necessary checks in this translator program may assure the generation of reliable programs. Furthermore, classical non-numerical optimization techniques applicable to compiler optimization (Rustin 1972) may assure efficiency of the machine executable programs (Ören 1978c, Ören, Zeigler 1979). Already several software systems exist to computerize the generation of simulation programs (Davies 1976, 1979, Howart 1975, Luker, Stephenson 1978, Mathewson 1974, Mathewson, Allen 1977, 1978) or for validation (Allen, Clema 1976).

3. USABILITY OF MODELS

Models can be used for communication, experimentation, and for algorithmic manipulations. Computer-assistance can be very valuable in all three categories of model use.

3.1 Communication

Computer-assisted model use can improve man/machine, machine/man, and man/man communication (Ören, 1978b, Ören, Zeigler 1979). Man/machine communication can be improved by having a program to guide the user to specify a model, or to modify a model which is already in a file of models. Computer-assisted modelling module can then automatically submit a user-specified model to a program module responsible for checking whether the model is acceptable with respect to a relevant modelling formalism (Ören, Zeigler 1979).

Machine/man communication can be drastically improved by different types of computerized documentation of models. Submodels or their interactions can be documented selectively on different types of output units including graphic display units. Selective documentation can answer questions like enumeration of all the inputs of a submodel and the names of the output variables of specific submodels from which they receive information. Computerized documentation of the sequence of structural change of a model with variable structure can bring new vistas in the simulation of large and complex systems. In this way, instead of generating numbers as a result of simulation study, one can study how the structure or the organization of the submodels evolve under complex operating rules.

Man/man communication can be improved by computerized editing (formatting) of mathematical models. Furthermore, if uniform modelling formalisms are used, one can have for every type of model, a framework to identify the elements of a model such as the inputs, states, outputs, parameters, constants, tabular functions, etc. Enhanced man/man communication implies improved comprehensibility and portability of models and hence their improved useability. Comprehensibility of models is paramount in the rational selection of models as bases for decisions in a participatory democracy (Ören 1977). Inspite of the existence of hundreds of simulation languages, due to the lack of appropriate modelling language(s), most large-scale models are incomprehensible (Nance 1977). Incomprehensibility of models reduce their chances of useability and hence their utility.

3.2 Experimentation

Experimentation is controlled observation. It has several aspects, i.e., (1) goals for experimentation, (2) elements of experimentation, and (3) algorithmic manipulation of experimentation specifications. Simulation is experimentation with models. In computerized simulation, necessary computations are done by a computer. On-line simulation and interactive simulation have been possible for some time. However, what is needed is computer-assisted simulation to have the assistance of computers with their non-numerical information processing abilities to explore systematically all aspects of experimentation. With the help of a specially designed program, a computer can direct a user to formulate an experimentation with a model. For example, since the state variables of a model are known to the computer, it can require that all of them have been properly initialized. Furthermore, if the ranges of the state variables are also specified as part of a model, the computer can also verify whether the initial values of the state variables are acceptable or not. Similarly, if the model specification involves a system coupling, input variables of all the subsystems which do not receive information from any subsystem of the model, are by definition external inputs to the model. In computer-assisted simulation, for a given system coupling, the computer can determine the external input variables and can require that the user specifies them.

Goals for experimentation. Experimentation is done either for decision making or to improve decision making capabilities which may be needed in the future. Improving skills for future decision making can be realized by simulation games or by training simulators. A simulation game is used at early stages of skill formation. It is useful in providing a synthetic situation where questions can be asked (and answered) within competitive and/or cooperative situations.

A training simulator is useful in late stages of skill formation to automate the response of a human in a possible operational situation. Simulation can be used in any type of decision making. The five types of possible decisions can be grouped in two categories, i.e., value-free decisions and value-dependent decisions. Value-free decisions or designative decisions consist of descriptive, explanatory, and predictive decisions (O'Shaughnessy 1972).

Value-dependent decisions are evaluative decisions (or appraisive decisions) and prescriptive decisions. An evaluative decision necessitates the determination of the relative worth of an alternative which may be a person or an object. "A prescriptive decision involves the selection of a course of action" (O'Shaughnessy 1972, p.16) . One has to distinguish conceivable, desireable, feasible and optimal alternatives.

The nature of the goal of experimentation delimits the scope of applicability of the results of the experimentation. And one should identify clearly the fact that the inherent draw-backs of a certain type of decision making technique are not necessarily the limitations due to computer assistance or to computerization at all. For example, rational decision making may imply that the decision makers have common goals and evaluative criteria. Especially in most multifaceted systems this assumption is not realistic. In most of the ill-defined, ill-structured problematique systems "decision-making invariably has political elements, where the emphasis is on reaching some agreement through the processes of persuasion, bargaining or straight exercises of power. ... legalistic decisions aim to stabilize conflict rather than to economize on means" (O'Shaughnessy 1972, p.182).

Elements of experimentation. The three basic components of experimentations are experimental frames, application of experimental frames to models, and generated/ observed behavior. Elsewhere in the literature this topic has been elaborated on (Zeigler 1976, 1978) especially on the value of the computer-assistance (Ören 1978d, Ören, Zeigler 1979).

Algorithmic manipulation of experimentation specifications. This can be done either for consistency checks or for documentation purposes (Oren, 1978b, d, Oren, Zeigler 1979, Zeigler 1976).

3.3 Algorithmic Manipulation of Models

This is an important area still to be fully explored for the computer-assisted use of models (Oren 1978b,d, Oren, Zeigler 1979, Zeigler 1976). One can identify the following goals for algorithmic manipulation of models: documentation, consistency checks, comparison, transformation, and analysis of models.

Computerized model-documentation involves algorithmic manipulation of models for the purpose of documentation and is mentioned in the section titled "communication."

Another important area for algorithmic manipulation of models is consistency checks of a model description with respect to a modelling formalism and is mentioned in the next section titled "acceptability of models."

Model comparison algorithms can be performed to detect homomorphism, isomorphism, state-equivalence, or input-equivalence of models (Wymore 1967).

Some of the goals for model transformation algorithms are: decomposition, simplification, minimization, coarsening, or elaboration of models (Zeigler 1976).

Some of the goals for algorithmic model analysis algorithms are: state, fault, or machine identification (Gill 1962) and checks to determine whether a model represents a controllable or observable system (Padulo, Arbib 1974).

4. ACCEPTABILITY OF MODELS

In computerized models, there are three important aspects of a model: 1) Its formal representation which consist of a parametric representation (or parametric model) and a set of values of the parameters of the model (or model's parameter set) (Oren 1978e, Oren, Dulk 1978); 2) A computer program which makes a formal representation of a model computer-readable (The program of a model augmented with the program of a conformable experimental frame constitues a simulation program); 3) Data which consists of real-system data and simulated data. Data of a real-system (existing in the case of analysis or control problems, or proposed, in the case of design problems) are necessary in the formulation of a model as well as in the assessment of the acceptability of data obtained from a simulation study.

As outlined in Figure 8, the assessment of the acceptability of a model, its program, and related data can be done based on conceived norms, real system, goal(s) of model use, modelling formalism, another or the same model, and the programming methodology.

Figure 9 represents different elements between real-system and simulated data, i.e., real-system data, model(s) and program(s). Observation, modelling, programming, and simulation are distinguished as processes which link two consecutive elements. Validation link simulated and real-world data. Groups of criteria for the assessment of the acceptability of model, program, and data are represented in Figure 9.

A detailed list of criteria for the assessment of the acceptability of model, program, and data is given in Figure 10.

Figure 8. Possibilities for the Assessment of the Acceptability of Models, Programs, and Data

Acceptability of:	With Respect to:	Conceived Normes	Real-System	Goal of Model Use	Modelling Formalism	Model (Same or Another)	Programming Methodology
Model	Parametric Model		✓	✓	✓	✓	
Model	Values of the Parameters of a Model		✓	✓			
Program	Representation					✓	✓
Program	Execution for Data Generation (Simulation)	✓					
Data	Real-System Data	✓	✓	✓			
Data	Simulated Data	✓	✓	✓			

Figure 9. Groups of Criteria for the Assessment of the Acceptability of
           Models, Programs, and Data

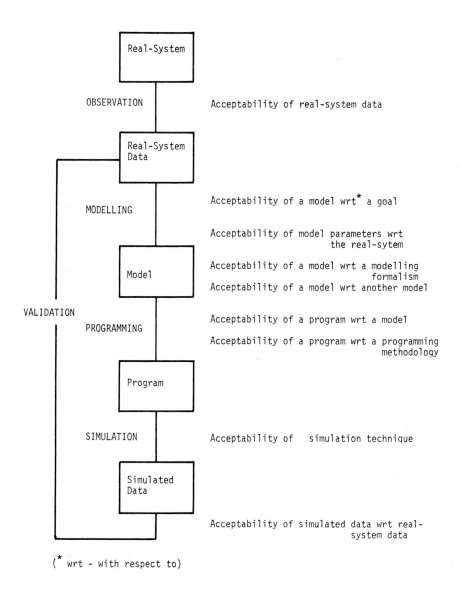

(* wrt - with respect to)

Figure 10.   Detailed list of Criteria for the Assessment of
the Acceptability of Models, Programs, and Data

- ACCEPTABILITY OF REAL-SYSTEM DATA
    - Measurement-noise assessment

- ACCEPTABILITY OF A MODEL WITH RESPECT TO A GOAL
    - Model evaluation
        - Model usefulness (utility)
            - Timeliness
            - Cost-effectiveness
        - Model credibility
        - Model relevance
            - Range of applicability of a model
        - Model comprehensibility
        - Model useability

- ACCEPTABILITY OF MODEL PARAMETERS WITH RESPECT TO THE REAL-
                                                   SYSTEM
    - Model fitting
    - Model calibration

- ACCEPTABILITY OF A MODEL WITH RESPECT TO A MODELLING FORMALISM
    - Model consistency
      (Mathematical correctness of a model)
        - Concordance of units
    - Model robustness

- ACCEPTABILITY OF A MODEL WITH RESPECT TO ANOTHER MODEL
    - Validity of model simplification
    - Approximation in model simplification

- ACCEPTABILITY OF A PROGRAM WITH RESPECT TO A MODEL
    - Program verification

- ACCEPTABILITY OF A PROGRAM WITH RESPECT TO A PROGRAMMING
                                              METHODOLOGY
    - Program reliability
        - Program correctness
            - Program debugging
            - Program testing
            - Program certification
        - Program robustness
            - Defensive programming
    - Program efficiency
        - Time   (programming/compilation/execution/program
                   modification/program documentation/...)
        - Memory (compilation/execution)

- ACCEPTABILITY OF SIMULATION TECHNIQUE
    - Determination of the run length
    - Determination of the data collection time period
        - Transient
        - Steady-state
    - Choosing integration algorithm
        - Discontinuity handling
        - Stiffness
        - Backcasting (Backword integration)
    - Choosing appropriate time advance
        - Time increment for discrete simulation
        - Integration step size
            - Error criteria

- ACCEPTABILITY OF SIMULATED DATA WITH RESPECT TO REAL-SYSTEM DATA
    - Model validity
        - Replicative validity
        - Predictive validity
        - Structural validity

5. REFERABILITY OF MODELS

The ability to use models, or modules of them, developed by other modellers is a crucial issue. Model transferability (or model portability) has two aspects, i.e. portability of the whole model or its submodels. Due to the lack of appropriate modelling language(s), most, if not all, of the computerized models are not portable. Computerized access to files of models expressed in terms of uniform modelling formalisms can increase the referability of models and hence can enhance the transferability of expertise vested in different modelling projects. In a recent article, files of models were elaborated on (Ören, Zeigler 1979). It is only by the standardization of the modelling language(s) that a framework to specify and manipulate system models based on different modelling formalisms can be realized.

6. REFERENCES

Agerwala, T. (1976), "Microprogram Optimization: A Survey," IEEE Trans. on Computers, Vol. C-25, No. 10, p. 962-973.

Allen, B., Clema, J. K. (1976), "Independent Verification/Validation Support Software," Proc. IEEE National Aerospace and Electronics Conference, Daytona, Ohio, pp. 276-281.

Asher, R. B., Andrisiani II, D., Dorato, P. (1976), "Bibliography on Adaptive Control Systems," Proc. of the IEEE, Special Issue on Adaptive Systems, Vol. 64, No. 8, pp. 1226-1240.

Bell, D. (1976), "Welcome to the Post-Industrial Society," SCITEC Bulletin, Oct. 1976, pp. 6-8, (Reprinted from Physics Today, Feb. 1976).

Davies, N. R. (1976), "A Modular Interactive System for Discrete Event Simulation Modelling," Proc. of the 9th Hawaii International Conference on Systems Science, pp. 296.

Davies, N. R. (1979), "Interactive Simulation Program Generation," In B.P. Zeigler et al. (Eds.) Methodology in Systems Modelling and Simulation, North-Holland Amsterdam, the Netherlands.

De Russo, P. M., Roy, R. J., Close, C. M. (1966), State Variables for Engineers, John Wiley, New York, 608 p.

Dijkstra, E. W. (1974), "Self-stabilizing Systems in Spite of Distributed Control," Com. ACM, Vol. 17, No. 11, pp. 643-644.

Elliott, M. B., Talavage, J. J. (1978), "A Theory for Three-Level Hierarchies," in T. I. Ören (Ed.), Cybernetics and Modelling and Simulation of Large-Scale Systems, International Association for Cybernetics, Namur, Belgium, pp. 101-125.

Elmqvist, H. (1978), A Structured Model Language for Large Continuous Systems, Thesis, Dept. of Automatic Control, Lund Institute of Technology, Sweden, 226 p.

Enright, W. H., Hull, T. E., Lindberg, B. (1975), "Comparing Numerical Methods for Stiff Systems of O.D.E.'s," BIT, 15, 1975, pp. 10-48.

Gill, A. (1962), Introduction to the Theory of Finite-State Machines, McGraw-Hill, New York.

Hartmanis, J., Stearns, R. E. (1966), Algebraic Structure Theory of Sequential Machines, Prentice-Hall, Englewood Cliffs, NJ.

Hogeweg, P., Hesper, B. (1979), "Heterarchical, Selforganizing Sim. Systems: Concepts and Applications in Biology," In B. P. Zeigler et al. (Eds.), Methodology in Systems Modelling and Simulation, North-Holland, Amsterdam, the Netherlands.

Howart, M. A. (1975), GRAFT - An Interactive Graphical Simulation Program Generator, M.Sc. Report, Imperial College, England.

Humphries, J. (1973), "Self-Reproduction and Chain Reactions," Kybernetes, Vol. 2, No. 3, pp. 157-159.

Hunt, B. R. (1978), "Large-Scale Systems Theories: Some Paradigms from Software Systems Engineering," in T. I. Ören (Ed.), Cybernetics and Modelling and Simulation of Large-Scale Systems, International Association for Cybernetics Namur, Belgium, 1978, pp. 141-150.

Klir, G. J. (1969), An Approach to General Systems Theory, Van Nostrand Reinhold, New York.

Klir, G. J., Uyttenhove, H. J. J. (1976), "Computerized Methodology for Structure Modelling," Annals of Systems Research, Vol. 5, pp. 29-66.

Locker, A., Coulter Jr.,N. A. (1975), "An Outline of Teleogenic Systems Theory," Proc. 2nd European Meeting Cybern. Systems Res., Vol 2, pp. 156-165.

Locker, A., Coulter Jr., N. A. (1976), "Recent Progress Towards a Theory of Teleogenic Systems," Kybernetes, Vol. 5, No. 2, pp. 67-72.

Luker, P. A., Stephenson, J. (1978), "Simulation without Programs," Proc., UKSC Conference on Computer Simulation, Chester, England, pp. 181-191.

Mago, G. A., Locker, A. (1975), "Towards a Mathematical Model of Goal Generation," Proc. 2nd European Meeting Cybern. Systems Res., Vol. 2, pp. 166-174.

Mahmoud, M. S. (1977), "Multilevel Systems Control and Applications: A Survey," IEEE Trans. on Systems, Man, and Cybernetics, Vol. SMC-7, No. 3, pp. 125-143.

Mathewson, S. C. (1974), "Simulation Program Generators," Simulation, Vol. 23, No. 6, pp. 181-187.

Mathewson, S. C., Allen, J. A. (1977), "DRAFT/GASP - A Program Generator for GASP," Proc. of the Tenth Annual Simulation Symposium, Tampa, FL, pp. 221-227.

Mathewson, S. C., Allen, J. A. (1978), "A Commentary on the Proposal for a Simulation Model Specification and Documentation Language," Proc. UKSC Conference on Computer Simulation, Chester, England, April 4-6, 1978, pp. 158-167.

Mesarovic, M. D., Macko, D., Takahara, Y. (1970), Theory of Hierarchical, Multilevel Systems, Academic Press, New York.

Minsky, M. L. (1965), "Matter, Mind, and Models," Proc. IFIP Congress, Vol. 1, Spartan Books, pp. 45-49.

McLeod, J. (1963), Simulation is What?, Simulation, Vol. 1, No. 1, pp. 5-6.

McLeod, J. (1970), Definitions, anyone?, Simulation, Vol. 15, No. 2, p. vi.

Nance, R. E. (1977), The Feasibility of and Methodology for Developing Federal Documentation Standards for Simulation Models, Final Report prepared for NBS, US Dept. of Commerce, Computer Science Dept., VPI & SU, Blacksburg, VI.

Ören, T. I. (1971), "GEST: A Combined Digital Simulation Language for Large Scale Systems," Proc. AICA Symposium on Simulation of Complex Systems, Tokyo, Japan, pp. B1/1-4.

Ören, T. I. (1974), "Deductive General Systems Theories and Simulation of Large Sclae Systems,"Proc. Summer Computer Simulation Conference, Houston TX, pp. 13-16.

Ören, T. I. (1975), "Simulation of Time-varying Systems," in J. Rose (Ed.) Advances in Cybernetics, Gordon & Breach Science Publishers Ltd., England, pp. 1229-1238.

Ören, T. I. (1977), "Simulation - as it has been, is and should be," Simulation, Vol. 29, No. 5, pp. 182-183.

Ören, T. I. (Ed.) (1978a), Cybernetics and Modelling and Simulation of Large Scale Systems, International Association for Cybernetics, Namur, Belgium, 1978, 191 p.

Ören, T. I. (1978b), "Rationale for Large Scale System Simulation Software Based on Cybernetics and General System Theories," In T. I. Oren (Ed.), Cybernetics and Modelling and Simulation of Large-Scale Systems, International Association for Cybernetics, Namur, Belgium, pp. 151-179.

Ören, T. I. (1978), "A Personal View on the Future of Simulation Languages," Proc. UKSC Conference on Computer Simulation, Chester, England, pp. 294-306.

Ören, T. I. (1978d), "Modelling, Model Manipulation and Programming Concepts in Simulation: A Framework," In G. C. Vansteenkiste (Ed.), Modelling, Identification and Control in Environmental Systems, North-Holland, the Netherlands.

Ören, T. I. (1978e), Reference Manual of GEST 78 - Level 1 (A Modelling and Simulation Language for Combined Systems), Technical Report 78-02, Univ. of Ottawa, Computer Science Dept., Ottawa, Ont., Canada.

Ören, T. I. (1979), Combined Continuous/Discrete System Simulation: Methodology and Software (Book in preparation).

Ören, T. I., Dulk, J. den (1978), Ecological Models Expressed in GEST 78, Report prepared for the Dept. of Theoretical Production Ecology, Dutch Agricultural Univ., Wageningen, the Netherlands.

Ören, T. I., Zeigler, B. P. (1979), "Concepts for Advanced Simulation Methodologies," Simulation (accepted for publication).

O'Shaughnessy, J. (1972), Inquiry and Decision, George Allen & Unwin, England, 200 p.

Padulo, L., Arbib, M. A. (1974), System Theory, W. A. Saunders Company, Philadelphia, MA, 779 p.

Palusinski, O. A., Wait, J.V. (1977a), Simulation Methods for Combined Linear and Nonlinear Systems, CSRL Memo: 306, Univ. of Arizona, Dept. of Electrical Engineering, 54 p.

Palusinski, O. A., Wait J. V. (1977b), "Simulation of Partitioned Linear/Nonlinear Systems," Proc. Simulation Symposium, Montreux, Switzerland, pp. 134-139.

Palusinski, O. A., Wait, J. A. (1978), "Simulation Methods for Combined Linear and Nonlinear Systems," Simulation, Vol. 30, No. 3, March 1978, pp. 85-94.

Pollin, J. m. (1969), Theoretical Foundations for Analysis of Teleological Systems Ph.D. Dissertation, University of Arizona, Tucson, AZ.

Pollin, J. M., Sanders, J. L. (1969), "Theoretical Foundations for Teleological Systems," Systems Engineering, Vol. 1, No. 1, pp. 57-89.

Roth, P. F., Gass, S. I., Lemoine, A. J. (1978), "Some Considerations for Improving Federal Modeling," Proc. Winter Simulation Conference, also in B.P. Zeigler et al (Eds.) Methodology in Systems Modelling and Simulation, North-Holland, Amsterdam, the Netherlands, 1979.

Rubinyi, P. (1975), "Project Planning and Management," Proc. Conference 75, Canadian Data Processing Institute, Ottawa, Ont., pp. 528-540.

Rustin, R. (1972),Design and Optimization of Compilers, Prentice-Hall, Englewood Cliffs, NJ.

Sammes, A. J. (1975), "Systems with state re-set," Computer Journal, Vol. 18, pp. 135-139.

Sela, Michael (1978), Private correspondence.

Shampine, L. F. (1977), "Stiffness and Nonstiff Differential Equation Solvers, II: Detecting Stiffness with Runge-Kutta Methods," ACM Trans. on Mathematical Software, Vol. 3, No. 1, pp. 44-53.

Shampine, L. F. and Gordon, M. K. (1975), "Typical Problems for Stiff Differential Equations," SIGNUM Newsletter (ACM), Vol. 10, p. 41.

SIGPLAN/SIGART (1977), Proc. Symposium on Artificial Intelligence and Programming Languages ,ACM SIGPLAN Notices, Vol. 12, No. 8, Aug. 1977, 179 p.

Tse, B., Bar-Shalom, Y. (1976), "Actively Adaptive Control for Non-Linear Stochastic Systems," Proc. IEEE, Special Issue on Adaptive Systems, Vol. 64, No. 8, pp. 1172-1181.

US GAO (1976), Ways to Improve Management of Federally Funded Computerized Models, Report LCD-75-111, Washington, DC.

Varela, F. J. (1975), "A Calculus for Self-Reference," International Journal of General Systems, Vol. 2, No. 1, pp. 5-24.

Wait, J. V. (1969), Proposed Simulation Methods Combining Linear Difference Equations and Numerical Integration, CSRL Memo 191, Univ. of Arizona, Dept. of Electrical Engineering, 13 p.

Wirth, N. (1973), Systematic Programming - An Introduction, Prentice-Hall, Englewood Cliffs, NJ.

Wymore, A. W. (1967), A Mathematical Theory of Systems Engineering - The Elements John Wiley, New York.

Zeigler, B. P. (1976)
    Theory of Modelling and Simulation, Wiley, New York, 435 p.

Zeigler, B. P. (1978), "Structuring the Organization of Partial Models," In T. I. Ören (Ed.), Cybernetics and Modelling and Simulation of Large-Scale Systems, International Association for Cybernetics, Namur, Belgium, pp.127-139.

Zeleny, M. (1977), "Self-Organization of Living Systems: A Formal Model of Autopoiesis," International Journal of General Systems, Vol. 4, No. 1, pp. 13-28.

METHODOLOGY IN SYSTEMS MODELLING AND SIMULATION
B.P. Zeigler, M.S. Elzas, G.J. Klir, T.I. Ören (eds.)
© North-Holland Publishing Company, 1979

WHAT IS NEEDED FOR ROBUST SIMULATION?

M.S. Elzas
Computer Science Department
Wageningen Agricultural University
WAGENINGEN, The Netherlands

This paper constructs a novel definition of "robust simulation".
This definition not only covers a type of "foolproofness" of
computational algorithms used during simulation but more
importantly automated means to assist in assuring model
description validity, program-integrity, program-structure,
program-readability, and bounds for permissible experimentation.
Therefore the three main parts of this paper reflect the needs
of robustness in terms of:

- simulation languages
- compiler "intelligence"
- computational algorithms

Items which will be handled under the language heading are
universality with respect to model classes (e.g. combined systems)
and structure with respect to model representation fidelity,
program quality and readability.
Under the compiler heading possibilities for "intelligent
compilation" are explored, where the intelligence is advocated
mainly for the modellers' convenience. Attention will be paid
to methods for checking the coherence of the model-description
at compile-time (e.g. automatic dimensional analysis, internal
decision table evaluation, completeness checks) and the dangers
of programming by exception will be set forth.
Specific computational algorithms that are predominantly used
are reviewed with respect to automatic error detection and
problem-adaptive characteristics. Furthermore the consequence
of accuracy requirements for basic algorithm design will be
commented upon.
As a last point the possible influence of emerging new computer
architectures on the above issues is briefly envisaged.

INTRODUCTION

A set of basic principles to which a user of simulation should adhere if he wants
his work to be useful where recently put forward by De Wit [1] and Elzas [2].
Based on these priciples a set of questions can be asked whenever a model of a
system to be simulated on a computer has been constructed. These questions are
(with no attempt at an order of priority):

1. Are the objectives and constraints perceived by the user the same
   as those present in the model?
2. Are fidelity measures, available in the implemented model and the
   experimentation process, adequate?
3. Is the sensitivity for change of the system correctly implemented
   in the model?
4. Have experimentation standards been explicitly stated in advance
   for the model and do they protect its integrity?
5. Are vital computational (sub)systems protected against failure?

6. Will the run-time image of the model accept and act upon signs of impending disastrous misbehaviour?
7. Have potential difficulties arising from the boundary relationships of submodels been adequately prevented?
8. Are the users who run experiments with the model actually performing according to the conditions inherent in the model-description and the experimentation plan?
9. Are the system boundaries chosen at modelling time and implemented appropriate for the intended validation and experiments?
10. Are adequate provisions available for simple updating of the model-description in view of results of validation, experimentation and the users' learning processes?

In the quest of answering these questions and following these principles the "robustness" of simulation devices becomes an important subject for the designer of new tools for this field.

AN ATTEMPT AT DEFINING "ROBUSTNESS"

Consider the area of computer utilisation by professionals not interested in computing for itself and/or whose experience lies mainly in fields quite remote from computing. In this context the author wishes to regard "robust" behaviour (1) of the computing entity (hardware, software, data and interactions) as a characteristic by which the user's attention can stay focussed mainly on his major field of interest: conducting (simulated) experiments in his discipline by means of a programmable computing device.
Elzas [3],[4] has asserted that the user should be freed (as much as possible) from hardware, numerical and economic constraints in this venture: this can be considered as a basic requirement in the "robustness" context. This requirement however does not make optimal use of the state of computing art to this date. More is known and has been implemented in several situations that can be used to the advantage of the user, especially if his application belongs to a particular class of computer utilisation like e.g. simulation, computer aided design or computer assisted instruction. Important basic ideas in this context have been reported by Spillers [5] and Weizenbaum [6].
Making use of these concepts mainly developed for other areas, the simulationist is entitled to expect some so called "intelligence" from his computing device.

Use of such possibilities implies that an excellent means of communication be provided in the first place. So "robustness" in this context leads to general requirements in the area of a "simulation" language. The language in itself should furthermore have characteristics which make it unequivocal on its own, and leave no room for free interpretation of the problems (models) described in it.
Once the language "barrier" has been broken, the user is entitled to expect an "intelligent" attitude from the computer that "digests" his model. In simulation such intelligence can be expected in the areas of rough verification of the model description, program integrity, etc. This is a second facet of robustness (Elzas [7]).
Once the model-description in the required simulation language has been converted to a computable form by the computer, utilisation of the model in an experimentation phase can start. In this situation the third facet of robustness leads to requirements as automated protection against numerical pitfalls, protection against experimentation out of the bounds inherent in the model-description, communication with the experiments at "source level".

---

(1) Webster's:robust: healthy, sound, wholesome
    Chambers':robust: stout, strong and sturdy

SIMULATION LANGUAGES

The concept of unified simulation languages that should be used for continuous and discrete simulation using various forms of computing hardware and independent of control engineering, electronic or numerical concepts, has reached a general acceptance in the past years (Ören [8] and [9]).
The advances in definition and acceptance of basic rules for well designed programs are beginning to show their influence in the basic structural properties of simulation languages (Elzas [10], Ören [11]). Even the concept of viewing a language by which a model is programmed on a computing device as a vehicle for transmitting model-descriptions between simulationists is gaining a status very close to formal acceptance (Ören [12]). Therefore it is to be expected that these three major concepts should be basic features of any new simulation language.

During the 10 years that the author's group has been doing research on simulation languages, the following basic ideas about simulation languages have emerged:

1. there should be one unified language for all simulation;
2. the simulation language should be strongly biased towards a structured approach compatible with structure elements in model- and experiment-discriptions.(This assures integrity and unequivocal model-experiment representation on a computing device);
3. inherent sequential and parallel processes should be easily representable as such in the language;
4. the language should be clear (transparent) and readable to assure its function as a communication vehicle for model-descriptions;
5. the operators and specification elements available in a simulation language should not form a restrictive set but be user-expandable. (This expansion should be possible using the language itself);
6. simulation languages should be independent as much as is possible of the underlying technology;
7. the language should not overly endanger simulation economy.

1. Unified language

Fortunately the recent past has clearly shown a move from rejection of this concept to the development and use of a great many products that all pursue this goal with differing degrees of success (Ören [9]). This change of attitude is motivated by the reality of the systems studied nowadays (Ören [8] and [12]). It must be remarked however, that this goal cannot be reached by "simple" addition of continuous elements to discrete simulation languages or vice versa as most "languages" referenced by Ören in [9] do. To obtain a useable approach the whole unification concept has to be built in from the theoretical foundations as recently proposed for example by Ören in GEST'78 [13],[14] under influence of Zeigler [15].
An important rationale at the basis of this remark is that explicit data flow between continuous and discrete submodels as well as implicit data transfer (e.g. for synchronisation of events with continuous time) has an important influence on the structure of the simulation program and thus may impose unnecessary restrictions on the freedom of model-description unless the language is carefully designed. This can be illustrated through the evolution in structure of the simulation language HL1 [10]. This evolution took place from a simple C.S.S.L. (Continuous System Simulation Language) for pure digital sequential computation (Fig. 1), to an augmented CSSL catering for parallel computation as well, and so providing object-code for hybrid systems (Fig. 2).
From there the language evolved into a multiple continuous process concept (Fig. 3) and finally became one of the early proposals for combined system simulation languages. The resulting structure (Fig. 4) was a first try at development of a combined (continuous/discrete) language as an enhancement to continuous simulation, allowing for multiple segment simulation programs. The multiple segment approach was deemed to be useful for large-model representation.

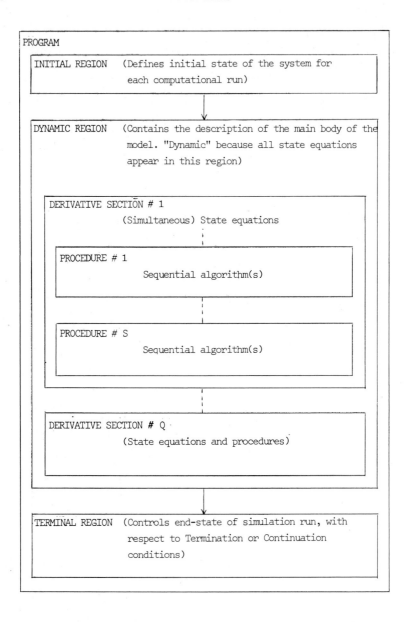

<u>Fig. 1</u>: Simple C.S.S.L. - Simulation program structure

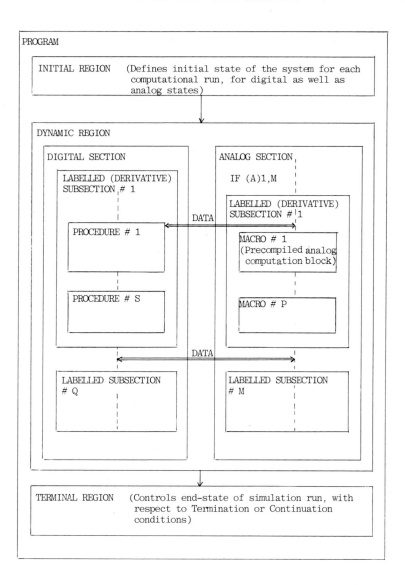

Fig. 2: Original HL1 (Augmented C.S.S.L.) program
         structure (e.g. for hybrid systems)
         (⟺ denotes permitted data transfer path).

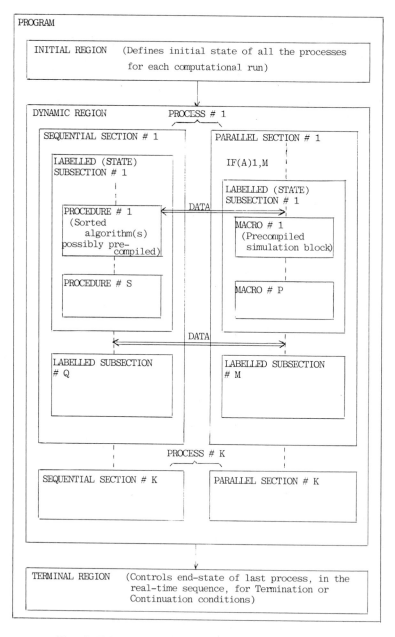

Fig. 3: Multiple continuous process concept in HL1,
         program structure.
         (⟺ denotes permitted data transfer path).

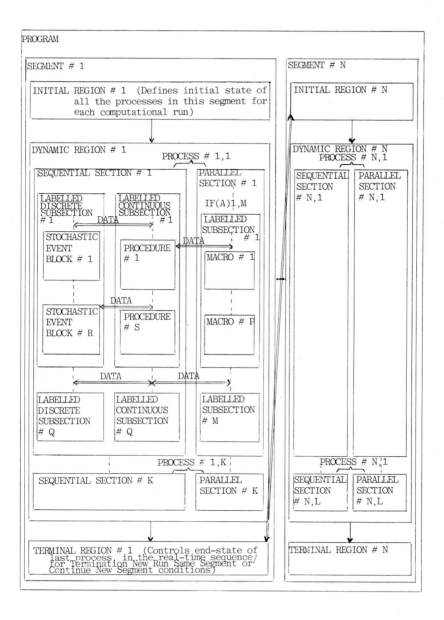

Fig. 4: Proposed "compound" (discrete/continuous)
Multi-segment program organisation in HL1 (ca. 1974)
(⟺ denotes permitted data transfer path).

Fig. 4 clearly shows one of the main shortcomings of this approach. Through the lack of appropriate implicit continuous/discrete process synchronisation provisions, data interchange was (by necessity) only allowed outside of major computational blocks. As a consequence extraneous model subdivisions were forced on the user by the restrictions of the simulation language. It is clear that this can be avoided by designing a language from scratch whose foundations take care of the mixed-concept from the start.

2. Structured model and experimentation specification

Much has been said about the value of structure in programs for different purposes. In the area of pure programming the concept of structure has been defended on grounds of:

    a. programming clarity (with reference to Dijkstra's [16] "informal correctness");
    b. systematic program construction (Wirth [17]);
    c. automated consistency checking (Wirth [17]);
    d. and possibly formal program verification (Luckham [18]);

If we consider simulation to consist of the following actions:

    e. system analysis;
    f. system definition;
    g. system/model-description;
    h. problem statement (e.g. statement of experiments to be undertaken with the model described);
    i. result description;
    j. result analysis;

it is legitimate to expect at least some of these items to appear as well in the structural framework encouraged by the simulation language. In this way such a language can by itself provide a contribution to "cleaner" modelling techniques. Let us take a closer view at points g. to i. and try to derive the implications. In the first place it is clear that model-description and experimentation with the model are intrinsically separate (though related) activities.
A robust language thus will logically separate these activities, and assign a priority (Fig. 5) (see also [13]).
This naturally does not negate the relationship between the two.

Within this framework further refinements should be:
        - the hierarchical ordering of system boundaries (e.g. input and output variables in GEST'78) with respect to the internal system structure (states in GEST'78, with reference to system theory principles). In the model-description this leads to declaration statements establishing a.o. connections between model and associated experiments. In the experiment the boundary variables acquire values through inputs (parameter sets (Fig. 6) and experimental frames (Fig. 7)) and produce outputs in RUNS (Fig. 7);

        - the model validity range: the great majority of models only has a limited range of validity. Outside of this range the integrity of the model is in danger, therefore experimentation outside of the validity space should not be allowed. Thus the language needs provisions to specify input, output and state variable ranges as well as room for the declaration of model parameters and their ranges (Fig. 6);

        - the possibility of a (hierarchical) submodel classification in a top-down approach establishing a hierarchy from conceptual main processes-description through a continuous series of refinements down to the level of simulation operators (Fig. 8). At every step down to a lower hierarchical level input/output declaration and range specification facilities should be available to allow systematic (modular) program construction, if need be from the bottom up.

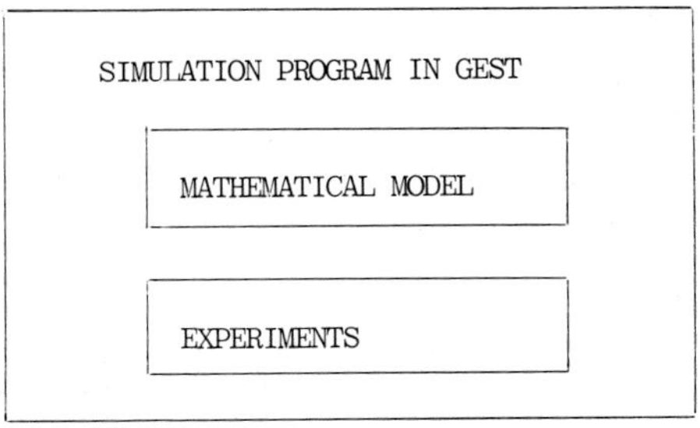

SIMULATION PROGRAM IN GEST

MATHEMATICAL MODEL

EXPERIMENTS

Fig. 5

MATHEMATICAL MODEL

PARAMETRICAL MODEL
DECLARATIONS:
- PARAMETERS
- STATES
- RANGES (OF STATES AND/OR PARAMETERS)
- VARIABLE TYPE (REAL, INTEGER, ARRAY, etc.)
- INPUTS (INPUT PARAMETERS, INITIAL INPUTS)
- OUTPUTS
- FUNCTIONAL RELATIONS
- TABULAR FUNCTIONS
- EXTERNAL (LIBRARY) FUNCTIONS
- CONSTANTS
_ VARIABLE DIMENSION (MASS, LENGTH, etc.)

SUBMODEL   # 1
- MODEL TYPE DECLARATION (CONTINUOUS, DISCRETE, etc.)
- OTHER DECLARATIONS (SEE ABOVE)
- STATEMENTS

....

SUBMODEL   # m
-
-
-

COUPLING OF ALL SUBMODELS
- INPUT OF SUBMODEL # i FROM OUTPUT
        OF SUBMODEL # j etc.

MODEL PARAMETER SET(S)

PARAMETER SET   # 1
- PARAMETER ASSIGNMENT
- PARAMETER COMPUTATION
- FUNCTION DATA

....

PARAMETER SET   # P
-
-
-

Fig. 6

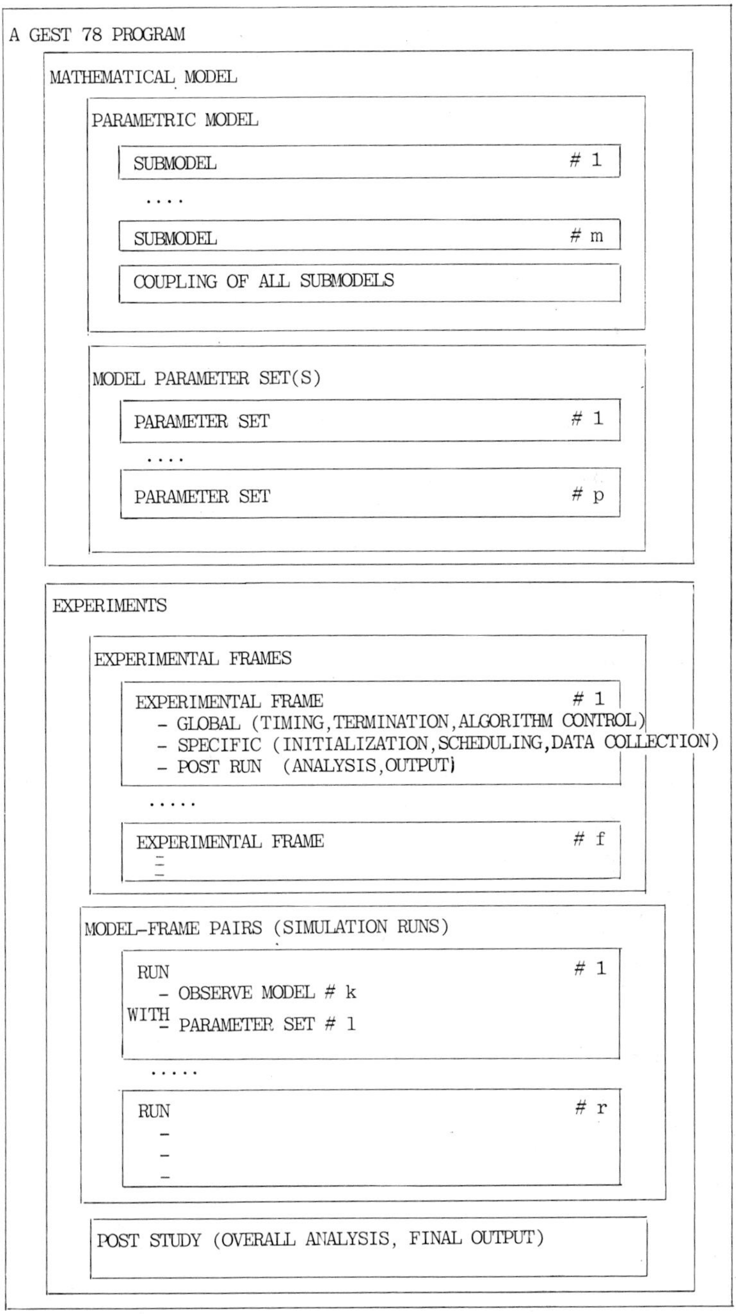
A GEST 78 PROGRAM
MATHEMATICAL MODEL
PARAMETRIC MODEL
SUBMODEL # 1
....
SUBMODEL # m
COUPLING OF ALL SUBMODELS
MODEL PARAMETER SET(S)
PARAMETER SET # 1
....
PARAMETER SET # p
EXPERIMENTS
EXPERIMENTAL FRAMES
EXPERIMENTAL FRAME # 1
- GLOBAL (TIMING,TERMINATION,ALGORITHM CONTROL)
- SPECIFIC (INITIALIZATION,SCHEDULING,DATA COLLECTION)
- POST RUN (ANALYSIS,OUTPUT)
.....
EXPERIMENTAL FRAME # f
MODEL-FRAME PAIRS (SIMULATION RUNS)
RUN # 1
- OBSERVE MODEL # k
WITH - PARAMETER SET # 1
.....
RUN # r
POST STUDY (OVERALL ANALYSIS, FINAL OUTPUT)

Fig.7

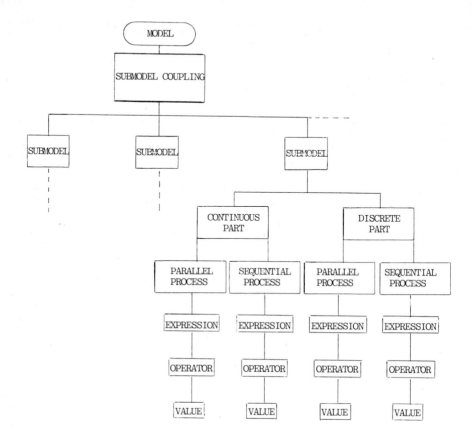

Fig. 8: Hierarchical program tree, model description part.

Apart from these points, the widely accepted features for structured programming
should preferably be available to allow points a. and b. above as a minimum.

3. Representation of sequential and parallel processes

When describing a sequential process for implementation on a computer, the
sequence itself introduces an ordering in expression (submodel) evaluation. This
is not the case for parallel processes. Indeed such processes could well lead to
an unsortable set which cannot be computed on a sequential device 2).
Although it is a task of the compiler (as will be pointed out later) to create
computable sets for such parallel setups, this restriction should not be inherent
in the language and imperil clear model-description.

4. Clarity and readability

Although no theoretical foundation has been published on this subject, there is
practical evidence enough to stress once more the great value of portability of
models and concepts versus portability of programs. A relevant remark on the
necessity of clarity of a programming language can be distilled from a remark
by Weizenbaum (p 152 of [6]):"Computer models have some advantages over theories
stated in natural language. But the latter have the advantage that "patching"is
hard to conceal. If a theory written in natural language is, in fact, a set of
patches and patches on patches, its lack of structure will be evident in its very
composition." Thus clarity of a modelling language should provide the same
revealing qualities.

5. Extendability

The gist of this remark is that in this way it becomes possible to adapt the
language to specific classes of simulation application without being obliged
to redefine a whole new language. A good example of such an approach is formed
by the "CLASS" concept in SIMULA, which has been reported in numerous publications
by Jacob Palme. The specific importance of this concept for general system
simulation applications has clearly been stated by Sol in [21]

---

2) Let us call W the set of all operator inputs and Z the set of all operator
   outputs of parallel processes to be represented on a sequential device. Let
   all these variables be partitioned into three disjoint subsets and let us call
   "system" the collection of all the described processes so that:
         a first set is   C: all operator outputs which are also operator inputs
                               ("connection variables", $C = W \cap Z$)
                               (-Note that operator outputs which are used
                                more than once are taken into account separately
                                every time they occur)
         a second set is   S: all "pure" system inputs ($S = W-C$)
         the third set is R: all "pure" system outputs ($R = Z-C$)

   Furthermore define the set Q which is the set $Q \subset Z$ of state operator outputs.
   The set $A = Z-Q$ which is left is then the set of algebraic operator outputs.
   Now a unique sequential computing procedure for the(se) parallel process(es)
   can only be constructed if all members $a_j \in A$ can be ordered in a sequential
   list such that all the inputs of each algebraic operator with output a_j
   is a member of either the set S, the set Q or the subset $P \subset C$, where P is
   the set out of C which represents inputs to $a_k \in A$ where all a_k precede a_j
   in the sequential list. For further details see Zeigler [15] pp 108-112
   (background), Giloi [20] and Elzas [19] (theory).

6. Technology independence, and 7. Simulation economy

The necessity for these requirements for applicability of a simulation language is so obvious by now that further elaboration seems unnecessary. A more precise economic motivation for a subset of simulation applications can be found in a relatively old speculation on the future of hybrid computation by Elzas[22].

COMPILER "INTELLIGENCE"

Most compilers for computer languages available today do some (marginal) program verification besides the translation from source-code to run-time object-code. Most of the checking, however, is limited to verification of adherence of the program to the semantic and syntactic rules laid down in the language definition, and restrictions added to them by the compiler limitations.

One could call this type of error detection text verification because it provides no information at all on the integrity of the program. The research on formal and practical program verification is in full swing, but to date has found only limited realistic application (for a review see Miller [24]). Some examples of such endeavours are known for limited purposes, e.g. some Pascal programs (Marnier [23]).

There is a consensus, however, that structure enforced by the programming language and implicit knowledge fostered by a limited application category improve the possibilities to help the user to construct sensible programs. They can do so by providing "conceptual" checking features in the compiler.

In order of increasing sophistication the following verification elements for program integrity checking can be mentioned:

  a. missing or unused variables and labels (most compilers);
  b. missing or faulty declarations (some Fortran and most Algol compilers);
  c. "weak" mode checking (e.g. integer, real etc.) (most compilers);
  d. infinite loop detection (though, in general, theoretically impossible and no automated examples are available, detection for limited domains may be feasible);
  e. "strong" mode checking (i.e. "weak" mode plus range checks and definition area) (Algol'68, some Pascal- and most hybrid-compilers);
  f. "dead" branch (or -process) or infeasible path detection (i.e. statements which are never reached) (some program checking routines e.g. for Pascal);
  g. control-structure or case-error detection (unknown if possible in most cases);
  h. hierarchical consistency verification (feasible, since Algol'60, but not often done);
  i. Boolean consistency verification (feasible in some cases, but not done yet);
  k. correctness/computability proofs.

Elements a. to c. are so common and down to earth that they can be considered as minimal requirements for any compiler.
   - Detection of type d. is mainly relevant if loop bounds are "dynamic".
       Until now the only way to check reliably for such a condition is to invoke (i.e. execute) the program with many sets of different input data until a loop termination error is found. This can only be attempted if the input data domain is known point by point and contains an instance of data that will cause this error condition. However, even in the case that the simulation language fulfils the conditions for separation of modelling and experimental phase outlined in the previous section, the input data domain for the model is only known as a range and the test would have to be carried out for every experiment-run, and specific statements to this purpose should be generated by the compiler. It is then questionable if carrying out such tests is worth while, especially if one takes into account that the exit from the space wherein the model is valid can be considered to be protected by the known range of the state variables, if the recommendations under ad 2, g. in the previous section are taken into account.

- Detection of type e.:"strong" mode errors,is implicit for a compiler
of a language which requests range specifications for model description.
Furthermore it is a proven technique, available and tested in most hybrid
compilers to date.
- The "infeasible path" (3) problem gives rise to different approaches in
the cases of parallel processes and sequential processes.
During the sorting process (which must take place for parallel processes to be
executed on a sequential device), statements which do not fall in the sequential
list are immediately suspect. Straight forward checking of the boundary relations
between these processing blocks, the system boundaries and the sequential process
interface can then reveal offenders.
Sequential processes which do not contain iterations nor dynamically variable
control structures can be tested for this condition by limited execution path
testing. This approach therefore is only valid for a restricted set of sequential
processes.
- Note that Howden [24] has stipulated that "path testing is a reliable
method to discover case errors if and only if a program with a case error
has input domains that do not intersect the input domains of the correct program."
This means only that programs with a priori limited input domains (like advocated
in the previous section) run a smaller risk for this type of error than programs
with unbounded input domains.
This conclusion only strengthens the case for specifying - and thus limiting - the
experimental range for a model. Without, however, providing any tool for realistic
verification.
- Hierarchical consistency (4) checking is only feasible for programs for
which a unidirectional hierarchical decomposition tree can be constructed
in which the relation between the hierarchically connected subsets can be defined
a priori in number and range. Simulation languages structured in the way recomended
in the previous section, paragraph 2, force this property on the model-description.
Verification of hierarchy is then feasible by sorting subsets with associated
variables in hierarchical order, and checking if the (associated) variables fit
in the hierarchy (and thus are only connected through the permitted boundary
relationships) (Fig. 8).
- It is clear that Boolean consistency (4) checks are only relevant for
two-valued (logical) variables which are related through a hierarchical
structure. A possible technique consists of:
1. compiling the stack(s) depicting the hierarchy of Boolean operations
2. constructing a truth table for every stack (neglecting all intermediate non-
Boolean expressions by forcing true and false conditions in such cases)
3. checking the truth table(s) for impossible conditions. Trivial example:

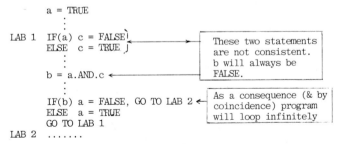

a = TRUE

LAB 1   IF(a) c = FALSE
     ELSE c = TRUE

     b = a.AND.c

| These two statements are not consistent. b will always be FALSE. |

     IF(b) a = FALSE, GO TO LAB 2
     ELSE a = TRUE
     GO TO LAB 1

| As a consequence (& by coincidence) program will loop infinitely |

LAB 2   ....

---

(3) The "infeasible path" is a set of program statements that is never reached
during program execution, e.g. because an erroneous GO TO statement prohibits
access to a specific program section.
(4) Consistency: (according to Webster: coherence, harmony of parts) is used in
this text for denoting the property of allowed and/or sensible relationships
between program statements and model-program homomorphism.

- It would be very valuable to be able to check conceptual model
consistency (4) for all possible model constructions. However, it is
obvious that this is not possible in many cases. Preliminary knowledge about
the fields in which simulation is used however, allow us to select a subset of
models for which such a verification is feasible. For all models of systems
to which conservation laws apply, a form of consistency checking has been
developed in the 19th century that we can use to advantage:dimensional analysis.
The basis for such an analysis is the notion of conservation of dimensions in
any expression. This means that if (and only if) dimensions of most variables,
all input and output data and parameters are declared in advance by the modeller,
the compiler can check consistency by checking dimensionality by statement (left
hand dimensionality=calculated right hand dimensionality) and throughout the
compiled computational sequence.
    - Although the research on computability for discrete as well as for
continuous automata has been going on for a long time (see rough outline
by Elzas in [19]), the results attained only cover a very limited of all situations
resulting from a model-description. The situation is even less advanced in the,
field of proving the correctness of a computational sequence (see e.g. Miller [24])
therefore - even for the restricted application area called simulation - proofs
of computability and/or sequence correctness cannot (yet) be automated.
    - Concluding one can remark that the state of the art permits to build
into a compiler (on top of standard features a, b, c above):
        1. "strong" mode checking
        2. infeasible path detection (especially for the
           parallel process parts of the model)
        3. hierarchical consistency verification (for sortable models)
        4. Boolean consistency verification (for sortable models)
    and 5. automated dimensional analysis.
There seems to be no reason not to include these features in every simulation
compiler for the benefit of the user and the robustness of the model-description.
This, however, at the extra cost of quite elaborate program texts. These are needed
to get a clear, explicit picture of what the modeller wants. "Programming by
exception", so making use of implicitly present views about the model in the
compiler has the advantage of allowing more compact program texts, it also has the
disadvantage that consistency is taken for granted and can not thoroughly be
checked.

An example:

To obtain some practical insight into some of the elements mentioned in the
previous two sections a simple case of simulation will now be studied. The example
technical system to be modelled consists of a simple servo-mechanism that serves
to bring a heavy body into the right position by rotation (e.g. a servo-
controlled ships-rudder). The system is illustrated in Fig. 9.

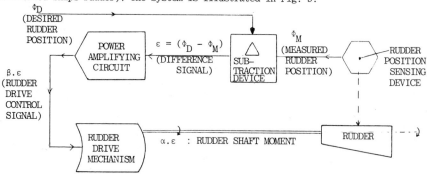

Fig. 9: System under study

The mathematical equation for this system is:

$$(I_R + I_D) \frac{d^2\phi_M}{dt^2} + D(\dot{\phi}_D - \dot{\phi}_M, \phi_D - \phi_M) \frac{d\phi_M}{dt} = \alpha(\phi_D - \phi_M) \qquad (1)$$

(Derived from moment equilibrium)

In this equation I_R: denotes the moment of inertia of the rudder, I_D: idem of the drive (Dimension: ML^2).

$D(\dot{\phi}_D - \dot{\phi}_M, \phi_D - \phi_M)$ is the damping ratio of the system, which has been taken to be a function of the change in angular position and rotational speed (Dimension: ML^2T^{-1})

ϕ_M is the measured angular rudder position

ϕ_D is the desired angular position

α is the overall tranfer function of the power amplifier/drive mechanism (Dimension: moment $\Rightarrow ML^2T^{-2}$)

If we transform equation (1) into an equation in the error signal $\varepsilon = \phi_D - \phi_M$ we obtain:

$$(I_R + I_D) \frac{d^2\varepsilon}{dt^2} + D(\dot{\varepsilon}, \varepsilon) \frac{d\varepsilon}{dt} - \alpha . \varepsilon = 0 \qquad (2)$$

Let us assume that in this equation D is of the form $K(p(1-\varepsilon^2) - q\dot{\varepsilon}^2)$ for the small values of $\varepsilon(\varepsilon \leqslant 1)$ to be studied.

In this expression $p(1-\varepsilon^2)$ represents a speed control term that decreases damping as the angle error signal is large and increases damping for small error signals with the intention to limit rudder overshoot. The term $-q\dot{\varepsilon}^2$ represents the decreased damping ratio due to lower oil viscosity in the drive mechanism caused by the temperature rise due to energy dissipation during a unit of time.

The dimension of K is ML^2, the dimension of p is T^{-1}, the dimension of q is T. This results in a dimension ML^2T^{-1} for D (change of inertia per time unit) and $ML^2T^{-2}*T$ for $Kq\dot{\varepsilon}^2$ (dissipated labour effort multiplied by time). For simplicity the values of p and q have taken to be 1.

If we normalize the resulting equation by substituting $a = \frac{K}{I_R + I_D}$ and $b = \frac{-\alpha}{I_R + I_D}$ then (2) reduces to:

$$\frac{d^2\varepsilon}{dt^2} + a(1 - (\frac{d\varepsilon}{dt})^2 - \varepsilon^2)\frac{d\varepsilon}{dt} + b\varepsilon = 0 \qquad (3)$$

(Now a is dimensionless, and b has dimension T^{-2}).

Then (3) is a known type of differential equation: the equation of Van der Pol.

In view of the ship's handling characteristics we are interested in the angle error correction rate as a function of the angle error (in terms of nonlinear differential equation theory this is known as a "phase-plot"), with b set at 1. The general organisation of CSSL languages (like CSMP) is shown in Fig.10. A CSMP-program for the simulation of this system, based on equation (3) is given in Fig.11.
Because CSSL's are based on the operation of integration, we have reorganised the equation into two statements involving integration operators.

In Fig. 11, we see more variables than in equation (3). These are essential for the solution of the problem.

Y and DERY represent ε and $\frac{d\varepsilon}{dt}$

YO and DERYO are the initial values of ε and $\frac{d\varepsilon}{dt}$ at the moment that we start to study the model. Arbitrarily this moment has been taken to be $t = 0$ (time zero).

FINTIM defines the extent of time that we want to study the model (0 to 12 seconds).

PRDEL is the sampling period, i.e. the intervals at which we want to see the values of the dependent variables (1 second intervals).

OUTDEL is our estimation for the maximum integration step in time that we want to allow, to obtain accurate results (step = 0.05 sec.).

The statement $Y = INTGRL(YO,DERY)$ is equivalent to $Y = \int DERY(t)dt + YO$

The TERMINAL REGION is in fact empty and leads directly to a new implicit INITIAL REGION changing only YO from 0.2 to 1.5

The next pass through the TERMINAL REGION leads directly to STOP.

INITIAL REGION	(Defines the initial state of the processes for each computational run)
DYNAMIC REGION	(Contains the description of the main body of the model. "Dynamic" because all differential and integral equations appear in this region. These equations describe the dynamics of the "real world" system under study. Also in this region "procedural" (or sequential) aspects of the model can be described in Fortran)
TERMINAL REGION	(Controls end-state of the process and/or New Run or Continue Conditions for the simulation experiment)

Fig. 10: Typical CSSL-program organisation.

```
TITLE V.D.POL  -  EQUATION

INITIAL
PARAM YO = 0.2, DERYO = 0.2, A = 2

DYNAMIC
Y = INTGRL (YO, DERY)
DERY = INTGRL (DERYO, DER2Y)
DER2Y= A*(1-Y**2-DERY**2)*DERY-Y

TERMINAL
TIMER FIN TIM = 12.0, PRDEL = 1.0, OUTDEL = 0.05

PRINT DELT, DER2Y, DERY, Y
PLOT Y(DERY) (20, 20)
METHOD RKS
END

PARAM YO = 1.5
END

STOP
```

Fig. 11: CSMP-program for the Van der Pol - equation

Now let us look at the same problem, but programmed in two ways in the language HL1. In the first case the equation is solved by INTEG-operators and in the second case with the DER-operators. The second solution will stop after 50 seconds. Then, in the TERMINAL-section the value of a is changed and again a second run is made. In total ten runs will be made.

```
1.   PROGRAM VD POL
       PARAM A, DYO, YO
       DATA  Y = (-10., +10.),
             DY = (-20.,   20.),
             DDY = (-30.,   30.)

       INITIAL
       A = 5.0
       DYO = -2.0
       YO  = 2.0
       END

       DYNAMIC
       ANALOG
         DDY = A*(1-Y**2-DY**2)*DY-Y
         DY  = INTEG (DDY,DYO)
         Y   = INTEG (DY, YO)
       END
         TITLE (14.5)
         PRINT (T, DY, Y)
         PLOT (DY, Y)
       END

       TERMINAL
       END
      **END
```

```
2.  PROGRAM VD POL
    PARAM A
    INTEGER PARAM I
    DATA A = 5. , I = 1
    DATA Y = (-10,10), T = (0.,50.)
    IC.. INITIAL
            DER 0(Y) = 2.0
            DER 1(Y) = 2.0
         END

         DYNAMIC
         ANALOG
           IF (T.GT.50) TERMINAL,COMPUT
            BEGIN COMPUT
             DER 2(Y)=A*(1-Y**2-DER(Y)**2)*DER(Y)-Y
            END
         END
           TITLE (14.5)
           PRINT (T, DY, Y)
           PLOT (DY, Y)
         END
         TERMINAL
         I = I+1
         IF (I.GT.10) ENDLAB, CONT
          CONT.. A = A+(-0.5)*I
                   GO TO IC
          ENDLAB
         END
    **END
```

Note that in comparison with CSMP, HL1 has extra features for structuring and program declarations, and control structures for experimental runs. In HL1 example 1 moreover, the ranges indicated with the variables allow "strong mode" checking. The block structure in Example 2 allows (rudimentary) hierarchical consistency verification, while the limited branching allowed by the HL1-control structure diminishes the infeasible path risks.

The model and experimentation underlying Example 2 are now expressed in GEST'78 version 2:

```
PROGRAM VD POL
 MODEL  RUDDER—CONTROL   IS   CONTINUOUS
  PARAMETERS (A)
  STATES (DY,Y)
  RANGE OF DY = REAL (>=-20.0, <=20.0)
  RANGE OF  Y = REAL (>=-10.0, <=10.0)
  DERIVATIVES
   DY' = A*(1.0 - Y**2 - DY**2)*DY - Y
    Y' = DY
 END MODEL RUDDER—CONTROL

 PARAMETER SET 1
  A = 5.0
 END PARAMETER SET 1
 PARAMETER SET 2
  A = 4.5
 END PARAMETER SET 2
             .
             .
 PARAMETER SET 10
  A = 0.5
 END PARAMETER SET 10

 FRAME 1
  GLOBAL
   SIMULATE UNTIL TIME 50.0
   INTEGRATE BY RUNGE KUTTA;
             MAX STEP = 0.05
  END GLOBAL
  MODEL RUDDER—CONTROL
   INITIALIZE STATES
    DY = -2.0
    Y = 2.0
   COMMUNICATE AT EVERY 1.0 TIME UNIT
   SAVE STATES
  END MODEL RUDDER—CONTROL

  POST RUN
   OUTPUT MODULE 1 ON PRINTER
   OUTPUT MODULE 2 ON PLOTTER
  END POST RUN

 RUN 1   TO OBSERVE RUDDER—CONTROL
   WITH PARAMETER SET 1 IN FRAME 1
 RUN 2   TO OBSERVE RUDDER—CONTROL
   WITH PARAMETER SET 2 IN FRAME 1
             .
             .
 RUN 10  TO OBSERVE RUDDER—CONTROL
   WITH PARAMETER SET 10 IN FRAME 1

 OUTPUT MODULE 1
  PRINT FOR FIRST PAGE 1 HEADING LINE
        STUDY OF RUDDER—CONTROL
  LIST TIME, DY, Y
 END OUTPUT MODULE 1

 OUTPUT MODULE 2
  PLOT  Y, DY
 END OUTPUT MODULE 2
END PROGRAM VD POL
```

Note that in comparison with the CSSL-compatible languages CSMP and HL1, the experimentation part is clearly separated from the model description, the language is well structured and requires clear declaration of variables and parameters. The language would however, benefit from the obligation to declare ranges (for "strong mode" checking of experiments versus model range). Another strong point is the imposed need to explicitly formulate all manipulations with the model, providing room in a future compiler for consistency checks of experiments with respect to models.

The languages presented could all provide better checking facilities if dimensions (e.g. length, mass, time) could be declared, allowing checking of expression consistency in the compiler. The reader is invited to do so for the presented programs and in this way verify himself which expressions can be thouroughly checked.
Although both HL1 and GEST do not require type declarations, they do allow them. If "weak" mode checking and logical consistency verification are to be carried out in the compiler, these declarations become mandatory. This might result in verbose program texts. On the other hand robustness in the program translation phase is engendered.

It is useful to remark at this point that as long as fully automated model (and/or program-) verification is not possible, post mortem error signalling referring directly back to the source text is invaluable. It should be noted that languages like HL1 and GEST'78 are well suited for this purpose. Especially the latter has a structure which should allow such a construction efficiently because the necessary crossindices will be needed at run-time anyway for reliable communication between the experiments and the model.

COMPUTATIONAL ALGORITHMS

We now apply the notion of robustness to computational algorithms which are provided by the compiler-run-time library to form the nucleus of the simulation run. This leads us to look for algorithms that:
- work for a large set of applications;
- are protected against (numerical) failure;
- give early notice of impending disaster (e.g. serious loss of approximation accuracy, instability etc.);
- are able to detect out of range, or incorrectly declared, input variables at an early stage;
- adapt their convergence characteristics to the problem at hand.

It will of course be noticed that all the fundamental matters needed for these features have been studied in the past. It is remarkable, however, that much of this knowledge has not been put to use in most of the available simulation run-time systems.
The largest majority of numerical algorithms in use on computers today to calculate solutions to mathematical equations described in our models, make use of approximation techniques and thus do not solve the problem at hand exactly. (Even if an exact solution would be available, computed results would be approximative because of the finite accuracy of representation on the computing device.)

At the basis of such approximations lie hypotheses about the nature of the solution of the problem at hand. Because of this, the approximations are only mathematically valid in a bounded subspace of the solution space considered, and in most cases require implicitly some a priori knowledge of the expected solution either for the choice of numerical method or for initialisation purposes.
Let us take, as an example, the problem of solution of simultaneous linear equations. Now consider the IMSL-subroutine library which provides the following possibilities in separate subroutines: inversion of band matrices, well conditioned

full matrices, well conditioned (positive) symmetric matrices, ill conditioned
full matrices, and ill conditioned (positive) symmetric matrices. The accuracy
bounds that a user requests can only be used internally as a safeguard against
cummulative round off errors.
The subroutines use: row equilibration and partial pivoting for (band) matrices;
Gaussian elimination (Crout algorithm) with equilibration and partial pivoting;
the Cholesky algorithm;or all previous methods followed by iterative improvement.
End accuracy estimates are based on residuals calculated per equation after
inserting the approximate solution in the original equation set. (Note that
therefore accuracy per solved unknown vector component can only be roughly be
estimated and only the overall solution accuracy per equation can be established,
the accuracy measure being the size of the residuals).

It is clear from this context that the available methods are appropriate for all
cases where the problem at hand is the inversion of a single matrix and where the
user has had the opportunity (and tools) to explore the nature of the matrix.
Very often users will tend to use the several inversion methods one after the
other until a "satisfactory" solution has been found.
Let us now put this approach in the context of (parametric) simulation.
In this situation the nature of the matrix can change considerably under the
influence of parameter changes.
Requesting the approach outlined for IMSL would mean requesting several changes
of the "experimental frame" (as defined for GEST'78) for every single run-
parameter set couple in order to evaluate the merits of the chosen numerical
method. This clearly puts the focus of the simulation experiment more on the
numerical mechanisation than on experiments with the model itself. This is
definitely not the intention of the simulation community.

The question then is what can be done to improve on this situation?
More explicitly: can "adaptable" methods be constructed which would obviate the
need for the above approach?
One way to look for the solution of this problem is by going back to the primitive
form of this type of problems: the general operator inversion problem.
This primitive form can be stated as:
      Find the vector $\underline{x}$ which is a solution to the equation $0(\underline{x}) = 0$
      in which 0 is any analytic vectorial operator (so not necessarily linear).

This general problem has been treated by the author extensively in [19].
It is shown there that:
      - a generalised iteration scheme can be constructed for
        solution of these problems;
      - convergence and stability domains for this iteration
        process can be established a priori;
      - under the restriction that the process is convergent, the
        solution found is unique.

(To the "old hands" among the readers it should be pointed out that this iteration
technique is a generalisation of schemes to obtain solutions for an implicit loop,
and fundamentaly represents a high gain inversion loop with variable - and
optimally controlled - loop gain.)

A set of linear expressions $0(\underline{x})=0$ can be written as $f(\underline{x},A,\underline{b})=0$ where the
operator f stands for $\underline{x}*A-\underline{b}$.
A program (HIGARR) written in Fortran 10 for a DECsystem-10 computer (Fig. 12)
was used to experiment with this approach. In these experiments the technique
was found to be competitive in accuracy and speed with the IMSL-routines for
the same system, and allowed the whole range of types of matrices to be covered.
A few samples of the results are shown in Fig. 13.
Note that the short form of the iteration formula used is:

PROGRAM HIGARR

```
1          DIMENSION OUT(10),OUT1(10),RIN(10),A(10,10)
2          DIMENSION GAIN(10),EPS(10)
3          ACCEPT 10010,M
4          ACCEPT 10020,((A(I,J),I=1,M),J=1,M)
5          TYPE 10050,M,A
6     10   ACCEPT 10020,(RIN(K),K=1,M)
7          DELT=1.E-6
8          N=0
9          DO 30 K=1,M
10         IK=K
11         GAIN(K)=-2./F(RIN,IK,A,M)
12         IF(ABS(GAIN(K)).GT.1.E+3)20,30
13    20   GAIN(K)=0.001*GAIN(K)
14    30   OUT(K)=GAIN(K)/ABS(GAIN(K))
15         TYPE 10050,M,(RIN(L),L=1,M),(OUT(K),K=1,M),(GAIN(KL),KL=1,M)
16    40   N=N+1
17         REL=0.
18         EPS1=0.
19         DO 70 J=1,M
20         KJ=J
21         EPS(J)=(RIN(J)+F(OUT,KJ,A,M))/2.
22         OUT1(J)=OUT(J)+GAIN(J)*EPS(J)
23         TYPE 10030,OUT1(J),OUT(J)
24         EPS1=EPS(J)+EPS1
25         REL=REL+ABS((OUT1(J)-OUT(J))/OUT1(J))
26         IF(ABS(REL).LT.1.E-6)60,50
27    50   F1=F(OUT1,KJ,A,M)
28         F2=F(OUT,KJ,A,M)
29         GAIN(J)=-2.*(OUT1(J)-OUT(J))/(F1-F2)
30         TYPE 10020,GAIN(J)
31         OUT(J)=OUT1(J)
32    60   IF((N.GE.1000).OR.(ABS(GAIN(J)).GT.1.E+30))90,70
33    70   CONTINUE
34         IF((REL.LT.1.E4).AND.(ABS(EPS1).LT.DELT))80,40
35    80   TYPE 10050,N,(OUT(L).L=1,M),EPS1,REL
36         GO TO 100
37    90   TYPE 10040
38         TYPE 10050,N,GAIN(KJ)
39    100  CALL EXIT
40    10010 FORMAT(I)
41    10020 FORMAT(F)
42    10030 FORMAT(1X,2(F12.4,1X),/)
43    10040 FORMAT(1X,'MISLUKT',/)
44    10050 FORMAT(1X,I6,5(E12.3,1X),/)
45         END
```

Fig.12: Matrix inversion program using generalized iteration.

Case 1

Matrix A = $\begin{pmatrix} 1 & 0 & 0 \\ 0 & 1 & 0 \\ 0 & 0 & 1 \end{pmatrix}$ known vector $\underline{b}$ = $\begin{pmatrix} 1 \\ 2 \\ 3 \end{pmatrix}$

Iterations 3

Solution vector $\underline{x}$ = $\begin{pmatrix} 1.0000 \\ 2.0000 \\ 3.0000 \end{pmatrix}$ max abs.error vs.analytical solution $< 10^{-5}$
calc.max abs.error: $0.9 * 10^{-7}$
calc.max rel.error: $0.2 * 10^{-4}$

Case 2

Matrix A = $\begin{pmatrix} 1 & 0 & 0 \\ 0 & 1 & 0 \\ 0 & 0 & 1 \end{pmatrix}$ known vector $\underline{b}$ = $\begin{pmatrix} 1.00000 \\ 0.02000 \\ 0.00003 \end{pmatrix}$

Iterations 4

Solution vector $\underline{x}$ = $\begin{pmatrix} 1.00000 \\ 0.02000 \\ 0.00003 \end{pmatrix}$ max abs.error vs.analytical solution $< 10^{-5}$
calc.max abs.error: 1.10^{-8}
calc.max rel.error: 1.10^{-6}

Case 3

Matrix A: $\begin{pmatrix} 1.005 & 1. & 1. \\ 1. & 1.005 & 1. \\ 1. & 1. & 1.005 \end{pmatrix}$ known vector $\underline{b}$ = $\begin{pmatrix} 6.005 \\ 6.01 \\ 6.015 \end{pmatrix}$

Iterations 682

Solution vector $\underline{x}$ = $\begin{pmatrix} 1.000 \\ 2.006 \\ 2.994 \end{pmatrix}$ max abs.error vs.analytical solution $< 6*10^{-3}$
calc.max abs.error: 2.10^{-7}
calc.max rel.error: 1.10^{-4}

Case 4

Matrix A: $\begin{pmatrix} 1.0025 & 1. & 1. \\ 1. & 1.0025 & 1. \\ 1. & 1. & 1.0025 \end{pmatrix}$ known vector $\underline{b}$ = $\begin{pmatrix} 6.0025 \\ 6.0050 \\ 6.0075 \end{pmatrix}$

Iterations 954

Solution vector $\underline{x}$ = $\begin{pmatrix} 1.006 \\ 2.006 \\ 2.984 \end{pmatrix}$ max abs.error vs.analytical solution $< 1.6*10^{-2}$
calc.max abs.error: 2.10^{-7}
calc.max rel.error: 1.10^{-4}

(It should be noted that the relevant IMSL routine was not able
to solve this last problem with comparable accuracy.)

NOTE: To date tests have only been carried out on small matrices
(max.size: 10 x 10)

Fig. 13

$$\underline{x}_{n+1} = \underline{x}_n + \underline{k} \cdot (\underline{b} + A * \underline{x}_n)$$

(where $\underline{x}_n$ is the approximation to the solution vector after n iterations)

where $\underline{k}$ is taken to be $= \dfrac{2*(\underline{x}_n - \underline{x}_{n-1})}{A*\underline{x}_n - A*\underline{x}_{n-1}}$ in accordance with the theoretical

convergence criterion in [19].

Note that the iteration count (N) and the magnitude of k (GAIN) are watched in the routine (at line 32) and serve as indicators for failure of the approach when they become excessively large. (In Dutch "failed"="mislukt")

The same approach has been tried for single nonlinear algebraric equations and has shown similar promise of universality in that field. Let us now go on to another class of problems central to continuous system simulation.

While the technique above might alleviate the task of solution of "single point" (static) inversion problems, does this help us in any way in the quest for general approaches for "functional"(dynamic) inversion problems?
One could argue that a continuum is just an infinitely dense set of points.
But then computing devices do not cater for infinity and punish us with cumulative errors when we try to approximate the continuum with a large set of point computations. In most cases the numerical core of these problems is the integration routine. Let us therefore take a closer look at the basic accuracy aspects of integration in a simulation context. The main sources of errors are:

- the choice of the basis function set used to approximate the unknown model trajectories;
- approximation error caused by the chosen (finite) order of the above approximating function (truncation errors);
- error per integration step caused by the finite wordlength of the computing device used (round-off error);
- cumulative error caused by the conjunction of the above errors propagated through many integration steps, dependent on the stability characteristics of the chosen integration method and on the transfer properties of the differential equations that are being solved;
- errors caused by non-analytic behaviour of the solution function (e.g. discontinuities);
- errors caused by wrong integration algorithm "strategy" with respect to computability criteria ("central" versus "distributed" integration).

Let us first handle the last point, and look at a continuous simulation problem to be solved on a digital computer described by m differential equations. Let us call $S_j[t_n]$ the j-th state variable in the set of equations, where $[t_n]$ indicates that S_j is known in discrete points (0, Δt, $2\Delta t$, ..., $n\Delta t = t_n$) on the interval $[0, t_n]$. Let us denote by F_j the functional relationship that is needed to calculate S_j, representing as well the terms of the j-th differential equation as the integration algorithm that is used to integrate the differential equation.
Now three main integration strategies can be distinguished. They can be described in the following generalized way:

- distributed integration: $S_j(t_{n+1}) = F_j(S_1[t_n], \ldots, S_{j-1}[t_n], S_j[t_n], \ldots, S_m[t_n])$
- ripple integration: $S_j(t_{n+1}) = F_j(S_1[t_{n+1}], \ldots, S_{j-1}[t_{n+1}], S_j[t_n], \ldots, S_m[t_n])$
- central integration: $S_j(t_{n+1}) = F_j(S_1[t_{n+1}], \ldots, S_{j-1}[t_{n+1}], S_j[t_{n+1}], \ldots, S_m[t_{n+1}])$

From the above formulation a few things are obvious.
- In the case of distributed integration a direct (extrapolating) computation of $S_j(t_{n+1})$ can take place in which the order of computation

is irrelevant. Every state variable at every timestep can be calculated
individually from the values of the other state variables in the "past". It is
the simplest integration strategy, applied for example in "Dynamo" using the
Euler algorithm. In general so called single-step methods (e.g. Euler, Heun,
Burlisch-Stoer) are applicable.
        - In the case of ripple integration a sequence for the computation of
        state variables has to be chosen. S_j 's are   updated in "ripple" fashion.
Almost all known integration algorithms are applicable, except predictor-
corrector methods.
        - For central integration some kind of iteration procedure is needed (e.g.
        the predictor-corrector technique) to obtain the value of all state
variables for the same value of the independent variable. The origin of this
strategy lies in predictor-corrector algorithms, of which e.g. the Runge-Kutta
algorithm can be shown to be a derivate (Elzas [29]).
Although experience has shown that the ripple-strategy yields good performance
for specific differential equations (Elzas and Zimmermann [27]) only the central
integration strategy fulfils completely the theoretical requirements for existence
of the solution of the set of differential equations. This applies especially to
one of the basic conditions for computability of the solution, namely that all
intermediate values of all state variables and their derivatives are available
for all values of the independent variable t   over the whole interval $[0, t_{n+1}]$
(see also [19] and [20]. So, although the other strategies have been shown
to be useable for many practical cases, central integration is the safer choice.

On the subject of the other points, excellent basic literature is available in
the form of Gear's book [25] and Hull's article [26] on the numerical solution
of differential equations. In the following part only those points will be
discussed for which viewpoints can be found that are not presented in the above
literature.
In discussing errors we should recall that these can-mostly-only be estimated
to lie within a closed region. So that the discussion about errors will have
to be limited to a discussion of estimated error bounds. These error bounds
can be of two kinds: a priori (depending only on knowledge of the differential
equation) and a posteriori (requiring knowledge of the properties of the solution).
For most practical cases occurring in simulation a priori error bounds are
- to say the least - extremely difficult to estimate and have to be recalculated
for every new model.
A posteriori error bound estimates are easier to handle in this respect, because
they depend mostly on the choice of approximating function and the behaviour of
the observed solution.
Note, however, that this observed solution is at its best a (close) approximation
to the true solution to the problem.

The integration methods available in the literature depend on two main families
for the basis of their approximating function: polynomials and rational functions.
This raises an interesting question: why these function families and not others?
It is well known that in the analytical treatment of the solution of (linear)
differential equations, harmonic and exponential functions play a predominant role.
To date no practical examples exist of the use of Fourier polynomials and exponential
functions as a basis for integration algorithms the evolution of computing practice
from mechanical calculators made the choice of polynomials or rational functions
natural. The ever increasing power of our computing devices opens new vistas on
experiments with other approximating functions although it is clear that the
universal approximating function will probably never be found.

The "goodness of fit" of the approximation is extremely important because
truncation errors and validity interval (integration step size) depend heavily
on this property. An automated choice of function family is therefore strongly
advocated. An example of such an approach can be constructed with the subroutine
DIFSUB (Gear, loc.cit., page 96) for one step integration by rational function
or polynomial approximation. One can modify the routine and switch MF-the

method flag-automatically with every approximation order tried from rational
approximation to polynomial approximation comparing estimates. Finally MF is set
to the method which yields the smallest error with the chosen step size. Such
subroutines have been tested by the author [29].

We will not dwell on a priori estimation of truncation errors, but rather take
a pragmatic (a posteriori) approach to truncation error size control, with
emphasis on reliability.
Once the approximation method has been chosen, there are three control
possibilities:
    a. step size control;
    b. variation of order of approximating function;
    c. variation of intermediate point location with weights.

To illustrate these techniques the Runge-Kutta family of formulae has been
chosen as an example.
    a. Step size control has always been the basic technique for estimating
      (and reducing) the truncation error per step. The reason for this is
obvious: all approximation techniques have the property that this error tends
asymptotically to zero when the step size goes to zero. The higher the order,
the stronger the tendency (c.f. Taylor series expansion remainder terms).
The problem with this type of a posteriori verification of step size lies in
the fact that comparison with the "ideal" solution is impossible and that
therefore criteria have to be used which are based upon comparison of approximated
solutions for several sizes of step in the unknown variable. In this way, at least
the calculation for two step sizes is needed, after which step size search schemes
like binary search and/or interpolation can be used to find the step size
corresponding to the approximated truncation error limits.
Note that this technique will have adverse effect on the problem of cumulation of
errors and pose synchronisation problems in the case of real-time simulation and
synchronisation of time-events resulting from associated discrete submodels.
In case the step size is comparatively large, and the difference of the two
approximate solutions mentioned above is within the truncation error tolerance,
the only statement that can be made about the solutions reached is that they have
the same degree of inaccuracy, while the size of the real truncation error remains
unknown. Naturally the case for this type of step size control becomes stronger as
more is known about the properties of the solution function (e.g. polynomial-
behaviour, absence of poles etc.).
Considering the above technique to be a best guess under the circumstances
(especially for "large" step sizes) other - similarly "arbitrary" - criteria
could be used as well. A technique that has been experimented with by the
author [27] is to take the estimated local radius of curvature of the observed
solution as a measure for step size control. This technique has - pragmatically -
been found to be more effective than the technique described above, although the
tendency to take very conservative estimates for the step size has been apparent.
Other techniques could be considered like the occurence of changes of shape
(convex versus concave) and saddle points.
They could also economize on the number of function evaluations to be carried out.
    b. An approach to variation of the approximation order (without necessarily
changing step sizes) has been advocated by Crosbie and Heyes [28] who compare
the results obtained by applying an (n+1)th order extension of an n-th order
Runge-Kutta formula and comparing the two results to get an estimate on the
truncation error for a given step length. The same approach can be found in Gear's
subroutine for polynomial/rational function extrapolation that was mentioned
earlier. Although this approach is in principle better than step size control, the
number of function evaluations needed and the problems posed by the finite
coefficient accuracy caused by finite word length put limits to the highest
order that would bring sensible improvements. Therefore also this technique often
needs to be augmented with step size control.

c. Several Runge-Kutta methods of the same order can be constructed. They only differ in the choice of the abscissa for the intermediate fitting points and the weights put to the function values at those points. An example of four such methods of order 4 is given by Runge-Kutta Kutta, Runge-Kutta Merson, Runge-Kutta Gill and Sarafyan-England formulae. Their construction is illustrated in Fig. 14. It is shown in a recent publication (Elzas [29]) that these methods show marked difference in their truncation error characteristics in tests which compared numerical results with an analytically known solution. This brought the author the idea to try to construct a technique which would automatically adapt the numerical coefficients in the approximating function that forms the base for a Runge-Kutta method. An example program for such a method is illustrated in Fig. 15. Experiments gave promising results, although further investigation has to take place.
The advantage of this technique is that the order of the approximation is fixed, so that highest order terms contribution will not be influenced too much by finite word length, while accuracy tolerance controls the coefficients of the approximating function within a fixed step length. At this stage, however, the technique still faces start-up problems.

In conclusion of these views about truncation errors one could say that techniques b. and c. seem preferable for several reasons and should be investigated in more detail.
     - Not much new can be said about round-off errors except that in the first place the integration formulae should be chosen in such way that the higher order terms should be allowed to contribute to the accuracy of the method and not be "drowned" by the finite accuracy properties of the computing device. These finite accuracy properties depend mainly on the floating point mantissa-word-length used in the computer at hand (about 6 significant digits in most minicomputers and about 8 significant digits on larger machines), because most types of floating point arithmetic units or routines do not provide any implicit round-off procedure. Let us take, as an example, the Sarafyan-England integration procedure as illustrated in Fig. 14, and compare results for typical differential equations (e.g. the harmonic equation and the Van de Pol equation). It can easily be shown that the difference in results between the 4th order and 5th order Sarafyan-England procedure is less than 10^{-8}. Application of the higher order procedure does not bring extra benefits.
Instead of this, repeated application of the higher order formula will contribute adversely to the accuracy of the end result by cumulation of errors through lack of proper round-off procedure during computation of intermediate steps.
As a consequence of the lack of implicit round-off procedures as pointed out above, accuracy improvement can be achieved by implementing these procedures explicitly (e.g. in the integration algorithm) by monitoring the number of significant digits left after a computation.
Techniques for maintaining significancy of the arithmetics in computation have been developed in the past, and were reported for example by Ashenhurst [30]. It has been shown (e.g. by E.O.Gilbert e.a. [33]) to be worthwhile to improve round-off behaviour (for example by implementing significant arithmetic) before resorting to higher order integration algorithms for better accuracy.

     - In most modern literature the formal a priori handling of cumulative errors in the numerical solution of differential equations is based on the integration method only, and does not really take into account as well the transfer properties of the system at hand.
An extension in this respect to the study of error propagation phenomena in sequential processes that can be found on pages 333 to 357 of Zeigler's book [15], would be most useful. The main problem with this type of error is that there are no appropriate a posteriori methods to evaluate these errors apart from either comparing the obtained results through complete new runs with variations in the integration method or effecting a sensitivity analysis on all state variables, the latter technique also requiring multiple runs (which approach is not often welcome).

*h↓	k_1	k_2	k_3	k_4
0	-			
1/2	1/2	-		
1/2	0	1/2	-	
1	0	0	1	-
order 4	1/6	1/3	1/3	1/6

KUTTA

	k_1	k_2	k_3	k_4
0	-			
1/2	1/2	-		
1/2	1/4	1/4	-	
1	0	-1	2	-
order 4	1/6	0	2/3	1/6

GILL

	k_1	k_2	k_3	k_4	k_5
0	-				
1/3	1	-			
1/3	1/2	1/2	-		
1/2	3/8	0	9/8	-	
1	3/2	0	-9/2	6	-
order 3	3/2	0	-9/2	6	0
order 4	1/2	0	0	2	1/2

MERSON

.	k_1	k_2	k_3	k_4	k_5	k_6
0	-					
1/2	1/2	-				
1/2	1/4	1/4	-			
1	0	-1	2	-		
2/3	2/27	10/27	0	1/27	-	
1/5	28/625	-1/5	546/625	54/625	-378/625	-
order 4	1/6	0	2/3	1/6	-	-
order 5	1/24	0	0	5/24	27/56	125/336

SARAFYAN/ENGLAND

Fig. 14: Different Runge-Kutta Methods of same order

PROGRAM INTHIG

```
 1              DIMENSION OUT(5(,OUT1(5)
 2     10       ACCEPT 10010,D,A,B
 3              DO 80 I=1,100
 4              RIN=A*I*D+B
 5              OUT(5)=-RIN
 6              DELT=ABS(RIN)*1.E-6
 7              FI=I
 8              N=0
 9              GAIN=1.
10              IF(ABS(GAIN).GT.1.E+3)20,30
11     20       GAIN=0.001*GAIN
12     30       CONTINUE
13     40       EPS=(RIN+F(OUT,D))/2.
14              OUT1(5)=OUT(5)+GAIN*EPS
15              TYPE 10040,N,RIN,GAIN,OUT(5),EPS,OUT1(5)
16              IF((ABS((OUT(5)-OUT1(5))/OUT1(5)).LT.1.E-5).AND.(ABS)EPS)1.LT.DELT))60,50
17     50       N=N+1
18              GAIN=-2.*(OUT1(5)-OUT(5))/(F(OUT1,D)-F(OUT,D))
19              OUT(5)=OUT1(5)
20              IF((N.GE.1000).OR.(ABS(GAIN).GT.1.E+30))90,40
21     60       TYPE 10020,FI,OUT(5)
22              TYPE 10020,OUT(4),OUT(3)
23              TYPE 10020,OUT(2),OUT(1)
24              DO 70 J=2,5
25              OUT(J-1)=OUT(J)
26     70       OUT1(J-1)=OUT(J)
27     80       CONTINUE
28              GO TO 10
29     90       TYPE 10030
30              CALL EXIT
31     10010    FORMAT(F16.6)
32     10020    FORMAT(1X,2(F12.4,1X),/)
33     10030    FORMAT(1X,'MISLUKT',/)
34     10040    FORMAT(1X,I6,5(E12.3,1X),/)
35              END
```

Fig.15: Adaptive 4th order numerical integration program,
using generalized iteration.

So no practical methods are available other than the calculation of the absolute value of the approximate size of the cumulated error by adding up the absolute values of all observed truncation errors per step as a very rough measure. That this measure is at best an indication of what is going wrong can be shown very simply by looking at the integral representation of the solution to the first order differential equation $y'=f(t,y)$ that has to be solved for every state variable. Expressing its value for any value T of the independent variable larger than the abscissa t_0 for which the independent value was last computed gives

$$y(T) = y(t_0) + \int_{t_0}^{T} f(\tau, y(\tau))d\tau$$

Assuming that the solution achieved during the interval $t < t_0$ has incurred an error $\varepsilon(t)$ and that the procedure to calculate f yields a small error $\eta(t)$ in $y(t)$ then this equation is transformed into:

$$\bar{y}(T) = y(t_0) + \varepsilon(t_0) + \int_{t_0}^{T} f(\tau, y(\tau) + \eta(\tau))d\tau$$

Through Taylor series expansion neglecting higher order terms the equation can be written as:

$$\bar{y}(T) \approx y(t_0) + \varepsilon(t_0) + \int_{t_0}^{T} f(\tau, y(\tau))d\tau + \int_{t_0}^{T} \eta(\tau) \frac{\partial f}{\partial y} d\tau$$

or:

$$\bar{y}(T) \approx y(T) + \varepsilon(t_0) + \int_{t_0}^{T} \eta(\tau) \cdot \frac{\partial f}{\partial y} d\tau$$

so quite a difference from the original.

Now let us choose such a small time step that f can be linearised so as to reduce above equation to the general form of a linear differential equation

$$y' = p(t) - q(t).y$$

and take ε and η to be equal and constant. Calling $\bar{y}$ the solution of the equation that we get under influence of errors for $t < t_0$ and y the exact solution, let us solve above equation analytically for the next time step. We then see that:

$$\bar{y}(t_0+h) = y(t_0+h) + \varepsilon e^{-\int_{t_0}^{(t_0+h)} q(\tau)d\tau}$$

Knowing this result it could be possible to try and estimate $q(t)$ for every variable and use it to compensate the result at every time step if ε is known. (Notice that $q(t) \approx \partial f/\partial y$, $\varepsilon(0)=0$ and η is the error per integration step).

    - Two problems have to be taken into account when discontinuities can occur in the simulation experiment.

The first one is fundamental: in principle, the integration algorithms that can be used have all been based on the assumption that the functions to be integrated are analytical, so a discontinuity will be approximated by a (steep) smooth function. E.G.Gilbert in [32] considers an example of the result of applying a unit step function at $t=0$ on the differential equation $y'= ay + br(t)$. ($r(0)=0, r(t)=1$ $t>0$) The analytical response is $y(t) = 1-e^{at}$ ($t > 0$), where RK4 responds with:

$$y(nh) \cong 1 - (1+\tfrac{1}{6}ah)[\,1+ah+\tfrac{1}{2}(ah)^2+\tfrac{1}{6}(ah)^3+\tfrac{1}{24}(ah)^4]^n$$

The reader can verify that if $a=1$, $t=0.4$, $h=0.1$ and $n=4$, though one expects an error of the order $h^4 (=10^{-4})$, the actual error is ≈ 0.025.
This indicates that a much smaller h is needed for reasonable accuracy. (Notice

that the expression within square brackets is a reasonable approximation of e^{anh}, so that the error is mainly caused by the term $(ah)/6$).

A technique to circumvent this problem by inserting an a priori correction into the integration algorithm has recently been proposed by Howe [34].

The second problem is the problem of synchronisation of the occurence of discrete events to the discretisation interval used for continuous simulation when a combined (continuous/discrete) system is simulated on a digital computer. In principle a matching technique has to be found such that the occurence of the discrete event is made to coincide with a mesh point of the continuous simulation integration interval.
This is not a trivial problem as the integration algorithm has to cater for two (often conflicting) requirements with respect to the step size: accuracy and precise "timing".
Originally this problem was solved by implementing a (binary) search technique that reduced the integration step size until a match was found. Apart from being time-consuming, this technique imperils the accuracy through the increase of cumulative errors.
A more detailed consideration of this problem by Cellier can be found elsewhere in this book [35].
Fortunately research on this subject has resulted in the development of relevant techniques which have either been published (e.g. Hay [31]) or are being implemented (e.g. by Kettenis/Ören at Wageningen University/University of Ottawa).

BRIEF NOTE ON NEW COMPUTER ARCHITECTURES

Much of the above knowledge has been gained through experimentation with conventional (though sophisticated) sequential digital computers.
A large part of the recommendations, especially the numerical ones, can only be implemented economically on today's (or tomorrow's) machines which provide high speed and ample storage at relatively low cost. The tendency to de-emphasize speed and cost of execution in favour of reliability of results will continue to grow while our equipment becomes faster and cheaper per unit of computation (E.O.Gilbert [36]).
Aside from this mainline tendency new types of computer architecture are coming of age. Associative computers, array-processors and digital parallel processing devices are commercially available now, and the associated application possibilities are developing steadily (Elzas [37]).
Most of these devices will make adaptations to the algorithms in use today, necessary to allow the simulation community to share the advantages offered by the new technology. Some of the algorithms mentioned here are amenable to such an adaptation (see Elzas [29]).

REFERENCES

[1] C.T.de Wit: On the usefulness of simulation models of ecological systems, Simulation of Continuous Systems (M.S.Elzas, editor, IMACS 1976) 126-130.
[2] M.S. Elzas: Results of a simulation: can they be "trusted"?, Modelling in Business (IFIP applied Information Processing Group IAG, 1978) 170-183.
[3] M.S. Elzas: Hifips, a proposed novel hardware/software system for sophisticated hybrid problem handling, SCI Hybrid Software Conference, Minneapolis (Aug, 1968).
[4] M.S. Elzas: Simulation languages and hybrid compilers, principles and techniques, Proc. Special Symposium on Advanced Hybrid Computing (U.S.Army Material Command, Alexandria, 1975) 77-79.

[5] W. Spillers: Some thoughts on automated design, Computer Aids to Design and
       Architecture,(Ed. N.Negroponte, Petrocelli/charter, N.Y. 1975) 61-72.
[6] J. Weizenbaum: Computer Power and Human Reason (Freeman & Co, San Francisco,
       1976).
[7] M.S. Elzas: Whither Simulation?, Proc. UKSC 1978 Conference on Computer
       Simulation, Chester (April 1978) especially pp 498-499.
[8] T.I. Ören: Digital simulation languages for combined systems:an overview,
       Proc. SCSC 1973, Montreal (August 1973) 346-353.
[9] T.I. Ören: Software for simulation of combined continuous- and discrete
       systems:a state of the art review, Simulation Vol.28, 2 (1977) 33-45.
[10] M.S. Elzas: HL1 or towards a unique language for all continuous system
       simulation, Proc. 7th AICA Conference, Prague (1973) 66-70.
[11] T.I. Ören: Modelling, model manipulation and programming concepts in
       simulation: a framework, Proc. IFIP Working Conference on Modelling and
       Simulation of Land, Air and Water Resources Systems,(Ed. G.Vansteenkiste,
       Ghent 1977) (In press at North-Holland, 1978).
[12] T.I. Ören: A personal view on the future of simulation languages, Proc. UKSC
       1978 Conference on Computer Simulation, Chester (April 1978) 294-306.
[13] T.I Ören: Technical Report 78-2, Computer Science Dept., University of Ottawa
       (June 1978) (produced while on sabbatical leave at Wageningen University).
[14] J.A.den Dulk and T.I. Ören: Ecological models in GEST'78, Technical Report
       Computer Science Dept., Wageningen Agricultural University (June 1978).
[15] B.P. Zeigler: Theory of modelling and simulation, (Wiley, New York,1976).
[16] E.W. Dijkstra: GOTO statements considered harmful, Comm.ACM, Vol.11 (March
       1968) 147-148.
[17] N. Wirth: An assesment of the programming language Pascal, IEEE Transactions
       on Software Engineering, Vol.SE-1, no.2 (June 1978) 192-198.
[18] D.C. Luckham: Program verification and verification oriented programming,
       Information Processing 77 (Ed. B.Gilchrist, North-Holland, 1977) 783-793.
[19] M.S. Elzas: Fundamental differences of discrete and continuous computing
       automata, Simulation of Continuous Systems (Ed. M.S.Elzas, IMACS, 1976) 39-55.
[20] W.K. Giloi: Principles of Continuous system simulation, (Teubner Verlag,
       Stuttgart, 1975).
[21] H.G. Sol: Simula(tion) software for the develoment of information systems,
       Modelling in Business (IFIP applied Information Processing Group IAG, 1978)
       especially 123-127.
[22] M.S. Elzas: Do automated hybrid computers have a credible future?, Proc.1973
       SCSC, Montreal (July 1973).
[23] E. Marnier: A program verifier for Pascal,   Information Processing 74
       (North-Holland, 1974).
[24] E.F. Miller Jr.: Program testing: art meets theory, Computer (July 1977) 42-51.
[25] C.W. Gear: Numerical initial value problems in ordinary differential equations,
       (Prentice-Hall, Englewood Cliffs, N.J., 1971).
[26] T.E.Hull, W.H.Enright, B.M.Fellen and A.E.Sedgwick: Comparing numerical
       methods for ordinary differential equations, SIAM Journal of Numerical Analysis
       Vol. 9, no.4 (Dec. 1972) 606-637.
[27] M.S. Elzas and A. Zimmermann: FORSIM/ESL, Applied Dynamics Europe internal
       report (September 1968).
[28] R.E. Crosbie and W. Heyes: Variable step integration methods for simulation
       applications, Applied Mathematical Modelling, Vol.1 (Dec. 1976) 137-140.
[29] M.S. Elzas: Old wine in new jugs, Proc. joint IMACS/SCI meeting (June 1978)/
       Research Report, Computer Science Dept., Wageningen Agricultural University
       (Dec. 1978).
[30] R. Ashenhurst: Number representation and significance monitoring,
       Mathematical Software (Ed. J.R. Rice, Academic Press, 1971) 68-92.
[31] J.L. Hay, R.E. Crosbie and T.I. Chaplin: Integration routines for systems with
       discontinuities, The Computer Journal, Vol. 17, no.3 (1974) 275-278.
[32] E.G. Gilbert: Dynamic error analysis of digital and combined analog/digital
       computer systems, Proc. Symposium on Hybrid Computation (Applied Dynamics
       Europe, 1966) 87-136.

[33] E.O.Gilbert, R.M.Howe and D.S.Bernstein: Numerical integration round-off error in the AD10, Technical Report (Applied Dynamics International, June 1978).

[34] R.M.Howe: A new method for handling discontinuous nonlinear functions in digital simulation of dynamic systems, Research Report (Dept. of Aerospace Engineering, University of Michigan, 1978).

[35] F.E. Cellier: Combined continuous/discrete system simulation languages, Methodology in Systems Modelling and Simulation (North-Holland, 1979).

[36] E.O. Gilbert and R.M. Howe: Design considerations in a multiprocessor computer for continuous system simulation, AFIPS-Conference Proc., Vol.47 (1978) 385-393.

[37] M.S. Elzas: An assessment of the architecture of some available multiprocessor computers, Technical Report 79-1, Computer Science Dept., Wageningen Agricultural University (Jan. 1979).

METHODOLOGY IN SYSTEMS MODELLING AND SIMULATION
B.P. Zeigler, M.S. Elzas, G.J. Klir, T.I. Ören (eds.)
© North-Holland Publishing Company, 1979

STRUCTURING PRINCIPLES FOR

MULTIFACETED SYSTEM MODELLING*

Bernard P. Zeigler†
Department of Applied Mathematics
The Weizmann Institute of Science
Rehovot, Israel

Abstract: Computer assistance in multifaceted system modelling
takes two related forms: organization of model/data bases and
provision of tools for operating on these bases. This article
adresses the former aspect by suggesting certain principles for
structuring the model and data bases and their interrelation.
The principles are compatible with the premise that computer
assistance   should support arbitrary composition and decomposition of systems. The resulting organizational structure recognizes that models encode empirical information through
parameter identification and facilitates the flow of this
information from model to model.

1. INTRODUCTION

The term "multifaceted system" has been coined to denote an approach to modelling
which explicitly recognizes the partial decomposibility of reality into *diverse*
but *related* aspects. Indeed, what are often called large scale systems are better
characterized by their multifaceted characteristics (Rosen, 1977, Courtois, 1977)
than by their scale which is purely incidental (both the biological cell and the
world ecosystem are multifaceted systems). Computer support of integrated modelling methodology has been proposed (Zeigler, 1978, Oren and Zeigler, 1979). Its
design must take into account (at least) the following considerations:

a) A multiplicity of models oriented towards different system aspects must be
    maintained in various states of validation.
b) The dynamic evolving nature of multifaceted modelling precludes a fixed base
    of primitive components for model construction -- both holistic and reduction-
    istic directions must be accomodated.
c) Multifaceted modelling is an on-going, history dependent, process (as opposed

*This research is supported by Grant No. DAERO-78-G-088 of the ERO, U.S. Army
†Final preparation of the manuscript was done while the author was a visiting
 professor of computer science at the Virginia Polytechnic Institute and State
 University, Blacksburg, Va.

to the start-from-scratch one-shot project that it is often implied to be in
methodological discussions (see for example, Shannon, 1975).

It follows that computer support of multifaceted modelling methodology must deal
with the organization and flexible manipulation of models around the questions
asked of them. A theory of modelling and simulation has been developed, which
offers concepts on which to base this organization (Zeigler, 1976, 1978). In this
paper, we extend the previous development in two ways.

1) We suggest a "substrate" of concepts to underlie the construction of models
   and experimental frames which includes unlimited composition and decomposition
   of entities and variables in keeping with considerations b) and c) above.
2) We suggest a widened set of structuring principles of model and experimental
   frame organization based on the above substrate.

Such structuring principles include:
-- linkage of variables, models and frames to entities
-- composition-decomposition hierarchies of entities
-- semantic structuring of variables
-- derivability ordering of frames
-- applicability of frames to models
-- model organization through parameter correspondences and morphisms.

The development of these concepts is preceeded by an example discussed in non-
theoretical terms. It concerns coin tossing -- a "two-faceted" system shown to be
multifaceted. This example and a larger one previously developed (Zeigler, 1979)
provide concrete illustrations of the concept application.

2. COIN TOSSING -- "MULTIFACETED SYSTEM"

As an illustrative example of the multifaceted modelling approach, we develop the
following hypothetical scenario. Starting from the model of coin tossing common
in probability text books, we develop a series of models which reveal various related facets of the coin tossing phenomenon. We wish to highlight in this
scenario, the fact that models encode empirical information through their parameter
settings and that progagation of parameter value information from model to model
is an important component of multifaceted system modelling.

After describing the system of coin tossing models, we shall continually return
to this description to illustrate the general concepts and structures of multifaceted system modelling.

The reader may wish to skim through this section and then return to relevant sub-
sections as these are needed in his subsequent reading.

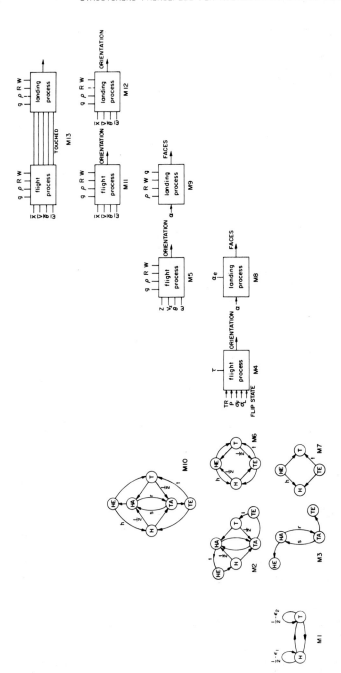

Figure 1 illustrates the family of models to be described

M1

Model M1 is the usual Markov 2 state model of coin tossing. For a fair coin, the probabilities of tails (T) following heads (H) and H following T are each 0.5. The parameters ϵ_1 and ϵ_2 represent deviation from fairness -- the probabilities of T following H and H following T are $\frac{1}{2}+\epsilon_1$ and $\frac{1}{2}+\epsilon_2$, respectively.

A sequence of trials on a certain coin was employed to identify the parameters ϵ_1 and ϵ_2 which were found to deviate from 0 to an extent not attributable to the finite sample size.

M2

Thus the question arose: how to explain these discrepancies. Two models M2 and M6 were proposed. M2 focuses on the behavior of the coin in flight. Once tossed, a coin can flip, from heads to tail and back, a number of times before landing.* If there were an asymmetry in the HA (heads in air) to TA (tails in air) alternation, then this might account for the discrepancies. Indeed, it can be shown that the probability of an H to T transition in model M2 is

$$P(T/H) = \tfrac{1}{2}+ \frac{r-s}{2(1-rs)}$$

and

$$P(H/T) = \tfrac{1}{2}+ \frac{s-r}{2(1-rs)} ,$$

where r and s are the probabilities of HA to TA and TA to HA transitions, respectively.

In this way, a parameter correspondence between M1 and M2 was established, viz.

$$\epsilon_1 = \frac{r-s}{2(1-rs)}$$

$$\epsilon_2 = - \epsilon_1 \text{ **}$$

(1)

This correspondence can be employed in two directions. Having values for ϵ_1 and ϵ_2 constrains the possible values of r and s . On the other hand, if r and

---

* For ease of subsequent exposition, the dummy states HE and TE have been added to M2. These have no effect on its behavior of interest.

** Note that even though M2 has two free parameters, r and s , the symmetries it assumes (for the H to HA , H to TA , T to HA and T to TA transitions) show up as a reduction in the number of free parameters in M1.

s can be independently identified, than ε_1 and ε_2 can be predicted. It is
the latter direction that leads to the sequence of models M3, M4, M5 and M11.

M6

Before discussing these latter models, let us return to the alternative proposal
for explaining the discrepancy. M6 focuses in on the <u>landing</u> of the coin. If the
coin is weighted more to heads than to tails, or conversely, it may have a bias
towards ending up in heads (or tails). Indeed, the parameter correspondence
from M6 to M1 is simply

$$\varepsilon_1 = \tfrac{1}{2}(t-h)$$

$$\varepsilon_2 = - \varepsilon_1$$

(2)

where t and h are the probabilities of the transitions TE (tails up when on
edge) to T and HE (heads on edge) to H . In the nonbiased case, t = h = 1 .
Once more, the desire to identify the new parameters independently, leads to the
development of a sequence of models -- M7, M8, M9 and M12.

Notice that M2 and M6 are not distinguishable on the basis of their relation to
M1. That is, the correspondences 1 and 2 permit pairs of values (r,s) and
(t,h) (within the range [0,1] for all values of $\varepsilon_1 \ \varepsilon \ [-\tfrac{1}{2},+\tfrac{1}{2}]$ and $\varepsilon_2 = -\varepsilon_1$).
Thus (on the basis of M1) both could be valid. Yet it is evident that the assump-
tions underlying each are mutually contradictory -- M2 assumes a bias in the
flight process which M6 denies, and conversely in the case of the landing process.

M10

In order to resolve this paradox, a "supermodel" M10 was constructed, combining
the flight hypotheses of M2 with the landing hypotheses of M6. This model is
essentially M2 except that the behavior from states HE and TE is taken from
M6. The correspondence between the parameters r , s , h and t of M10 and
those of M1 can be shown to be

$$\varepsilon_1 = \tfrac{1}{2}\frac{1}{1-rs} \ [s-r+h(r-1)(s+1)+t(1-s)(r+1)]$$

$$\varepsilon_2 = - \varepsilon_1 \ .$$

(3)

It can be checked that (3) reduces to (1) and to (2) when t = h = 1 and r = s ,
respectively. Presumably, when neither of the latter conditions holds, the correspondence (3) is more likely to be correct than are either (1) or (2). This
conclusion is obvious given the motivation for constructing model M10. As they

stand, however, the correspondences appear to be inconsistent and the question raised is: how can the computer help detect such inconsistencies and suggest ways of resolving them.

M3

Let us return to the models M3, M4, M5 and M11. Model M3 represents that part of M2 which has to do with the flight behavior of the coin. M4, M5 and M11 are successive elaborations of M3, i.e., they encorporate more and more details which may be relevant to any asymmetry in the coin's alternation in flight.

M4

M4 represents the flight process as decomposed into two components -- the spin and the translation process, respectively. The spin process recognizes two events -- flipping from heads to tails and the reverse. To schedule events, let τ_{HA} be the time the coin will spend in the HA state (recall, this means heads while in the air). Then a flip event from HA to TA is scheduled in τ_{HA} seconds. This event will schedule a flip from TA to HA in τ_{TA} seconds. The period of oscillation is thus $P = \tau_{HA} + \tau_{TA}$ and the ratio $\tau = \frac{\tau_{HA}}{P}$ is considered to be characteristic of the coin. Asymmetry in spinning is thus represented by a value of $\tau \neq \frac{1}{2}$.

The translation process representing the coin's non-rotational travel through the air, is realized simply in M4 by a travel time TR between the TOSS event and the LAND event. At the occurence of the LAND event, the state of the spin process, HA or TA , is reported as the state at landing HE , or TE , respectively.

The period P and travel time TR are considered to be input variables of M4, determined by the manner in which each particular toss is made. M4 is a discrete event model with state variables σ_F (time left to next flip), σ_L (time left to landing), and flight process ORIENTATION with range {HA, TA, HE, TE} (see Zeigler, 1976, chapter 6).

Following the strategy of linkage evident in the case of models M1, M2, M6 and M10, we seek a parameter correspondence between M4 and M3, i.e., between τ and the pair (r,s). However in this case, the correspondence depends on the constraints placed on the input variables P and TR . Assuming that TR is exponentially distributed with mean λ and P is constant, it can be shown that

$$r = e^{-\lambda P \tau}$$

$$s = e^{-\lambda P(1-\tau)}$$

(4)

Thus, depending on the distributions of the input variables, there may be more than one correspondence between τ and the pair (r,s) . Also for some distributions, a correspondence may not be demonstratable. Indeed, a correspondence exists just in case M3 is a homomorphic image of M4 (Zeigler, 1966, chapter 8)*. The basic problem is that M3 represents the transition between successive elements in sequences over HA , TA , HE and TE as Markovian (independent of the past) a property which may not hold for ensembles of such sequences generated from certain distributions of input variables in M4. Under such considerations, the modeller may decide to revise the model M3 and subsequently M2 in which it is embedded. However, in the present scenario, this has not been done. The correspondence (4) now enables an independent estimate of the parameters r and s once the parameter of the coin τ and the parameters of the input variables distribution P and λ have been identified.

M5

Model M5 continues the sequence of refinements of the flight process. It models the translation process as the motion of the coin's center of gravity in the vertical direction. The spin process is considered to be a rotation about the horizontal axis. Thus there are four state variables: Z (height), V_Z (velocity), θ (angle of rotation) and ω (angular velocity). Employing elementary Newtonian dynamics of a free falling body, a differential equation for coin's center of gravity is set up. Similarly, the angular accaleration of the coin can be related to the torque experienced due to gravity acting on its mass distribution. Under rigid body assumptions, we obtain an expression of the form: $\frac{d^2\theta}{dt^2} = F(\theta)$. If the time spent by θ in the heads region $(-90°\leq\theta\leq90°)$ is in constant proportion to the time spent in the tails region, independent of initial state (θ,ω) , then a parameter correspondence between model M5 and model M4 exists and is of the form:

$$\tau = F(g,\rho,R,W) \qquad\qquad\qquad (5)$$

where g is the gravitational constant, ρ the mass density distribution in the coin of dimensions R (radius) and W (thickness).

Note that in contrast to M4, M5 is autonomous (has no input variables). The manner of tossing is reflected in the initial state (Z,V_Z,θ,ω) assigned to the

---

* More precisely, a correspondence exists just in case, for each model of type M4 there is a model of type M3 which is morphic image of it, where model of type Mi means a model obtained by assigning particular parameter values to parameters of Mi (see Section 8).

process. Under certain conditions the input variables of M4 are readily related
to the initial state of M5. Indeed, the time taken for the center of gravity to
reach the surface is readily computed in terms of the initial vertical velocity
and height above the surface. Neglecting the disposition of the coin upon striking
the surface, this time can be taken to be the travel time TR of model M4. Also,
if to a first approximation, the angular accelaration is zero, then the period P
is inversely related to the initial angular velocity.

In general, however, M5 must be simulated to study its trajectories and relate
them to those of M4. The efficiency with which this is done may be greatly increased by constructing a discrete event model intermediate between M4 and M5
which keeps track of the state variables of M5, but updates them only at event
times (crossing of the heads-tails boundaries in θ-space). Zeigler 1978 discusses
this modelling procedure.

M11

Model M11 conceives of the coin's flight in six degrees of freedom -- translation
and rotation in three dimensions each. Such a model would be a natural stopping
point to the refinement sequence initiated with M3. Its use here is conceptual
rather than operational. Thus it has not been specified, would probably not be
justified by the additional information it could provide about the original question of interest -- the asymmetry in spinning. Nevertheless, it will serve a
fundamental role in enabling us to pinpoint the assumptions made in constructing
the simpler models of the sequence.

The end result of such a study would be the knowledge of a correspondence of the
form given in (5). When pushed through correspondences (4), we obtain estimates
of the asymmetry parameters r and s in terms of the static parameters of the
coin (density, distribution, dimensions). When pushed through correspondence (1)
we have estimates of the discrepancy parameters ε_1 and ε_2 in terms of these
(static) parameters. *Of course these composed correspondences apply only under
the simultaneous holding of the conditions justifying the component correspondences.*

M7

We now follow the parallel sequence of models M7, M8, M9 and M12 which relate to
the landing behavior of the coin. Model M7 isolates the landing behavior from
M6 in an analogous manner to the isolation of M3 from M2.

M8

Model M8 takes as input α , the angle made by the coin with the surface upon impact. It compares α with α_e , a threshold-like parameter; if $\alpha < \alpha_e$, then the output is H (heads); otherwise, it is T (tails). Thus $\alpha_e = 90°$ characterizes an unbiased tendency to fall. Just as in the case of M4 and M3, the correspondence between parameter α_e of M8 and parameters t and h of M7 depends on the distribution of input values. Assuming α is uniformly distributed we obtain the correspondence

$$(h,t) = \begin{cases} (\dfrac{\alpha_e}{90} , 1) & \text{if } \alpha_e < 90° \\ (1 , \dfrac{180 - \alpha_e}{90}) & \text{otherwise.} \end{cases} \qquad (6)$$

M9

Model M9 provides a justification for the threshold structure of M8. It considers the coin to be pivoted to the surface. Using elementary mechanics, the coin falls to the right when leaning right if its center of gravity lies to the right of the pivot. Otherwise, the coin will right itself and continue on to fall to the left. The angle at which the center of gravity is directly above the pivoted edge -- the angle of static equilibrium -- is identified as the angle α_e in M8. Since the static parameters (density distribution, dimensions) of the coin determine the position of the center of gravity as a function of angle, there is a correspondence of the form

$$\alpha_e = F(g,\rho,R,W) . \qquad (7)$$

Analogously to the case of the flight process, the correspondences 6 and 7 can be composed to yield estimates of the parameters t and h . When farther pushed through correspondence (2), we obtain discrepancy parameters values ε_1 and ε_2 , this time on the basis of the landing process bias. Of course employing the estimates of r , s , t and h in correspondence (3), we also obtain discrepancy parameters values ε_1 and ε_2 , on the basis of the combined flight process -- landing process model M10 (but see M13 below).

M12

Model M12 serves a conceptual role analogous to M11. In contrast to M9, which is limited to statics, it formulates the dynamics of the coin falling -- essentially it takes into account the effect of the coin's non zero translational and angular velocities at time of contact on its subsequent fall.

M13
===

Model M13 is the base model (i.e. the most comprehensive model) of the set of
models so far described. That is, represents the coin at the finest level of detail present in any of the models. Being composed of M11 and M12, it is intended
for conceptual use. Its formulation raises the problem of interfacing the compon-
ent flight and landing processes. It is clear that the state $(\bar{X},\bar{v},\bar{\theta},\bar{\omega})$ of the
coin at the end of its flight must become the initial state for the landing pro-
cess. However, the model must be able to decide when the flight process has terminated in order to initiate the landing process from this state. To do this we
introduce the variable TOUCHED which has initial value NO and becomes (and remains)
YES when the coin first makes contact with the surface. Thus we consider landing
to begin on the first bounce of the coin. Note that TOUCHED is not an instant-
aneous function of the dynamic state $(\bar{X},\bar{v},\bar{\theta},\bar{\omega})$ of the coin (this is obvious when
one considers that knowing the latter state does tell us whether the coin has
bounced or not, the very information needed to initiate the landing process).
Thus TOUCHED is a state variable of the model M13 (see Zeigler, 1976, chapter 6).

M13 enables us to formulate the problems involved in recombining information derived by isolating its components. We noted that the estimates of pairs (r,s)
and (t,h) obtained by isolating the flight and landing processes respectively,
could be applied to the combined model M10. However, to do so, the assumptions
made in constraining the variables which were freed by the isolation of a component, must in fact hold when the components are interconnected in the undecomposed
base model. For example, the assumption made that the impact angle α is uniformly distributed must be (at least approximately) true as determined by the
final state of the flight process. In principle these conditions can be checked
by simulating model M12, or better, a feasible simplified version of it.

3. THE SUBSTRATE OF MULTIFACETED MODELLING
==

Having developed a minature example of multifaceted modelling, we turn to an exposition of the concepts and structuring principles proposed for computer support
of this modelling approach. We begin with the "substrate" -- a universe of entities and variables from which to construct models and experimental frames. The
universe evolves dynamically under the definitional control of the modeller or the
modelling team. Concepts of entity composition (Bunge, 1977) and decomposition,
of variable construction and semantic structure, are developed in line with this
view of modelling as an on-going history preserving process.

The overall effect is to add more dynamism to world views already available in
such simulation languages as SIMSCRIPT (the entity concept) and SIMULA (the process class concept).

ENTITIES

At any time, the universe of discourse consists of a set of <u>entities</u>. An entity
has a <u>name</u> which uniquely identifies it for formal purposes. The entity may be
<u>hypothetical</u> or <u>detectable</u>, i.e., capable of real existence. In the latter case,
there is given an associated <u>detector</u> which, when embedded in a real system of
interest, is capable of detecting the presence or absence of the entity.

The entities of the example are displayed in Figure 2. COIN, FLIGHT PROCESS and
LANDING PROCESS are detectable, the latter two being segmented by the first bounce
of the coin. Detection of the spin and translation processes is possible through
reduced speed motion picture playback for example.

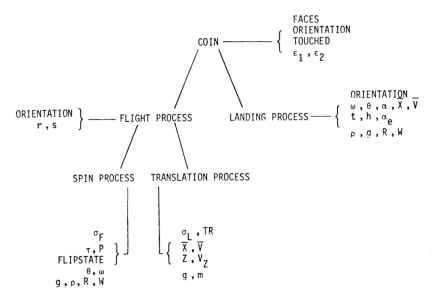

Figure 2. Entity Structure

VARIABLES

A variable has a <u>name</u> and a <u>range-set</u> (the set of values it can assume). Variables
may be called descriptive variables or parameters depending on their role in a
given context. Descriptive variables may further be categorized into classes such
as input, state and output variables. The variables employed in our example are
listed in Figure 3, categorized as descriptive variables or parameters according

to the use they have received thus far. Definitions of the different types of variables will be given later.

descriptive variables	range (units)	meaning
FACES	{H,T}	H - coin has landed finally on heads T - coin has landed finally on tails
FLIPSTATE	{HA,TA}	HA - heads up in air; TA - tails up in air
COIN ORIENTATION	{H,T,HA,TA,HE,TE}	H,T - same as FACES HE - heads up with edge on surface TE - tails up with edge on surface
LANDING PROCESS ORIENTATION	{H,T,HE,TE}	H,T,HE,TE - same as COIN ORIENTATION
FLIGHT PROCESS ORIENTATION	{HA,TA,HE,TE}	HA, TA - same as COIN ORIENTATION
α	{0,180} degrees	angle of impact of coin with surface with $\alpha=0$ meaning heads up, $\alpha=180$ meaning tails up
σ_L	{0,∞} seconds	time left to land
	{0,∞} seconds	time left to next flip
P	{0,∞} seconds	period of spin
TR	{0,∞} seconds	travel time of coin in air
θ	{0,360} degrees	angle of coin with horizontal plane, measured counter clockwise with $\theta=0$ meaning heads up
ω_z	(-∞,∞) degrees/sec	angular velocity of coin rotation about vertical axis
$\bar{\theta}$	$(0,360)^3$	vector of rotational angles about the roll, pitch and yaw axes
$\bar{\omega}$	$(-∞,∞)^3$	vector of angular velocity about roll, pitch and yaw axes
ωx	(-∞,∞) degrees/sec	angular velocity of coin flipping about horizontal axis
$\bar{Z}$	(0,∞) cm	height of coin above surface
V_z	(-∞,∞) cm/sec	velocity of coin in vertical direction
$\bar{X}$	(-∞,∞)×(-∞,∞)× (0,∞) centimeters	position coordinates of coin (x,y,z)
TOUCHED	(YES,NO)	indicates whether coin has touched ground or not

parameters	range (units)	meaning
$\varepsilon_1, \varepsilon_1$	[-½,½]	deviation of heads to tails (tails to heads) transition probability from ½
r,s	[0,1]	probability of flipping from heads to tails (tails to heads)
t,h	[0,1]	probability of transition from TE to T (HE to H)
ρ	real value distribution over coin volume g/cm^3	density distribution of coin
m	(0,∞) grams	mass of coin
α_e	(0,180) degrees	static equilbrium angle of coin lean
τ	(0,1) seconds	fraction of period of spin spent with heads up
R	(0,∞) cm	radius of coin
w	(0,∞) cm	width of coin

Figure 3. List of Variables

ENTITIES AND VARIABLES

Every variable belongs to one and only one entity. Thus the entities partition
the set of variables. When an ambiguity exists due to identical names of variables, it may always be removed by prefixing the names of the entities. Figure 2
displays the entity partition of variables in our example. Note that ORIENTATION
is disambiguated by reference to COIN.ORIENTATION, FLIGHT PROCESS.ORIENTATION
or LANDING PROCESS.ORIENTATION depending on the entity in question.

The requirement that a variable belongs to a unique entity is natural under the
interpretation that a variable refers to the possible readings of a measuring instrument applied to that entity. Thus a variable is directly observable if, and
only if, it belongs to a detectable entity and there is given a measuring instrument such that at least a subset of its readings is in one-one correspondence with
the range set of the variable.

Often parameters with the same meaning and value belong to more than one entity
(e.g., the gravitational constant g in our example). In this case, the modeller
may employ name equivalencing (see below) to constrain the different parameters to
have the same value (e.g., LANDING PROCESS.g = FLIGHT PROCESS.g)

SEMANTIC STRUCTURE OF VARIABLES

The set of variables is organized by an underlying semantic structure. By this
we mean that the modeller may attach _a priori_ meanings to the variables which cause
them to be logically dependent. In effect, the modeller agrees to abide by these
conventions throughout the course of model building or until he decides to alter
them. This is the spirit of the meaning postulates of Carnap (1960).

The semantic structure for our example is shown in Figure 4.

The general form of a meaning convention is as follows:

Let A with range R_A and B with range R_B be variables. The modeller establishes a meaning convention between A and B by supplying a relation
$R \subseteq R_A \times R_B$.
The effect of such a convention is to make legal those and only those, assignments
of values to variables A and B which satisfy the relation. A typical use of
such a convention is the case in which the relation R is given by a function,
say $f : R_A \rightarrow R_B$. Then whenever variable A takes on a value a , say, the value
of variable B is determined and equal to $f(a)$.

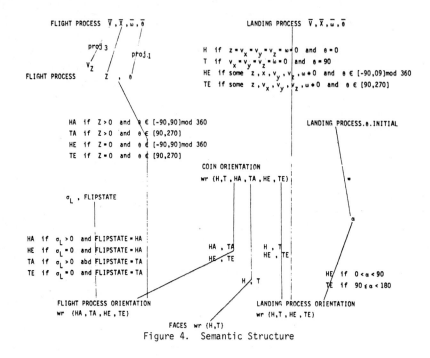

Figure 4.   Semantic Structure

Usually such a relation is of direct, easily understood, and simply computable nature. The following are two general types:

Range reduction

The range of B is identified with a subset of that of A . In this case the relation R is given by the identity function mapping R_B into R_A . For example, the range of FACES is a subset of that of COIN.ORIENTATION (Figure 4).

A special case of range injection is <u>name equivalencing</u>. Here, by writing A=B , two variables A and B with the same range but different names are declared to be the same variable. For example, α is declared to mean the same as LANDING PROCESS INITIAL.θ in Figure 4.

Note that two variables may have identical ranges, say the real numbers, but unless a meaning convention is declared to bind them, they will be logically independent variables. Of course some model constructed using these variables may make them dependent. Similarly, they may be empirically correlated as evident in the analysis of some data set. But in either case, the dependence will only hold in the context of the model or data and not necessarily in all contexts.

Range coarsening

The range of A is coarsened into equivalence classes which are identified with elements in the range of B . In this case, the relation R is specified by a function $f : R_A \rightarrow R_B$. For example, α and LANDING PROCESS ORIENTATION are related in this way, the function f being defined in Figure 4.

The semantic relation R does not reduce to a function in case one defines a covering of R_A with overlapping blocks (Klir, 1976).

CONSTRUCTION OF VARIABLES -- AGGREGATION, REDUCTION

At any time the modeller may wish to introduce new variables into the universe. Often new variables are defined in terms of existing variables -- as in an aggregation procedure. Conversely, old variables may be redefined in terms of new ones -- as in a refinement or reduction procedure.

To deal with such extensions of the universe, we employ the cross product operation as follows:

Let $V_1, \cdots, V_n$ be variables in the universe with ranges $R_1, \cdots, R_n$ respectively. Then the composite variable, denoted <u>composite</u> $(V_1, \cdots, V_n)$, may be added to the universe. It has range $R_1 \times \cdots \times R_n$ and is related by projections, proj_i : $R_1 \times \cdots \times R_n \rightarrow R_i$ to each of the component variables, V_i .

The meaning of composite $(V_1, \cdots, V_n)$ is that it is assigned a value $(r_1, \cdots, r_n)$ whenever it is simultaneously true that V_1 has value r_1, V_2 has value $r_2, \cdots,$ and V_n has value r_n . For example, composite (A,B) has value (a,b) when A has value a and B has value b ; $\text{proj}_1(a,b) = a$ and $\text{proj}_2(a,b) = b$.

Note that the family of projections automatically establishes the meaning conventions linking the composite with the existing variables.

In our example, $\bar{V}$, $\bar{X}$, $\bar{\omega}$, and $\bar{\theta}$ are composite variables. The range of $\bar{X}$ is a three dimensional real vector space with typical element (x,y,z) .

To introduce a new variable defined in terms of existing variables $V_1, \cdots, V_n$ we give the variable a name, say V , a range R_V , and we provide a meaning convention linking V with composite $(V_1, \cdots, V_n)$. V is then said to be <u>constructed</u> from $V_1, \cdots, V_n$.

In our example, FLIGHT.PROCESS ORIENTATION is constructed from Z and θ (Figure 4).

Aggregation, in various forms, is well described by the variable construction process. In the simplest form, summation, the variable TOTAL.V is constructed from the variables $V_1, \cdots, V_n$ by specifying the coarsening function f :
$R_{V_1} \times \cdots \times R_{V_n} \to R_V$ given by $f(r_1, \cdots, r_n) = \sum_{i=1}^{n} r_i$. For example, TOTAL POPULA-
TION is a summation of POPULATION.IN.CITY.1, $\cdots$, POPULATION.IN.CITY.N .

To refine an existing variable, we introduce new variables and link the old variable to the composite of the new ones via a meaning convention. When this convention takes the form of a function mapping the range of the cross product onto
the range of the old variable, we say that the old variable has been reduced to
the new ones. We shall discuss this further in the context of entities.

DERIVABILITY OF VARIABLES

Let V and $V_1, \cdots, V_n$ be variables. We say that V is derivable from
$V_1, \cdots, V_n$ (with respect to the semantic structure) if there is a subset of value
assignments to $V_1, \cdots, V_n$ which uniquely determine, via the semantic structure,
the values of V , and all the values of V are determined in this way. More
formally, there is a partial function f which maps $R_{V_1} \times \cdots \times R_{V_n}$ onto R_V ,
such that $(v_1, \cdots, v_n, f(v_1, \cdots, v_n))$ is compatible with the semantic structure
for all $(v_1, \cdots, v_n)$ in the domain of f . If such a function exists, it is
expressible as a composition of relations of the semantic structure. For example,
in Figure 4, FACES is derivable from COIN.ORIENTATION and also from the LANDING
PROCESS variables $\bar{X}$, $\bar{V}$, $\bar{\omega}$, and $\bar{\theta}$.

Note that we explicitly allow non-minimality of source variables, i.e., if V is
derivable from $V_1, \cdots, V_n$ it may also be derivable from a subset of $V_1, \cdots, V_n$.

VARIABLES AND MEASUREMENTS

A variable may be observable or non observable. An observable variable may be
directly observable or inherit its observability through the semantic structure.
Recall, a directly observable variable is one for which there is given a measuring
instrument applicable to a detectable entity whose readings are matched with its
range set. An indirectly observable variable is one which is not directly observable but which is derivable from a family of directly observable variables which
can be simultaneously observed. For example, suppose that TOTAL.V is summation
aggregation of $V_1, \cdots, V_n$ and that this is the only existing semantic relation.
If each V_i is directly observable and they all can be measured simultaneously,
then if TOTAL.V is not directly observable, it is indirectly observable. But note
that it can happen that though each V_i may be directly observable individually,
they may not be measurable simultaneously. Then TOTAL.V would not be indirectly

observable although it could be directly observable. An example to the following:
A microscope has a limited field of view. The number of bacteria within a sufficiently small subregion of a container may thus be counted, but since the numbers in each region are not simultaneously observable, the total number of bacteria in the container cannot be observed by microscopic means. However, the
total number of bacteria may be observable through other means, say by means of
their total light production (Varon and Zeigler, 1978).

As often noted, quantum mechanics implies certain constraints on simultaneous
measurement.

We return now to discuss the handling of entities.

COMPOSITION AND DECOMPOSITION OF ENTITIES

Entities may be composed to form composite entities. If A and B are names of
entities, the composite is named A·B . We allow equivalencing of names: if C
and D are names, then C = D will cause both to refer to the same entity. Of
course checks must be made to see that if both C and D are names of detectable
entities, then the detectors are compatible. No order is attached to composition,
i.e., A·B = B·A (refer to the same entity).

If A , B and C are entities, then C = A·B indicates that C has been expressed as the composition of A and B . For example, COIN = FLIGHT PROCESS ·
LANDING PROCESS, indicates that the coin is to be considered as consisting of two
components: the flight and landing processes.

An entity may be given more than one decomposition. Thus it is possible that
$C = A_1 \cdot B_1 = A_2 \cdot B_2$ so that A_1 and B_1 are components of C in one decomposition,
and $\{A_2, B_2\}$ in another. Such a possibility for multiple decomposition is extremely
important in modelling, where one may represent the same system in more than one
way. For example a system may be expressed as a composition of its activities
(flight and landing processes in the coin case), or a composition of its structural
elements (e.g., COIN = HEADS HALF·TAILS HALF).

Entities receive a hierarchical ordering induced by the subordinates relation:
" A is a component of B " (provided that we disallow an entity to be a component
of itself either directly or more than once removed). The minimal elements in
such a hierarchy are called atomic entities. These are the entities that have
not yet received a decomposition. Maximal entities in the hierarchy are the entities that have not yet been composed with other entities.

However, considering the modelling process to be a dynamic one, we desire not to constrain the modeller to a fixed set of atomic entities. In principle, entities should be <u>decomposable without limit</u> in terms of new atomic entities. Likewise there should be <u>no intrinsic limit</u> on the ability to <u>compose</u> entities to form new maximal elements. Of course, the actual degree to which composition and decomposition are carried out at any time will depend on the specifics of the modelling situation.

ENTITY COMPOSITION AND VARIABLE CONSTRUCTION

While each variable <u>belongs</u> to exactly one entity, a variable <u>pertains to</u> (may be referenced by) all entities which dominate its owner in the entity hierarchy. Thus when two entities A and B are composed, the composite A·B may reference all the variables pertaining to the components A and B . In addition, the modeller may assign variables to belong to A·B which are constructed from, and only from, its pertinent variables.

In our example, FLIGHT PROCESS ORIENTATION is constructed from Z and θ, variables belonging to the components TRANSLATION.PROCESS and SPIN.PROCESS of FLIGHT.PROCESS (Figure 4).

Finally, the modeller may assign new variables to belong to A·B which are not constructed from pertinent variables. In our example, the variable TOUCHED belonging to COIN.PROCESS is not constructed from variables of its components. We call such variables <u>holistic</u> since they may describe properties of an entity which are in principle essential to model, but which are not reduceable to properties of its components. Variables which have to do with interactions of components which are not meaningful for the components in isolation are of this nature.

When an existing entity C is decomposed into new entities A and B , the modeller must decide upon the status of the variables of C . Some may be reassigned to belong to A or B ; some may be reduced to variables of A or B ; and some may remain holistic. Reducibility in our sense is clearly dependent on the availability of finer variables, the knowledge of the modeller (to make the reduction) and his decision as to its utility (the costs versus benefits in carrying out the reduction, where the costs might be measured in terms of extra experimentation required, extra parameters introduced into models, extra labor, etc. and the benefits might include increased insight, increased experimental control, etc.).

4. QUESTIONS

We take it that a model is basically a device to answer questions about a real
system (Minsky, 1965). Since models are calibrated to existing data, any answers
they provide derive from this data. However, a "model base system" is not merely
a data base system (which is capable of answering certain questions "or queries")
in that models embody hypotheses about aspects of the real system (e.g., boundaries, behavior, structure) which go beyond the available data and thus enable
rational extrapolation from it (prediction, forecasting). Moreover, model building interacts with questioning and data collection -- questions lead to modelling
and data gathering; these two activities stimulate new questions, etc.

Thus, question asking is central to the modelling enterprise and serves to provide
a focus for the collecting of relevant data and the constructing of relevant
models. In a multifaceted systems modelling approach, the "structure" of the
questions determines the organization of the data and models. By "structure" we
mean the following informally defined relation: let questions A and B be
related if the answer to A also helps to answer B . Figure 5 depicts the
helps to answer relation for our example. We shall not try to formalize the
question structure directly but will formalize the structural organization of
experimental frames which it motivates.

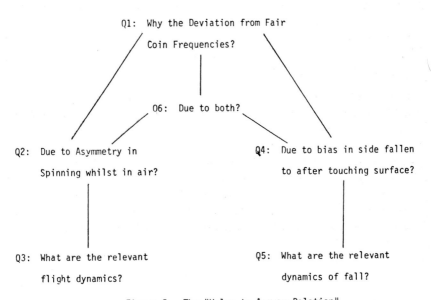

Q1: Why the Deviation from Fair
     Coin Frequencies?

Q6: Due to both?

Q2: Due to Asymmetry in
     Spinning whilst in air?

Q4: Due to bias in side fallen
     to after touching surface?

Q3: What are the relevant
     flight dynamics?

Q5: What are the relevant
     dynamics of fall?

Figure 5. The "Helps to Answer Relation"

5. EXPERIMENTAL FRAMES

An experimental frame specifies a restricted mode of data acquisition and reduction. The data in question derives either from a real system or from a model.
The frame may limit access to the data by selecting observational modes relevant
to a behavior of interest at a desired level of abstraction. It may also control
the acquisition of potential observable data by passing only that data meeting
specified tests. Such control specifications arise in two ways which are best
explicated in the context of model handling.

1. Isolation of Components

Here the modeller wishes to isolate a component of a model from the other components, (i.e., he is interested in observing the behavior of the component untainted
[as far as possible] by the behavior of the rest of the system). The isolation of
the flight (or the landing) process in Section 2 is an example. To do this he
may try to scrutinize the input and output of the model so that only input-output
pairs attributable to the component of interest are passed. (Zeigler 1976,
Chapter II; Zeigler 1979). Let us note that the scrutinizing tests can be expressed independently of a particular model -- as the control specification of an
experimental frame. (Of course, the success of the frame in bringing about the
intended isolation will depend on the model.) The role of isolation in parameter
identification will be discussed later (Section 9).

2. Formalization of Simplifying Assumptions

Here the modeller wishes to formally express the assumptions under which he believes (or can prove) that a model he has constructed is valid. As pointed out
in our coin example, this involves specifying a finer (base) model and conditions
under which the base model reduces to the original (lumped) model. These conditions are of two kinds: structural and behavioral. Structural conditions take
the form of parameter constraints on the base model which we discuss later (Section 9). Behavioral conditions can be thought of as defining a subspace in the
state space of the base model, called an operating space. As long as the base
model trajectories lie within the operating space, it is supposed to reduce to
the lumped model. In the kinetic theory of gases (base model) the behavioral
assumption that the molecular positions and velocities are uniformly distributed
at all times is sufficient to establish Boyle's Law (lumped model). Other examples
are given in Zeigler (1977, 1978, 1979). Finally, as in the isolation case, note
that the operating space can be expressed independently of the base model, as
the control specification of an experimental frame.

Experimental frames can be thought of as the formalized counterparts of the "questions" mentioned in Section 4. A frame specifies a mode of data aquisition appropriate to the answering of the question. (Frames are like queries in data base terminology.) A frame always relates to some entity and thereby to its component entities. An experimental frame (Zeigler, 1976, 1978; Oren and Zeigler, 1979) specifies the following:

a) The descriptive variables of the entity to be treated as input variables.
b) The input segments applicable as stimuli in experimentation on the real entity or its model.
c) The entity descriptive variables to be treated as output variables*.
d) The conditions under which experimentation may be initialized and terminated.
e) The statistical processing to be performed on the data acquired in the frame. (We shall not consider further this aspect here.)

Note that a) and b) may be omitted as would be appropriate if the entity is to be regarded as a closed (or autonomous) system.

The observational part of the frame comprises items c) and e). The control part comprises a), b) and d).

The control conditions d) may be further elaborated as follows (although it is not clear at present whether the following formulation includes all of the cases of interest):

d') The frame specifies a set of entity descriptive variables called control variables. The control variables are disjoint from the input variables but not necessarily disjoint from the output variables. The frame also specifies two sets of contraints whose effect is to specify two subspaces of the composite control space: the first initialization space is included in (or equal) to the second specified subspace called the continuation space.

The effect of these specifications will be that all experiments start with the control variables' values within the initialization space and end as soon as the control variables try to leave the continuation space. The complement of the continuation space is called the termination space. It embodies the conditions under which an experiment is terminated. The continuation space was called the range of validity by Zeigler (1976), since it corresponds to the informal concept relating to the range in which a model is expected to be valid.

---

* These are called compare variables by Zeigler (1979), which considers a more restricted study of frames.

A set of frames for our example is presented in Figure 6. As indicated in Figure 7 (soon to be explained) E6 , E7 and E9 are isolation frames for the landing process entity; E3 , E4 , E5 and E8 play the same role for the flight process; E1 , E2 and E10 are frames for observing coin behavior. Note for example that in E8 , there is one control variables, TOUCHED, which stops the collection of data relevant to the flight process when it leaves the NO value (becomes YES).

Frame	Input Variables	Input Segments	Output Variables	Control Variables	Initialization Conditions	Continuation Conditions
E1			FACES			
E2			COIN.ORIENTATION			
E3			FLIGHT PROCESS ORIENTATION	TOUCHED		TOUCHED = NO
E4	TR , P	TR is exponential with mean λ ; P is constant	FLIGHT PROCESS ORIENTATION	TOUCHED		TOUCHED = NO
E5			FLIGHT PROCESS Z, V_2, θ, ω	TOUCHED		TOUCHED = NO
E6			FACES	TOUCHED, LANDING PROCESS $Z, \bar{V}, \omega$	$\omega = 0$	TOUCHED = YES $\bar{V}=0$, $Z=0$
E7	α	α is uniform in range [0,180]	FACES	TOUCHED, LANDING PROCESS $Z, \bar{V}, \omega$	$\omega = 0$	TOUCHED = YES $\bar{V}=0$, $Z=0$
E8			FLIGHT PROCESS $\bar{X}, \bar{V}, \bar{\theta}, \bar{\omega}$	TOUCHED		TOUCHED = NO
E9			LANDING PROCESS $\bar{X}, \bar{V}, \theta, \omega$	TOUCHED		TOUCHED = YES
E10			FLIGHT and LANDING PROCESS $\bar{X}, \bar{V}, \bar{\theta}, \bar{\omega}$ TOUCHED			

Figure 6. Set of Frames

5.1 Experimental Frames and Data

Let E be an experimental frame with input variables $X_1, \cdots, X_n$ and output variables $Y_1, \cdots, Y_n$. Further, let X and Y be the composite $(X_1, \cdots, X_n)$ and composite $(Y_1, \cdots, Y_n)$, respectively. (R_X, T) and (R_Y, T) denote the set of all segments over time base T , $\omega : <t_0, t_1> \rightarrow R_X$ and $\rho : <t_0, t_1> \rightarrow R_Y$, respectively. Data elements acquirable in E take the form of input-output pairs (ω, ρ) . If there are no input variables, data elements are output segments ρ . A data set in E is a collection of data elements acquirable in E , [so that it takes the form of an IORO (Input-Output Relation Observation) (Zeigler 1976, Chapter 9)].

An experimental frame is realizable in a real system if all the variables it specifies are observable in the real system.

When a frame E is realizable in a real system, an input-output pair records the
result of an experiment performed on the real system. That is, ω is a time
series of input stimuli applied over a time interval, and ρ is a time series
of output variables observed as the system response over the same interval. Both
the real system and the frame constrain the possible pairs which actually are
collectable in E (see Section 5.3).

5.2 The Experimental Frame Derivability Relation

The basic structuring principle for experimental frames is the derivability relation which is a partial order. Essentially, a frame E' is derivable from a
frame E (written E' $\leq$ E) if all questions posable in E' are posable in E
as well.

More precisely, for E' $\leq$ E the following kinds of requirements must be satisfied:

1) The restrictions on data acquisition in E' are over and above those of E.
2) Data acquired in E can be processed to yield the data acquirable in E' .

Note that 2) does not follow from 1). Indeed, 2) requires that the additional
restrictions specified by E' be ascertainable from the data acquired in E
without access to the original data source.

To formalize the derivability relation on frames we shall employ the derivability
relation developed for variables (Section 3). Let V be derivable from
$V_1, \cdots, V_n$ through function f .

If ω is a segment in $(R_{V_1} \times \cdots \times R_{V_n}$, T) then f , applied pointwise, carries
this segment into a segment ω' in (R_V, T) (i.e., $\omega'(t) = f(\omega(t))$ for all
t ϵ dom(ω)). We call this mapping, the induced segment mapping.

Now we define E' $\leq$ E (E' is derivable from E) if:

a) The input variables of E' are derivable from the input variables of E .
b) The input segments of E' are images of those of E under the induced seg-
    ment mapping.
c) The output variables of E' are derivable from the output variables of E .
d') The control variables of E are included in the control variables of E' and
    the (initialization, continuation) constraints placed by E' on these vari-
    ables are identical to those of E ; any additional control variables speci-
    fied by E' are derivable from the input and output variables of E .

The conditions in d') apply to frames specifying control variables and imply the
more general conditions:

d) Any experiment initializable in E' is initializable in E ; its being con-
    tinued in E' implies its being continued in E as well.   Whether to ter-
    minate (or disallow starting) an experiment in E' must be checkable from
    data acquirable in E .

Notice that d) and d') require not only that the initialization (continuation)
space $S_{E'}$ of E' be included in the initialization (continuation) space S_E of
E , but also that the relative complement S_E-$S_{E'}$ be specified via control variables derived from observable variables in E . That is, the additional termination conditions must be checkable from the data in E . This is in keeping with
the informal requirements 1) and 2), given before.

The ordering induced by the derivability relation in our example is shown in
Figure 7. Note that E10 is the maximal frame. E9 , E7 , and E6 are successively

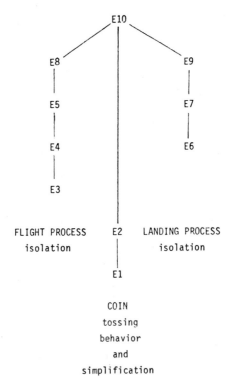

Figure 7. Frame Derivability Relation

derived from it by omission and coarsening of variables (reduction of observation potential), and by adjoining more and more stringent control conditions. For example, in addition to requiring that TOUCHED = YES, E7 requires that the coin be touching the surface and initially at rest. Clearly E7 is more restrictive than E9 which allows experiments from all initial conditions of the landing process.

(We should note that E4 is shown as derivable from E5 employing discrete event processing, which we have not discussed here.)

5.3 Processing Induced by Frame Derivability

Let E and E' be frames, with $E' \leq E$. Then every data element (ω, ρ) in E is mapped to a single element (ω', ρ') in E' . We call this mapping the E-to-E' processing, since it expresses the processing which may be applied to data acquired in E to map it to data acquirable in E' . We refrain from sup- plying a formal definition of the E-to-E' processing, but the reader can assent to its existence by noticing that (see Figure 8) :

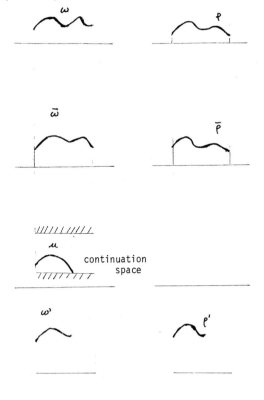

Figure 8. E to E' Processing

a) (ω,ρ) is mapped into a pair $(\bar{\omega},\bar{\rho})$ of E' by the induced segment mappings known to exist due to the derivability of the {input-output} variables of E' from the {input,output} variables of E .

b) (ω,ρ) is mapped into a control segment μ within the additional control variable space by the induced segment mapping (known to exist due to the derivability of the additional control variables from the observable variables of E). (The additional control variables are those in E' but not in E .)

c) (ω',ρ') is the null element if μ does not begin within the initialization space of E' ; otherwise (ω',ρ') is obtained by retaining only that part of $(\bar{\omega},\bar{\rho})$ during which μ remains in the continuation space of E' .

Now consider frames E , E' and E" where E" $\leq$ E' and E' $\leq$ E . Then it can be shown that E" $\leq$ E . Hence derivability is a transitive relation and partially orders the frames. Moreover, let (ω,ρ) be a pair in E , and let (ω'',ρ'') be the pair in E" obtained from (ω,ρ) via the E-to-E" processing. Then it can be shown that (ω'',ρ'') is also obtained by composing the E-to-E' processing with the E'-to-E" processing. This latter result relies upon condition d') and formally substantiates the desired criteria for derivability expressed in requirement 2) above.

5.4 Experimental Frames and the Real System

The real system is formalizable as an assignment to each realizable frame E of a data set within E . This assignment, call it R(E) , represents the data potentially aquirable within E . At any time t , only a subset $R^t(E)$ will have actally been acquired. The fact that data acquired within E must be in accordance with its acquisition restrictions, shows up in the following consistency constraint:

For every pair of frames E and E' , it E' $\leq$ E then the result of applying the E-to-E' processing to $R^t(E)$ is included within $R^t(E')$.

This condition, together with the transitivity of derivability and the associated frame-to-frame processing, guarantees that the data sets $R^t(E)$ resident in the data base at any time t are all mutually consistent. The collection of data sets constitutes the observable knowledge of the real system, acquired until time t .

6. MODELS

In the multifaceted approach there may be many models of a real system, expressed in various formalisms and at various levels of specification (Zeigler, 1976, 1977, Klir, 1977). Figure 9 demonstrates this fact for our example. In the present

formulation, a model is always a _model_ of some entity, i.e., models are partitioned by entities. An entity may be modelled by any number of models, which relate
to it at various levels of abstraction. Figure 10 lists the entities and their
models in our example.

Model	Formalism	Autonomous (input free)	Level of specification	Other comments
M1	Markov	Yes	state	stochastic
M2	Markov	Yes	state	stochastic
M3	Markov	Yes	state	stochastic
M4	discrete event	No	structured state	deterministic
M5	differential equation	Yes	structured state	deterministic, aerodynamic, Newton's Laws
M6	Markov	Yes	state	stochastic
M7	Markov	Yes	state	stochastic
M8	instantaneous	No	input-output relation	deterministic
M9	instantaneous	No	input-output relation	deterministic, mechanical statics
M10	Markov	Yes	state	stochastic
M11	differential equation	Yes	structured state	deterministic, aerodynamic, Newton's Laws
M12	differential equation	Yes	structured state	deterministic, Newton's Laws
M13	differential equation	Yes	network	deterministic, dynamic Newton's Laws

Figure 9. Model Classification

A model is specified by its components, descriptive variables and parameters, and
its rules of interaction. The components of a model of an entity are models of
the components of the entity. The descriptive variables and parameters of the
model are selected from the variables pertaining to the modelled entity (which
include those belonging to the component entities). The distinction between
descriptive variables and parameters is that the former are employed in the description of the model state, while the latter represent a certain degree of freedom in specifying the model which must be reduced by calibrating it with real
system data. In effect, each vector of parameter settings specifies a new model,
and the search for a valid model is being waged within the parameterized family
of models. The rules of interaction determine the time trajectories of the descriptive variables.

i - input variable
s - state variable
o - output variable

entity	model	component	descriptive variable	parameter
COIN	M1:	COIN	$FACES^{S,O}$	e_1,e_2
	M2:	COIN	$COIN\ ORIENTATION^{S,O}$	r,s
	M6:	COIN	$LANDING\ PROCESS\ ORIENTATION^{S,O}$	t,h
	M10:	COIN	$ORIENTATION^{S,O}$	r,s,t,h
	M13:	COIN	$ORIENTATION^{O}$	ρ,R,W
			$TOUCHED^{S}$	
		SPIN PROCESS	θ^{S},ω^{S}	g,m
		TRANSLATION PROCESS	$\bar{X}^{S},\bar{V}^{S}$	
		LANDING PROCESS	$\bar{X}^{S},\bar{V}^{S},\theta^{S},\omega^{S}$	
FLIGHT PROCESS	M3:	FLIGHT PROCESS	$ORIENTATION^{S}$	r,s
	M4:	FLIGHT PROCESS	$ORIENTATION^{O}$	
		SPIN PROCESS	$FLIPSTATE^{S},\sigma_F^{S},p^{i}$	τ
		TRANSLATION PROCESS	σ_L^{S},TR^{i}	
	M5:	FLIGHT PROCESS	$ORIENTATION^{O}$	g
			θ^{S},ω^{S}	ρ,R,W
			Z^{S},V_Z^{S}	m
LANDING PROCESS	M7:	LANDING PROCESS	$ORIENTATION^{S,O}$	t,h
	M8:	LANDING PROCESS	α^{i}	
			$FACES^{O}$	α_e
	M9:	LANDING PROCESS	α^{i}	
			$FACES^{O}$	ρ,R,W,g
	M12:	LANDING PROCESS	$X^{S},V^{S},\theta^{S},\omega^{S}$	
			$ORIENTATION^{O}$	ρ,R,W,g

Figure 10.　Entities and Models

Input variables are those descriptive variables over which the model has no control. Output variables are those of potential interest external to the model, i.e., for comparison with the real system or when the model is composed with other models to form a composite model. A set of state variables carries all the information about the past history of the model sufficient to uniquely determine along with the input variables) its future behavior under the rules of interaction. (See Zeigler, 1976, Chapter 3, for a more detailed exposition.) Figure 10 provides a categorization of the variables in our example.

6.1 Operations Structure of Models Set

As implied above, models may be composed or decomposed just as the entities they model may be composed or decomposed (Oren, 1977). In addition, models may be coursened (simplified) or refined (elaborated) (Zeigler, 1976, Chapter 2). The sequence of application of these two classes of operations on model descriptions provides an underlying structure to the set of models resident in the model base, called the operations structure. As illustrated in Figure 11, for the models of our example, there are two types of relations linking the models. Vertical lines represent composition-decomposition; diagonal lines represent simplification-elaboration. Thus M1 was refined to construct M10. M10 is composed of M3 and M7; the latter are refined by M4 and M11, and M8 and M12 respectively. M13 is a composition of M11 and M12 and refines M10.

Construction sequences need not always be up-building (i.e., go in the composition/elaboration direction). For example, M13 might be simplified to yield M14, which might not be a refinement of M10 and contain traces of both M11 and M12. Indeed, the modelling tools provided should facilitate the construction of feasible-to-simulate models from comprehensive models even though the latter may be only partially specified. Thus the modelling system should help the modeller to construct the base or most comprehensive model implied by the extant models only to the point that this model is required for constructing new models for simulation and validation in response to new experimental frames.

7. APPLICABILITY OF FRAMES TO MODELS

The applicability relation constitutes the basic structuring principle for linking frames and models. The applicability of a frame to a model formalizes the notion that the questions embodied in the frame can be meaningfully posed to the model. Of course, whether the model can successfully answer the question is another issue (see Section 8). The most extensive frame applicable to a model is called its scope. All frames applicable to a model are derivable from its scope.

122   B.P. ZEIGLER

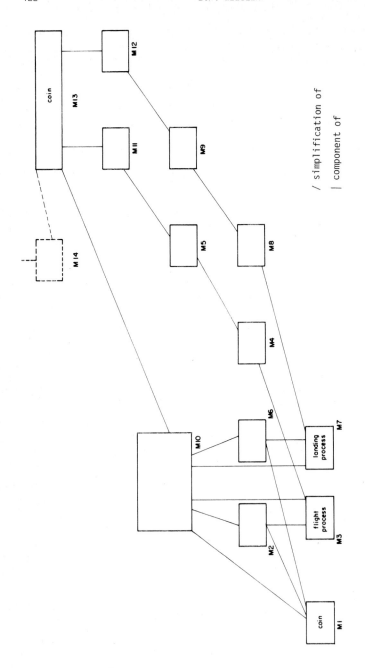

Figure 11.   Operations Structure of Models

More formally, the scope E_M of a model M is the frame with input variables, input segments, and output variables equal to those of M respectively; E_M has no control variables.

We say that a frame E is applicable to a model M if E is derivable from E_M . The following structuring principle follows directly from the definition of applicability and the transitivity of derivability:

E is applicable to M and $E' \leq E \rightarrow E'$ is applicable to M .

Applicability is a strong concept and we explain in Section 9.1 why it is necessary to weaken it. In the weaker version we allow a frame to be applicable to a model even if not all of the control conditions demanded by the frame can be checked in the model. The idea is that as long as a model has the necessary output variables requested by the frame, it may provide useful information, despite the fact that it lacks the requested input or control variables.

A frame E is partially applicable to a model M if the output variables of E are derivable from those of M and either (or both)

a) Input overspecified: The input variables of M are derivable from (e.g. included in) those of E , and the input segments of E are mapped into input segments of M by the induced segment mapping.

b) Control overspecified (not checkable): The control variables of E are not all included in the output variables of M .

Figure 12 displays the applicability relations (full and partial) for the frames in our example. Let us exemplify some of the entries of the applicability table.

E1 is (fully) applicable to M_1 because from Figure 6, E1 specifies only one output variable, FACES (and no other constraints), which from Figure 1 is seen to be an output variable of M1.

E1 is applicable to M10 and M13 since FACES is derivable from COIN ORIENTATION according to the semantic structure of Figure 4.

E1 is in none of the three applicability relations to M3 since FACES is not derivable from FLIGHT PROCESS ORIENTATION.

E3 is partially applicable to M2 because its requested output variable FLIGHT PROCESS ORIENTATION is derivable from COIN ORIENTATION, the output variable of M2, but it requested control variable TOUCHED is not included in the output variables of M2.

Frame - applicable to - model   partially
                 (fully) : A   -   input overspecified : 0
                                  - control not checkable : N

Frame \ Model	M1	M2	M3	M4	M5	M6	M7	M8	M9	M10	M11	M12	M13
E1	A	A				A	A			A			A
E2		A								A			A
E10													A
E3		N	N	N	N						N		A
E4		N,O	N,O	N	N						N		A
E5					N						N		A
E8											N		A
E6		N				N	N	N	N	N	N	N	A
E7						N,O	N,O	N	N		N	N	A
E9												N	A

Figure 12. Applicability Relation

E4 is partially applicable to M_2 since the input variable set of M_2 (being empty) is derivable (trivially) from the input variables of E_4, (while E_4 is identical to E_3 in the remaining features).

7.1 The Results of Applying Frames to Models

When an admissible input segment is applied to a model started in a specified state, it generates an output segment. The input-output relation R_M of a model M is the set of all such input-output pairs (Zeigler, 1976, Chapter 9). We define the behavior of M within the scope frame E_M to be the data set R_M . The behavior of M within any frame E applicable to it, written R_M/E , is just the result of applying the E_M-to-E processing to R_M (Figure 8). Thus $R_M/E_M = R_M$. Note that both the real system and the models produce data sets. Once such data sets are generated and the frame to which they are attached is identified, further processing is given the same uniform treatment in both cases.

In the case of partial applicability, the result of applying a frame to a model is slightly more complex to define. Let E be partially applicable to M . We shall construct a frame $\bar{E}$ which is applicable to M . $\bar{E}$ is specified as follows:

a) the input variables of $\bar{E}$ are those of M ,
b) the input segments of $\bar{E}$ are the images of those of E under the induced
    segment mapping (known to exist due to the derivability of the M input
    variables from the E input variables),
c) the output variables of $\bar{E}$ are those of E ,
d) the control variables of $\bar{E}$ are those of E which are also derivable from
    output variables of M . The (initialization, continuation) constraints of
    $\bar{E}$ are those of E which apply to the retained control variables.

Now $\bar{E}$ is applicable to M , so we have a data set $R_M/\bar{E}$ in $\bar{E}$. The behavior
of M in E is obtained by substituting for the input segments in $R_M/\bar{E}$ their
inverse images. Formally,

$$(\omega,\rho) \in R_M/E \;\leftrightarrow\; (\bar{f}(\omega),\rho) \in R_M/\bar{E} ,$$

where $\bar{f}$ is the induced input segment map.

Knowing how to obtain the behavior of a model in a frame (partially) applicable
to it, we can discuss model validity.

8. MODEL VALIDITY -- PARAMETER IDENTIFICATION

A model M is _valid_ in a frame E for the real system R if $R_M/E = R(E)$,
i.e., the input-output segments generated by M in E are exactly those potentially observable for the real system in E . While this definition of validity
is the most basic one, there are related stronger and weaker concepts (Zeigler,
1976, Chapter 1, O'Neill et al, 1977, Corynen, 1975). Here we shall limit discussion to what we shall call partial validity: M is _partially valid_ in frame E
for real system R if $R_M/E \supseteq R(E)$.

It is easy to show that the following structuring principle holds:

Let a model M be (partially) valid for a real system R in frame E . Then for
any frame E' derivable from E , it is the case that M is (partially) valid
for R in E' .

A partially valid model cannot be relied upon for prediction, since it may generate responses which are not observable in reality. However, any input-output
pair produceable by the real system is produceable by the model as well (within
the frame). This implies, as we shall see, that while partial validity is not
approprite to prediction, it is appropriate to model calibration.

Let P be the parameter space of a model M . Then $\{M(p) \mid p \in P\}$ is the _family_
of models of type M and $\{p \mid p \in P , R_{M(p)}/ E \supseteq R(E)\}$ is the set of parameter

settings which specify partially valid models in E . Parameter identification
is the process of discovering the latter set, which hopefully is not empty and
ideally is a singleton. Notice that every data element collectable from the real
system can, and must, be employed in this process. We apply an input segment ω
and observe a response ρ of the real system in E . We search for those parameter settings which specify models which can reproduce ρ when given ω . Thus
at any time t , our best estimate of parameters is $\{p \mid p \in P , R_{M(p)}/E \supseteq R^t(E)\}$.
Notice that had we chosen the validity concept $R_{M(p)}/E \subseteq R^t(E)$ (reverse
inclusion), we would not know, for any given real system data element (ω, ρ) ,
whether to employ it or not, to calibrate the model.

We shall return to the connection between partial validity and parameter identification (Section 9.1) after discussing the parameter correspondence structure.

9. PARAMETER CORRESPONDENCE

The development of the coin example showed that models can be organized by means
of parameter correspondences. We shall now treat this idea more generally. Recall that the semantic structure organizes the variables of the universe of discourse, whether they be descriptive variables or parameters. However, we can
expect that the burden of the semantic structure relates to descriptive variables,
while parameters are only sparsely related by meaning conventions. A more powerful organizing principle for parameters lies in the parameter correspondences
developed between models. Structurally, parameter correspondences are set up and
behave like meaning conventions, with the differences to be given below. The
parameter correspondence structure of our example is shown in Figure 13.

The differences between meaning conventions and parameter correspondences arise
from the prior definition of the former -- meaning conventions are laid down and
upon this basis and depending on the models and frames employed, parameter correspondences are established. Thus while a meaning convention holds uncondition-
ally, the conditions under which a parameter correspondence holds form an integral
part of its specification.

The operations structure of models underlies the parameter correspondence structure as follows: 1) Whenever one model is a component of another, the parameters
of the first constitute a subset of the parameters of the second. Thus vertical
lines in the operations structure are labelled by identity correspondences in the
correspondence structure. 2) When one model is a simplified version of another,
the parameters of the first may relate to those of the second. Thus diagonal
lines in the operations structure may be labelled by non-trivial parameter correspondences in the parameter correspondence structure.

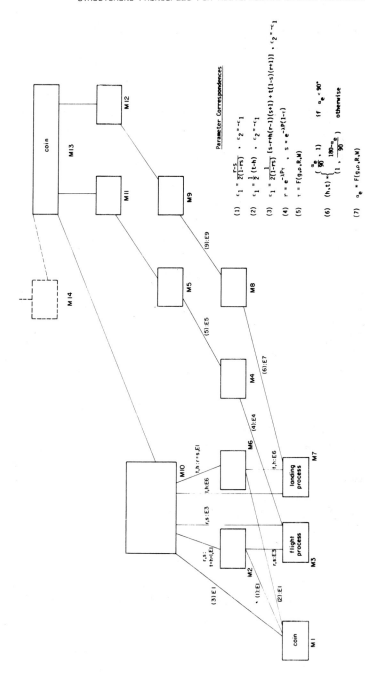

Figure 13. Parameter Correspondence Structure

Neither of these types of correspondences are unconditional. Both must be specified relative to an experimental frame -- in the first case to specify the isolating conditions for the submodel, in the second case, to specify the conditions
under which the simplification is valid. Note that carrying over the parameter
values identified for a component to the model, is justified to the extent that
the conditions under which the component was calibrated in isolation, are compatible with those it experiences within the model.

9.1 Experimental Frames for Parameter Correspondences

We see that prior to the establishing of a parameter correspondence between two
models M and M' , is the specification of an experimental frame E . The following situations arise:

1) <u>E applicable to M and M'</u>. In this case we may aim for complete validity.
    We say that M and M' are <u>equivalent</u> in E if $R_M/E = R_{M'}/E$, where
    R_M is the input-output behavior of M , and recall that R_M/E is R_M after
    it is "windowed" by frame E (Section 7.1). In the case of simplification,
    we say that M' is a <u>valid simplification</u> of M in frame E if M and M'
    are equivalent in E .

2) <u>E applicable to M , E partially applicable to M'</u>. There are two non-
    exclusive possibilities (Section 7):

a) <u>The control variables of E are not checkable in M'</u>. In this case we cannot
    expect complete validity as in Case 1, since trajectories may be stopped by
    the termination condition in M , but may continue on in M' due to the
    inability to test the termination conditions. Nevertheless, we may try for
    <u>partial validity</u>. M' is <u>partially valid</u> with respect to M if $R_M/E \subseteq$
    $R_{M'}/E$, i.e., any input-output pair allowed by E in M is also allowed by
    E in M' . Note that not being able to check the control condition in M'
    may permit it to generate behavior not found in M -- thus the inclusion
    goes only one way.

    As indicated before, this case commonly arises when M' is either a component
    or a simplified version of M and the control conditions relate to the isola-
    tion constraints or the assumptions justifying simplification respectively.
    While M' cannot be employed to <u>predict</u> the behavior of M , it may neverthe-
    less be settable so as to reproduce already available behavior of M . It can
    thus be employed for parameter identification. In order to clarify this con-
    cept, we shall anticipate the discussion of Sections 9.2 & 9.3 with a simple
    example.

Consider the Lotka-Volterra model of prey-predator interaction specified by the differential equation:

$$\frac{dx}{dt} = ax - \alpha \; xy$$

$$\frac{dy}{dt} = -by + \beta \; xy \qquad\qquad (A)$$

This model has four parameters to identify: a, α, b, β. However, a and b can each be independently identified. For example, if predators are absent ($y \equiv 0$), then the model reduces to the prey-only form:

$$\frac{dx}{dt} = ax \qquad\qquad (B)$$

Prey growth curves can easily identify a.

Formally, we have a frame E with output variable PREY, control variable predator and initialization/continuation condition PREDATOR = 0 . Note that E is applicable to model A, but only partially applicable to model B (which has no PREDATOR variable). Real system data collected in E is employed to identify the parameter of model B. Since it is easily shown that model B is partially valid with respect to model A in frame E , the estimate of B's parameter a, carries through to A's parameter a. Note that partial validity means here that if predators should enter the experimental system or model A, the experiment would stop. Model B, however, would continue to produce uncontaminated prey growth.

b) The input variables of M' are overspecified by E. In this case, it is still possible that M and M' generate the same set of output segments in E . To see this, consider that it is possible to start with a model M and to add input segment generators to all of its input variables. Let M' be the composition of M and the generators. Then a frame E in which all the input and output variables of M appear, is applicable to M and partially applicable to M' . Now let (ω, ρ) be a pair in R_M . Then there is a state of M' which generates ω internally and, feeding it to its internal copy of M , generates ρ . Conversely, every segment ω generated internally by M' can be considered to be an input segment to M so that every output segment of M' is also an output segment of M . Thus M and M ' are output segment equivalent (generate the same set of output segments). The usefulness of this form of equivalence is further discussed in Section 9.3.

3) <u>E applicable to M but underspecifies its input, E applicable to M'</u>. In this case expand E to Ē by adding the input variables and input segments of the scope E_M . Then $E \leq \bar{E} \leq E_M$. Now Ē is applicable to M and Ē overspecifies the input variables of M' . Thus we have case 2b, and we can expect that M and M' generate the same set of output segments in Ē . Since $E \leq \bar{E}$, this equivalence transfers to E as well.

4) <u>E is not applicable to either M or M'</u>. In this case E cannot be employed for justifying parameter correspondences between M and M' .

9.2 Parameter Correspondence Structure

Recall that the modeller may label the lines of the operations structure of models with identity parameter correspondences in the case of the composition relation and with correspondences of his choice in the case of the simplification relation. A parameter correspondence between M and M' in effect pairs up models of type M with models of type M' (recall Section 8). More formally, let P and P' be the parameter spaces (the crossproduct of parameter range sets) of M and M' respectively. Let (p,p') be an element of the parameter correspondence $C \subseteq P \times P'$. Then (M(p),M'(p')) is the pair of models <u>placed into correspondence</u> by C , where M(p) is the model of type M obtained by fixing the parameters to setting p . For example, correspondence (1) (Section 2) between M1 and M2 places each of the symmetrical versions of M2 (r = s ≠ 1) in correspondence with the fair coin version of M1 $(\varepsilon_1 = \varepsilon_2 = 0)$. The model M2(1,1) (r = s = 1) is not placed into correspondence with any model of type M1 (this parameter setting does not allow the coin to land).

Let E be a frame applicable, at least partially, to both M and M' (cases 1, 2, 3 above). We say that a parameter correspondence C is <u>justified for complete validity</u> in E if each pair of models it places into correspondence are equivalent in E , i.e., $R_{M(p)}/E = R_{M'(p')}/E$ for all (p,p') ∈ C .

We say that C is <u>justified for partial validity</u> in E if M'(p') is partially valid with respect to M(p) in E , i.e., $R_{M(p)}/E \subseteq R_{M'(p')}/E$ for each (p,p') ∈ C .

The following structuring principle holds:

If a correspondence is justified for (partial, complete) validity in E , then the same holds true for any frame E' derivable from E .

The form assumed by the <u>parameter correspondence structure</u> can now be given. It has as its underlying graph the operations structure labelled by correspondences (Section 6.1). Associated with each correspondence C is an experimental frame

E and a confidence index which indicates the degree to which C is believed to
be justified for (complete, partial) validity in E . Full confidence is possible
in the case where an analytic proof is employed to establish model equivalence or
inclusion. Partial confidence may be obtained via simulation and comparison of
model trajectories. The basis for either approach is the morphism concept which
we mention below.

9.3 Morphisms and Justification of Parameter Correspondences

Morphisms as relations for structuring the organization of models have been extensively discussed (Zeigler, 1976, Chapter 10; Zeigler, 1978). Essentially, a
morphism is a correspondence between models which preserves their structure and
behavior. The basic concept can be understood by restricting attention to morphisms at the state level, called homomorphisms.

Let C be a parameter correspondence between M and M' , and let E be a frame
in which C is to be examined for justification. For example, let E be applicable to M and partially applicable to M' . It is easy to see that the input
(output) variables of M' are derivable from the input (output) variables of M .
In addition, the modeller may set up a mapping from the state space of M to
that of M' . The triple of mappings from input, state and output of M to the
respective objects in M' is called a model structure correspondence, denoted
H . If H is preserved under all parallel state transitions of M(p) and
M'(p') which remain in the continuation spaces of M and M' , it is called a
homomorphism from M(p) to M'(p') . It can be shown (Zeigler, 1976, Chapter 10)
that if a homomorphism holds from M(p) to M'(p') , then M'(p) is partially
valid with respect to M(p) in E . Thus, demonstrating that a homomorphism
exists provides an analytic means for justifying parameter correspondences.

The case of models M1 and M2 of our example illustrates justification for complete
validity. The derivability of FACES from FLIGHT.PROCESS.ORIENTIATION establishes
the output correspondence. The state correspondence is here taken to be the same
as the output correspondence. A homomorphism requires that when M1 can make a
single step transition from H to T , then M2 can make a parallel multistep
transition from H to T ; and similarly for the H to H , T to H and T to
T transitions. Since for $r \neq 1$, $s \neq 1$, these conditions do in fact hold, we
have a homomorphism between corresponding models. Hence the set of output
sequences over {H,T} generated by corresponding models is the same, i.e., they
are equivalent in frame E1 (output variable = FACES). More than this, however,
we can require that the probabilities assigned to these sequences be the same for
corresponding models. Thus correspondence (1) is justified for this probabilistic
behavioral correspondence.

The case of models M7 and M8 illustrates justification in the case of partial
applicability. Here frame E7 is partially applicable to both M7 and M8. Both
the input and control variables of E7 are overspecified for M7, while only the
control variables are overspecified for M8. Since E7 places a uniform distribution on the input variable α (appearing in M8 but not in M7), it can be easily
shown that the correspondence (6) associates models which generate sequences over
{H,T} with the same probabiltiy.

The case of models M9 and M12 is also an example of partial applicability, this
time in which the control variables are at issue. The control variables and constraints of E7 check for zero translational and angular velocity upon landing and
maintainance of zero translational velocity upon falling. These can be checked
in model M12 but not in model M9. As long as the control conditions are satisfied,
M9 generates the same behavior as M12 under the same parameter settings. However,
if M12 should enter the termination space (develop some translational velocity
while falling), then this cannot be detected in M9 and M9 will continue to generate a trajectory which is not generated by M12. Thus the identity parameter correspondence is justifiable for partial validity in frame E7.

Finally we note that often, as in the case of M4 and M5, simulation may be employed to establish and/or assess parameter correspondences between models. In this
case, a pair of models with related parameter settings is chosen and the models
are set into initial states which correspond under the structural correspondence
H . Then corresponding input segments are applied and the state and output trajectories generated are tested for correspondence. To the extent that correspondence is maintained over many initial states and lengths of run, and for many pairs
of corresponding models, we can gain confidence in the justification of the parameter correspondence in the given frame E .

A software package has been implemented which aids in this validity assessment
process (Hopfeld, 1978). It enables the modeller to describe a pair of models,
the correspondence between them to be checked, and the initial states from which
checking is to begin. The system then translates the models into a simulation
language (SIMSCRIPT) and carries out the simulation and the trajectory compairson.

9.4 Parameter Identification and the Parameter Correspondence Structure

We can now tie together the various threads related to parameter identification
and organization which we have developed. The central concepts are:

1) The Parameter Correspondence Structure, which overlays the model operations
    structure (Section 6.1) with parameter correspondences, the frames in which
    they are believed to be justified and a confidence rating for their justifica-

tion (Section 9.2, Figure 12). The frames embody isolation conditions and
simplifying assumptions (Section 2, 5 and 9.1).

2) Parameter Identification employs the concept of partial validity to find
parameter values consistent with observed real system data (Section 8).

3) Parameter Value Propagation. When parameter values are identified for one
model, they may be propagated to other models via the parameter correspondence
structure (the coin example in Sections 1 and 9.3 provides an extensive discussion of this). The concept of (partial) applicability of frames to models
(Sections 7 and 9.1) and parameter correspondences justified for partial
validity (Sections 8 and 9.2) are the basic concepts in this enterprise.
Parameter values identified for a model M' via comparison with the real system in frame E may be propagated to model M via a correspondence which
is justified for partial validity of M' with respect to M in E .

10. THE SEMANTIC AND PARAMETER CORRESPONDENCE STRUCTURES BROADLY VIEWED -
MODELLING AS CONSISTENCY CHECKING

The development we have been following leads us to view modelling as a never ending process of consistency checking. There is first of all the logical consistency of the semantic and parameter correspondence structures to be checked with
every modification (Friedman, 1976; Mackworth, 1977). Assuming that these structures are logically consistent, they set up multiple pathways for the cross-check-
ing of empirical data.

The directly measurable variables of the semantic structure can be viewed as terminals which can be connected to data sources. These sources are either the real
system or its models. Data being fed into multiple terminals flows through the
multiple derivability paths of the semantic structure. When two or more streams
meet, they invoke a comparison. Thus, real system data arriving from different
frames is checked for consistency (Section 5.4), model data is compared against
real system data (Section 8), and behavior of one model is compared against the
behavior of another model (Section 9.3).

Likewise, the parameters of the parameter correspondence structure can be viewed
as terminals which can be connected to sources. These sources are the outputs of
the parameter identification procedures. Parameter information arriving at multiple terminals flows through the correspondence structure. When such streams meet
and are compared, the consistency of the various parameter estimates with each
other is being tested.

Recent philosophy of science (Suppe, 1976) has emphasized that scientists test
a new theory not only by comparison with experiment (the older view), but also

by its compatibility (either explicitly determined or intuitively apparant) with
existing theories. Thus a primary goal of science is the self-consistency of
its whole system of theories and data. Our formalization of multifaceted modelling methodology is consistent with this view.

REFERENCES

Bunge, M. and A.A.L. Sangalli (1977), "A Theory of Properties and Kinds", Int. J. Gen. Sys. Vol 3, No. 3, p. 183-189.
Carnap, R. (1960), "Meaning and Necessity". University of Chicago Press, 1956.
Corynen, G. (1976), "A New Look at the Concept of Model", Proc. South East. Sys.
Courtois, P.J. (1977), "Decomposability", Academic Press, New York.
Friedman, G.J. (1976), "Constraint Theory: An Overview", Int. J. Sys. Sci. Vol. 7, No. 10, p. 1113-1151.
Hopfeld, A. (1978), A Software System For Checking Model Simplification, Masters Thesis, Weizmann Institute of Science.
Klir, G.J. (1977), "On The Representation of Activity Arrays", Int. J. Sys. Sci. Vol. 2, No. 3, p. 149-168.
Mackworth, A. (1977), "Consistency of Networks of Relations", A.I. Vol. 8, No. 1, p. 99-118.
Minsky, M. (1965), "Models, Minds and Machines", Proc. I.F.I.P.S. Conference A.F.I.P.S. Press, Montvale, New Jersey.
O'Neill, R., et.al. (1977), "The Importance of Validation in Ecosystem Analysis", in New Directions in the Analysis of Ecological Systems, Ed. G. Innis; Simulation Councils, Pat 1, p. 63-72.
Oren, T.I. (1977), "Revised Report on GEST: A Simulation Language for Large Scale Discrete/Continuous Systems", Univ. of Ottawa, Computer Science, Technical Report.
Oren, T.I. and B.P. Zeigler (1979), "Concepts For Advanced Simulation Methodologies", Simulation (In press)
Rosen, R. (1975), "Complexity and Error in Social Dynamics", Int. J. Gen. Sys. Vol. 2, No. 3, p. 145-148.
Shannon, R.E. (1975), Systems Simulation: The Art and Science, Prentice Hall, N.J.
Suppe, F. (1977), "Afterword" in Structure of Scientific Theories, 2nd Ed., Univ. of Illinois Press, Urbana.
Varon M. and B.P. Zeigler (1078) "Bacterial Prey-Predator Interaction at Low Prey Density", J. App. Env. Microb. (38) p. 11-17.
Zeigler, B.P. (1976), Theory of Modelling and Simulation, Wiley, New York.
_____ (1978), "Structuring the Organization of Partial Models". Int. J. Gen. Sys., Vol. 4, No. 1.
_____ (1979), "Multi-Level, Multi-Formalism Modelling: An Ecosystem Example", in Theoretical Ecosystems, Ed. E. Halfon, Academic Press, New York.
_____ (1979), "Simplification of Biochemical Systems", in Continuum Models in Biology, Ed. L. Segel, Cambridge University Press, Cambridge.

SECTION II
MULTIFACETED MODELLING:
APPROACHES AND ISSUES

METHODOLOGY IN SYSTEMS MODELLING AND SIMULATION
B.P. Zeigler, M.S. Elzas, G.J. Klir, T.I. Ören (eds.)
© North-Holland Publishing Company, 1979

INTEGRATED MODELLING SYSTEMS: APPLICATION TO
ENERGY SYSTEMS AND CONSIDERATIONS OF
SOFTWARE ENGINEERING DESIGN

Warren T. Jones
Associate Professor
Applied Mathematics and Computer Science Department
Speed Scientific School of Engineering
University of Louisville
Louisville, Kentucky 40208

Integrative modelling is a concept which suggests that the
modelling and simulation enterprise can best be advanced
within a framework which provides a strong theory and software domain for dealing with many models at many levels of
specification at once. The importance of this approach and
its evolution is discussed as analogous to that of database
management systems. The role of software design methodologies in the development of individual integrated modelling
systems is presented. Finally, energy systems are given as
an example context where an integrated approach can provide
needed benefits. The integrated modelling system approach
to large-scale systems modelling and simulation provides
many needed advantages, not the least of which is a
self-imposed "higher level" of systems thinking in modelling
efforts.

INTRODUCTION

This paper is the combined product of experience, research and observation in the
field of simulation methodology. Experience and observation have demonstrated
that the practice of simulation model development has not been as effective as it
could be. In particular, model development has not contributed effectively toward
the development of cumulative science. For example, investigations into
large-scale systems such as energy, environment or education have taken the form
of development of a large number of models, each with a slightly different view
of the system and its differing assumptions. At the implementation level the
diversity is even greater with discrete and continuous approaches, deterministic
and stochastic as well as a diversity of special purpose simulation languages.
This diversity of methods has distinct advantages since the questions which the
model is designed to answer can often be most naturally expressed in terms of
properties of the processes involved.

However, if one takes a more comprehensive view, as will be attempted in this
paper, it can be noted that certain difficulties arise when, for example, comparing the results of the output of one model with another. It is also often the
case that the problems which require a simulation approach also demand the urgent
presentation of recommended strategies (e.g. energy problems). To develop a
large-scale model in response to such demands without making use of previous knowledge and existing models seems a waste, but this is more often the case than not.
A methodology is needed which will permit the cumulative development of a collec-
tion of interrelated models |1|. These models would constitute the cumulative
efforts toward an increased understanding of the system under study. This proposed approach would not preclude the existence of more than one such collection
which may be developed within the context of another paradigm in the Kuhnian
sense |2|.

Perhaps the easiest way to envision the proposed modelling concept is that of communicating researchers, sometimes referred to as an "invisible college" sharing
the same integrated collection of models even though from slightly differing viewpoints. This approach would, of course, require the use of interconnected computer networks or the management of the logistics of updating copies of the same
simulation on several computers. As it has been in the last few years that
time-sharing and distributed computing, when viewed from a higher-level, predominantly derive their merit from the possibility to share knowledge expressed in
data and models.

This integrative modelling concept suggests that the modelling and simulation
enterprise can best be advanced within a framework which provides a strong theory
and software domain for dealing with many models at many levels of specification
and application at once. This approach to modelling is important for several reasons. First, and perhaps most important, the approach will provide for a more
cumulative development of the modelling and simulation enterprise, and thus indirectly more cumulative development of the ultimately desired knowledge and
understanding of large-scale complex systems. Secondly, once the required computer software is designed and implemented, the system of integrated models should
enable the development of additional related models for a lower total "cost" in
terms of total effort which includes model formulation and validation processes.
The lower cost is due to the fact that new models will also be integrated and
therefore, derived from earlier available ones. Third, this approach, when implemented, will tend to centralize the data and models of a particular large-scale
system thus making it more readily available in responding to the needs of public
policymaking.

A simulation model of a real system is most often developed because the real system itself is too complex to study directly. Thus, given that complexity motivated the development of the simulation at the outset, the model itself does not
incorporate all aspects of the real system and hence constitutes a *partial model*
of the system of interest. Each of these partial models is capable of answering
only a limited number of questions about the real system. Zeigler |3,5| charac-
terizes limited sets of questions of this type as *experimental frames*.

Zeigler has also provided a theoretical framework within which to countenance the
organization of a plurality of partial models as a coherent system of views of the
same large-scale real system. What remains is further development of the theory
and also the need to address the software requirements for implementing this inte-
grative approach. The former has been begun in |5| and this paper is an attempt
to address the latter.

In the following section the basic concepts of integrative modelling are presented. In the discussion parallels are made with the problems of data integration
within an organization in the development and implementation of a data base management system. Section 3 highlights the necessity as well as utility of
recently developed software engineering techniques for facilitating the design of
an Integrated Modelling System (IMS). The potential need for the development of a
special purpose language for communication and manipulation of an IMS is also discussed. Section 4 presents some concrete examples of ongoing large-scale modelling in the energy area where the need for the proposed approach is readily
apparent and recognized. Section 5 presents specific approaches which should be
taken in the design of any IMS.

INTEGRATIVE MODELLING CONCEPTS

The concepts and notation presented in this section are based upon Zeigler |5|.
The following definitions are needed:

Experimental frame – embodies a restricted set of questions by specifying the
                    restrictions on experimental access to the real system suffi-
                    cient to answer them. A single frame is denoted E, a collec-
                    tion of frames E.

Input-Output relations – a collection of data sets determined by an experimental
                    frame and denoted D(E) such that D ϵ D(E) is an
                    $a\ priori$ possible result of complete data acquisition
                    within frame E.

Real system – represented by the specific data which could be collected by experi-
              ment and denoted R. Thus R associates with each experimental frame
              E a unique set R(E) ϵ D(E).

Domain of possible models – assumed to be transition systems which can be speci-
                    fied at various levels of structure and behavior and
                    within various formalisms such as the sequential
                    machine, discrete event, differential equations, etc.

Thus the data acquired in frame E at any time t will be some subset R^t (E) $\subset R$(E).
Moreover, not all of the possible models in M will have been considered at time t.
Hence, the triple (E^t, R^t, M_t) reflects the situation at time t. As t increases,
the results from models in M^t will likely raise new questions that will in turn
suggest the introduction of new experimental frames and stimulate the development
of new models.

The parallels between the proposed integrated modelling approach and the recent
development of database management concepts are worth noting since the basic philosophy is identical. Prior to the appearance of the database management system
(DBMS) machine-readable files of data were resident in the computer for separate
departments within an organization. These files often contained some of the same
data which was in the files of other organizations. Each organization had its
own programs which accessed and processed its files. In addition, each organization was responsible for updating its files. These updates did not always occur
at the same time thus sometimes led to conflicting reports and problems. More
importantly, the value of reliable data to the management of the organization has
been increasingly recognized. However, planning information on an ad hoc basis
is difficult since each individual organization has programs that generate results
which could be obtained relatively quickly but only by crossing these organizational data barriers. Often, however, it is denied this capability since the
data is "locked" in different file formats and is even sometimes accessed by pro-
grams in different languages. The desirable situation here |6| is a software
system which will support:

    (a) The structuring of a centralized database and the accessing
         mechanisms thereof and

    (b) A high level language capability within which requests for
         data can be made on routine as well as an ad hoc basis.

The parallels between DBMS philosophy are clear when one considers the information
requirements for each department within the organization need only deal with a
$partial\ view$ E of the database which is referred to in DBMS terminology as a
$subschema$. The $schema$ of the DBMS is used to represent the structure of the entire database. Thus, the evolution of management information systems has proceed-
ed from individual departmental files and their associated application programs
to an integrated environment wherein the problems of incompatible files are eliminated. Moreover, in this integrated context, the user is developing applications
programs in the context of the entire organization.

The proposed IMS system is envisioned as an analogous endeavor with an integration of several models of the same large-scale system, each with a slightly different perspective on the same overall system. For example, in the field of education perhaps one is only interested in the financial aspects of the educational system and not immediately concerned with the learning dynamics in the classroom even though this may be of vital concern to those more involved with academic side of the system. Such a financial model could be developed within the same context as other models and taking advantage of modules which are already established within the IMS.

To initiate the development of an IMS the parallel with the DBMS implementation at a given organization can be a helpful guide. The process begins with the establishment of a data dictionary which is a catalog of the data elements in the database. A similar initial step would be necessary in the IMS development, however, the interpretation in this case would be a reference base of primitives from which to develop interrelated models. The utility of a design description language is discussed later.

The problems and complexities of development of the database management systems are well known and will not be reviewed further here. Suffice it to say that the problems involved with their development are similar and can perhaps be used to estimate the magnitude of the problems involved in IMS development. It would appear that the complexities may be greater since we must interrelate models and not simply data. Hopefully this is a somewhat extreme statement since many of the concepts of database can be mapped into the IMS development domain as guiding techniques.

To recognize the complexities of the IMS as a software system is to recognize the need for application of software engineering techniques in its design and development. Several of these design methods are suggested as applicable in the following section.

UTILITY OF SOFTWARE ENGINEERING TECHNIQUES IN THE DESIGN OF AN IMS

The term software engineering has recently (1969) emerged not so much as a collection of design techniques, but was born of the recognition that few such techniques existed as attested by the increasing cost of software and its decreasing reliability. Since that date, there has been a rapidly growing interest in the development of engineering design techniques for software. Those approaches which are seen to be appropriate for application to the design of an IMS are briefly reviewed here.

Consider first the goals of this IMS software system. First, we wish to be provided with an environment within which models in M can be developed that are interrelated. That is, the development "cost" of a new model will hopefully be minimized by the fact that at least some of the relationships needed are already available in the system and can be, through standard conventions, mapped into the new model. This process corresponds to Zeigler's concept of introducing new frames and model and is analogous to the database counterpart of developing another applications program which will answer a new set of questions about the large-scale system under study. There are many ways that a new model may be related to other models in the system. For example, a collection of models may correspond to regional partitions of some country with respect to energy systems. The regional groups of electrical utility companies in the United States |7| provide a specific example. A new model may be desired which is capable of countenancing the electrical systems of the entire country, but not of the detail existing in the regional level. Thus an aggregated version of the collection is required.

A second goal of an IMS is the capability for easily extending existing models.
Hence, it is clear that a key concept in the underlying design philosophy of the
IMS software system is modifiability. This concept is a difficult one for which
to develop software design techniques. Presumably models are (or certainly can
be) constructed in terms of a primitive collection of functions which are interrelated in such a manner as to constitute the model. This collection can be view-
ed as a functional nucleus or similar to Zeigler's base model |3|. At this level
the detail should be considerable with anticipated future model development to be
either expanded at the same level or aggregations at higher levels. It is noted
that it is also possible to adopt the convention of model development in the opposite direction of disaggregation by beginning at a high level of aggregation.
These two approaches would correspond to the well-known bottom-up and top-down
software design philosophies |4|.

The traditional mode of software development has evolved as a bottom-up procedure
where the lowest level units are coded first, unit tested and prepared for integration (see Figure 1). Data definitions and interfaces between units tend to be

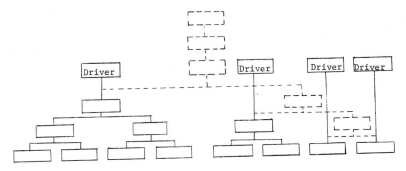

Traditional Bottom-Up Approach

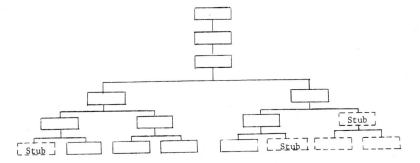

Top-Down Approach

Figure 1. Design Approaches

simultaneously defined by each of the programmers, including those involved with the lowest levels of code, and consequently, are often inconsistent. Management control is frequently inaffective during much of the traditional development cycle because there is no coherent, visible product until integration.

Top-down is a management strategy aimed at reducing these problems by sequencing the development of modules in the order in which those modules are to be executed. A program module is coded only after the unit that calls for its execution has been coded. Hence, top-down development assumes some form of hierarchical structure and proceeds from a single starting point while conventional implementation proceeds from as many starting points as modules in the design.

The second IMS goal of extending existing models can be best considered in the context of the traditional bottom-up approach. However, the first and primary goal of providing an environment for the integrated development of many models can be usefully considered in the context of a combination of three development methodologies, the above two and Dijkstra's layered design approach |8|. In the layered approach, model development can take place from the nucleus outward by the aggregation of entities and variables which constitute lower levels. In this way, new models can be developed that are derived from earlier existing ones. In the strictest sense of layered design, each layer i of the "onion" can communicate with only the layer i-1 found by peeling away layer i. This criterion is probably too severe for IMS design applications and should be somewhat relaxed to require only the clear definition of these layers and the entities and variables which constitute them.

Thus a combination of the above three well-known software design methodologies is proposed as appropriate for an IMS. The development is envisioned as beginning with some central model which is highly disaggregated. Modular design is also essential at this level since more aggregated related models will be developed utilizing this infrastructure. New models will be defined and developed in this integrated system in a top-down fashion, but not completely so, for the system is to be integrated thus requiring some bottom-up philosophy be included. Figure 2. depicts the structure of a proposed IMS.

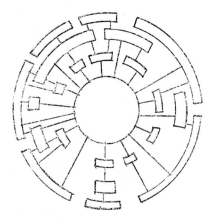

Figure 2. Architecture of An IMS

Elements of all three design methodologies are apparent. Models are shown as boxes with a central model at the center. Models found along radii from the center are increasingly more aggregated as one moves outward. Experimental frames can be defined in terms of the layers and the entities and variables which constitute them. The associated model specifications for structuring the layers can be model schemas defined by Zeigler |9| having the form <components, input variables, state variables, possible output variables> where components are entities and the mentioned variables are the descriptive variables of the components satisfying certain restrictions.

Since model development and use will rely heavily on a clear understanding of how the models are interrelated and at what levels the relationships are present, it is also appropriate to consider a special purpose design oriented language for inquiry and manipulation purposes. The development of design languages as tools for software development has been viewed recently with increasing importance. Several such languages have been developed for this purpose |10, 11|. The function of such a Language for IMS (LIMS) would be primarily for description and manipulation of the experimental frame schemas (defined as entities, input variables, compare variables, control variables) and the above defined models schemas. By use of these descriptions, the IMS use can readily determine if, for example, there is a resident model which can answer a given set of questions about which the user is interested with respect to the large-scale system represented by the IMS. If not, the descriptive knowledge available in the system will enable the user to make an assessment of the effort needed to derive a model to meet his needs. Ideally this type of interaction with the system should take place in an interactive mode at a terminal.

In summary, the software engineering design techniques which seem appropriate to the IMS design are the layered software, bottom-up design, top-down design and interactive use of a design description language for model development and inquiry in IMS software design should focus on the use of software specification and design languages, their limitations and capabilities in describing the interrelationships between model schemas, the description of experimental frame schemas and procedures for deriving new model schemas, taking into account that uniform model description formalisms are being developed now and will be generally accepted for the IMS concept.

The need for this integrated approach to modeling has been discussed in a very general context. In the following section, a specific large-scale system problem of recent critical interest, that of energy systems, will be reviewed.

ENERGY SYSTEMS: A CANDIDATE LARGE-SCALE SYSTEM FOR IMS APPLICATION

There is perhaps no better example of a complex large-scale system for which there is a demonstrated need for a greater understanding than that of energy systems. The recent "energy crisis" has dramatized this need and stimulated the development of several simulation modeling projects throughout the United States. In terms of level of detail as discussed in the last section, the modeling efforts can be classified into two primary groups: national and state level models. These efforts are obviously responding to the need to develop energy policies at these two governing levels. Some models have also been developed for use at the regional (e.g. a group of coal producing states) and international levels.

The current approach to energy planning involves projecting scenarios based upon an array of underlying assumptions, then analyzing the energy requirements of the scenario. Basically a scenario is a statement of present understandings of a system and expected future conditions. The primary purpose of a scenario is to establish energy demands over the analysis period. Then, based upon assumed types of energy facilities, requirements are projected for the various energy facilities. Also the capital, manpower, and materials to construct and operate the facilities are estimated. Although most scenarios represent a particular point of

view of the total system, they are somewhat complex in reality. The following is
a partial listing of major scenarios:

> Department of Interior scenarios
> Ford Foundation scenarios
> Nuclear Electric Economy scenarios
> Joint Committee on Atomic Energy scenarios
> Atomic Energy scenarios
> National Academy of Engineering scenarios
> Environmental Protection Agency scenarios
> Shell scenarios
> National Petroleum Council scenarios
> Council on Environmental Quality scenarios
> Project Independence scenarios

In conjunction with these scenario definitions, models are developed to analyze
the scenario and to determine its requirements. The models then reflect the
particular point of view of the scenarios. Of course it is also often the case
that the models developed are of greater generality than that of the motivating
scenarios. It should be apparent to the reader that the scenario is simply
another name for our experimental frame within an energy systems context. Thus
the design descriptions for an energy IMS can be scenario oriented with new models
derived in terms of new proposed scenarios.

As can be seen from the listing of example scenarios given the collection of
points of view is diverse indeed. Some, such as the Atomic Energy scenarios, are
relatively narrow in scope. Others such as those of the Environmental Protection
Agency are more general and represent problems of second order effects. Project
Independence was a study undertaken by the Federal Energy Administration and
resulted in an elaborate econometric analysis. The following examples of scenarios used in the Project Independence study are given to illustrate the scenario
concept:

1. *business as usual*, in which the government takes a minimum of
   new initiatives;
2. *accelerated supply*, in which the government adopts a number of
   measures intended to stimulate domestic energy production, including accelerated leasing of federal lands onshore and offshore,
   diversion of the naval petroleum reserves to commercial production,
   and introduction of regulatory changes where current practice
   impedes production (as in nuclear licensing delays);
3. *conservation*, in which the government adopts mandatory regula-
   tions or offers incentives that will lead to a reduction in
   energy demand growth, a concept also referred to as demand
   management;
4. *emergency preparedness*, in which the government takes steps to
   insure against severe economic disruption in case of interruptions
   in imported energy supplies, basically through a stockpiling
   program plus provisions for standby authority to curtail demand
   and allocate supply.

The above examples make more explicit the equivalence of the scenario and experimental frame concepts. The manner in which the experimental access to the real
system is restricted is made explicit in a scenario.

The models which have been developed with which to investigate scenarios are also
diverse in their implementations as well as underlying methodology. Among the
methodologies are econometric, input-output and mathematical programming approaches.

An econometric model contains variables and relationships which are derived from economic theory. Parameters in such models are usually obtained from carefully designed statistical estimating procedures which use past economic data as a basis. This fact tends to increase the validity of these models, however, it also limits the scope and level of resolution since economic historical data is readily available only for certain aggregate phenomena. Thus, the aggregation layers in an IMS including econometric models would tend to correspond to the local government units, state, regional and national for example since the aggregated data would be available.

Input-output models are a highly structured form of an econometric model based on a table of relationships between the input and output (in dollar terms) of each sector of the economy. For example, a nation's economy can be structured into producing-industry sectors, final demand categories and sources of primary imports. Dollar flows among these groups can be used to measure the degree of interdependency in the economy. Input-output models have been used to study regions of smaller size than the entire nation and even organizations such as hospitals. Since the level of aggregation of an input-output model is a function of the structuring of the economy, again the IMS layers would tend to correspond to the geographic scope of the model.

The mathematical programming approaches include linear and non-linear programming and their optimized solutions where they exist. These models are highly structured and are generally applicable in well-defined cases where resources are allocated to predefined processes or facilities. Thus these approaches can be applied at any level of aggregation.

Each of the above methodologies have their advantages and disadvantages with respect to some point of view on the system under study. In particular, each of the above three example methodologies can represent an optimum methodology for a model with which to investigate proposed experimental frames. Hence it can be seen that many of the existing models can be viewed as separately developed views of the same energy system. Had the conceptual structures as well as software design methods been brought to bear on the problem of IMS design and implementation earlier many of these models would very likely have been developed using this approach. However, since such was not the case and the models are related but individually developed, the next best approach would be to investigate how these models can be coupled or perhaps at least share a common database. This conclusion has been reached by at least two research centers (Argonne National Laboratory and Brookhaven National Laboratory) as a result of comprehensive studies |12,13|. The recommendations of these investigations were to maintain a library of individual models from which to obtain required results and to investigate the problems of coupling individually developed models, respectively. In the latter case above Brookhaven has successfully coupled two models of the optimization and input-output type by developing a compatibility interface |14|. A brief overview of this model coupling situation is given in the following paragraphs as an example.

There are drawbacks to the isolated use of either the single period energy optimization model or the input-output model for forecasting the future. The input-output model is limited because, although it provides for alternative energy technologies, the model cannot internally determine the extent to which each is utilized in some future planning year. Conversely, the optimization model is weak in macroeconomic content. Supplemental information on the Gross National Product and its components or on inter-industry demand for energy cannot be easily related to the specification of energy demands which drive the model.

The coupling of these two models reduces the *a priori* judgements that must be made before running either model. With the input-output model one no longer must specify the coefficients which describe fuel flows and conversions into end-use products. These quantities are determined from an optimization model solution.

With the optimization model one no longer must specify total demands for energy products since only those portions which comprise final demand in an input-output sense (i.e. included in a compilation of the GNP) must be determined. Inter-industry (sectorial) input demands for energy products is obtained through solution of the input-output model. The interface between the two models consists of two sets of equations. The first set calculates demand constraints for the linear programming model from the total outputs determined by the input-output model. The second set computes the input-output coefficients from the solution to the energy optimization model.

The interfacing of existing models in general will not be as easily accomplished as was the above described case for the mathematical programming. The assumptions underlying two or more models will often differ widely. In an IMS context these assumptions can be taken into account in the initial design description and included in the experimental frame and model schemas. When it is apparent that a new related model is required, the new model schema can be derived from the existing schema.

A key to success in designing an IMS seems to be the development of guidelines and techniques for facilitating the modifiability of the software in the system. Progress in this development can perhaps be best achieved by concentrating on the potential utility of design description languages. A summary of this and other recommendations is given in the following section.

CONCLUSIONS

This paper has been an attempt to present some of the more pragmatic issues involved in the development of an integrated modeling system. The approach and philosophy has been shown to be quite similar to that of the well-known database management system. Moreover, it is likely that some of these similarities can be exploited for adoption in an IMS design. It is also clear that the design task is complex and requires the application of all appropriate software engineering tools available. The need for an integrated approach to modeling has been recognized by the energy systems community as demonstrated by the example. This interest in the coupling of existing models indicates the need for the development of a more comprehensive approach.

The following list summarizes the methods and techniques which are believed to be appropriate for incorporation into the design and development of an IMS.

1. Use of a design description language - this implies that research is needed both in the area of design languages and their evaluation, which is a relatively new area in software engineering, and the representation of the required IMS functions in these languages.

2. Definition of model and experimental frame schemas - this work has already begun with Zeigler |5|. Methods for deriving new frames and models in these terms is also needed.

3. Layered Software Approach - this technique is included to indicate the need for clear structuring of the design whether developed bottom-up or top-down.

4. Modifiability - guidelines should be developed for a highly modular design and partitioning of functions for flexibility of derivation of new models.

Earlier in the paper it was noted that the modeling approach discussed herein did not preclude the development of other IMS's with respect to another paradigm. In conjunction with the consideration of other possible modeling systems, one can

raise the design question of what the criteria should be for the initiation of
other related systems. At some point it may be that the criteria are determined
by computer resources. In this event it is not difficult to envision a
"distributed processing" version of an IMS wherein the functions of the system
are partitioned out to separate but interconnected processors. Thus conceptually
the integrated approach can be extended through an ever expanding collection of
interconnected computer systems. Thus the IMS approach to modeling and simulation
of large-scale systems provides many needed advantages not the least of which is
a self-imposed "higher level" of systems thinking in our modeling efforts.

References

|1| Jones, W.T. and Jones, B.J., "Computer Simulation Using Hierarchical Models",
    Proceedings of Pittsburg Conference on Modeling and Simulation, 2 (1975)
    759-760.
|2| Kuhn, T., *The Structure of Scientific Revolutions*, University of Chicago
    Press, Chicago, 1962.
|3| Zeigler, B.P., *Theory of Modelling and Simulation*, Wiley-Interscience,
    New York, 1976.
|4| McGowan, Clement L. and Kelley, John R., *Top-Down Programming Techniques*,
    Petrocelli-Charter, New York, 1975.
|5| Zeigler, B.P., "Structuring the Organization of Partial Models",
    International Journal for General Systems, 4 (1978) 81-88.
|6| Date, C.J., *An Introduction to Database Systems*, Addison-Wesley, Reading,
    Massachusetts, 1975.
|7| Jones, W.T. and Dyre, M., "META: An Algorithm for the Analysis of Multilevel
    Energy Transport Systems", *IEEE Transactions on Systems, Man and Cybernetics*,
    4 (1977) 243-247.
|8| Goos, G. and Hartmanis, J. (eds.), *Software Engineering: An Advanced Course*,
    Springer-Verlag, New York, 1975.
|9| Zeigler, B.P., "Structuring Principles for Multifaceted System Modelling",
    (this volume).
|10| Riddle, W.E., "Modelling and Simulation in the Design of Complex Software
    Systems", (this volume).

|11| Yeh, R.T., *Current Trends in Programming Methodology: Software Specification
    and Design*, Vol. 1, 1977.
|12| Regional Energy Modeling: An Evaluation of Alternative Approaches, Argonne
    National Laboratory, ANAL/AA, 1975.
|13| Annual Report 18984-R. Brookhaven National Laboratory, 1974.
|14| Tessner, R., Hoffman, K. and Marcuse, W., "Coupled Energy Systems - Economic
    Models and Strategis Planning", Brookhaven National Laboratories, 1975.

METHODOLOGY IN SYSTEMS MODELLING AND SIMULATION
B.P. Zeigler, M.S. Elzas, G.J. Klir, T.I. Ören (eds.)
© North-Holland Publishing Company, 1979

ECOSYSTEMS MODELING METHODOLOGIES:
DESIRABLE COMPUTER ASSISTANCE

George Innis
Department of Wildlife Science
Utah State University
Logan, Utah 84322

A modeling system that would speed ecosystem model production,
make modeling these systems more accessible to less experienced programmers and produce results in a form readily
usable by ecosystem researchers and managers is described.
This system would take advantage of recent developments in
computer hardware and software as well as knowledge of
ecosystem structure and function to lighten the modelers'
load even more than general simulation languages. The
system described does not exist but would be nice.

INTRODUCTION

The modeling of ecological systems is, like many other human pursuits, a mixture
of the original, creative and exciting with the dull, repetitive and boring.
Modeling is the decision maker's heaven because in a modeling activity there may
be more overt decisions/hour than in any other activity. The modeler must choose
objectives, hypotheses, formulations, implementations, data, experiments, parameters, parameter values, verification, validation and sensitivity analysis
exercises, etc. Many of these are unique to the problem being solved while others
apply to entire ranges of problems. The unique decisions allow the modeler to
exercise his art; the repetitive ones cause him to consider other careers or to
seek automata that will make them for him.

In this paper I will distinguish these two types of decisions as they appear to
me in the current ecosystem modeling milieu. There is a great deal that computers
can do for us as regards the repetitive decision, and I will suggest a modeling
structure/computer system that would promote this effort. The implications of
such a system are argued for on several grounds.

First, the human use of human beings (Wiener 1950) demands that we allow the
computers to do their things and relieve the human of the mechanical tasks.

Secondly, a number of bright young ecosystem modelers have chosen to pursue other
fields largely because the repetitive processes are so terribly boring.

Thirdly, the repetitive processes often require the greatest programming skill to
make them operate efficiently. By reducing this task we will allow less skilled
programmers and better skilled ecologists to construct their own ecosystem models.

Finally, in many studies, time is of the essence. The repetitive part of a job
may be much more time consuming than the interesting part and may result in the
product being too late to meet the objective of the effort.

But don't simulation compilers do exactly this? Of course, but not to the extent
that they could. Any higher level programming language is designed to achieve
these same goals. I will argue below that they fall short of their potential in
ecosystem simulation and that special language development is not necessarily
expensive.

ECOSYSTEM MODELING

Identification of a modeling procedure (well defined collection of steps to be
taken in constructing a model) facilitates communication of that process, speeds
model development and adds to the sense of repetition in model construction.
Many such procedures have been published, but I will refer to one which is
described elsewhere (Innis 1975). Briefly, the steps of the procedure are:

1. Identification of model objectives.

2. Statement of the hypotheses thought be be needed to achieve objectives.

3. Formulation of these hypotheses as mathematical statements.

4. Implementation of the mathematical statements as computer statements.

5. Model experiments to determine if the model meets the objective.

6. If the model meets the objectives, write a report and stop.

7. If the model does not meet objectives, one must decide why. It could be
    that A) the data on which the hypotheses and formulations were based and
    the associated parameter estimates are inadequate. If the data are
    wanting, design and conduct the appropriate experiments to refine the
    hypotheses and formulations and following the experiments return to steps
    2) and 3) above. One may find that the point of return is 1) because the
    new information gained from the modeling exercise and experiments have
    exposed weaknesses in the objectives. B) The model contains serious
    errors. If the model is wanting, analysis and refinement of the formu-
    lations and implementation lead to further model experiments (step 5).

Objective setting is original, creative and difficult. It is a step that is often
given too little emphasis. It is underestimated because we often proceed on other
activities with rather broad general objectives. However, the myriad of decisions
required in modeling obviate this ability to work with loosely stated objectives.

Hypothesis choice for achieving objectives has both original and repetitive
components. The choice of an asymptotic relationship between nutrient uptake rate
and concentration is original (in this one problem), but if each of several
nutrients is to be treated similarly, the choice becomes repetitive.

Formulation is largely repetitive. Surely there are infinitely many continuous
functions that have a sigmoid shape, but only a few of these are commonly used
and, in a given modeling exercise, perhaps only one or two. Formulation is also
a step that is difficult for many who would construct ecosystem models. Formulation involves parameter identification but in a restricted sense vis-a-vis the
general problem. We generally work with only a few alternative formulations for
a given hypothesis. The one of this small list that best represents a given
hypothesis may depend on data or on the modeler's concept of the mechanism being
represented.

Implementation is highly repetitive. At the level of coded statements, many
organisms look alike. Indeed this is one of the strengths of the modeling
process; by reformulating a problem in a new language we see similarities that
are lost in the original jargon (see Bunge 1967). Simulation languages have
concentrated on relieving the modeler of this burdensome step. By allowing the
formulation to be more directly expressed in a coding language, these tools
assume many of the repetitive tasks.

Model experiments are repetitive and, for computer models, almost completely done
by a machine.

Determination of objective obtainment is creative or dull, depending on the
character of the objectives and the structure of the model. If the objective

contains a quantitative, determinable criterion for model success, then this step is dull. Whatever data or other information is needed to measure attainment is invoked and the result observed. The process is dull, but the results may be very interesting and lead to new insights and studies. If the objective is too general, then determining if the objectives have been met can be a very creative and exciting exercise (witness the series of world models, Forrester 1973, Meadows et al. 1972, Mesarovic and Pestel 1974).

Experimental design and conduct are clearly creative processes.

Analysis and refinement of mechanisms is both repetitive and creative. Location of the source of the errors in a large model can be difficult. The highest levels of creative and analytical skills may be required to locate a simple misplaced statement or an order of magnitude error in a parameter estimate. On the other hand, the analysis process may involve searching for dimensional (unit) problems that can be quite tedious and dull.

I should emphasize the point that the dull and repetitive are so because of the experience of the modelers themselves and the gradual development of a consensus about the structure and function of ecosystems. Ten years ago few ecosystem modelers existed and still fewer found their work repetitive. Early modeling exercises tried quite different hypotheses and formulations for the most fundamental processes and, when done for the first time these are unique (Patten 1971, Watt 1966). Experience led to the elimination of all but a few of these approaches and that reduction encodes a developing consensus on ecosystem structure and function. Books by Patten (1975), Hall and Day (1977), and Levin (1974) illustrate this consensus, particularly when compared to the books cited just above; but the journals, meetings and informal discussion provide stronger evidence. Early in the 1970s ecosystem modelers often spent hours on difference vs. differential equations or linear vs. nonlinear discussions. What I am calling for is a modeling structure that promotes this consensus and makes its nature more directly observable and usable.

AVAILABLE COMPUTER ASSISTANCE

Computers are peerless when doing certain repetitive tasks. "Higher level languages" like FORTRAN, COBOL, ALGOL, etc. eliminate much of the repetitiveness from certain calculations by, for example, doing memory accesses for you. Thus the FORTRAN

$$X = Y + Z$$

replaces the machine language coding sequence for adding two numbers and storing the sum. These languages differ in the kinds of repetitive tasks that they do well because they are designed for different applications: FORTRAN eases equation writing, COBOL has good input/output features for business applications, ALGOL facilitates algorithm implementation.

Simulation languages carry this process of relief from overhead programming a step further for their special applications. For example, DYNAMO (Pugh 1961) simplifies input and output for a differential equation description of a state-space system. Data are input in the form of FORTRAN-like equations so that from one run to the next, only the parameters that are changed need be specified. Graphical and printed output are specified with a few simple commands and formatted for convenient use. Generalized functions are provided, e.g. TABLE and CLIP, that allow for the implementation of a wide variety of hypotheses.

With the SIMCOMP series, Gustafson and I (1973) carried the DYNAMO ideas further. First, by deciding on a first order ordinary forward difference equation representation of a model in the form

$$X_{T + DT} = X_T + DT * (AVERAGE\ RATE)$$

we eliminate the necessity of writing these equations. If the AVERAGE RATE is
specified in such a way that it is easily associated with a state variable
(done by indexing), then the compiler can write the equation. Second, one of the
greatest sources of programming errors was the multiple specification of variables. With FORTRAN a dimensioned variable that must be read in, communicated
to a subroutine and printed out must be spelled identically the same way in each
calling statement. In SIMCOMP, all input variables are automatically put into a
COMMON block that is inserted in each subprogram. This not only reduces the
number of times the variable must be spelled correctly but speeds execution by
shortening the list of call parameters for many subprograms. Linking programs
together and the use of library functions, however, requires special care.

It is not necessary to extend this list. The points to be made are:

First, that as a process becomes repetitive in the programming of a problem,
changes to the compiler can reduce the repetitive effort.

Secondly, the nature of the compiler is specific to the task under consideration.

Third, the simultaneous development of compilers and programs of a given type
can make better use of both men and machines.

It is reasonable to ask about compiler writing languages as a solution to this
problem. While they hold promise, our present compilers are beyond the capability of compiler writers. It will be some time before compiler writing
languages could generate a next generation of SIMCOMP (Jon Gustafson, personal
communication).

Compilers aren't the only limitation. Alternate forms of input and output and
better man/machine communications are also needed. An excellent review of the
state of the art is found in Benyon (1976).

SCENARIO

In this section I will describe the use of a system that I would like to have.
This will set the stage for the description of many of the features to be presented in the sequel.

A modeler, having set his objectives carefully, including a criterion for model
acceptance, calls for the ecosystem simulation package. The system and the
modeler's responses are given below:

    S: Is your system terrestrial or aquatic?
    M: Terrestrial.
    S: What latitude (range)?
    M: 40°N.
    S: What elevation above sea level?
    M: 600 M.
    S: What soil type?
    M: Unknown.
    S: Assume sandy loam. What rainfall?
    M: 45 In.
    S: Give percentage each season.
    M: 40, 10, 30, 20.
    S: The plant types I have for this system are: Warm season grass, cool
        season grass, annual grass, deciduous tree, conifer, broad leafed ever-
        green...
        Which plants do you wish to include?
    M: Deciduous trees, conifer, annual grass.
    S: What biomass density (KgDW/M**2)?
    M: 2000, 4000, 400

```
S:  Do you wish to add producers?
M:  No.
S:  The consumers I have for this system are...etc.
        :
        :

S:  System specification is complete.  You may specify changes.
M:  Change Hypothesis 24 of Nitrogen Model as follows:
        :
        :

S:  OK.  You may specify changes.
M:  Change deciduous tree loss to consumer to 20%.
S:  Inconsistent.  Repeat to override.
M:  Change deciduous tree loss to consumer to 20%.
S:  OK.  May result in imbalance of carbon model.  You may specify changes.
M:  Run.
S:  Time interval?
M:  20 Yr.
S:  What data sets for parameter estimation?
M:  MYDATA1, 24, STD set 5.
S:  The goodness of fit achieved met acceptance criteria.  Which parameters
    do you wish to see?
M:  Deciduous tree nominal death rate.
S:  .3
M:  NO.  Try PO2 (parameter optimization routine level 2).
S:  Deciduous tree nominal death rate = .03.
M:  Run.
S:  All parameter values printed on output unit 3 (line printer).
    What data sets for validation?
M:  MYDATA 2, STD set 5.
S:  RUNNING.
S:  Which output do you wish to see?
M:  Graph, deciduous trees/time.
S:
```

```
Dec.
tree
            Time
```

```
(modeler, expletive deleted).
M:  Print, plot.
S:  OK.
S:  Output available at the computer center.  Do you wish to modify this
    model?
M:  Not at this time.
S:  Save or remove this model?
M:  Save.
S:  Saved as MYMODEL 1.  BYE.
```

1. Objective setting is an art that at present defies automation. The modeler
    comes to this system with his objectives set. He translates these objectives
    into a general structure and with the computer begins to build up a model of
    the system designed to achieve the specified objectives.

2. The abiotic environment is specified. All required inputs are either

provided by the modeler or default values produced by the machine. This part
of the process is excellent training for the modeler. He sees how much information is needed to achieve his objectives, and, by changing the values of
defaulted parameters sees the effect of his ignorance on the system.

3. Information is provided to the system in many distinct units. These are
    converted internally into a consistent set.

4. Given a specification of the environment, the system can identify some of the
    plant types that could exist in that environment and which are coded into the
    system. The modeler can choose the plant types he wishes to include and
    their densities. The system might check for consistency and, if, for example
    the GRASS were more than a preset value, comment to that effect (Innis 1975b,
    and Innis and Clark 1977).

5. Before proceeding with the consumers (in ecosystem jargon, producers refer to
    green plants, consumers refer to animals, in the broad sense, from microscopic
    to large) the modeler is given the opportunity to insert additional producers
    into the system. He might elect to code from the terminal in "discussion"
    with the computer, he might exit the discourse mode and write a model in a
    convenient language, or he might call a new model in from another source (e.g.
    card or disk file).

    One of the uses of this system would be to construct a driver (submodel to
    provide input) for a subsystem that the modeler wishes to develop. When
    making producer, consumer, nitrogen, water or other subsystems, one begins by
    using constant or simple varying (sinusoid) default values for inputs to the
    subsystem from other parts of the system. At the next level of development,
    the interactive nature of the relationship between the subsystem and the
    remainder of the system becomes important. Parameter values and behavioral
    properties depend heavily on the way the larger system responds to the
    subsystem. At this state in the development of a subsystem model, the
    investigator is often diverted into a several day or longer process of con-
    structing a simple but interactive system to drive his model. This is dull
    work and often not done very well because it is peripheral to the main inter-
    est. The desired computer system would allow the development of substantial
    drivers tailored to the needs of the study.

6. The consumers would be added based on the abiotic environment and producers.
    Options for forcing the inclusion of certain consumers could occur (see
    Haefner 1977).

7. Once the system is complete in the sense that all trophic levels have been
    included and all the unresolved inputs provided, changes to the hypothesis
    structure of the model are considered. First, though consider the unresolved
    inputs. A producer model, e.g. DECIDUOUS TREES, might require as input SOIL
    PHOSPHORUS. This input might be unresolved because of the failure to provide
    a soil phosphorus model or because the soil phosphorus model did not have an
    output labeled SOIL PHOSPHORUS. The modeler might resolve the problem by
    entering the phrase DEFAULT SOIL PHOSPHORUS, in which case a value for soil
    phosphorus would be provided by the DECIDUOUS TREE model; or he might enter
    IDENTIFY SOIL PHOSPHORUS WITH SOLUTION PHOSPHORUS if the latter occurred as a
    state variable in the model. Whenever default values are used, that infor-
    mation should be recorded for output in permanent form.

    Changing the hypothesis structure of a model in a relatively painless way
    would really help to find a collection of mutually compatible hypotheses
    whose joint operation would be consistent with observation. I can envision
    modeling sessions in which experts in specific disciplines like nitrogen
    cycling, primary production, consumer bioenergetics, etc. changed one hypo-
    thesis several times to investigate its effects on one or more output

variables, then moving to another hypothesis and doing the same thing.

The hypothesis change procedure may result in inconsistencies between the loss of material at one level and its uptake at the next, or other such problems. These will be flagged. If the modeler wants them left alone, he simply repeats the command.

8. After he has the model structured as he wishes, he issues the command RUN. The machine requests run control information. Two kinds of output are produced. First, material selected is displayed on the modeler's terminal. This material should be stored in pages with rather free access to these pages under operator control. The other output is much more complete and goes to a disk file. If the run was useful either as a final product or for diagnosis, the disk file is processed for output. This processing might be simply printing hard copies of the material displayed on the screen plus identification data for the run. It might also consist of statistical analysis of variables, computation of other output variables (such as sums over groups of variables or dates on which certain events occur).

Another form of model output would be the storing of each of a selected collection of variables on disk at selected times. For example, we might wish to store all state variables at each time step. This way we could do an extensive analysis of the results of a simulation without repeating it. This form of output would require a further program to analyze these data and present the results.

9. Data and communication with the data base would be a feature of the proposed system. Data might be used to produce hypotheses (as with a TABLE Function), to estimate parameters, to select functional forms, and for validation in the broad sense of comparison of model output with data. To be used in this way the data must be extraordinarily clean and the system accessing the data must be able to change units (within reasonable limits, e.g. meters to feet, not acres to kcal).

This data structure could be used to facilitate communication between submodels. If each submodel addressed a data structure for input values and placed its output in a data structure, the communication among submodels would be simplified, changes of units as needed would be routine, and input and output would be from this structure. A system like this has been developed for use in the aerospace industry and it was very expensive to make (Filippa 1973). Physicists also developed PATCHY at CERN for subroutine interfacing. Perhaps a later and less expensive system would be feasible.

DISCUSSION

This system differs radically from that described by Benyon (1976) because we have different objectives for our systems. A general purpose simulation language as discussed by Benyon must be like a general purpose automobile--good at many things but lacking in many ways. Such languages are improvements over FORTRAN for simulation work but do not take advantage of features of the system being modeled or of the approach that the modeler has elected to take. A general purpose automobile cannot take advantage of the types of roads on which it will be used or the fact that it will, for some user, be driven more frequently at 50 mph than 30 mph. This is evidenced well by Benyon's (1976) discussion of procedural vs parallel structuring. He points out that the problem of parallel model structure disappears if one uses a state variable description of the system. By retaining a strict procedural language one can accomplish the variable indexing and array handling that Benyon wants rather easily (Gustafson and Innis 1973). A problem of indexing appears immediately; however the compiler must organize the data and arrays in an efficient fashion and assign indices accordingly. For example if one has a model nicely structured with 10 consumers and decides to

expand to 12, the data and array restructuring can be messy. Without such restructuring the automatic indexing features are confounded. Thus, the modeler
needs a system that allows him to speak in parameterized terms (CONSUMER(1) -
CONSUMER(LAST CONS)), while the compiler guarantees that the memory space use and
indexing are efficient.

Work with SIMCOMP convinces me that the solution to the simulation language problem does not lie in "general" or "standard" languages but in structures from which
special purpose languages can be readily developed. "General" implies features
that are expensive and are often not used. "Standard" means, among other things,
that almost all input/output and file manipulation are placed outside the language because these are patently nonstandard. Yet input/output and file manipulation are often critical to the implementation of an effective tool. Thus, for
large scale simulations I think we need to encourage experimentation with special
purpose languages like SIMCOMP and their development with tools like compiler
writing languages.

In point of fact, this last statement may expose the real issue in simulation
languages - scale. I will venture the following table:

Scale	Language Choice
Small = few man weeks	Whatever the modeler is familiar with
Medium = few man months	General and standardized simulation language
Large = few man years and up	Special purpose language

A small scale modeling effort hasn't the flexibility,nor is there time to learn a
new language, much less develop one. A medium scale effort would surely benefit
by having efficient tools, even if a fair amount of learning were required to use
those tools effectively. In a large scale effort, efficiency of use of men and
machines becomes something deserving of attention. The development of a language
and other features that are tailored to the job at hand is probably warranted.

Communication of results is certainly made more complicated in this modern Tower
of Babel. But that need not be the case. If the community agrees (and agrees
with the U.S. Government) to produce the final product in ANSI Standard FORTRAN,
then one of the features of the special purpose language development would be
either a FORTRAN intermediate stage (as with some extant languages) or a translator that converted the special purpose language into FORTRAN (perhaps achieving
only 90% translation with the remainder being done by hand).

CONCLUSION

A computer system for ecological modeling has been described. This system is an
ideal that may be beyond reason at the present. I would argue that progress
toward that system will be cost effective in that improved tools for ecosystem
simulation will derive. There are ample precedents for the combined tool/appli-
cation development of simulation languages and computer structures.

LITERATURE CITED

Benyon, P. R., (1976) Improving and standardizing continuous simulation lan-
     guages. Presented at the SIMSIG Simulation Conference, Melbourne. Excerpts
     in Simulation 27 vii-ix.

Bunge, M., (1967) Scientific Research. Vol. 1. The Search for a System. Vol.
     2. The Search for Truth, Springer-Verlag, New York.

Felippa, C. A., (1973) Finite element and finite difference energy techniques for the numerical solution of partial differential equations. pp. 1-15 In Proceedings of the 1973 Summer Computer Simulation Conference, Vol. 1, Simulation Councils, Inc., La Jolla, CA.

Forrester, J. W., (1973) World Dynamics, 2nd Ed., Wright-Allen Press, Cambridge.

Gustafson, J. D. and G. S. Innis, (1973) SIMCOMP Version 3.0 User's Manual, USIBP Grassland Biome Technical Report 218, Colorado State University, Ft. Collins, CO.

Haefner, J. W., (1977) Generative grammars that simulate ecological systems. pp. 189-211 In George S. Innis (ed.) New Directions in the Analysis of Ecological Systems, Soc. for Comp. Simulation. Simulation Councils Proc. Series Vol. 5, La Jolla, CA.

Hall, C. A. S. and J. W. Day, Jr., (1977) Ecosystem modeling in theory and practice: An introduction with case histories, John Wiley, New York.

Innis, G. S., (1975a) The use of a systems approach in biological research. pp. 369-391 In G. E. Dalton (ed.) Study of Agricultural Systems, Applied Science Pubs. Ltd., London.

Innis, G. S., (1975b) One direction for improving ecosystem modeling. Behavioral Sci. 29(1):68-74.

Innis, G. S., and W. R. Clark, (1977) A self-organizing approach to ecosystem modeling. pp. 179-187 In George S. Innis (ed.) New Directions in the Analysis of Ecological Systems, Soc. for Comp. Simulation. Simulation Councils Proc. Series Vol. 5, La Jolla, CA.

Levin, S. A. (ed.), (1975) Ecosystem analysis and prediction, Soc. for Industrial and Appl. Math., Philadelphia, PA.

Meadows, D. H., D. L. Meadows, J. Randers, and W. W. Behrens, III, (1972) The limits to growth, Universe Books, New York.

Mesarovic, M., and E. Pestel, (1974) Mankind at the turning point, Dutton, New York.

Patten, B. C. (ed.), (1971, 1975) Systems analysis and simulation in ecology. Vols. 1, 3, Academic Press, New York.

Watt, K. E. F. (ed.), (1966) Systems analysis in ecology, Academic Press, New York.

Wiener, N., (1950) The human use of human beings: cybernetics and society, Houghton Mifflin, Boston, MA.

METHODOLOGY IN SYSTEMS MODELLING AND SIMULATION
B.P. Zeigler, M.S. Elzas, G.J. Klir, T.I. Ören (eds.)
© North-Holland Publishing Company, 1979

STRATEGIES FOR STANDARDIZATION
IN SOCIO-ECONOMIC MODELING

Siegfried Dickhoven

Institut für Planungs- und Entscheidungssysteme (IPES)
Gesellschaft für Mathematik und Datenverarbeitung (GMD)
PoB. 1240, D-5205 St. Augustin-1, F.R. Germany

We understand standardization as a necessary premise to improve the
transfer of models and modeling know-how, to diminish the gap between
model builders and model users and to put the art of modeling on a
scientifically consolidated basis. This paper tries to evaluate, whether
and how new concepts and strategies for (1) the modeling products (the
models), (2) the modeling tools, and (3) the modeling people (builders
as well as users) can contribute to standardization.
Some remarks on limits in socio-economic modeling are added.

INTRODUCTION

Regarding recent activities in the modeling scene with emphasis on socio-economic modeling it may be stated that a 'new generation of models' has arisen. Though
this expression was originally used for models covering very long time horizons
[1], I would like to apply it to a broader spectrum of new modeling activities.
These activities are going into two main directions. The first and more experimental direction can be described by following characteristics:

- The taking off into enormous dimensions of size;
- The application of more sophisticated methods;
- The combining of different approaches (eclectic approach).

The second direction, often combined with experimental modeling activities, in-
tends to lead towards consolidation in modeling and can be shaped by slogans like:

- Model evaluation;
- Model review and comparison;
- User participation (education);
- Implementation research.

These activities differ essentially from former modeling activities by their tendency towards division of labor and their growing preoccupation with meta modeling
questions - both caused by the inevitable trend to large scale modeling as a consequence of the need to manage the growing complexity of problems and of the modeling process itself. This tendency towards division of labor to integrate the
available modeling know-how is necessary because these new (socio-economic) modeling activities presently afford more substantial, methodological, instrumental
and managerial knowledge than one person or one team is able to overview.

This division of labor concept requires a substantial improvement of model- and
modeling knowledge transfer. And as there many different groups are involved, it
is necessary to set up special agreements, guidelines ('standards') for this
transfer process.

Thus understanding standardization as an aid to improve the transfer of models and
modeling know-how, this paper reflects on some possible standardization concepts
and strategies for

(1) the modeling-products (the models),

(2) the modeling tools and

(3) the modeling people (builders as well as users).

Though the discussed concepts and strategies seem to be quite general, it must be said that they rely on our special experiences with the development of an integrated Model Base System (MBS) [2] for two ministries of the German federal government, which were faced with the problem of setting up and using complex socio-economic models, which were developed in different methodologies and by different (and generally outhouse-) groups [3]. (Its general concept is very similar to the Integrated Modeling System, discussed by Jones [4] in the volume.) Due to this special (and more technical) background the following concepts are based (or eventually biased) on technical concepts of the underlying software instruments for modeling. Furthermore there exist many interconnections among the different concepts, though they are discussed separately.

THE MODELING PRODUCTS (THE MODELS)

With the increasing scale and complexity of socio-economic models the application of the modularization concept becomes more and more appropriate. This well regarded technique to manage large and complex systems is finding increasing application in the field of modeling.

Well known applications of this technique like the Mesarovic/Pestel World Model (having regional and sectoral modules combined with a hierarchical layer concept of interactions [5] and the Project LINK [6] or the INFORUM approach [7] (regional modules) use the modularization concept in a substantial way. This is probably the most important method to control model complexity available to model builders as well as to model users.

With this substantial concept, for example, it will be quite convenient to combine complementary or to compare similar model parts (modules), which are written in the same language (e.g. DYNAMO) and belong to the same methodology. But there will arise severe technical problems like respecification, translation and manual data transfer, if the to-be-combined or to-be-compared modules are of different types or languages as they will rather often be in 2nd generation modeling. Consider for example the intended dynamic linkage of an econometric growth model, an input/output world trade model and a system dynamics solid resources model in project FUGI [8], the model comparison and assessment projects of the National Bureau of Economic Research (NBER) [9], or the Energy Modeling Forum (EMF)-Project [10].

To avoid these technical problems it is often suggested that a new language or modeling system should be created that combines the advantages of all (or some) different systems and methodologies and meets all their different needs. Though some modeling systems that we have analysed actually had started with this objective, I have doubts about this approach, because it will probably lead to a general purpose language like FORTRAN or PL 1 (and is completely useless). x)

A better way to overcome these technical difficulties with 2nd generation modeling activities is an approach that integrates some already existing modeling languages of different methodologies for the specification of the (partial-) model's structure, while other tasks (such as model interfacing, scenario generation, run specification, reporting, documenting and analysing) get a new and largely uniform kind of treatment. The realization of this task oriented concept in our (interactive) Model Base System roughly looks as follows:

---

x)
   see the article by Innis in this volume for more discussion on this point
   (editor)

The basic elements (elementary modules) of socio-economic models are so called 'partial simulation operators', which generally are time dependent and which are special forms of general operators that transform a set of input quantities according to its transformation prescriptions into a set of output quantities for one and only one (time) step. This reduces the building process of models to the construction of the model 'core' (the model structure). General tasks like data transfer, dialog handling, run and time loop control are performed by a central simulation processor.

These elementary modules are generated by MBS from socio-economic models, speci-fied in modeling systems such as DYNAMO III-F [11], the econometric systems IAS [12] (developed by the Institute for Advanced Studies, Vienna, and also used for project LINK at Bonn University), MEBA [13] (used in Bundesministerium fuer Wirt-schaft), MASH [14], a microanalytic system and the general purpose language FORTRAN IV. Under control of the MBS-user, and more or less supported (depending on the elementary modeling systems) by special preprocessors, there will be pro-duced a special information block for each elementary module that contains the description of the module's data interface for the meta construction level.

On this meta construction level the modules can be linked together, and can be connected with special reporting (or analysing) modules or with data elements of a data base, using a special linkage language. Special model runs - including conditions and systematic search - may also be specified on this level.

This formal concept of modularization preserves the preferred construction en-vironment of experienced model builders on its basic construction level, provides an additional level for construction and supports the computer aided transfer of models (especially models of different approaches). By supporting a largely uni-form approach to meta construction and run specification (especially for 'pro-duction runs') we hope that MBS will facilitate better user participation and more know-how transfer among modeling groups.

The existence of a model base makes already existing models technically but not intellectually available: The transfer of modeling knowledge requires further-more sufficient documentation of the available models. Though this need is widely accepted as one of the greatest in modeling [15], there don't exist any general or obligatory guidelines (or standards) for documentation. Great progress, how-ever, can be expected from the current Model Standards Program of the National Bureau of Standards, reported by Roth/Gass/Nance in this conference [16].

THE MODELING TOOLS

Comparing the trends in socio-economic modeling with the existing software tools it becomes evident that many modeling tasks and especially those which require the tranfer of models are not supported by the existing instruments. This again requires a new generation of modeling tools, which we think will go in the direc-tion of integrated modeling systems and which will cover more modeling tasks as well as more modeling approaches. While these integrated modeling systems are under development we should ease the transfer of models by linkages (software interfaces) between already existing modeling tools.

This software interfaces concept involves linking of different modeling tools, especially of central tools (model construction systems) with peripheral in-struments like data base systems, report generators and analysis packages with-in one computer (or operating system) environment. This does not create severe problems for a computer scientist, but for a model builder it is really a great problem and therefore very seldom applied. By developing interface (bridge) pro-grams the situation would become much better.

Two kinds of such interfaces are possible in relation to the kind of linking:

(1) Direct interfaces between modeling instruments:

This kind of linking is more important for combinations of central (contruc-
tion) systems with peripheral tools like report generators, because most
central modeling systems are rather poor in these peripheral (post pro-
cessing) tasks. This direct interfacing has been done, for example, by the
Urban Institute, Washington D.C., where the microanalytic modeling system
MASH [14] has been linked for analysis purposes with the time series package
TSP of the Brookings Institution and with the report generator TPL of the
U.S.Department of Labor to produce tables ready for printing [17]. This Table
Producing Language again is based on an already existing data base system and
linked with the statistical package SOUPAC from the University of Illinois.

(2) Interfaces into a general purpose high level (HL) language:

Though direct interfaces are also possible between central modeling tools
they are not recommendable for these purposes. While interfaces between
central and peripheral systems only have to provide (numerical) data-trans-
fer between different instruments, an interface between central systems
generally has to transfer programs. This makes it much more complicated
and too expensive for only one connection. The detour into high level
languages like FORTRAN here has the advantage that the generated code can
also be used by those people, who work with this high level language as
their modeling tool. This kind of interface can be developed either as an
input-interface (able to adopt programs of that HL-language) or as an out-
put-interface (producing HL-language programs) or as both. Applications of
this interface type exist for example for the DYNAMO-F-language (type: input-interface) and for the Viennian econometric system IAS (type: output-
interface), both having FORTRAN-interfaces.

We think that efforts into this direction will also amplify the model builder's
impetus towards standardization, because the use of linked modeling tools will
lead to much more technical transfer problems, which may be overcome by the
setting of interface standards.

As already mentioned the development of new modeling tools will go in the direction of integrated modeling systems. These systems are understood here as modeling instruments that support the set up and processing of models (multi-task),
which are of different methodological approaches (multe-type); for example:
system dynamics and econometric or microanalytic and econometric models. The
linking of one central with one or several peripheral modeling tools is not such
an instrument, though it broadens the spectrum of working with models (but only
with models of one approach).

Besides some microanalytic systems like MASH [14] and MOVE [18], which have been
combined with econometric modeling nearly from their beginning, there exist only
some rudimentary systems like SIMA [19] or RSYST [20]. While SIMA is one of the
systems that has created an overall concept for econometric and system dynamics
models, the RSYST-System comprises different, but newly developed subsystems for
each approach. Our Model Base System is similar to that last system, but it provides already existing subsystems for the different approaches. It also has a
two level concept that makes clear differences between elementary construction
(1st generation activities) and meta-construction (2nd generation activities).
Such multilevel systems for modeling activities including additional levels
(if it cannot be done on the elementary level) for the writing of methods by
its users will probably become the modeling tool of the future, as it is indicated by experimental systems like the ACOS SYSTEM [21] or the KARAMBA Concept [22].

As large scale modeling often deals with models developed by others and as integrated modeling systems are the specific tools for this kind of modeling active-
ties, they will facilitate the transfer of models as well as accelerate the process of consolidation and experimentation in modeling. They will also contribute
to a standardization in modeling and to a harmonization of software development
for modeling purposes. While the technical problems will encourage the users' wish
for standardization, the high barrier of development costs will automatically lead
to a concentration in development of such modeling tools.

THE MODELING PEOPLE

The third and probably the most critical factor in the transfer process of modeling know-how are the people involved. Besides the current 'classical' communication problems between model builders and decision makers there will arise additional communication problems arising from the trend towards division of labor in
modeling. These latter problems will be the greater the more interdependent the
divided parts of work will become.

As this area has not yet been sufficiently explored, we can only begin to suggest
concepts for dealing with these problems. Nevertheless there exist some helpful
ideas for possible strategies to improve the transfer process. These strategies
are directed towards the users' (esp. the clients') receptivity for modeling know
how, towards the modeling process itself, and towards the institutional environment.

Seeing the main reason for the model utilization crisis in an educational gap of
the clients (the decision makers), it was said that this crisis will end with the
next generation of decision makers. Therefore the obvious recipies to overcome
these problems were a more formal (modeling oriented) education in most sciences
and (eventually) basic modeling courses for the clients. Though both strategies
are helpful, they don't take care of the fact that a politician has to behave po-
litically which is often different from (scientific-) rational behaviour; and even
formally well skilled politicians will behave like this (if they are good politicians), because models will (and should!?) only describe the more rational and
more quantitative components of their real decision environment. Therefore it
seems to be necessary to bring decision makers and modelers together for better
learning and understanding of the constraints and conditions, under which they
have to do their jobs. A successful example for this kind of communications is the
Energy Model Forum (EMF)-project [10] where ad hoc working groups of politicians,
businessmen and scientists perform comparative studies of a range of energy models around a specific issue to which many existing models can be applied.

Another possible approach to diminish the transfer problems caused by the people
involved refers to an improvement of the modeling process itself by an improved
internal organization. An important attempt in this direction is the GAO-proposal
[15] for a five-phases model development concept. The subsequent documentation
effort of the U.S.-National Bureau of Standards [16] belongs also to this kind of
approach. But it must be stated that a special modeling management science is
still to be developed, and especially the communication problems within the modeling group arising from large scale modeling efforts have not yet been considered.
(Experiences with very large software development teams have shown [23] that under
certain circumstances up to 95 percent of the total efforts had to be spent on
communication and management.)

Closely related to this management approach regarding the internal organization
are some ideas to improve the institutional environment of modeling efforts. Such
ideas range from the foundation of Model- and Data Banks to make the existing
knowledge available (clearinghouse function) [24] to the organization of institu-
tionalised model reviews (either as external expertizes [25] or as comparative
studies [10] to the creation of special transfer agencies [26] between model biul-
ders and decision makers.

On account of the softness and complexity of human factors there are no simple and cogent solutions for this problematique; and nearly all cited concepts seem to make sense, unless they are applied altogether. They all intend to provide better communication among the people involved, a need that will increase rapidly with the inevitable tendency towards division of labor in modeling. On the other hand, the concepts discussed for models and modeling tools seem to be much more evident. Even in these fields human factors are of great importance, but the chances for success haven risen, because division of labor necessitates it, and because the problem has been recognized by the sponsoring agencies.

SOME FINAL REMARKS ON LIMITS IN SOCIO-ECONOMIC MODELING.

There are several dimensions of limits in socio-economic modeling. First and easiest to overcome is the technical dimension. Though special simulations with large models (e.g.: stochastic optimization of large econometric models [27] or microanalytic models with large data bases [28] are still very expensive and time and storage consuming, this limitation is diminishing (unless models will (for-ever) grow in size). Other, (solvable) technical difficulties, will arise with the management of large scale modeling efforts.

More severe problems are to be expected with the theoretical dimension of socio-economic modeling. In this field, theoretical deficiencies in modeling methodology itself seem to be(-come) less crucial than deficits in social (economic) theory and/or availability and correctness of data. While modeling methodology can be im-proved by new concepts for modular, hierarchic and focussed modeling, by review methodologies and complexity reduction methods, and by other consolidation orien-ted, works, the theoretical or formal deficiencies of the social sciences will (and should!) set up limitations to modeling applications by their very nature. I strongly believe that formalistic modeling will never cover the field of social research and therefore should restrict itself to the more formalistic aspects of social science.

Finally, the modeling people should be concerned with ethical limitations of their work. Besides the duty of all participating parties in modeling to be honest, they all should consider how to prevent and how to react to abuses of models. Above all they should respect the limits and caveats which are set by the basic human rights (esp. the right for privacy) as well as by human deficiencies, especially when they are investigating 'challenging' social problems in a large scale, highly disaggregated, and integrated manner. These works could easily mutate into 'big brother'-concepts.

REFERENCES:

[1] H. Chestnut, Th. B. Sheridan (eds.): Modeling Large Scale Systems at Natio-nal and Regional Levels, Report of a Workshop held at Brookings Institution, Washington D.C., February 10 - 12, 1975 (MIT-Project granted by National Science Foundation under grant no.: ENG 75 - 00738)

[2] S. Dickhoven; W. Kloesgen: Grundlagen fuer die Entwicklung eines Modellbank-systems, in: Öffentliche Verwaltung und Datenverarbeitung, 6(11) p. 325 - 333, (1976) Stuttgart 1976

[3] see: P. Hoschka, U. Kalbhen (eds.): Datenverarbeitung in der politischen Planung, Campus Verlag, Frankfurt 1975; H. Schmidt: Das Sozialformations-system der Bundesrepublik Deutschland, adl-Verlag, Eutin 1977; and: W. Schmidt: Probleme bei der Übernahme fremder Modelle, in: S. Dickhoven (ed.): Modellierungssoftware, IPES-Report 76.102, Bonn 1976, p. 425-430

[4] W.T. Jones: Software Engineering Problems in the Design of an Integrated Modeling System (this volume)

[5]   M. Mesarovic; E. Pestel: Mankind at the Turning Point, E.P. Dutton & Co, Inc., New York 1974

[6]   R.J. Ball (ed.): The International Linkage of National Economic Models, North Holland Publ. Comp., Amsterdam 1973 and: J. Waelbroek (ed.): The Models of Project LINK, North Holland Publ. Comp., Amsterdam 1976

[7]   D. Nyhus, C. Almon: The INFORUM International System of Input/Output Models and Bilateral Trade Flows, paper presented at the fifth Global Modeling Conference, Sept. 26 - 29, 1977, IIASA, Laxenburg 1977

[8]   Y. Kaya et al.: Future of Global Inderdepencence, Report on Project FUGI; Addenda-Part 3, chapter 3.7, paper presented at the fifth Global Modeling Conference, Sept. 26 - 29, 1977, IIASA, Laxenburg 1977

[9]   D. Kresge: Energy Model Integration and Assessment, in: National Bureau of Economic Research (NBER) - 57th Annual Report, New York Sept. 1977, p. 88

[10] W.W. Hogan, S.C. Parikh: Comparision of Models of Energy and the Economy, Energy Modeling Forum (EMF)- Report 1, Appendix D, Stanford University, Stanford Ca. May 1977

[11] A.L. Pugh III: DYNAMO User's Manual, fifth edition, The MIT Press, Cambridge Mass. 1977

[12] K. Plasser: IAS-Benuetzerhandbuch, IAS-Level 2.7, Institute for Advanced Studies, Wien 1976

[13] Bundesminister für Wirtschaft (ed.): Ökonometrische Methodenbank, BMWi-Studien-Reihe 22, Bonn 1978

[14] G. Sadowski: MASH - A Computer System for Microanalytic Simulation for Policy Exploration, URI 17600, The Urban Institute, Washington DC 1977

[15] U.S. General Accounting Office (GAO): Ways to Improve Management of Federally Funded Computerized Models, Report to the Congress: LCS-75-111, Washington D.C. August 23, 1976; and: G. Fromm, D. and W.Hamilton: Federally Supported Mathematical Models - Survey and Analysis-, Washington D.C., 1975

[16] P. Roth, S. Gass, R. Nance: On the Development and Management of Computerized Models; paper prepared for the Symposium on Modeling and Simulation Methodology, Weizmann Institute of Science, Rehovot, August 14 - 18, 1978 (see also article by Roth et al., this volume).

[17] U.S. Dept. of Labor: The Development and Uses of Table Producing Language, U.S. Govt. Printing Office, Washington DC 1975

[18] R. Brennecke: Das MOVE-System, Ein Prozessor fuer oekonometrische Anwendungen in: Datascope, Vol. 5 (1974) 13 Frankfurt 1974

[19] H. Maier/ Computer-Simulation mit dem Dialogverfahren SIMA, Birkhaeuser-Verlag, Basel 1976

[20] R. Ruehle: RSYST I-III, experience and further development in: Atomkernenergie, Vol. 26 (1976), Stuttgart 1976

[21] National Bureau of Economic Research (NBER): ACOS - An Overview, NBER Report DO082, Cambridge, Mass. 1974

[22] R. Hueber; P.C. Lockemann; H.C. Mayr: Architektur von Methodenbanksystemen,

Internal Report, Universitaet Karlsruhe 1976

[23] F.P. Brooks: The Mythical Man-Month. Essays on Software Engeneering, Addison
      Wesley, Reading (Mass.) 1975

[24] see: "What is Modeling and Simulation Methodology?", introduction to this
      volume, or: the NSF-Study by Fromm et.al. (see reference [15], and private
      consulting companies like Data Resources Inc. (DRI) or Chase Econometrics

[25] e.g.: The Evakuation Studies (on behalf of the Bundesministerium für For-
      schung und Technologie) of the 'Pestel: Deutschland-Modell', done by Gesell-
      schaft für Mathematik und Datenverarbeitung (GMD), Bonn and Science Center
      Berlin (West) (both forthcoming)

[26] M.Greenberger, M.A. Crenson, B.L. Crissay: Models in the Policy Process,
      Russell Sage Foundation, New York 1976

[27] H.P. Galler: Optimale Wirtschaftspolitik mit nichtlinearen ökonometrischen
      Modellen, Campus Verlag, Frankfurt 1976

[28] R.F. Wertheimer: Methodical Issues in Building Microsimulation Models,
      IPES-Report 78.207, Bonn, 1978

METHODOLOGY IN SYSTEMS MODELLING AND SIMULATION
B.P. Zeigler, M.S. Elzas, G.J. Klir, T.I. Ören (eds.)
© North-Holland Publishing Company, 1979

SOME CONSIDERATIONS FOR IMPROVING FEDERAL MODELING*

Paul F. Roth, National Bureau of Standards
Saul I. Gass, University of Maryland
Austin J. Lemoine, Control Analysis Corporation

There is a growing awareness of the need for better communication and management techniques to improve the process of developing computerized models. Modeling is defined as the process of solving complex system problems and making
decisions using experimental data generated by a computerized conception of the
system. The process is subtle and sophisticated and often misunderstood.

The Government is a large underwriter of models. The Government's modeling
process has been criticised by several GAO Reports. Some critical issues
involved in improving the Government modeling process are disclosed, evaluated
by expert practitioners, and ranked according to importance: better communication
between users and developers of models is important; special bureaucracies to
address modeling is unimportant. The responsibility for sponsorship of research
is addressed.

1. Introduction

A recent issue of SPECTRUM (IEEE) has an
editorial (Christensen, 1978) concerning
"systems thinking." In alluding to the
state of mind of a "systems thinker" it
notes the shortcomings of systems science
and engineering in terms of the specific
area of modeling. The editorial refers
to issues in the modeling process identified by an MIT workshop on systems
sciences, including shortcomings in validity demonstration, adequacy of data
bases, model compatability, model documentation, and model incorporation of
stochastic effects and uncertainties.
These shortcomings are discussed in the
context of the workshop recommendation
that a "National System Center" be established to conduct needed basic research in the general areas of large
scale systems analysis, modeling methodologies, and tools for policy analysis
and synthesis.

A second recent article, (Swain, 1978)
in "Natural History," on the computerized
ecosystem, describes the increased understanding brought to the Great Lakes
through the use of mathematical modeling
and simulation. The article depicts the
tool of mathematical modeling as very
meritorious but "widely misunderstood by
both scientists and lay persons." The
reason given for the misunderstanding is
the absence of a "common language" about
modeling. The language of modeling is
portrayed as being "unknown...to many managers and even to a good proportion of the
scientific community." Therefore the sub-

ject of modeling seems to "generate
feelings of apprehension and distrust."
The article goes on to characterize a
mathematical model as being more than a
simple organization of coefficients,
equations, programs, etc., but a dynamic
conceptual framework. It generalizes
that a model incorporates a vision of
the future enabling the analyst to evaluate current states and the interrelation of states, and to look ahead to
envision future states. It makes the
very strong point that modeling is a
dynamic process and that as the model
becomes better developed and more data
become available, the model and the information which it generates grow.
Therefore the process is iterative and
dynamic.

An interesting inference drawn from
these articles and others is that there
is growing awareness by the technical
community of the complexity and sophistication of modeling and a growing concern for better understanding of the
process and identification of the problems. What is equally interesting is
that these concepts have been enunciated
in the same month in two very disparate
publications with two very disparate
viewpoints: 1) system science research,
and 2) a particular application. We
may presume that there is a growing preoccupation with the process of modeling
itself and the questions pertaining to
the uncertainties and imperfections in
the process, both in the practitioner
aspect and the interpretation aspect.
Hopefully, we can surmise that we are

---

* Reprinted from the Proceedings of the 1978 Winter Simulation Conference with
  permission of the Institute of Electrical and Electronic Engineers.

at the beginning of a period of intro-
spection and self-examination concerning
the very nature of this tool which has
been developed and become increasingly
sophisticated in the last 15 or 20 years.

Another publication lending further impetus to the awareness of the defects and
shortcomings of the process is the 1976
Report to the Congress from the General
Accounting Office ("Ways to Improve
Management of Federally Funded Computerized Models," August, 1976), which attempts to describe and quantify the loss
to the Government in the poor use, mis-
use, and nonuse of the modeling process,
and which makes recommendations that
certain management practices be developed to improve the Government's approach to modeling. Among these recommendations are:

  "....formulate standards for, and....
  develop and provide guidance to Federal Agencies for improving management of computerized models."

which can be seen to overlap the concerns expressed by both the MIT workshop and the ecological systems application paper.

What we appear to have, therefore, is an
emerging concern with perfecting the process of modeling: that better practices
be invented; that guidelines for managing the process be synthesized, that
all-in-all, the newly emerging, rampan-
tly growing modeling organism be transformed from its adolescent stage into
one of recognizable useful maturity. In
other words, that professionalism and
discipline be given to the technique.

On the other hand, what is not identified as a problem in the literature, or
elsewhere, is any hint of imperfections
in the computerized tools of modeling.
Quite to the contrary, the development
of modeling languages and software can
be perceived as having far outstripped
the capabilities for using these tools.
Therefore, we are confronted by well-
engineered user programs being employed
in a rather undisciplined fashion by
practitioners. What is indicated is that
research and development must be focused
on achieving a catch-up in the maturity
of the application process, not in the
tools themselves. Questions of who
should sponsor such research and who

should perform the research are ones
which must be solved. It is hoped that
this paper will galvanize those who are
concerned with the problem into identifying the sources of resources for such research efforts.

Our concern then is with large scale
computer-based decision making models and
the problem of how to improve their utility. The discussion in this paper deals
with models employed throughout the Government; however, conclusions drawn for
Government modeling can in general be
applied more broadly. The article discusses the extent of Government modeling
activities, indicates how to assess the
utility of large scale models, and summarizes the opinions of various developers and users of large scale models.
It then discusses problems encountered
in defining the research environment
needed for improving model utility.

2. Government Modeling

The U.S. Government is the largest sponsor and consumer of models in the world.
Estimates have indicated that over one-
half billion dollars is being spent annually on developing, using, and maintaining models in the Federal Government.
Large scale models are employed in the
decision-and-policy-making functions of
Government. They can be classified by
type: e.g., mathematical, simulation,
econometric; or by end use: e.g., war
gaming, engineering, resource allocation,
futures projection. Government models
are developed both by the Government and
by contractors such as universities and
private organizations, both profit and
non-profit. The application of complex
decision making models at all levels of
Government and industry is increasing,
due to better trained analysts and development and refinement of analytic decision making methodologies. However, the
final utility of these models has been
questioned, especially in the Federal
Government.

Consider the utility of a model as some
function of its being useful, useable,
and used. ("Computer Simulation Methods
to Aid National Growth Policy," 1975)
A model can be considered useful if it
shows promise of attaining its stated
objectives. It can be considered useable
if it is understandable and plausible to
both technicians and decision makers,

economic to run on a computer and accessible to those who wish to use it. If a model is useful and useable, it has a chance of being used. What are models? Distilling the opinions from a number of experts, modeling can be defined as the process of solving complex system problems and making system decisions using data generated by a computerized conception of the system. The conception can be mathematical; it can be functional; it can be symbolic. The process of modeling is complex and imperfect. It is initially conducted at a high level by sophisticated analysts who are familiar with the problem and who are supported by statisticians, programmers, etc., and who may or may not use problem-oriented languages to implement their model on the computer. The entire process can be partitioned, according to Emshoff and Sisson (1970) into various tasks or phases:

1. Defining the problem,
2. Analysis of data requirements and sources,
3. Formulation of subsystem models,
4. Implementation and debugging,
5. Model validation,
6. Design of simulation experiments,
7. Analysis of results and presentation to management.

Management of the modeling process is similar to management of engineering projects and requires, as does engineering practice, the existence and use of management metrics: specifications, guidelines, standards, and objective planning techniques. The subtleties of modeling, however, have contributed to the lack of development of these metrics. As a result, the Government and others have experienced waste in many steps of the modeling process.

In addition, and as indicated by one of the previously referenced articles, there is a problem in the communication between management and technologists in the modeling process itself. There is no common language of understanding. Therefore, the models and the process are frequently misunderstood. The General Accounting Office (GAO), in its report to Congress, points out that the Government-sponsored models are largely 1) misused, 2) unused, 3) not validated and 4) not documented. GAO's solution to this problem is to request that a body of standards be developed to make the process more manageable and cost-effective. It also proposes that the Government adopt a formal lockstep phased approach to model development. As part of its solution, GAO requested that the Department of Commerce undertake the necessary work to develop the appropriate body of standards. It did not indicate what resources are available to do this. The main questions to confront the Government developer of better practice are:

1) can such standards actually be developed,
2) if developed, can they be deployed and enforced?
3) should the modeling improvement responsibility be organizationally centralized or decentralized even and
4) basically, is this a problem which the Government alone should solve or indeed participate in the solution of at all?

That standards or guidelines could be developed to improve some of the problems alluded to by the GAO, e.g. documentation, seems reasonable. That good management practices can be infused into the Government modeling process in a formal or quasi-formal sense is questionable.

3. Software is Not the Problem

It should be noted here that an obvious target for the application of standards is software. However, in the modeling context, and particularly the simulation application, software shows the least apparent need for standardization. Problem-oriented languages for simulation have, in general, been developed responsibly. They have been engineered. They have internal diagnostics. They exhibit practices of good software engineering. Also, a large majority of programming language applications in the modeling field do not use problem-oriented languages, but FORTRAN, a language currently being subjected to a standardization process. Simulation languages themselves have attained a degree of reliability and sophistication far in advance of their wide use, as indicated by the very slow process of change in simulation language developers over the years. The language GPSS, for example, over the last ten years (which encompasses GPSS 360 through GPSS V, IBM's versions) has not incurred a great deal of change in basic structure; but upwardly compatible enhancements have been made. We see the same pattern in other simulation languages such as SIMSCRIPT and SIMULA, leading to the conclusion that the software has taken its "great leap forward" and is now waiting for the process of modeling to catch up. This situation is tantamount to having built a modern motor car in the year 1800 where the state of the civilization and technology was obviously not ready for the application of such a product! It is quite clear, from reading recent GAO reports and articles, that nowhere does a software problem seem to manifest itself. Therefore, we must look outward from software to the management of the modeling process for a place to begin to implement professionalism and discipline.

4. Expert Opinion Concerning Modeling Improvements

In order to clarify some of the issues involved, a study has been performed which has accumulated, quantified and weighted opinions of various experts, model practitioners, and theorists as to the major issues confronting the subject improvement of the modeling process. (Control Analysis Corporation, 1978) This was done to identify the main issues confronting the Government in model use, but the comments and opinions may largely be generalized to include the entire modeling community.

Opinion briefs were obtained from a group prominent in modeling. The group could be characterized by affiliation:

>   University
>   Non-profit company
>   Profit company
>   Government

or by application area:

>   Analysis
>   Simulation
>   Economics

A good cross-mix was achieved. This group voted on their degree of agreement, or non-agreement, with certain propositions. The results reported here are from partial raw

data, due to publication time constraints. The complete report will have been finished by the publication date of this volume.

Using rudimentary weighting of raw scores, the propositions can be ranked and numbered on a scale of 1 to 18, indicating a range between highest and least agreement averaged over all opinions. Of 18 basic propositions, those showing significantly high agreement (descending ranking) were:

1. Documentation Plan and Guidelines

   Model developers should specify a documentation plan at the beginning of the project and have it approved by the Government Contract Officer's Technical Representative (COTR). The Government should develop documentation guidelines for computer models, similar to the NBS FIPS PUB 38, "Guidelines for Documentation of Computer Programs and Automated Data Systems." (NBS, 1976)

2. RFP (Request for Procurement) Statement of Work

   The RFP statement of work should indicate the technical and management aspects the developer must follow, including specification of the analytical procedures, data to be used, reports required, and briefings that must be given. If a contract is to advance the state of the art or a research area, then it should be so stated in the RFP and a procedure be established to ensure user-developer interaction to ascertain final model specification.

3. Relationship Between Developer and User

   The scope of work should stipulate who the ultimate user(s) will be, and that meetings between the developer and user(s) be held to enhance developer-user concurrence.

4. Modeling Forums of Users and Developers

   The Government should establish modeling forums that deal with

specific application areas and/or
methodologies that are of concern
to Government model sponsors and
users. Whenever possible, a modeling forum should be organized with
the support of the appropriate pro-
fessional organizations and industrial groups.

5. Data Availability

Prior to the issuance of an RFP, most
modeling projects should undergo a
preliminary data availability and
costing assessment, where this assessment would be used by the sponsor to continue or stop the effort.
The RFP should require an explicit
data collection effort to be conducted by the model developer or other
designated group. The availability
of suitable data, as measured at certain milestones, should be a basis
by which the sponsor and developer
determine whether or not the modeling project objectives can be attained.

6. User Training

All modeling projects should explicitly address the training issue.
If formal training is required by
the contract, then training should
include how to use and analyze the
results of the model, along with data
maintenance and program change procedures. If training is not required,
the model specifications should
state why it is not needed.

7. Verification and Validation

A detailed verification and validation test plan should be required of
most modeling projects. The project
reports should describe the results
and their implications to the future
use of the model. Exceptions to a
detailed plan should be based on a
model's complexity and proposed use.

8. Phased Management Approach

The Government should develop the
idea of a phased approach to modeling.

The GAO report suggests a five-phase
approach to the management of a model
development project, with each phase

having a checkpoint that must be
passed before the next phase can be
initiated. Under this approach,
the Government could contract for
all or some of the phases, but
would reserve the right to cancel
the project if the Government felt
that criteria for passing a checkpoint had not been met. The suggested phases are: problem definition, preliminary design, detail
design, evaluation, and maintenance.

9. Definition of Large Scale Models

The Government should develop a
basis for classifying models. Any
future modeling standards or management procedures would then be
applied based on the level of a
model's classification.

Those propositions showing significant
disagreement (ascending ranking) were:

18. Model Testing, Verification, and
     Validation Center

The Government should establish a
center at which certain classes of
models would undergo verification
and validation testing by an independent staff of analysts.

17. Post Review

The Government should establish a
post-review panel that would evaluate a model and provide guidance
to potential users as to the model's strengths and weaknesses.

16. Government Modeling Research
     Center

The Government should create a new
agency, namely a Government Modeling Research Center (GMRC), that
would coordinate and direct some
of the Government modeling research activity. The research activities at the GMRC could include
developing model research goals,
validating present models, developing software and documentation
standards, conducting training
programs, organizing standard
data bases, testing algorithms,
comparing methods, organizing a
library of models, etc. Much of

this research could be done by contractors, but the GMRC would be responsible for coordinating and directing these efforts.

Conclusions can be drawn by characterizing the propositions. There is agreement that improvements are needed in the process of model project initiation (propositions 1, 2, 5); in developer-user communication (propositions 1, 2, 3, 4, 6, 9); and in model quality control (proposition 7, 8).

Negative opinions can be projected into almost total rejection of the concept of increased bureaucratic intervention (proposition 18, 17, 16).

Some interesting assessments may be obtained by examining results categorized by respondent affiliation grouping. The table indicates, by group, the first and second highest, and lowest, ranking propositions.

Respondent Affiliation	Overall Ranking	Proposition Subject
University	1	Documentation plan & guidelines
	4	Modeling forums
	16	Government Modeling Research Center (GMRC)
Non-profit	1	Documentation plan & guidelines
	2	RFP statement of work
	17	Model post-review panel
Profit	8	Phased management approach
	5	Data availability
	18	Model test verification and validation center.
Government	8	Phased management approach
	1	Documentation plan & guidelines
	18	Model test verification and validation center

Particularly interesting is the Government category which indicates favoring a more rigid approach to both the development process and to documentation and disfavoring the establishment of a centralized research function.

In summary, the experts have expressed a consensus opinion indicating concern with both the starting of the modeling process and the communication between model users (decision makers) and developers (model builders).

5. Conclusions

The goals for improvement of the modeling process are clear. They are:

1. Provide tools to improve communication between model users and developers.

2. Improve the model development process through better preliminary definition specification and initiation of modeling.

3. Improve the modeling process through establishment of quality control techniques.

The path to reach these goals is not so clear. Since the early 1970's the goals have been enunciated repeatedly by GAO reports, research reports and the opinions of expert practitioners.

At least three times within that period, the National Bureau of Standards has attempted to secure appropriated funds to underwrite research on improvement of the large scale modeling process and has·failed. The most recent and most successful attempt was an NBS modeling standards initiative in the Fiscal Year 1979 budget report which successfully traversed the budget process but failed final Congressional budget review. The NBS program proposal included the development of guidelines for model documentation, model development, and model procurement, which would have gone far in satisfying the above goals.

At this point, the prospects for near-term unilateral Government agency sponsorship of the undertaking appear dim unless the attention of Government policy and decision makers is focused on the problem. Unfortunately, a problem

must not just be important - it must appear to be important. And the desired
attention can only be focused with a
large expression of concern from the constituency of the problem requiring attention. A possible medium for organizing
the constituency into a unified voice
might be a cooperative research organization such as the "National Systems
Center," proposed by the MIT workshop.
(Christensen, 1978) Certainly such a
center, in its quest for better understanding of systems would more than likely be able to subsume the goal of improving large-scale modeling, and provide the leverage necessary to gain
support for the problem.

6. References

1. "On Systems Thinking," D. Christensen,
Spectrum, Vol. 15, No. 8, August 1978.

2. "The Computerized Ecosystem,"
W. H. Swain, Natural History, August-
September 1978.

3. "Ways to Improve Management of
Federally Funded Computerized Models,"
B115369, U.S. GAO, Washington, D.C.
August 23, 1976.

4. "Computer Simulation Methods to Aid
National Growth Policy," Committee on
Merchant Marine and Fisheries, 94th
Congress, U.S. GPO, Washington, D.C.,
1975.

5. "Design and Use of Computer Simulation Models," J. R. Emshoff and
R. L. Sisson, MacMillan, N.Y. 1970.

6. "A study for Assessing Ways to
Improve the Utility of Large Scale
Models - Task 3 Report: Reviewer
Opinion and Discussion Report of Alter-
nate Improvement Possibilities,"
Control Analysis Corporation, Palo
Alto, Calif., June 15, 1978.

7. "Guidelines for Documentation
of Computer Programs and Automated
Data Systems," Federal Information
Processing Standards Publication
(FIPS) 38, National Bureau of Standards, Washington, D.C. February 15,
1976.

SECTION III
SOFTWARE TOOLS
FOR MODEL CONSTRUCTION

METHODOLOGY IN SYSTEMS MODELLING AND SIMULATION
B.P. Zeigler, M.S. Elzas, G.J. Klir, T.I. Ören (eds.)
© North-Holland Publishing Company, 1979

INTERACTIVE SIMULATION
PROGRAM GENERATION

N. R. Davies*
Industrieanlagen-Betriebsgesellschaft mbH
Department SOP
8012 Ottobrunn bei Muenchen
West Germany

This paper explains how an interactive questionnaire can
be used to extract the essential information which describes
a model of a queueing situation. The character strings
which are supplied in response to the questionnaire are
used to construct syntactically correct statements in
various high-level simulation programming languages.
These interactive program generating systems reduce
considerably the time taken to produce a simulation
program, at the expense of some loss of flexibility in
the approach to modelling. A system is described which
provides alternative questionnaires in the styles of the
commonly available simulation programming languages.

1. INTRODUCTION

Folklore has it that the first high-level programming language was a simulation
language. Programs to simulate complex dynamic queueing situations have
certainly been amongst the first applications to be put onto the computer, and
it is not surprising to find that many individual workers devised their own
"language" or "package" to aid the writing of simulation programs. At least
thirty simulation systems are reported in the literature of the early 1960's
(see Tocher (21) for an early review), of which about seven are widely
available and in use today, usually in greatly enhanced versions.

It did not take long to recognize that certain features of simulation programs
are unique to the particular model being implemented while other features are
common to all models, notably, the time-advance mechanism. For, of course, we
are dealing with the simulation of situations which involve some degree of
randomness. The central problem of discrete system simulation is scheduling the
execution in correct chronological sequence of sections of program which
represent the occurence of random phenomena. It would be nonsensical to have to
write and debug quite complicated routines to handle this dynamic sequencing
every time we implement a simulation model. Hence most workers are more than
happy to utilise a well-proven commercially available simulation programming
system.

---

*Formerly of Portsmouth Polytechnic Computing Centre, Hampshire Terrace,
Portsmouth PO1 ZEG, England.

But let us be quite clear what this involves. Each particular simulation programming language (SPL) provides the skeleton of a program, principally the time advance mechanism, together with a series of routines and data structures which we may use to add the flesh to describe the characteristics of our own models. The SPL provides high-level concepts to help us articulate the unique features of our model, but at the same time it imposes a rigid structure within which to define the dynamic behaviour of the elements of the model. There are in fact some half dozen different skeletal structures, each of which dictates apparently quite different forms of (user-supplied) flesh (see Appendix 1). Clearly, the originating author of each simulation programming system had a particular type of application in mind, and his way of looking at situations is reflected both in the skeletal structure of his system and in the terminology incorporated into his simulation language. To give some idea of the range of nomenclature found in the various SPLs, a selection of terms is presented in Table 1.

CLASS	FILE	ACTIVITY
ENTITY	LINK	BLOCK
FACILITY	LIST	EVENT
STORAGE	QUEUE	LIFE CYCLE
TRANSACTION	SET	PROCESS

Table 1: Sample of Terminology used in Discrete Event Simulation.

The next step in the evolutionary progress was to recognise that all simulation programs do perform certain basic functions, even though these are often concealed by the jargon of individual SPLs. In essence, time-consuming activities which take place simultaneously in the real world are represented by the sequential execution of event routines on a serial computer. Usually, we are dealing with situations where there is competition for scarce resources. What emerge as the essential features of a particular model are:

1. the identities of the resources involved in each activity or event,
2. where these resources are to be found,
3. the length of time for which resources interact with each other or are committed to an activity,
4. and what happens to the resources when a period of cooperative activity terminates.

This information is built into every simulation program, be it a one-off FORTRAN program or a sophisticated program written in a high-level special purpose simulation programming language.

Clementson (4) was the first to design an on-line questionnaire to solicit this essential information about a model, and to use it to generate automatically a syntactically correct simulation program in the target simulation programming language (in this case, ECSL). Since then, a number of interactive program generating systems have followed, notably Mathewson's DRAFT (6, 13, 15) which generates programs in a number of different target SPLs, and Davies' modular interactive system, MISDESM (5, 6), which extends the user interface by providing alternative questionnaires and so covers the range of terminologies and skeletal structures inherent in the commonly available SPLs.

2. INFORMATION CONTENT OF A MODEL

In discussing the various simulation programming languages and the way they affect the user, we tend to loose sight of the fact that, ultimately, all these systems are aids to modelling the same type of queueing situation. Given a particular situation, each of the SPLs can be used to produce programs which are very different in structure and syntax and yet, when run, produce identical results by executing event routines (explicitly or implicitly) which make the necessary changes to the data structures representing the state of the components of the model.

What are the features of the real-world situation that are reflected in the models described by the various SPLs? What information about the situation must be incorporated into the finished model, regardless of the SPL in which it has been described? For of course this information is essential to the process of interactive program generation.

2.1 A Language-Independent Description

There is a strong conceptual link between the event which starts an activity and the event which terminates it. They are separated in time by the duration of the activity; and the entities which must be available to start the activity are the ones which are released when it terminates. Furthermore, the activity is a direct representation of what consumes time in the real world. This inherent unity of the activity concept makes it an ideal building block for defining discrete event simulation models.

This information, which uniquely describes the static and dynamic inter-connectedness of a model quite independently of the terminology or structure of any particular SPL, can be summarised in a set of descriptive units consisting of

1. At least one, and preferably all three of
   a) the activity name
   b) the name of the event which initiates the activity
   c) the name of the event which terminates the activity.
2. For each entity which engages in the activity
   a) the entity name
   b) the name of the queue or set in which the entity is to be found waiting
   c) the number of entities of this type involved, and
   d) the name of the queue or set to which the entity is to be sent when the activity terminates.
3. The duration of the activity

A definition in Backus-Naur Form is presented in Figure 2.

```
<descriptive unit>  ::=  <name list>,<entity list>,<duration>


<name list>  ::=  <start event>,<activity>,<termination event>|
                  ,<activity>,<termination event>|<start event>,,<termination
                                                                        event>|
                  <start event>,<activity>,|<start event>,,|,<activity>,|
                  ,,<termination event>
<start event>  ::= <identifier>
<activity>  ::= <identifier>
<termination event> ::= <identifier>


<entity list>  ::= <entity group>|<entity group>,<entity list>
<entity group>  ::= <identifier>,<from queue>,<how many>,<to queue>
<from queue> ::=  <identifier>|INF
<to queue> ::=  <identifier>|INF
<how many> ::=  a non-negative integer
<identifier> ::=  usual definition


<duration>  ::=  any arithmetic expression, including the usual distribution
                 functions NORMAL, UNIFORM, ERLANG, etc.
```

Figure 2: BNF definition of the Descriptive Unit.

This information is of course essential to any interactive program generator,
part of whose on-line questionnaire must solicit the information from the user.
A file of such descriptive units plays a central role in the Modular
Interactive System for Discrete Event Simulation Modelling (5,6) developed at
Portsmouth Polytechnic. The system is illustrated in outline in Figure 3. An
important feature is that the user has a choice of questionnaire through which
to describe his model.

1. B DEFINE allows the user to define a set of (bound) events in the style of
   SIMSCRIPT or GASP. What entities are released? In what other activities can
   they now take part, and in conjunction with what further resources? Should
   these extra resources not be available, what queue or pool of entities does
   each released resource join?
2. ACD DEFINE asks questions about an activity cycle diagram in the style of
   HOCUS. Name an activity. What entities are needed to start it and from which
   source queue? How long does the activity take to accomplish? To which
   destination queue are the resources subsequently released?
3. B AND C DEFINE is a questionnaire which assumes the user thinks of his
   model in terms of bound events and conditional events to fit into the
   three-phase structure of SIMON. Name an event; is it bound or conditional?
   If bound, what entities are released? If conditional, what entities are
   required?

Each of these questionnaire modules assembles an identical set of descriptive
units for a given situation.

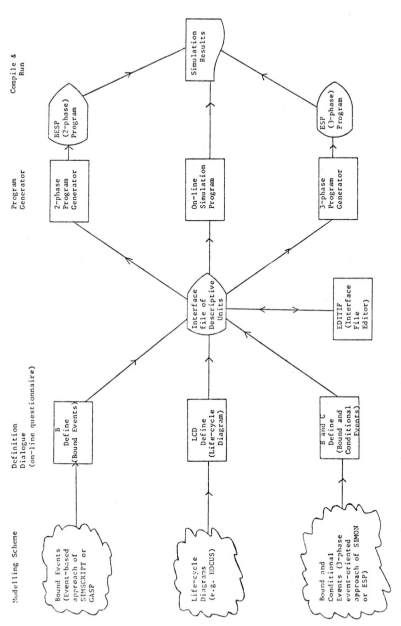

Figure 3:   Modular Interactive System for Discrete-Event Simulation Modelling

2.2 Language- and Run-Dependent Information

A file of descriptive units defining a particular situation probably contains enough information to generate the outline of a simulation program, but a lot more information needs to be supplied if the target simulation program is going to be useful. Much of it must, of necessity, depend on the simulation programming language in which the program is being generated. For example:

1. The descriptive unit identifies entity types but does not mention what attributes they may have. Some of the attributes may well form part of the essential fabric of the model, others may be closely bound up with the queue disciplines, while yet others are dependent on what statistics one individual run is designed to collect.
2. Source and destination queues are named in the descriptive unit, but the user may want to change queue disciplines between runs.
3. Another factor to be considered is the number of each type of entity in the model. Languages such as SIMSCRIPT and SIMULA, which incorporate dynamic storage allocation, may not need this information. ECSL and SIMON require the number of each type of entity to be declared at the head of the program.
4. SPLs vary in the ways they deal with multiple or simultaneous occurrences or instances of an activity.

There is a set of information which defines the parameters of a particular run, the objectives of which will probably dictate

5. the initial state of the model,
6. the queue disciplines,
7. what statistics are to be collected,
8. the (simulated) time for which the model is to be run,
9. seeds for the pseudo-random number generators, and
10. what trace facilities should be produced.

Although much of this information could be classified as language-independent, nevertheless it seems sensible to divide the on-line questionnaire into two phases, the first of which seeks to establish the fabric of the model and to build up a file of descriptive units, while the second phase solicits the details for a particular run, including the features which depend on the target SPL.

There are three major advantages of this split.

1. A simulation experiment often requires many runs which do not change the essential fabric of the model. It would be time-consuming to have to re-enter this information for each run. (This comment ignores the possibilities of editing the target simulation program).
2. The choice of on-line questionnaire, and hence of the terminology used to define the model, can be made quite independently of the SPL in which the target program will ultimately be generated. Thus a two-phase event-based (SIMSCRIPT-like) definition phase can be followed by a process-oriented second-phase to generate a program in SIMULA. This feature is particularly useful in a pedagogical environment.
3. The split emphasises the clear distinction that exists between the model and experimentation with the model.

2.3 A Working System

A modular system to aid the teaching of simulation methodology, as well as to perform some serious simulation experiments at Portsmouth Polytechnic, is illustrated in Figure 3. A file of descriptive units is used as an interface between the two phases of the on-line questionnaire of the program generating system. The three definition modules, which solicit the information needed to build up the descriptive units, have been described in Section 2.1. Four modules use the interface file.

1. An on-line simulator, SIMULATE, gives an event-by-event trace for the model, and is useful for checking the correctness of the descriptive units.
2. EDITIF allows changes to be made to individual descriptive units which have been incorrectly defined. This module has proved invaluable.
3. A two-phase event-oriented program generator, BESPIPG, uses the interface file of descriptive units to produce a simulation program with the skeletal control structure of SIMSCRIPT (Figure 4). The actual target SPL is a two-phase version of the ALGOL-Based package ESP (19). (This reflects the non-availability of compilers on the Polytechnic's computing facilities.)
4. A three-phase event-oriented program generator for ESP, which has the skeletal control structure of SIMON (Figure 5).

The function of the program generators is to construct character strings which are synactically correct statements in the target SPL and into which have been inserted the names of entities and queues which are unique to the particular model and which have been supplied by the user in response to the generator's on-line questionnaire. This is not as complicated as it sounds, since there is a high degree of symmetry in most simulation programs as a result of the pre-defined control structure of the chosen SPL (see Figures 4, 5, 7 and 8 in the Appendix).

In the Portsmouth system, MISDESM, (Figure 3), the three-phase program generator involves two passes of the interface file of descriptive units. The first pass extracts the names of the queues from which entities are needed to start each activity in order to construct the conditional event routines. The activity durations are also extracted and built in to the SCHEDULE statement to trigger the events which terminate the activities. The second pass constructs bound event routines which release the entities into their destination queues.

A descriptive unit requiring a single entity from an INFinite source queue indicates an exogenous event. These are ignored by the first pass, but the second pass inserts an extra statement to SCHEDULE this same event after the inter-arrival time (duration).

The two-phase program generator is slightly more complicated, since multiple scans of the file of descriptive units must be made. The event routines which must be generated for the two-phase structure are of course those which release entities. For each descriptive unit, the set of released {entity, destination queue} pairs is matched with the set of {entity, source queue} pairs required to start each activity. An overlap or union of these two sets indicates that this activity might be able to start if certain extra resources were available. The set of extra resources, for which tests of availability must be built in to the current event routine, is the set of required entities less the entities being released. Only if a particular entity cannot engage in an other activity is the code generated to make it join its destination queue.

The MISDESM system has evolved over a number of years, and figure 3 illustrates the current state of development. The major difficulty in implementing the system has been the lack of interactive facilities other than at assembler level. This has enabled very compact programs to be written, but portability has suffered. A BASIC version for the ICL 2904/2960 computer system has been started, and new modules are planned to generate ECSL and SIMULA programs, pending the installation of these compilers.

3. FUTURE DEVELOPMENTS

There can be no doubt that interactive program generation has brought simulation techniques within the bounds of confidence and competence of a large number of engineers and managers who have, in the past, been daunted by the problems of implementing a simulation model. The activity cycle diagram approach has done much to bridge this communication gap that existed between the man who had the problems and the computer. It is not surprising to find that HOCUS, Clementson's CAPS/ECSL and Mathewson's DRAFT systems have all adopted a common input style based on the activity cycle diagram.

There are, however, certain limitations to this approach. Activity cycle diagrams can only express the simplest logic, AND-ing the availability of entities needed to start an activity. And yet, complex decision rules are the essence of many simulation models. In mitigation it is argued that 90 % of a program can be written successfully by a program generator and that it is up to the user to edit in the peculiarities of his particular model.

Although activity cycle diagrams adequately represent entities and their set membership, the attributes of entities are not represented, nor are more complex data structures such as would be required to represent networks. A step in the right direction is CAPS/ECSL's automatic assignment of attributes needed to define queue disciplines.

Not every situation lends itself to modelling as an activity cycle diagram, and it is often necessary to introduce artificial entities to make a situation fit the modelling scheme - in the process, masking the essential simplicity of the activity cycle diagram approach. A model of road vehicles flowing through a set of traffic lights is an example of a situation which is more readily modelled from an even-based point of view. Much of my own work has been aimed at providing the user with alternative questionnaires, to match his own conceptual background and world view. I should like to see an extension of the constructs available to the user, such as the "facility" introduced by Thesen (20) into his extension of GASP, with program generators to match.

An area of major importance in the development of simulation program generating systems is the statistical design of experiments (Mathewson (14)). The statistical techniques are well established (see for example Kleijnen (11)) and it seems a logical step to generate a program with the appropriate statistical devices built in automatically. De Carvalho and Crookes (7) point out that partitioning a model into "cells" can reduce the amount of computation required to perform a factorial experiment. Those cells or parts of a model, which occur before the first cell in which a particular factor changes, will behave identically from one run to the next. By recording once the behaviour of the earlier cell, a reduction in run-time can be achieved by using the "output" of this early cell's simulation as the "input" when running simulations of the cell in which the factor changes. There is no reason why the cell or activity which is affected by each particular factor cannot be identified automatically (say, from a file of descriptive units), and the model partitioned into cells and the factorial experiment conducted with a minimum of runs. The trade-off is with the storage space required to hold the behaviour patterns which describe the interaction between one cell and the next.

Simulation programming languages, and the program generators for which they are the target language, are already highly specialised in the type of problem for which they can usefully be employed. This is inevitable: as the level of the language and concepts used to describe a situation become higher, so the range of applications narrows. It can be argued that current SPLs are sufficiently high-level for the simulation of most applications with, at the pinnacle, GPSS, HOCUS and the various program generating systems described in this paper. Nevertheless, there are areas where extensions to current systems are desirable. One is the automatic generation of programs to simulate situations which involve networks, for example, the design of tracked document conveyors in a building or the movement of people through a hospital or airport. Small changes to the network or path through the building make necessary substantial edits to the simulation program. A specialised program generating system, able to deal with the data structures which describe a network, would allow a much more flexible approach to the design process.

Another field, in which there is sufficient interest and potential to make it worthwhile investing in highly specialised simulation tools, in the evaluation of hardware/firmware/software trade-offs in the design of computer systems, especially in this age of microprocessors and very large scale integrated circuitry. Here again, data structures to represent networks will be an important consideration.

Simulation program generation is the half-way house between the do-it-yourself high level language and the packaged program. It gives the advantage that ninety or more percent of the program will be generated in a standard, syntactically correct form and guaranteed to run correctly, with the flexibility to define the unique features of a situation by editing in the peculiarities of the model. These developments must and have enhanced the availability of simulation methodology. No longer need we simulate only as a last resort.

Appendix

APPROACHES TO MODELLING WITH THE MAJOR SIMULATION PROGRAMMING LANGUAGES

There are three reasons for examining certain simulation programming languages
in some detail. The first is that the skeletal control structure of an SPL has
a marked influence on the way a user must view his model. It dictates the
questions he should be asking of the situation, and the order in which those
questions should be posed. Secondly, the terminology appropriate to a particular
SPL should be reflected in the style of questions which form the on-line
questionnaire. After all, the questionnaire section of a program generating
system must lead the user through the modelling process, and should therefore
use terms and concepts which are familiar to the user. The third point is rather
subjective: the user should have some feeling for the syntax and structure of
the target language, even though he may be using the program generator to avoid
getting involved in such details. Nevertheless, it is quite likely that he may
wish to edit the source code of the target program, to incorporate some special
feature not covered by the program generator's questionnaire. Some knowledge of
the target language is, surely, highly desirable.

There are no doubt dozens of simulation languages in use around the world, but
only a limited number is widely known, well maintained, well documented and
readily available.

To exemplify the different approaches to modelling, six high-level simulation
systems have been selected for comparison. The survey by Kleine (12) and the
comments by Palme (17) both confirm the feeling that users like what they get,
and so the major languages GPSS (8), SIMSCRIPT (10), and SIMULA (2) must be
included. Structurally, GASP II (19) is so similar to SIMSCRIPT that it has
been omitted from discussion, inspite of the availability of a program
generator (15) and a rehashed version of the language by Thesen (20).

ECSL (4) is the extended version of the Control and Simulation Language (3) and
is a powerful stand-alone language with many novel statements; and of course, it
has the support of the CAPS program generator (4). It is widely used in the
United Kingdom (Bailey (1)). SIMON (9) has been given a new lease of life by
Mathewson's program generator DRAFT (13) and, furthermore, is an example of a
three-phase event-based SPL. Finally, the life-cycle approach of HOCUS (18) has
contributed so much to the design of on-line questionnaires in this context and
to the popularisation of simulation techniques in general.

Discrete system simulation is concerned with the representation on a serial
machine, of time-consuming activities which take place simultaneously or in
parallel in the real world. No attempt is made to monitor continuously the
progress of each activity from its inception to its termination; only the
transitions from one activity to another are represented explicitly. These
instantaneous transitions, during which some facet of the system under study
changes, are called events. Ultimately, every simulation program consists of
event routines which must be executed in chronological sequence. The irony of
the situation is that what consumes time in the real world - the activities -
are not explicitly represented in a simulation program, and so consume no
computer time, whereas the instantaneous (zero time-consuming) transitions
between activities - the events - must be represented by sections of program,
and so do consume computer time.

A.1 SIMSCRIPT

SIMSCRIPT (10), currently in version 2.5, follows closely the idea of "event routines executed in chronological sequence", as illustrated in figure 4. The user supplies the event routines, and SIMSCRIPT provides the time-advance mechanism. Execution normally follows two alternating phases:

1. The time-advance phase
2. The event execution phase.

(In order to isolate the essential structure, no attempt is made to show how SIMSCRIPT deals with such things as simultaneity, nor is a loop shown which would give multiple runs.)

The time-advance mechanism consists of a calendar of future events, together with routines for scheduling the occurrence of some particular event at some future time, and a mechanism for searching the calendar to find which event is due to take place next, so that control can be passed to the appropriate event routine.

What are the implications of this two-phase skeletal structure for the user? From figure 4, we see that he must supply appropriate event routines, and an initialisation section. Experience suggests that the questions he should be asking about the situation he is modelling are:

1. Which events release resources?
   These will normally be the termination of activities, for the duration of which various entities have been co-operatively engaged or committed, or the arrivals of entities into the system from outside. In answer to this question, a list of events can be drawn up, for each of which an event routine will have to be written.
2. What entities need to be represented?
3. When defining each event routine, what precisely can happen to each released entity?
   The event routine might interrogate the availability of "other" resources to ascertain whether a newly-released entity can be re-utilised immediately in performing the next activity or whether to pass it to wait in the next queue. If the necessary "other" resources are available, they must be seized and committed for the duration of this next activity, the terminating event of which should appear in our list from question 1. There may be several possible activities in which the newly-released entity can engage, including a repetition of the activity which has just terminated.

A.2 Three-phase event-based structure of SIMON

The commencement of an activity causes changes to the state of the model, and is therefore just as much an event as the termination of the activity. SIMON (9) distinguishes between bound events which depend for their occurrence only on the passage of time, and conditional events which only take place if certain resources are available. Typically, the starts of activities are conditional, and the terminations of activities are bound to occur some time after starting - releasing any committed resources.

SIMON requires the user to write separate bound event routines and conditional event routines, these latter being preceded by the set of conditions (test of availability of resources) which must be satisfied if they are to take place.

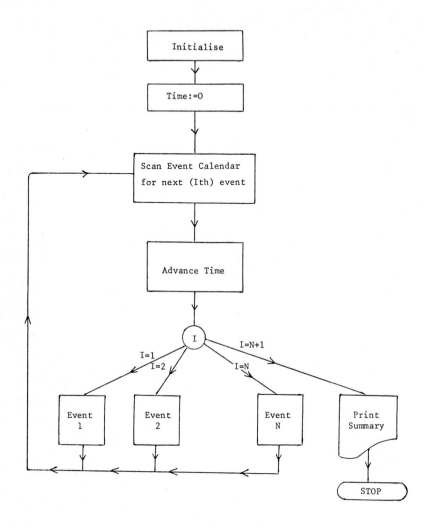

Figure 4:   Control Structure - SIMSCRIPT

The skeletal control structure (figure 5) shows three phases:

1. the time-advance phase,
2. execution of the next bound event, which releases resources,
3. and execution of any conditional events for which the requisit resources are available.

This third phase is often referred to as an activity scan, since its main function is to start activities if or as soon as their required resources become available.

From the modellers point of view, SIMON implies that certain questions be asked.

1. What events take place?
2. Which are bound and which conditional?
   As previously indicated, events which start activities tend to be conditional; those which terminate activities or make resources available tend to be bound.
3. What entities and sets (queues) need be represented?
   SIMON provides appropriate data structures to represent entities with attributes in sets. In writing the event routines (in ALGOL 60 or FORTRAN, since SIMON is a package of subroutines),
4. each bound event releases its resources into the next queue, while
5. conditional events test the availability of resources in the various queues in order to start an activity. Starting an activity involves seizing the requisit resources and scheduling the occurrence of the event which terminates the activity after an appropriate time interval.

We note that the possibility exists, under SIMON's three-phase structure, for an entity to be released into a queue by a bound event and removed from the queue by a conditional event without time being advanced.

A.3 Activity or Entity Cycle Diagrams: HOCUS

A system which has the same three-phase structure as SIMON is the Hand of Computer Universal Simulator, HOCUS (18). This is not a "language" a such, but a program which automates the movement of resources through a sequence of active and idle states as represented on an Activity Cycle Diagram. The life of each type of resource or entity is represented on the diagram as an alternation of queues (idle or waiting state) and activities; arrows are used to link the activities and queues, and so trace the path or life cycle of each entity type through the model. An activity will of course be common to the life cycles of the various entities which come together or co-operate in the activity. An example of an activity cycle diagram is shown in figure 6.

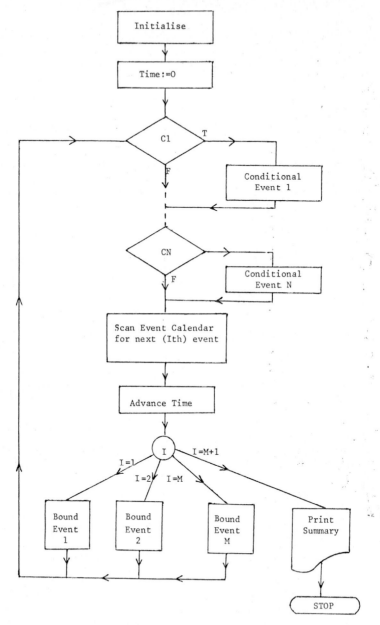

Figure 5:   Control Structure - SIMON, ESP

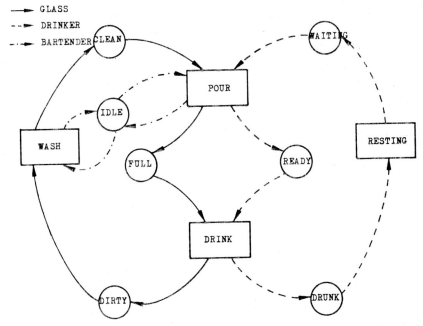

Figure 6:   Activity Cycle Diagram for Pub.

If counters are placed on the diagram to represent the individual entities, a hand simulation can be performed by moving the counters according to the following three rules:

1. Try to start each activity.
   Examine the queues which feed the activity. If the requisit resources are available, the counters are moved from the source queues into the activity (seize the resources) and the time at which the activity will terminate is calculated and noted. When no more activities can start,
2. Advance Time to the termination time of the activity which is due to terminate soonest.
3. Terminate all activities due to terminate at this new time by moving the counters (releasing the resources) into the next queues.

The steps in building a HOCUS model can be summarised:

1. Identify the resources.
2. In what queues and activities does each resource spend time?
   Bear in mind that we can usually predict the length of time a resource will be tied up in an activity, whereas it may have to wait in a queue until something unpredictable occurs, such as the release of some other resource which is needed to start the next activity.
3. Draw an Activity Cycle Diagram.
   It is often said that the solutions to many problems are obvious once the diagram has been drawn. The final step will often be
4. Simulate, on the computer or by hand.

A.4 Extended Control and Simulation Language

ECSL (4) has essentially a two-phase event-oriented skeletal control structure, though the preferred approach to modelling follows the activity cycle diagrams of HOCUS.

Confusingly, ECSL describes its event routines as ACTIVITIES, each of which consists of a "chain" of conditional statements followed by the statements which change the state of the model. If the conditions test the passage of time only, the ensueing instructions represent a bound event; if they test the availability of resources they represent a conditional event such as the start of an activity. This is quite sensible if we think of time as a scarce resource like any other.

The control structure of ECSL can be thought of as shown in figure 7, in which the two phases are:

1. the time-advance
2. and the activity scan.

No attempt has been made to show how the flow of control from one activity to the next can be altered by such statements as REPEAT, ADVANCE or RECYCLE.

ECSL includes a number of novel, high-level statements for manipulating sets (queues), of which the conditional FIND and the construction.

FROM <set name 1> INTO <set name 2> AFTER <time interval> are the most noteworthy.

The FROM ..... INTO ..... AFTER ..... construction exposes ECSL's hidden third phase, and clearly reflects the development of the language in the direction of HOCUS's activity cycly diagram. But the main reason for adopting this modelling approach is the existence of the Computer Aided Programming System (CAPS), a program generator for ECSL for which the on-line questionnaire interrogates the user about an activity cycle diagram, as described in section A.3 on HOCUS.

A.5 SIMULA 67

SIMULA (2) is a sophisticated general purpose programming language offering simulation facilities such as set handling and time advance as natural extensions of ALGOL 60, which forms a subset of the language.

It is usual to refer to SIMULA as a process-oriented SPL, a process being a description of the attributes, events and activities in which an object takes part. The important new idea in SIMULA is that a PROCESS routine defines both the static and the dynamic aspects of the object it describes. Local (static) data structure declarations are followed by statements of two sorts: those which manipulate the data structures, and those which determine the time dependance of the process. A process is said to be "passive" when it is waiting - for example, for an activity to terminate or for some other resource to become available - and "active" when changes are being made to the state of the object - in other words, when an event is taking place. A (dynamic) statement such as

HOLD < time interval >

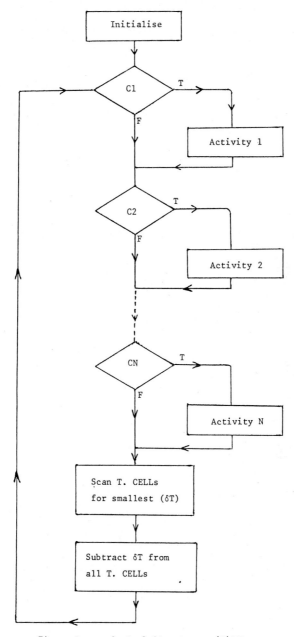

Figure 7:   Control Structure - (E)CSL

causes execution of the process routine to be interrupted until simulated
time has advanced by the specified interval, whereupon execution resumes from
the point in the process at which it was broken off. Interaction between
processes, for instance when a process releases an entity on termination of an
activity, is achieved by ACTIVATE-ing any other process which may be passively
waiting for this entity.

Many different processes and many instances of the same process may co-exist
in various stages of execution. SIMULA schedules the transfer of control between
the process routines and executes the various sections of the processes in the
correct chronological sequence.

Figure 8 shows the essential structure of SIMULA to be two phase,

a) the time advance phase, followed by
b) the active phase of a process.

Note the multiple re-entry points to the processes, and the possibility of
transfer of control to another process addressed by the ACTIVATE statement.

From the users point of view, SIMULA offers considerable flexibility to mimic
the structures of SIMON and HOCUS, ECSL or even SIMSCRIPT.

A.6 General Purpose System Simulator

GPSS (8) is perhaps the highest-level simulation system in existence. It is
perhaps worth commenting that the higher the level of the language, the less
flexible it becomes and the narrower the range of problems for which it can be
used. But this does not detract from its excellence as a simulation programming
language, which is well supported and widely used.

Basically GPSS is a process-oriented SPL and, like HOCUS, it has a diagramatic
representation. Unfortunately, it has a vocabulary all of its own, and familiar
terms such as activity, entity, set and even event are foreign to it.

GPSS views models in terms of "transactions" (material-like entities) which
travel through a set of "facilities" and "storages". A facility can accomodate
only one transaction at a time, and hence a queue may form in front of it; a
storage can accomodate many differnt transactions at one time, up to a maximum
which may be specified. The life-cycles of transactions are traced through a
succession of functional "blocks", which describe the generation, movement,
queueing, statistic-collecting and termination of transactions. Some of these
blocks clearly involve a time delay, which may be predictable (ADVANCE) or
unpredictable (QUEUE in front of a SEIZE block). The language processor provides
the necessary code and structures to sequence the movement of transactions
through their life cycles at the appropriate chronological moment.

An important feature of GPSS is that each block has a unique geometrical
symbol which translates into a single language statement, facilitating both
the modelling and programming process.

The control structure of GPSS is essentially three-phase. Several chains or
lists of transactions are maintained, the structurally important ones of which
are the current events chain, the future events chain and the various delay
chains. Current transactions are those which are due to move on at this moment
in simulated time, or were blocked at an earlier time. Future transactions will
be held up in some block until some time in the future. Delay chains are

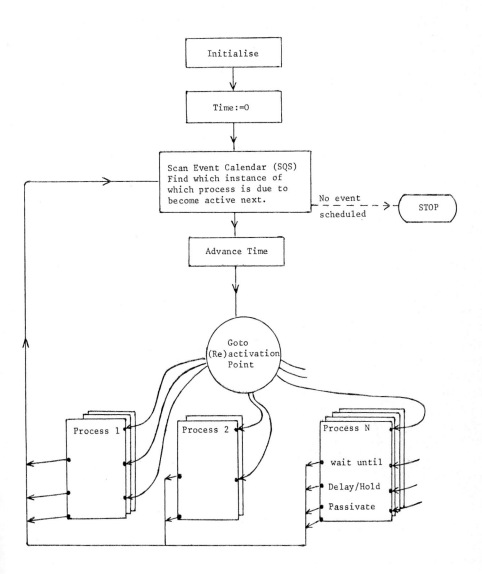

Figure 8:   Control Structure - SIMULA 67

associated with each facility, storage or logic switch, and contain those
transactions held up by this particular modelling element.

Broadly speaking, and ignoring transactions which have been pre-empted or are
being matched, GPSS proceeds as follows:

1. Time advance phase.
   The transaction(s) at the head of the future events chain (which has the
   lowest future time-of-occurrence) is transferred to the current events chain.
   For each current transaction,
2. Attempt to move the transaction forward to or through the logically next
   block.
3. On leaving a facility, storage or logic switch, thereby changing its status,
   transfer entities from the associated delay chain to the current chain.

In the sort of terminology we have been using to describe the other SPLs, the
release of a transaction resource from an activity-like facility enables the
transaction to attempt to start the next activity, and enables other
transactions to reutilise the facility, by making them current.

REFERENCES

1. Bailey J. E.: Survey of Digital Simulation Languages and their Machine
   Implementation. M. Tech Dissertation, Brunel University, Uxbridge,
   Middlesex, England (1973).

2. Birtwhistle G. M.: SIMULA BEGIN. Petrocelli Charter (1973).

3. Buxton J. N., Laski J. G.: Control and Simulation Language. The Computer
   Journal. Vol. 5 p. 194 (1962).

4. Clementson A. T.: Extended Control and Simulation Language/Computer Aided
   Programming System. Lucas Institute for Engineering Production, University
   of Birmingham (November 1977).

5. Davies N. R.: A Modular Interactive System for Discrete Event Simulation
   Modelling. Proceeding of the Ninth International Conference in Systems
   Sciences, Hawaii (January 1976).

6. Davies N. R.: On the Information Content of a Discrete Event Simulation
   Model. Simulation Vol. 27 p. 123 (October 1976).

7. De Carvalho R. S., Crookes J. G.: Operational Research Quarterly. Vol. 27
   No. 1i p. 31, (1976).

8. Gordon G.: System Simulation (Second Edition) Prentice-Hall (1978).

9. Hills P. R.: SIMON 75 Manual. P. R. Hills (Consultants) Ltd., England (1975).

10. Kiviat P. J., Villanueva R., Markowitz H. M.: The SIMSCRIPT II Programming
    Language. Prentice Hall (1968).

11. Kleijnen J. P. C.: Statistical Techniques in Simulation Part I and II.
    Marcel Dekker (1977).

12. Kleine H.: A Second Survey of Users' Views of Discrete Simulation Languages.
    Simulation Vol. 17 p. 89 (August 1971).

13. Mathewson S. C.: Simulation Program Generators, Simulation Vol. 23 p. 181
    (December 1974).

14. Mathewson S. C.: Technical Comment, Simulation Vol. 28 p. 96 (March 1977).

15. Mathewson S. C., Allen J. H.: DRAFT/GASP - a Program Generator for GASP.
    Proceeding of the Tenth Annual Simulation Symposium, Tampa, Florida (1977).

16. Mathewson S. C., Beasley J. E.: DRAFT/SIMULA Proceedings of the Fourth
    SIMULA Users Conference. National Computer Conferences (1976).

17. Palme J.: Technical Comment. Simulation Vol. 17 p. 94 (August 1971).

18. Poole T. G., Szymankiewicz J. Z.: Using Simulation to Solve Problems.
    McGraw-Hill (1977).

19. Pritsker A. A. B, Kiviat P. J.: Simulation with GASP II. Prentice Hall
    (1969).

20. Thesen A.: The Evolution of a New Discrete Event Simulation Language for
    Inexperienced Users (WIDES). SOFTWARE Practice and Experience. Vol. 7 p. 519
    (1977).

21. Tocher K. D.: Review of Simulation Languages, O.R.Q. Vol. 16 No. 2, p. 189
    (June 1964).

22. Williams J. W. J.: The Elliot Simulator Package. The Computer Journal.
    Vol. 6 p. 328 (1964).

METHODOLOGY IN SYSTEMS MODELLING AND SIMULATION
B.P. Zeigler, M.S. Elzas, G.J. Klir, T.I. Ören (eds.)
© North-Holland Publishing Company, 1979

COMBINED CONTINUOUS/DISCRETE SYSTEM SIMULATION
LANGUAGES --- USEFULNESS, EXPERIENCES AND
FUTURE DEVELOPMENT

François E. Cellier
Institute for Automatic Control
The Swiss Federal Institute of Technology Zurich
Physikstr. 3
CH-8006 Zurich
Switzerland

Combined system simulation is a relatively new technique for the simulation of a class of systems
having properties suitable to both continuous system simulation and discrete event simulation, two
techniques well known to the simulation community.
This combined technique has first been proposed by
Fahrland [13,14].

This paper surveys the techniques and methodology
involved in this simulation approach. Major aspects
considered are numerical behaviour and information
processing. It is shown that this technique is
applicable to a much larger class of problems than
originally suggested by Fahrland [13,14].

I) INTRODUCTION:

The term "combined simulation" is not yet sufficiently well understood in the literature to mean one and only one specific methodology or problem class. For example, one can find references where
combined simulation is used as synonym for hybrid simulation. This
term, therefore, first requires some definition to clarify how it is
going to be used in this paper.

If one speaks of simulation as a technique one usually thinks of a
specific solution tool (digital simulation, analog simulation,
hybrid simulation). On the other hand the term "system simulation"
refers to a specific class of systems under investigation (continuous system simulation, discrete system simulation). However, as
early as 1967 Kiviat [17, p.5] stated that it is common to find the
terms "simulation" and "system simulation" used interchangeably. In
this paper we do not have primarily a specific simulation methodology in mind, but rather the simulation of one specific class of
problems which we call combined systems. However, restricting ourselves to fully digital solutions only, simulation of this class of
problems does suggest the use of a specific simulation methodology
which we are going to discuss in detail. Although the term "combined
system simulation" is thus appropriate, we will use the term
"combined simulation" as well for simplicity.

It remains to define what the term "combined systems" means precisely. It can be paraphrased as follows:

*Combined systems are systems described, either during the
whole period under investigation or during a part of it, by
a fixed or variable set of differential equations where at
least one state variable or one state derivative is not continuous over a simulation run.*

Using this definition the famous pilot ejection study (which is probably the best known test case for continuous system simulation)
will also fall into this class of problems, since the acceleration
of the ejector seat and the first time derivative of its angular
position (both state derivatives in the system's definition) are
discontinuous at the moment when the ejector seat is disengaged from
the mounting rails.

The most comprehensive volume on combined simulation published to
date [27] cites two examples of continuous systems -- the above mentioned pilot ejection study and an analysis of a slip clutch. Both
do belong to the class of combined systems according to our new
definition. This shows that the definition used here is not entirely
in accordance with the "common" use of this term (as a matter of
fact, a proper definition for this term has never been given!). We
must redefine the term "continuous system" as well to keep it consistent with our definition for combined systems.

*Continuous systems are systems described by a fixed set of
differential equations with state variables and first state
derivatives both being continuous over the whole simulation
run.*

This definition restricts the term "continuous system" to a more
narrow sense than is commonly used.

According to these definitions for combined and continuous systems,
most of the more complex "continuous" systems belong to the class of
problems under investigation. The motivation for this definition
will be given in due course.

As the traffic control example presented in [8] shows, one and the
same physical system may be modeled either in an entirely continuous, entirely discrete, or in combined fashion. The term
"combined system" is, therefore, rather an attribute of the selected
model than of the underlying physical process, and we should thus
better talk of "combined models" than of "combined systems". As
explained in [34], one should in any event not expect to simulate a
physical system but rather a model derived from the physical system
via an experimental frame (within which data can be collected representing the behaviour of the real system under specified experimental conditions). Hopefully, under novel experimental conditions
the constructed simulation program will produce data representative
of the data to be observed when the real system is observed under
these new conditions. However, since we will disregard the problem
of modeling in this paper entirely, we shall use the terms "model"
and "system" interchangeably.

II) HISTORICAL DEVELOPMENT:

Surveying simulation languages and packages for combined system simulation available on the software "market" we may find that most of them are extensions of existing "pure" discrete simulation languages/packages. As examples we may mention:

```
GASP-II          --> GASP-I .
SIMSCRIPT-II.5   --> C-SIMSCRIPT
SIMULA-67        --> CADSIM.
```

The reason for this is the following: Although a numerically well performing package for continuous system simulation is much more difficult to achieve than one for discrete system simulation, the language structures for the latter are much more complex than for the former. Thus, extending discrete simulation languages to encompass combined problems is a much easier task to achieve than extending a continuous simulation language for that purpose.

Extensions of discrete simulation packages have been performed in most cases either by the original designers or at least by former users of the original software. However, these people usually having a background in operations research, normally do not consider the requirements of systems' analysts for continuous systems from either the numerical or information point of view. For example, one may find that one specific integration algorithm has been coded into the control routine, or that no provision has been made for parallel structures, adequate run-time control procedures, etc..

We have already seen that most so-called "continuous" systems are really "combined" systems according to our definition. On the other hand there exist many systems, which may be conveniently described by purely discrete simulation elements. As a result there is a much greater impact of combined simulation on the treatment of continuous processes than of discrete processes. This allows for the conclusion that the state-of-the-art of combined system simulation languages is by no means satisfactory yet.

III) USEFULNESS OF COMBINED SYSTEM SIMULATION:

In references [14,15] it has been shown, that there exist problems which cannot be modeled in a proper way by either purely discrete or purely continuous simulation elements. (The author believes, however, that for most problems it is possible to find either an entirely discrete or an entirely continuous formulation, but the work required for converting the problem may be considerable and even may not be desirable.) Examples given in the above references include a steel soaking pit and slabbing mill and also a chemical batch process. The arguments given in these references to justify the new combined approach to these processes are certainly correct. However, we shall show that the needs for combined simulation languages are even more evident and elementary than explained in these references.

Reference [30] describes control of the motion of trains by SCR's (silicon controlled rectifiers). For this purpose a current has to follow a sine wave within a prespecified tolerance range. This was achieved by controlling the SCR's in a way that the current always switches back and forth between basically two different models where reaching the tolerance bound is the condition for switching to the other model. A simulation using CSMP-S/360 required approximately US $400 for one half period of the sine wave, whereas a simulation using GASP-V required only US $13 for two whole periods of the sine wave. (The CSMP program was coded using only standard features offered by the CSMP language.) Thus the reduction factor in computing time was more than 100 for this example. The reason for this is that CSMP utilizes the step size control mechanism of the integration algorithm for event location which is a rather inefficient method for location of state-events and an extremely inefficient method for location of time-events. (For the definition of the terms state- and time-event see [27].) Any other continuous simulation language would behave in the same way as CSMP, so this is not a shortcoming of the particular language CSMP, but a problem of the inadequate underlying solution technique applied.

Fig.1 depicts schematically what happens to the integration step size at event times.

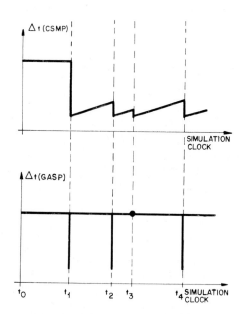

Fig.1: Integration step sizes of continuous and combined languages versus the simulation clock.

Using a continuous simulation language the step size will be reduced heavily when a discontinuity is encountered since the integration algorithm is unable to compute a step properly over discontinuities. This fact has been used in these languages to "localize" event times. However, since the program cannot know that there is a discontinuity taking place (no language element is provided to explain this to the system) the algorithm will "think" that the set of equations suddenly became extremely stiff and reduce the step size to cope with the new situation.

Having located the event, the algorithm will carefully explore the possibility to make the step size larger again, but is not allowed to enlarge the step size immediately since this would lead to instability behaviour in a real stiff case.

A combined simulation language, on the other hand, will provide for a language element to describe discontinuities. Now the step size will be reduced by an event iteration procedure inherent in the program to locate the unknown event time in case of state-events (events 1, 2 and 4 on Fig.1) taking place, whereas the step size will simply be reset to the known event time in case of a time-event (event 3 on Fig.1) taking place. After accomplishment of the event the integration algorithm will be restarted and use the internally provided (and hopefully efficient) algorithm to obtain a good guess for the new "first" step size to be used. Note, that the independent axis on the graph denotes the simulation clock and not CPU-time.

Considering the numerical aspects of the problem one should describe the above mentioned control system rather by combined simulation techniques than by purely continuous simulation. This gives the required motivation for our redefinition of combined and continuous simulation stated in the introduction. (Note that the example would definitely belong to the class of continuous systems according to the "common" use of the term.) Considering the aspect of information processing CSMP obviously offers modeling elements (such as a switch function) which it is unable to preprocess into properly executable code. Again this is not a problem of the language CSMP, but holds for all CSSL-type languages. Some of the languages (like DARE-P [18]) are somewhat more modest in the facilities they offer in this respect, for which they are blamed by many users. The author, however, believes that it is more faithful to offer few facilities than to write a beautiful manual offering many nice features, which effectively cannot be properly used. On the other hand the reaction of these users proves that the facilities are useful and needed. For this reason the author feels, that in a future revision of the CSSL specifications [36] combined simulation facilities should be taken into account, opposing herein to the opinion expressed in [1]. (A revision of the CSSL specifications will be necessary anyway, if for no other reason, the original definition contains over 40 syntactical errors as shown in [22].)

As can be seen from the previously stated: The problem of combined simulation can be subdivided into the numerical aspects (executability of the run-time system) and into the aspects of information processing (definition of the descriptive input language). These two problems are now to be considered more carefully.

IV) NUMERICAL ASPECTS:

IV.1) Structure of the run-time package:

Experience has shown that for the execution of combined simulation the following concept is to be used: A combined problem may be sub-divided into

a) a discrete part consisting of all elements used for discrete simulation,

b) a continuous part consisting of all elements used for continuous simulation and

c) an interface part describing the conditions when to switch from (a) to (b) and vice-versa.

During the execution of a combined simulation we are, therefore, either performing entirely discrete simulation (with its well known properties) or entirely continuous simulation (with its also well known properties), whereas execution of simultaneously combined continuous and discrete simulation does not exist. Thus a combined simulation run-time package must be composed of

a) a discrete simulation run-time package,

b) a continuous simulation run-time package and

c) some algorithms describing the activities to be taken, when branching from (a) to (b) or vice-versa is required.

The numerical requirements for the subsystems (a) and (b) are both well known and discussed on many occasions and, thus, need not be considered here again. An excellent survey of the major simulation systems for problem class (a) is [19], whereas the problem class (b) is surveyed in [1,4,20]. Once this structure has been understood we can restrict ourselves merely to combining previously developed software for discrete and continuous simulation to obtain a good run-time package for combined simulation as well.

In the following we will restrict our view on subsystem (c).

A) Conditions for changing to continuous simulation when executing discrete simulation:

Let the simulation clock be advanced to event time t1. Executing discrete simulation means that the system is about to perform event-handling at time t1. We have to execute discrete simulation until all events scheduled for time t1 have been performed. We have then to switch to continuous simulation if there are differential equations currently involved in the combined simulation (for some intervals of time there may be none). Otherwise we reset the simulation clock to the next event time and continue with executing discrete simulation until again there are no events left to be performed for this new event

time. Therefore, no special algorithms need to be developed for
this case. After event handling being performed the integration
algorithm needs to be restarted. This is especially important in
case multi-step methods are being used.

B) Conditions for changing to discrete simulation when executing
continuous simulation:

Continuous simulation has to be performed either up to the next
scheduled event time (for time-events) or until a state-
condition is met triggering execution of a state-event,
whichever comes first. In both cases the step-size control
mechanism of the integration algorithm has to be disabled. In
the former case (handling of a time-event) the step-size simply
has to be reduced down to the scheduled event time, in the
latter case (handling of a state-event) a new step-size control
algorithm must be activated for iteration of the solution to the
unknown event-time. Again these algorithms are not really new.
Any good iteration procedure (like Newton-Raphson) can solve the
problem. The author recommends a combination of the inverse
Hermite interpolation (fast convergence) with Regula-falsi (unlimited convergence range). This iteration scheme has been described in detail in [7].

C) Selection of the initial subsystem:

Having discussed the conditions for branching from (a) to (b)
and vice-versa it remains to determine the subsystem to be used
first at initialization time t0. The rule is simply to start
with the discrete subsystem. This will then check whether there
are any events to be treated at time t0 and if not transfer
control to the continuous simulation package in which case the
taken activity would be none (this under the assumption that
differential equations form part of the system's description at
time (t0+dt), otherwise proceed as described above).

IV.2) Unsolved problems:

In the author's opinion the best among existing coded packages
utilizing the ideas above is GASP-V [7,8]. Many problems have been
tested using this software and the results were quite promising.
There are still two unsolved problems:

A) Taking the definition of Pritsker [27] for event times:

*"An event occurs at any point in time beyond which the
status of a system cannot be projected with certainty"*

it is clear that an infinite density of events must not occur.
This may, however, happen in at least the following two cases:

a) A system is modeled by a set of PDE's and discontinuities
   exist. In this case the discontinuity may "walk" through
   space with the time and can no longer be localized in the

way proposed in section IV.1. As an example let us consider
a long electric wire where a current is imposed at one end
which suddenly (at time t1) changes its value. This discontinuity will remain in the system for some time and
"flow" through the wire. If there are effects of reflection
assumed at the other end, it may even remain in the system
forever. Thus, in this example we will find, that for any
instant of time t t1 the system will be discontinuous at one
particular point in space (x1) which is moving around.

b) The behaviour of the continuous subsystem is stochastic in
nature. The spectrum of a random number stream has infinite
frequencies which has the effect that it is nowhere differentiable. If such a random number is superposed to the
input of an integrator, we face the problem mentioned above.
This holds of course only for stochastic behaviour of the
continuous subsystem and not for the discrete subsystem.
Stochastic interarrival times of customers to a queue, for
instance, will not effect the numerical behaviour of the
system, since new samples for the random numbers are only
computed at event times, whereas in between these variables
are constant.

Zeigler [34, Chapter 9] has shown that the existence of an infinite density of events always results in an illegitimate
model. In the case of (a) it is, theoretically, always possible
to respecify the model so that the new equivalent model is no
longer illegitimate. In this new formulation the propagation of
discontinuities will follow the axes of the coordinate system.
This is well known as the "method-of-characteristics". In the
case of the linear wave equation we know that the characteristics are straight parallel lines and the required variable
transformation is easy to achieve. For complex situations (nonlinear cases), however, to find the characteristics of the
problem (which are now curves bended in time and space) is almost equivalent to solving the entire problem. Thus while we can
solve the problem (a) theoretically, in practice the required
computations for obtaining the variable transformations are extremely tedious and may prevent us from doing so.

Therefore, we usually find another solution for this problem: In
using the method-of-lines approach [3,6] we found that the integration over time is not much effected in most applications by
these discontinuities whereas the computation of the spatial derivatives is heavily disturbed. Therefore, we first try to
identify (for each step) in which discretization interval the
discontinuity is situated at the moment, then we split up the
region and compute the spatial derivatives independently for the
two parts lying to the left and to the right of the discontinuity. This procedure can easily be expanded for several space
dimensions as well.

The case of (b) is in principle more difficult to treat.
Zeigler's characterization of illegitimate models was developed
only for discrete event models. However, his discussion of the
intrinsic limitation of the class of continuous systems which
can be simulated by digital computers [34, Chapter 5] and [35]

may be applied to the present problem. According to this analysis, there must always be a non zero interval separating computer updates of the model's state. Thus the computer must guess
what the behaviour of the model is in the interval separating
computational instants -- the problem of "bridging the gap".
Since the computer is given a description of the model components and their coupling it can guess correctly only if certain
conditions enabling perfect interpolation in the gap hold. Polynomial trajectories, commonly assumed in integrating differential equation models, serve this purpose.

In the case of stochastic continuous models it is not easy to
justify the assumption of polynomial trajectories. For example,
if the model contains a white noise component then no means of
bridging the gap exist in principle. This is because, by definition, the correlation between sample values, however closely
spaced in time, is zero. Even if the noise is not white, current
numerical methods are not geared to exploiting autocorrelations
specified by the model for optimum choice of integration step.
As a result, most step size control algorithms will produce extremely pessimistic guesses for the step sizes to be used, resulting in high computational costs.

We found the following approach useful in many applications:
First we compute one run by setting the noise to zero using
variable step integration (the continuous subsystem is now deterministic). In this run we collect statistics (histogram) of
the utilized step sizes dt. From the cumulative frequency curve
we select the 0.1 level point (10% of the step sizes fall below
this point). Now we compute a new run, this time with inclusion
of the noise, where we keep the step size dt fixed at this 0.1
level point. A disadvantage of this solution is, of course, that
we now have no measurement for the quality of the approximation.
We must thus be very careful in the interpretation of results
obtained in this way. Furthermore, the proposed method can be
applied only, if the signal/noise ratio is high. For a low
signal/noise ratio we do not know any good numerical technique
to go round this problem.

B) As explained by Elzas [12] the user of a simulation package
wishes either to obtain reliable results or have a "bell" ring
when an algorithm is unable to perform proper work. Under no
circumstances does he want to obtain results which are wrong. So
far this can be guaranteed with a high confidence in the case of
ODE problems only, whereas for PDE problems numerical difficulties need not necessarily be detected by the package, resulting
in inaccurate or even entirely wrong results. More research
needs to be devoted to this problem. It arises from the fact
that we always use a fixed grid for the spatial discretization
and thus have no control on the error resulting from this discretization. Adaptive algorithms would be required (similar to
the variable-step algorithms for numerical integration) to solve
this problem. Some attempts have been made in several places to
find a solution, but these have not been successful so far. Also
in this respect, packages like GASP-V [7,8] or FORSIM-VI [3] may
prove very useful, since they allow to design new experiments in
a very flexible and simple manner.

V) ASPECTS OF INFORMATION PROCESSING:

So far we have discussed the numerical behaviour of a run-time system able to perform combined simulation. Now it remains to question
what is the easiest and most convenient way for the user to formulate combined problems to the computer, so that the computer will be
able to produce properly ececutable run-time code. For this we will
have to identify the structural elements of combined simulation languages.

V.1) The elements of the language:

A combined simulation language will primarily consist of the well
known elements of continuous and discrete simulation languages.
There are few additional elements required to weld these two subsystems together.

A) The state-event:

The only essential new element is the state-event describing
conditions of the continuous subsystem status when to branch to
the discrete subsystem. A typical situation is illustrated in
the following:

> *When the angular velocity of a DC-motor crosses a
> threshold of 1500 RPM in the positive direction, the mo-
> tor has to be loaded.*

This situation could, for example, be coded using a 'CONDIT'-
statement in the continuous subsystem:

```
CONTPROCS
    ...
    CONDIT EV1: OMEGA POS CROSS
           1500.0 TOL=1.0E-3 END;
```

and the reaction to this would then be coded by an event description in the discrete subsystem:

```
DISCPROCS
    EVENTS
        EV1: TL := 200.0 END;
    ...
```

(the torque load (TL) is to be reset to 200.0). The CONDIT-
statement is similar to a CSMP FINISH condition, except that the
time of the crossing is iterated until a prespecified tolerance
is met (TOL=1.0E-3), and in that the simulation run is not terminated, but control is handed over to the discrete simulation
system. After event handling as described by the discrete subsystem (DISCPROCS) control is returned to the continuous subsystem (CONTPROCS) where the new value of TL will be used somewhere
on the right hand side of the equal sign.

B) Operations of the continuous subsystem on the discrete subsystem:

There are none.

C) Operations of the discrete subsystem on the continuous subsystem:

It is most commonly found that not only parameters of the continuous subsystem (as the torque load TL above) change their values at event times but that some of the equations are replaced by others. This situation can be taken care of by the following language elements:

  a) The "one out of n" situation:

  There are n possible "models" out of which one is always active. This situation can best be expressed by a CASE-statement:

                    CASE NMOD OF

  where NMOD is an integer number pointing to the currently active model. This language element is used in general to describe n different functional ways of behaviour of one model component.

  b) The "k out of n" situation:

  Another frequently found situation is illustrated by the following example:

      There are n cars in a system, out of which k are moving around and (n-k) are parked somewhere.

  This situation can be represented by the following syntactical construct:

                    FOR I:=1 TO N DO
                      IF CAR[I] THEN

  where CAR is a boolean array with the values "true" for cars moving around and "false" for parked cars.

  For n = 1 this case degenerates to a simple IF clause.

  c) Example:

  Let us consider a mechanical system with a dry friction force (TFR) modeled somewhere in the system. This can be shown by the following graph:

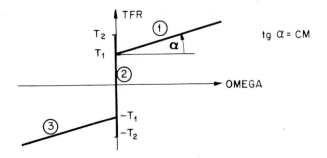

Fig.2: Dry friction force graphed versus velocity

In this example we face the typical "one out of n" situation, where n = 3 are the three continuous branches of the
discontinuous TFR-function. Each of them is represented by a
different equation and by a different set of state conditions.

This situation could thus be coded as shown in Fig.3. Using
this formalism for describing a combined system, the resulting description is not much more complicated than for
normal CSSL-type languages, but allows the preprocessor to
produce properly executable run-time code.

V.2) Requirements of the language:

When developing a new language for combined system simulation the
following points should be remembered:

a) The language should provide for flexible structures.

b) It should be extendable (open-ended operator set).

c) Both syntax and semantics of the language should be easy to
learn and to remember.

d) The language should contain as few elements as possible but as
many as are required.

e) Models should be codable by as few elements as possible.

f) The preprocessor should contain provisions for faithfully detecting coding errors.

```
SYSTEM
  CONTPROCS
    ...
    ...
    MODEL DRYFRICTION (TFR <- T, OMEGA)
    (* COMMENT: <- SYMBOLIZES A LEFT ARROW AND IS USED TO
                SEPARATE INPUT FROM OUTPUT VARIABLE LISTS *)
      CASE NL OF
        1:   TFR = T1 + CM*OMEGA;
             CONDIT MOD2: OMEGA NEG CROSS + 0.0 TOL=1.0E-3 END
             END;
        2:   TFR = T;
             CONDIT MOD1: T POS CROSS + T2 TOL=1.0E-3 END;
             CONDIT MOD3: T NEG CROSS - T2 TOL=1.0E-3 END
             END;
        3:   TFR = -T1 + CM*OMEGA;
             CONDIT MOD2: OMEGA POS CROSS + 0.0 TOL=1.0E-3 END
             END
      END (* DRY FRICTION *)
    ...
  END (* CONTINUOUS SUBSYSTEM *)
  DISCPROCS
    EVENTS
      MOD1:  NL := 1 END;
      MOD2:  NL := 2; OMEGA := 0.0 END;
      MOD3:  NL := 3 END
    END (* TIME EVENTS DESCRIPTION *)
  END (* DISCRETE SUBSYSTEM *)
END (* SYSTEM DESCRIPTION *)
```

Fig.3: Combined description of a dry friction force

g) The language must contain all elements required to enable the preprocessor to produce numerically well-conditioned run-time code.

Some of these requirements are contradictory. For example: If we want to enable the preprocessor to detect as many errors as possible, made by the user when formulating his model, the language must contain some redundancy. This certainly competes with the wish to have as short user's programs as possible.

Flexible structures: We want to obtain a universality of the program's applicability. This problem has been considered carefully when designing the CSSL-type languages which are in existence and also in designing the SIMULA-67 language for discrete simulation [10]. So we need not discuss this here again.

Extendability: Two different points must be considered: On one hand, we want to enable the user of such a language to extend it for his personal needs. This requires a superposed macros-structure [4,9]. This, however, need not necessarily form an integral part of the language definition, as it is interpreted by the preprocessor. The macros-handler can as well be realized by a stand-alone program preceeding the normal preprocessing [5]. This would then allow for more flexible macros-structures (interpretative macros-language) without calling for too high core memory requirements for its realization. On the other hand the system's specialist should also be given the possibility to extend the basic language definition itself. For this purpose the preprocessor should be constructed in such a way that it can be easily augmented to accommodate new ideas. For this purpose the most recent compiler building techniques employing structured programming and structured data representation should be applied, as described for instance by Wirth [33]. The author suggests, there- fore, to design the language as much as possible as a determi- nistic, one pass, left to rigth language (DLR-1 language), for which the syntax is to be described formally either by use of a BNF-nota- tion or by use of syntax diagrams. Compilers for such languages can be written in a straight forward manner and are, thus, easily readable.

Ease of learning syntax and semantics: Two main goals are to be achieved: It should be easy to write programs by one's self and it should as well be easy to read programs coded by somebody else. These two goals tend to compete with each other. To meet the former goal we want the different elements of the language to use the same syntactical constructs as much as possible. To meet the latter goal we want to be as flexible as possible in choosing appropriate mne- monics and close to conversational english constructs.

To give an example for the competing nature of the two goals let us consider once more the dry friction example given above (Fig.3). To meet the second goal we introduce the '=' symbol in the notation of equations of the parallel section and the ':=' symbol in the nota- tion of statements of the procedural section. By these means the in- herent difference between parallel and procedural sections is clari- fied, in that, for instance,

$$I := I + 1;$$

is a meaningful statement, whereas

$$I = I + 1;$$

is a meaningless equation. This rule will thus help to improve the readability of programs, it will at the same time, however, compli- cate the writing of programs since it simply introduces an additio- nal not necessarily required syntactical construct to remember.

Few language elements: The language should be constituted of as few elements as possible to make it easily learnable. On the other hand we require many language elements to obtain short user's programs. If there are not enough primitives offered by the language, the co- ding of complex problems becomes very difficult and the resulting source program will be long. This problem is best solved by provi- ding a hierarchical structure of both language and documentation. By means of this the user can first read an introductory manual which teaches him how to utilize the basic features required for modeling simple problems. This can be learned in a short time. Later on, when he realizes that his problem is more intricate than he thought in the beginning he may study another manual which enables him to use advanced features of the language. The user must be able to code simple situations in a simple manner, but should be assisted when coding more complex situations as well.

<u>Short users' programs</u>: The user should not be bothered by being asked to provide unnecessary information (like typing FORTRAN COMMON-blocks). This point must be balanced against:

<u>Provisions for error detecting</u>: Some redundancy should be left to the program for error detecting purposes. The author feels that a modern simulation language should for example require from the user that all variables he is going to utilize are declared in the beginning. This enables the preprocessor to detect many typing errors. This statement has been mentioned on many occasions (e.g. in the development of PASCAL [16]). It is, therefore, amazing, that none of the CSSL-type simulation languages to our knowledge has adopted this idea, and that this fact is even praised by many developers of brand new simulation software.

<u>Well-conditioned run-time code</u>: This point has been discussed in section V.1 already.

VI) DISCUSSION OF EXISTING SOFTWARE:

The existing software for combined simulation has recently been reviewed in an excellent survey by Oren [23,24]. He has collected information on about 30 different simulation systems for combined system simulation. Due to the fact that combined simulation never before had been properly defined, some of the surveyed languages/packages lie on a somewhat different line from what has been presented here. Furthermore, some of the programs described have never been released. Considering only those programs being implemented at several different installations and being widely used by different people, just a very few of them remain. Among these, GASP-IV [27] which is an ANSI-FORTRAN-IV coded subroutine package, has by far the largest distribution. Together with its descendent, GASP-V this program follows the ideas mentioned in section IV. The numerical behaviour of the GASP software is, in our experience, the best of all existing run-time packages for combined simulation. Unfortunately there is no provision in GASP for a user-oriented input definition (no preprocessor is involved). Therefore, none of the ideas presented in section V is realized in GASP.

From the information processing point of view, pioneer work has been done in the definition of the language GSL [15] following the original ideas of Fahrland [13,14]. GSL has the best language structure of all the languages so far published. Unfortunately GSL has never been released and must therefore be considered to be a collection of new ideas rather than a simulation language. However, GSL also has severe shortcomings both in the underlying run-time structure and in the language definition itself (the recent developments in the field of information processing and especially compiler building, as presented in section V, have not sufficiently been taken into account). A new simulation language COSY [2] (standing for COmbined SYstems) is under development by the author. It will involve a PASCAL-coded preprocessor which translates a new input definition language (following the ideas developed in section V of this paper) into GASP-V executable code. GASP-V thus will be used as target language for COSY. This language, however, will not be released before the end of 1979. Another new language under development which has some resemblance to COSY is GEST [21]. The language definition will soon be replaced by the newer version GEST'78 [25,26].

An entirely different approach has been taken in the definition of SMOOTH [32], which uses a network approach combining GERT networks for discrete simulation [28] with STATE networks for the continuous subsystem. A new program of this class will be released in 1979 by Pritsker combining Q-GERT [28] with GASP-IV/E (extended version of GASP-IV, released in 1978 by Pritsker). This approach certainly results in extremely short application programs, at least for such applications for which there are elements provided in the language. However, a network approach cannot be as general as a structured language. Benyon [1] states: "Such a diagrammatic approach to modelling can be very useful in some instances, but experience with the continuous languages has been that the diagrams soon grow too complicated to be enlightening, once one advances beyond quite simple models".

Moreover, it is often stated that network languages are easier to learn compared to equation oriented structured languages. The author would deny this statement for two reasons:

a) The number of language elements of such block oriented languages required to obtain at least a certain degree of flexibility is much larger than for equation oriented languages. GPSS-V [31], for example, consists of 41 building blocks and Q-GERT [28] offers 24 of them (Q-GERT requires a smaller number of blocks for an even higher degree of flexibility, because the single building blocks are more decomposible and recombinable). All of these building blocks must be understood before a truly complex program can be written.

b) Since the single building block describes a rather complex entity compared to a simple event description the semantics required to describe such an element are much more complex. (A complex situation can either be expressed by a complex syntax consisting of many "small" building blocks with primitive semantics, or by a simple syntactical construct consisting of few "large" building blocks with complex semantics, but never by both simple syntax and semantics.) This can be illustrated with a simple example. Considering the GENERATE-block of GPSS, it seems first, that the meaning of this block is very easily explained.

GENERATE A, B

means that a new transaction is to be generated with a uniform distribution in the interval [A-B,A+B]. The novice user of GPSS will take this definition and let this block be followed by a SEIZE-block to have the transaction occupying a facility. In practice, if this facility is already occupied when the transaction is born to the system, this transaction will stay in the GENERATE-block and inhibit the generation of new transactions. This shows, that the semantics required to describe this simple situation properly, are much more complex than might be thought. Very commonly, an error occurs due to the fact that semantics are involved which have not been reported to the user in the introductory manual.

This last example unveils another weak point of network languages: The program as specified above will "work", which means that there will be output produced, although this output will be wrong. With a high probability the user will thus never detect that his program is erroneous. The reason for the inability of GPSS to detect the error lies in the fact, that hardly any redundancy has been left in the code which could enable the system to detect errors in the source program.

Much more promising, it seems to us, is a new network approach proposed by Elmqvist [11] and by Runge [29]. Intended for continuous systems, it could also be extended to encompass discrete systems as well. This new approach has its background in the equation oriented languages. The single network element consists of a set of equations programmed by the user. Different modules are connected by special elements, called cut- and path- elements in DYMOLA [11]. This new approach can be thought of as an extension of the earlier macros-constructs of CSSL. It is even more general than the classical CSSL-type language, since this latter forms a subset of the new network language. In this approach the user does not need to describe in advance which connecting variables of his submodel (variables which are visible from outside) are input and which are output variables of the module. Formulae manipulation algorithms are used in DYMOLA to obtain a computational set of statements, whereas MODEL [29] uses implicit integration techniques to go round the difficulty. Some problems may arise from the fact that these languages are no longer context-free. They are on the contrary extremely context-dependent. This sometimes may result in ambiguities which have to be resolved. The possible set of executable statements is not necessarily unique. The author believes, however, that it will be worthwhile devoting more research to this approach.

VII) ACKNOWLEDGMENTS:

The author would like to express his deep indeptedness towards Prof. A. Alan B. Pritsker. Many of the ideas expressed in this article originated from the pioneer work in combined simulation done by Pritsker and his group, and also resulted from personal discussions with them. He would furthermore like to thank Prof. Tuncer Oren for the many good ideas he suggested in several long discussions of the subject. He is also very grateful to Prof. Bernard P. Zeigler for the stimulating comments obtained as a reaction to the initial abstract sent.

218   F.E. CELLIER

[1] P.R.Benyon: (1976) "Improving and Standardizing Continuous
     Simulation Languages". Proc. of the SIMSIG Simulation
     Conference, Melbourne, Australia, May 17-19, 1976;
     pp. 130 - 140.

[2] A.Bongulielmi: (1978) "Definition der allgemeinen Simulations-
     sprache COSY". Semesterwork, Institute for Automatic Control,
     The Swiss Federal Institute of Technology Zurich. To be ob-
     tained on microfiches from: The main library, ETH - Zentrum,
     CH-8092 Zurich, Switzerland. (Mikr. S637).

[3] M.B.Carver: (1978) "The FORSIM-VI Simulation Package for the
     Automated Solution of Arbitrarily Defined Partial and/or
     Ordinary Differential Equation Systems". Form: AECL-5821.
     Atomic Energy of Canada, Ltd.; Chalk River Nuclear Laborato-
     ries, Mathematics & Computation Branch, Chalk River, Ontario,
     Canada KOJ 1J0.

[4] F.E.Cellier: (1975) "Continuous-System Simulation by Use of
     Digital Computers: A State-of-the-Art Survey and Prospectives
     for Development". Proc. of the SIMULATION'75 Symposium,
     Zurich. To be obtained from: ACTA Press, P.O.Box 354,
     CH-8053 Zurich, Switzerland; pp. 18 - 25.

[5] F.E.Cellier: (1976) "Macro-Handler for Simulation Packages
     Using ML/I". Proc. of the 8th AICA Congress on Simulation of
     Systems, Delft, The Netherlands. Published by North-Holland
     Publishing Company (Editor: L.Dekker); pp. 515 - 521.

[6] F.E.Cellier: (1977) "On the Solution of Parabolic and Hyper-
     bolic PDE's by the Method-of-Lines Approach". Proc. of the
     SIMULATION'77 Symposium, Montreux, Switzerland. To be obtained
     from: ACTA Press, P.O.Box 354, CH-8053 Zurich, Switzerland;
     pp. 144 - 148.

[7] F.E.Cellier: (1978) "The GASP-V Users' Manual". To be obtained
     from: Institute for Automatic Control, The Swiss Federal
     Institute of Technology Zurich, ETH - Zentrum, CH-8092 Zurich,
     Switzerland.

[8] F.E.Cellier, Blitz A.E.: (1976) "GASP-V: A Universal Simu-
     lation Package". Proc. of the 8th AICA Congress on Simulation
     of Systems, Delft, The Netherlands. Published by North-Holland
     Publishing Company (Editor: L.Dekker); pp. 391 - 402.

[9] F.E.Cellier, Ferroni B.A.: (1974) "Modular, Digital Simulation
     of Electro/Hydraulic Drives Using CSMP". Proc. of the 1974,
     Summer Computer Simulation Conference, Houston, Texas, U.S.A.;
     pp. 510 - 514.

[10] O.J.Dahl, Nygaard K.: (1966) "Simula; A Language for Program-
     ming and Description of Discrete Event Systems". Oslo, Nor-
     wegian Computing Center.

[11] H.Elmqvist: (1978) "A Structured Model Language for Large Continuous Systems". Form: CODEN LUTFD2/(TFRT-1015)/1-226/(1978). Ph.D Thesis. Lund Institute of Technology, Dept. of Automatic Control, Lund, Sweden.

[12] M.S.Elzas: (1978) "What is Needed for Robust Simulation?" Article in this volume.

[13] D.A.Fahrland: (1968) "Combined Discrete Event / Continuous System Simulation". MS Thesis, Systems Research Center Report SRC-68-16, Case Western Reserve University, Cleveland, Ohio.

[14] D.A.Fahrland: (1970) "Combined Discrete-Event Continuous System Simulation". Simulation vol. 14 no. 2 : February 1970; pp. 61 - 72.

[15] D.G.Golden, Schoeffler J.D.: (1973) "GSL - A Combined Continuous and Discrete Simulation Language". Simulation vol. 20 no. 1 : January 1973; pp. 1 - 8.

[16] K.Jensen, Wirth N.: (1974) "PASCAL User Manual and Report". Lecture Notes in Computer Science, Springer Verlag.

[17] P.J.Kiviat: (1967) "Digital Computer Simulation: Modeling Concepts". Form: RM-5378-PR, The Rand Corp., Santa Monica, CA, U.S.A..

[18] G.A.Korn, Wait J.V.: (1978) "Digital Continuous-System Simulation". Prentice Hall.

[19] W.Kreutzer: (1976) "Comparison and Evaluation of Discrete Event Simulation Programming Languages for Management Decision Making". Proc. of the 8th AICA Congress on Simulation of Systems, Delft, The Netherlands. Published by: North-Holland Publishing Company (Editor: L.Dekker); pp. 429 - 438.

[20] R.N.Nilsen, Karplus W.J.: (1974) "Continuous-System Simulation Languages - A State-of-the-Art Survey". Annales de l'Association Internationale pour le Calcul Analogique (AICA), No. 1, january 1974; pp. 17 - 25.

[21] T.I.Oren: (1971) "GEST: A Combined Digital Simulation Language for Large Scale Systems". Proc. of the AICA Symposium on Simulation of Complex Systems, Tokyo, Japan, September 3-7, 1971; pp. B-1/1 - B-1/4.

[22] T.I.Oren: (1975) "Syntactic Errors of the Original Formal Definition of CSSL 1967". Technical Report TR75-01 (IEEE Computer Society Repository No. R75-78), Computer Science Dept., University of Ottawa, Ottawa, Canada.

[23] T.I.Oren: (1977) "Software for Simulation of Combined Continuous and Discrete Systems: A State-of-the-Art Review". Simulation, vol. 28 no. 2 : February 1977, pp. 33 - 45.

[24] T.I.Oren: (1977) "Software Additions". Simulation, vol. 29 no. 4 : October 1977, pp. 125 - 126.

[25] T.I.Oren: (1978) "Reference Manual of GEST'78 - Level 1 (A Modeling and Simulation Language for Combined Systems)". Technical Report 78-02, Computer Science Dept., University of Ottawa, Ottawa, Canada.

[26] T.I.Oren, den Dulk J.A.: (1978) "Ecological Models Expressed in GEST'78". Technical Report Prepared for the Dept. of Theoretical Plant Ecology, Dutch Agricultural University Wageningen, The Netherlands.

[27] A.A.B.Pritsker: (1974) "The GASP-IV Simulation Language". John Wiley.

[28] A.A.B.Pritsker: (1977) "Modeling and Analysis Using Q-GERT Networks". John Wiley.

[29] T.F.Runge: (1977) "A Universal Language for Continuous Network Simulation". Form: UIUCDCS-R-77-866. Ph.D Thesis. University of Illinois at Urbana-Champaign, Dept. of Computer Science, Urbana, Ill., U.S.A..

[30] H.Schlunegger: (1977) "Untersuchung eines netzrueckwirkungsarmen, zwangskommutierten Triebfahrzeug-Stromrichters zur Einspeisung eines Gleichspannungszwischenkreises aus dem Einphasennetz". Ph.D Thesis, no. DISS.ETH.5867: The Swiss Federal Institute of Technology Zurich, Switzerland.

[31] T.J.Schriber: (1974) "Simulation Using GPSS". John Wiley.

[32] C.E.Sigal, Pritsker A.A.B.: (1973) "SMOOTH: A Combined Continuous/Discrete Network Simulation Language". Proc. of the 4th Annual Pittsburgh Conference on Modeling and Simulation. Pittsburgh, Penn., U.S.A., April 23-24, 1973, pp. 324 - 329.

[33] N.Wirth: (1976) "Algorithms + Data Structures = Programs". Prentice Hall, Series in Automatic Computation (Chapter 5). or: N.Wirth: (1977) "Compilerbau". Teubner Studienbuecher, Informatik.

[34] B.P.Zeigler: (1976) "Theory of Modelling and Simulation". John Wiley.

[35] B.P.Zeigler: (1977) "Systems Simulateable by the Digital Computer". Logic of Computers Group Report, University of Michigan, Ann Arbor, U.S.A..

[36] (1967) "The SCi Continuous System Simulation Language (CSSL)". Simulation, vol. 9 no. 6, December 1967; pp. 281 - 303.

METHODOLOGY IN SYSTEMS MODELLING AND SIMULATION
B.P. Zeigler, M.S. Elzas, G.J. Klir, T.I. Ören (eds.)
© North-Holland Publishing Company, 1979

HETERARCHICAL, SELFSTRUCTURING
SIMULATION SYSTEMS:
CONCEPTS AND APPLICATIONS
IN BIOLOGY

P. Hogeweg and B. Hesper
Bioinformatica
University of Utrecht
de Uithof, Utrecht
Netherlands

1. INTRODUCTION

In this paper we outline a framework for modelling and simulation, which emphasises local specification of entities, dynamic generation of entities, heterarchical control and self structuring properties of simulation models. Two examples of programs using these principles are discussed in some detail. They are:
- MIRROR, a program for modelling $\underline{M}$oving, $\underline{I}$nteracting, $\underline{R}$eproducing and $\underline{R}$etiring $\underline{O}$rganisms, and
- MICMAC, a program, exploring "micro/macro" relations among processes for self structuring.

In the discussion on MIRROR we emphasise the local specification and heterarchical control using DEMONs; in the discussion on MICMAC we employ self structuring properties through the use of EXPERTs.
Both programs are implemented in SIMULA/67 (Dahl, Myhrhaug and Nygaard, 1970). We found the CLASS concept of this language useful, although the strict hierarchical structure of CLASSes is less suitable for our purposes.
In the more technical part of the paper, familiarity with SIMULA/67 is assumed, but the concepts introduced should be understandable without such background knowledge.

"global state considered harmful"
2. LOCAL SPECIFICATION OF ENTITIES
C. Hewitt (1973)

2.1 Parts and wholes

Biology, like physics, is simple if and only if viewed locally. This implies that an entity is to be described in terms of information available to the entity itself, without reference to an outside observer. For example, "a free falling particle is to be observed from the inside of a free falling rocket with the windows closed" (Misner, Thorne and Wheeler, 1973).
Likewise, the behaviour of a cell is to be viewed in relation to its neighbours only, not in terms of the global position it occupies in an organism.

Even in models which represent such a "localness" explicitly there often remains some implicit global variable or control structure. This was true for Newton's description of moving bodies (absolute space and time remained global variables). A rather similar situation is seen in contemporary cellular models for biological systems. While the system is subdivided into cells, supposedly autonomous units, time remains a global variable: cell transitions are globally synchronised in cellular space systems (von Neumann, 1960; Ulam, 1962; cf Burks, 1970) and L-systems (Lindenmayer, 1968a,b; Herman and Liu, 1973; Herman and Rozenberg, 1975)(See also Herman et al, 1974; Lindenmayer and Rozenberg, 1975). This global synchronisation not only affects drastically the results in all practical applications as shown by Hogeweg (1978a,b), but also lacks any biological rationale.

The artificiality of the global synchronisation of the transformations of the cells is at once apparent if viewed from the standpoint of a cell as an autonomous unit. However, from the standpoint of an outside observer, who seeks to model the trans-formations of an entire system (subdividing it into (arbitrary) units) imposing a global synchronicity seems quite acceptable. Indeed all differential and difference equation models, as well as all automatic theoretic models employ such global syn-chronicity implicitly. Likewise most simulation strategies for such models enforce global synchronicity (on a sequential computer).
Such simulation models typically contain a global control structure (monitor) to run the program and use global data structures (arrays, etc.) to store the varia-bles and parameters. Thus, although, for example in continuous system languages (CSMP, LEANS, etc.), the user decomposes the system into a set of standard blocks, and specifies the interconnections, the system translates this into the global transformation of a vector of system variables.

Implicitly global formulation can be very dangerous, in particular if the behaviour of the entities is not known (or not modelled) in detail. This danger lies in the fact that the entities are formulated as 'parts of a whole' instead of as autonomous units. Simulation models make it feasible to study how a set of (dynamically) in-terrelating entities can generate seemingly (i.e. in the eye of the observer) 'emergent' properties. This obviates metaphysical discourses on emergence, unless entities are implicitly controlled by the 'whole' whence all this metaphysics 're-emerges'.

In many interesting problems in biology, both the overall system and the subsystems are supposed to be an incomplete representation of parts of the biological system studied. If such is the case the subsystems should be regarded as autonomous units, and should be represented in the (simulation) model as such.

Thus we envisage a system in which the behaviour of a subsystem is entirely speci-fied within the subsystem itself, using only its own local variables (no global data structures) and those of other subsystems with which it is acquainted. Thus the information and the information processing capabilities of the subsystem are stated explicitly in the model. It is important to endow the subsystem only with capabilities and concerns which are reasonable with regard to the object it is supposed to model (Note that we do not shun from using anthropomorphic terminology in describing our entities: it protects us from the far more serious fallacy of implicitly global control).

Such a simulation methodology is envisaged by us in the first place because of the usefulness in biology (our own research is mainly in modelling morphological de-velopment and modelling ecosystems). Besides it has many attractive properties from the simulation point of view: complete local specification of the modules in-creases the flexibility, extendability, transferability and last and least the readability of the model.

The maxim above this paper is "global state considered harmful". Indeed in our type of simulation methodology, the global state is not readily available even to the user (or output unit). In fact an output unit is, like the rest of the system, a locally defined entity, observing the system from its own local viewpoint through inspection of its acquaintances. As is the case for other units, the set of acquaintances may change dynamically also for the output unit, thus extracting the output on different parts of the system. Only as a special case can we endow the output unit with the power to observe all the local variables.

> "an ant viewed as a behaving system is quite simple,
> the apparent complexity of its behaviour in time is
> largely a reflexion of the complexity of the en-
> vironment in which it finds itself ....."
> "a man viewed as a behaving system is quite simple,
> the apparent complexity of his behaviour in time is
> largely a reflexion of the complexity of the en-
> vironment in which he finds himself ...."
>   H.A. Simon (1969)

2.2 Sample model:
Moving, Interacting, Reproducing and Retiring Organisms (MIRROR)

Features of the foregoing simulation methodology will be elaborated further in the context of a specific example: the behaviour of individual organisms in a plane. The organisms are viewed as information processing entities whose behaviour is determined (possibly stochastically) by features of the environment they can directly observe, and by their memory structure. Because they can move about, are born or die, the set of entities with which they interact varies with time. We endow our organisms with only the information actually available to them at any time. This makes it possible to study the amount and kind of information needed to perform a certain behaviour.

Examples of simple problems we can tackle with the system are:
- Formation of spatial patterns due to interaction between organisms. For example patterns due to density dependent influence on germination, growth and death of plants, or due to changes in moving direction of animals in meeting other animals etc.
- Influence of behaviour parameters of the prey on the density of prey needed by various typed of predators to survive. For example, the influence of trooping of prey on predators hiding in a bush, or predators moving about in certain ways.

In short we want to be able to model the structuring of the environment by physical processes, plants and animals, and the influence thereof on the behavioural patterns of the organisms. Thus we hope to gain further insight into the simplicity and complexity of behavioural patterns following the paradigm of Simon as stated above this chapter.

We desire that the simulation system be structured so as to allow for easy modelling of quite simple problems (such as those mentioned above) and of much more complex models, while providing a smooth path from the former to the latter.

In designing such a system it becomes apparent that the most obvious choice of information processing primitive, i.e. individual organisms, does not suffice. Organisms should be rather modelled as dynamic colonies of information processing units. This is so because:
1. We are interested in modelling a number of partially independent processes taking place in an individual. Different information is relevant to each of these, they take place in different time scales, and they interact only once in a while. Even in the simplest cases, we might need an information processing entity which handles movement of the organism step by step, one which handles the daily cycle of waking and sleeping and one which handles the reproduction cycle. These processes are partially independent (an animal does not have to check at every time of the day whether the night is falling), but also should modify each others behaviour at certain times.
2. In order to obtain the relevant information for its behaviour, each information processing unit may moreover need to 'expand' itself to check on potential interaction partners for important events which trigger its own activity. For this purpose we implemented DEMON-like structures in our system (Charniak, 1972; see also Bobrow and Winograd (1977) on "active programming").

This twofold need for organism decomposition is again a manifestation of the maxim above: describing an organism by its global state is harmful.

Moreover such a decomposition amplifies greatly the strength of Simon's paradigm.

2.3 DEMONs

Activities of animals may have to be triggered by other events. Take as an example an ambush predator (LION) hiding in a bush, waiting for a prey to come along. It may have to wait for a long time, but when the prey arrives it should be fast catching it. Implementing this by frequent checking of the environment would be very inefficient. Instead the LION should be triggered by the arrival of the prey. However the prey is prohibited to call the LION explicitly upon entering its sur-rounding by our demand that the knowledge of each entity is to be confined to know-ledge it may reasonably possess. Instead it should ward the LION without knowing it (compare the snapping of a branch). Such a trigger can be achieved by using DEMONs.

DEMONs provide the time driven heterarchical interconnection and activation structure of our simulation system. They may be seen as generalised and localised 'wait until'. They combine the features of process (time) oriented activation systems and (conditional) event oriented systems. They reduce to either of these for particular parameter settings, and go beyond these for other parameter set-tings.
In most applications DEMONs are entirely transparant except at the time of their creation.
DEMONs have four parameters: TIE, TARGET, DT and FLAG. TIE is a reference to the entity which is to be 'revived' by the DEMON (which is often but not necessarily the same as the one which created the DEMON); TARGET is a reference to an entity to be watched by the DEMON; DT the time delay after which the DEMON activates it-self and FLAG is a parameter to check for obsolescence of the DEMON in relation to its TIE.

REVIVAL of an entity involves:
1. providing the entity with a reference to the cause of its revival, i.e. to TARGET.
2. providing the TIE with a MESSAGE[A].
3. revival of the DEMONs associated with the entity to be revived.
4. reactivation of the entity to be revived.
REVIVAL is only executed if the DEMON is not obsolete, otherwise the DEMON is de-leted from the system.

All entities in a DEMON-driven system possess a list for DEMONs and a pointer called REVIVER to receive the cause of the revival. Entities include INTEGER,REAL, BOOLEAN,ARRAY1,ARRAY2.....ARRAYn,LISTS and DPROCESS and any user defined entity which is a subCLASS of DLINK or DPROCESS. All entities can be member of a list. Convenient side-effects of the representation of the variables in such a form in-clude stacks and dynamic allocation as standard facilities for all variables.

DEMONs take the role of REACTIVATION clauses in SIMULA/67 (there implemented through entities calles EVENT NOTICES) if TARGET is NONE and the FLAG of an entity is increased upon DEMON creation to a level above the one given to DEMONs up to that time (a REACTIVATION clause in SIMULA supersedes all previous ones whether

---

[A] MESSAGES are an important concept in Hewitt's ACTOR systems, which are related to our approach (Hewitt, 1973). A MESSAGE mechanism is included in our system (each entity has a SCRATCHPAD for this purpose) but is so far little used.

the scheduling is for earlier or later times). If on the contrary the FLAG is increased upon REVIVAL of the entity beyond the value given to all previously generated DEMONs the scheduling for the earliest time supersedes all the others. Likewise other FLAG manipulations can provide us with the latest scheduling or the
first generated scheduling. Moreover any amount of interaction among DEMONs of the
same TIE or the same TARGET may be achieved.

We will use the above mentioned LION as an example of a DEMON-driven animal in our
implementation of a system for modelling moving, interacting, reproducing and re-
tiring organisms.
The space in which the organisms live is subdivided into PATCHes, discrete spatial
units within which the environment is supposed to be homogeneous. Consider the
space divided in a fixed set of PATCHes, of fixed size (actually we implemented an
extended patch structure in which the number of PATCHes and their sizes are determined by the needs of the system itself, using a way of structuring similar to the
one discussed in sections 3.2 and 3.3). PATCH possesses a number of environmental
variables and a list of organisms inhabiting it (REF (DHEAD) BIOTA). The distinction between environmental variables and organisms is defined in such a way that
the set of environmental variables for each PATCH is fixed during the simulation,
while the set of organisms will vary.

ANIMAL is a subCLASS of ORG which possesses, besides the observation procedures
which are shared by all ORGs, procedures to move about. The moving procedures enter
the ANIMAL in the appropriate BIOTA list. In such a context LION is defined as a
subCLASS of ANIMAL. The LION sets up DEMONs watching over the surroundings (i.e.
the BIOTA lists of nearby PATCHes). Moreover the LION sets up a DEMON to warn it
when it is hunting time (i.e. when it has grown so hungry that it has to go searching for its prey). An outerloop of the CLASS body of LION defines the hunting behaviour, i.e. moving about and setting up new DEMONs in its new surroundings; it
increases the FLAG of the LION to render obsolete old DEMONs watching over areas
now out of view. The innerloop of the CLASS body defines the behaviour while
hiding in the bush: upon REVIVAL, when the REVIVER is the surrounding area, it
tries to find a prey there; if it finds a prey it kills it, goes fast asleep and
wakes up again to watch over the surrounding area; if no prey is found (false
alarm by entry of a non-prey species into its surroundings) nothing happens. The
LION goes hunting when the hunger level is high (REVIVED by the hunger-DEMON); if
the hunt remains unsuccessful for too long it dies from starvation.

2.4 Some simple simulation results from this system

We will mention some results which became immediately apparent while working with
the system, but which, at least to us, were not obvious beforehand.
- In case of the above mentioned LION its viability is crucially dependent on
  the social behaviour of its prey: if the prey animals have only a slight tendency to clump (i.e. moving preferentially in the direction in which a fellow-
  species is observed) a much larger prey population is needed to maintain the
  LION, especially if it is not to go out searching when hungry. While such an
  effect was expected, its size was rather surprising to us.
- If entities interact by changing direction of movement when meeting, this only
  leads to specific spatial distribution patterns, if an entity "can see beyond
  its nose". By this we mean that its sensors provide it with information about
  a larger area of space than it covers in one 'step', i.e. the area it covers
  without updating its information.
  This result seems to have profound implications for the simulation methodology
  which we advocate.

3. THE MICMAC STRUCTURE OF MODELS

3.1 Discrete event formalism and the selection of variables and events

In formulating a simulation model we have to select 'relevant' features of the
system. What is considered to be a relevant feature of a subsystem is dependent on
other subsystems, which observe some features (and react on them) and do not observe or react on other features. The user, as output unit, is just one such a subsystem, as argued above, and selection of relevance by the user is analogous to selection by other subsystems. Selection of interesting features occurs very markedly
in discrete event modelling, in which the underlying conceptualisation of continuous simulation models "everything is changing all the time" is replaced by "once
in a while something interesting is changing somewhere in the system". The latter
reduces, in theory but never in practice, to the former if "everything is considered interesting all the time". Obviously in any implementation on a digital computer some selection of interest is necessary: in fixed time step simulations the
global state if considered interesting at fixed regularly spaced points in time;
in variable time step integrators the selection is on the basis of calculability
of the next global state. The selection of interesting features of the system on
the basis of calculability is also important in discrete event systems, but the
calculations are for local states as opposed to global states, and interesting
points in time will be different for different subsystems. Moreover the selection
of an interesting point for a subsystem A may be due to the fact that some other
subsystem B needs A's local state to compute its own next state (with respect to
timing or value or both). For example A may be a 'micro' entity operating on a fast
time scale and B a 'macro' entity operating on a slower time scale (or vice versa).
In fact, if no part of the model is directly or indirectly interested in the state
of a subsystem, there is no purpose in computing it.

It should be emphasised that the selection of interesting features of the model by
other components of the model contrasts with the global control denounced above.
It involves only selection, not control: the 'micro' entity does not know the
'macro' entity. All that happens is that it receives a request to provide it with
its state at a certain time, or to provide it with the time in which it has a certain state. Recall the LION requesting to be warned when a prey enters its range
of vision; as programmed, using DEMONs, the prey warns the LION without being
aware of it itself. In the case of the LION this resulted only in a reduction of
the amount of calculation done by the LION; the prey kept moving step by step,
choosing its moves on the basis of local circumstances. Below we describe a system
in which such requests reduce the amount of computation more drastically.

3.2 Selfstructuring in simulation models

It is well known that formulating the dynamics of a model in terms of small
changes in small time steps, obviates the incorporation of indirect interactions
between variables. However, the goal of compounding these direct interactions
postulated for the model in a simulation run is exactly to gain insight in the indirect dependencies between variables on a longer time scale. Aids for obtaining
such an insight are: simulation of the entire system and inspection of the output,
analytical solutions of subsystems, simulation of subsystems in isolation, fixing
of variables etc.

Zeigler (1977) showed how longer time scale information gathered for subsystems
once and for all, could be repeatedly used to improve the efficiency of the simulation of the composite system. Thus it is useful to incorporate partial knowledge
in the model itself. Furthermore it is useful to endow the simulator with "self
awareness" so that it can itself gather such partial knowledge about itself and
use it to improve the simulation. However, because such knowledge is partial, it
is bound to be wrong once in a while. Thus the usage of partial knowledge should
be accompanied by self criticism, and the ability to retreat to the basic formula-

tion of the dynamics supplied to it. As an example of such a simulation methodology we discuss our implementation and elaboration of the model for predator prey inter- actions in a patchy environment proposed by Zeigler (1977).

3.3 Example: predator/prey in a patchy environment

The model is about a field of discontinuous patches in which predator and prey species interact. The interaction between predator and prey is supposed to be go- verned by Lotka-Volterra dynamics with a self-limiting term for the prey. Such dy- namics give rise to a stable equilibrium; if however the population numbers in this equilibrium (or anywhere along the trajectory to the equilibrium) are very small, extinction will occur in practice. Such extinctions are often observed in homoge- neous experimental settings of predators and prey. The model is designed to dis- cover whether in an environment of extinction prone patches, both predator and prey may survive. In the model proposed by Zeigler, migration between patches occurs if and only if food shortage occurs, i.e., the prey migrates if it has reached its carrying capacity in a patch devoid of predators, and the predator migrates if it has exhausted the prey in a patch. It is shown by Zeigler (1978) that this system will indeed give rise to continued existence of predators and prey for a reasonable large number of parameter settings.

Straightforward simulation of this system consists of continuous simulation of the predator/prey interactions of all the patches, checking all the time for the con- ditions of migration, and, if these are fulfilled, establishing the migrations as discrete events. This exhaustive approach is quite unfeasible if we use many patches. However, we are interested only in occurrence of both populations and not in exact population numbers. Moreover migrations take place only at specific stages of the within-patch system. Thus we can simplify matters considerably by selecting interesting events only. Such a simplification can be carried out if we know the following partial solutions of the model: 1) how long it takes a prey to reach carrying capacity after it has colonized a patch, and how large the population is at that time; 2) how long it takes for a predator to exhaust its food, when migra- ted to a patch with a certain number of prey, and how many predators there will be at that time; 3) how long a skeleton predator population survives in a patch devoid of prey. This knowledge can be obtained partly by analytical solution and partly by simulating the trajectories of the predator/prey system in isolated patches. How- ever this knowledge is not always applicable because intermediate migrations may disturb the path which was considered to be isolated in these calculations. Zeigler implemented the model providing it with the above mentioned knowledge, approxima- ting the effect of intermediate migrations in the predator/prey system by inter- polation, and changing the scheduled events for emigration from the prey patch.

We implemented Zeigler's model within the above simulation framework, emphasising local specification of the units of the system. Moreover because we wanted the system to be applicable (or at least easily extendible) to different models of interaction of populations in patches (e.g., competition or symbiotic relations) and to different numbers of interacting populations (e.g., two predators predating on one prey species), it seemed to be begging the question to require an extensive a priori knowledge of the isolated patches. Therefore we designed the system so as to gather the necessary knowledge by itself, while it could use the basic formula- tion of the model when no specialised knowledge was (yet) available.

3.4 EXPERTs

The selfstructuring properties are implemented using EXPERTs. EXPERTs have exper- tise in carrying a process from a specified initial state to a (partially) speci- fied next state or over a specified time stretch. In the former case an EXPERT supplies the complete specification of the next state and the time at which it occurs; in the latter case it supplies the next state. Thus an EXPERT is a triple (INISTATE, NEXTSTATE, DT). EXPERTs exist in a local state space of initial condi- tions. EXPERTs have to know when their expertise is applicable. For this they have

to know the similarity structure (metric) of the surrounding state space. This similarity structure may be quite different in different regions of the state space
and in different directions. Thus such knowledge is only locally applicable by the
EXPERT itself. The EXPERT has to build up this knowledge from experience within the
framework of its data structure. EXPERTs may consult other EXPERTs and the knowledge of nearby EXPERTs may be used as first approximation to its own knowledge.
EXPERTs are eager to learn and apply their knowledge. As soon as a situation arises
within their capabilities, they apply their knowledge to it. As soon as a situation
arises which is just outside their capabilities they generate hypotheses about it.
They may also be so doubtful about their capabilities that they do not solicit for
a job, but are nevertheless consulted. In both the last mentioned cases they play
it safe: they test their hypotheses by direct simulation.

An EXPERT is endowed with a number of alternate hypotheses it may generate. Example
hypotheses are: "my own NEXTSTATE may be used for initial conditions sufficiently
like my own INISTATE" (and what is sufficiently alike, it will learn), or, "the
NEXTSTATE for initial conditions intermediate between my own INISTATE and that of
other EXPERT(s) can be derived by linear interpolation of our NEXTSTATEs" (weighted
for different directions in state space for higher dimensional state spaces). The
hypotheses are tested and given a confidence rating. This rating extends to all
situations closer to the EXPERT than the ones which were successfully tested. If
the confidence rating is high enough the hypothesis is applied. If an EXPERT is not
consulted often enough, or its knowledge is never successfully applied, it is de-
leted from the system. EXPERTs react on REQUESTs. REQUESTs exist in the same state
space as EXPERTs, and a REQUEST is also a triple (INISTATE, partially specified
NEXTSTATE, DT). It requests the missing information not given in the triple. When
a REQUEST enters the state space, the EXPERTs nearby examine it; if they can handle
it, they supply it with the needed information and reactivate it at the appropriate
time to report back to the process which generated the REQUEST. If the REQUEST is
not handled thus, it resorts to direct simulation. Upon reaching the partially
specified NEXTSTATE, the EXPERTs which generated hypotheses for the REQUEST, test
these and finally the process which generated the REQUEST is reactivated. If an
interrupt occurs before the condition on the NEXTSTATE is fulfilled, the REQUEST
deletes the previously generated hypotheses: failure should in this case not decrease the confidence rating. The REQUEST then checks whether EXPERTs are available
to handle the new situation (etc.). If upon reaching the NEXTSTATE, none of the
hypotheses proves to be correct (or no hypotheses were available), the REQUEST
generates a new EXPERT, supplying it with the now attained missing information.
This newly generated EXPERT is initialised with a region in state space, about
which to generate hypotheses, of the same shape as that of its nearest neighbouring EXPERT. Otherwise, if a correct hypotheses was generated upon arrival of the
REQUEST, and is now validated, no new EXPERT is generated but the system relies in
the future on this hypothesis.

3.5 MICMAC, program for simulating interacting popoulations in a patchy environment

MICMAC (micro/macro) explores the use of EXPERTs in the context of the patch
structured model of interacting populations. The basic entities distinguished in
MICMAC are:
CLASS PATCH(X,Y);
PROCESS CLASS MICDYN (micro dynamics);
PROCESS CLASS MACDYN(PTCH) (macro dynamics); see fig. 1.
The patches are hexagonal and arranged in a toroidal FIELD, by a procedure local to
the CLASS PATCH, which provides the pointers to neighbouring PATCHes. MICDYN
possesses the definition of the population interactions within a PATCH, in the procedure DYN(T,POP,DPOP) and the associated parameters. MICDYN takes care of direct
simulation of the population interactions as defined in the differential equations
of the procedure DYN. It uses variable time step integration using the extrapolation method of Bulirsch and Stoer (1966). The timestep is adjusted on the basis of
calculability (as predicted by the extrapolation method) and on the basis of interest: a check is performed to see whether an overshoot would occur with regard

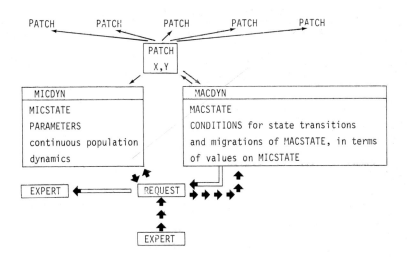

Fig. 1 Relationships among entities in MICMAC
    ⟶ pointer
    ⟹ generation of a new entry
    ➠ activation of an entity
For further explanation, see text.

to an interesting state using the predicted time step. If such an overshoot would
occur, the time step is repeatedly halved (down to a certain minimum), using the
intermediate calculations of the integration until the exact timing and the value
of the interesting state are determined. After computing timing and next state,
the process (MICDYN) is delayed for the calculated time stretch and afterwards
assumes the new state. If this new state is equal to the goal state requested by
MACDYN then REQUEST (and subsequently MACDYN) is reactivated and MICDYN is suspended until reactivated by a new REQUEST from MACDYN. However, MICDYN may be interrupted during its idle time (after computing timing and value of the next state)
by an unexpected new REQUEST from MACDYN, which is sent because of migrations.
Therefore MICDYN has to check whether it is revived on schedule (i.e, by itself)
or by such an interrupt. In the latter case it has to recompute its next state
after readjusting the time step according to the interrupt time.

The structure of the PROCESS CLASS MACDYN is quite different from MICDYN because it
has no definition of a time dependent next state function. Instead the next state
function is expressed in terms of conditions on the state of MICDYN. MACDYN
possesses the procedure CONDITION which defines the conditions to be fulfilled for
transition to the next state given a certain initial state. For example, if the
present state is GROWTH PHASE the next state is the PREY MIGRATION PHASE and is
entered upon reaching the carrying capacity (i.e., a certain number of prey).
MACDYN should be warned at the time that these conditions are fulfilled. For this
purpose MACDYN files a REQUEST, and then is passivated. Upon reactivation it sets
its next state and files a new REQUEST, etc. Without self organising properties
the REQUEST simply reactivates MICDYN and attaches itself as a DEMON to the FLAG
COND, to be reactivated if this is set true, and reactivates MACDYN in its turn.
The importance of the REQUEST structure lies in its use in the self organising
properties of the model: EXPERTs may take over control and delay the REQUEST, so
that it does not reactivate MICDYN. The EXPERT supplies in that case the REQUEST
with the missing information, so that it knows when to reactivate MACDYN and what
should be the next state of MICDYN and MACDYN at that time (see section 3.4).

3.6 Discussion of MICMAC

We will examine the flexibility of the above described model, and its rigidity and
specificity in turn.

1. Flexibility
MICMAC easily adapts to changes in the model, which can be made by very local modifications. For example the population dynamics of the patches in isolation can
be changed by modifying the procedure DYN. Likewise the dynamics of the macroprocess can be changed by local modifications: for example by modifying the conditions for migration or the migration process itself.
Moreover the model can be generalised to any number of interacting populations. In
this case both the micro and the macro-dynamics are changed, but the changes are
confined to model-specific changes. A very interesting generalisation of the model
seems to be to change the definition of the PATCHes.
The model may be embedded in MIRROR (universe of mobile interacting ANIMALs), see
section 2.2. The PATCHes are then identified with such ANIMALs and the migration
neighbours are determined as the neighbours of ANIMALs of the appropriate species
using the procedure which ANIMALs have for this purpose.

2. Rigidity and specificity
The system as it stands, although flexible (i.e. able to support a large class of
models), is limited in several respects, which are not inherent in the framework
of modelling and simulation which we have in mind.
In particular, it is simplified in the sense that there is a sharp and fixed distinction between micro and macro processes. The micro process has a time controlled next state function (here even with infinitely small time steps in the definition), while the macroprocess has an entirely condition controlled next state
function and the time to reach the next state always exceeds the time step of the

microprocess. Although the terms micro and macroprocess may suggest such a relation, it is not what we have in mind: MICMAC is a relation which can exist between
processes at certain times, a relation however which may change dynamically during
model simulation (it may for instance be reversed). Generalisations in that direction are presently being investigated.

4. DISCUSSION AND CONCLUSION

1. "Parts and wholes"

In this paper we outlined a framework for modelling and simulation stressing as de-
siderata local representation and heterarchical control.
The implementation of the models here described has indeed achieved these deside-
rata to a considerable extent. Although the two systems were developed independently, they may each serve as a subsystem for the other.
The way in which MICMAC can be a subsystem of MIRROR has already been mentioned:
the fixed patches of MICMAC can be replaced by the mobile organisms of MIRROR. As
in the original MICMAC the surrounding of the unit (patch or organism) accessible
for migration is given in this unit. This generalisation is useful for modelling
epidemic processes in relation to interaction patterns. The other way around,
MIRROR can serve as a subsystem of MICMAC. In this case the local dynamics of the
interaction within a patch is modelled by a MIRROR-type system of interactions
between individuals. In this case, such a MIRROR-like system operates in each
patch of MICMAC. Note that the MICDYN is not necessarily a continuous formulation
as seen in this example. EXPERTs may again find regularities in the so modelled
dynamics to speed up the simulation of the distribution of populations over the
field. Moreover, both these generalisations may be true in the same system, i.e.
MICMAC serves as a subsystem of MIRROR and MIRROR serves as a subsystem of MICMAC
at the same time (and these patterns may be nested). In this case we are modelling
individual moving organisms which interact with each other and which are infected
by several other species. The interactions between these species are again modelled
in term of the behaviour of the individuals. This mutual embedding of the two
systems in each other is the best proof for achieving heterarchical control.

2. Mixed mode, continuous and discrete event simulation

Mixed mode continuous and discrete event simulation systems proved to be very useful if continuous simulation is employed in those cases in which the discrete
event simulation breaks down because of unexpected interrupts. Continuous simulation should then be employed until the system arrived at a recognizable state from
which the larger time scale discrete event simulation can be picked up again. In
the worst case this strategy results in a simulation similar to an analogous, entirely continuous, simulation (if interrupts occur very often). This manifests a
use of mixed mode simulation opposite to the one usually advocated. In the latter
use (e.g. Cellier, this volume) continuous simulation is employed unless it
breaks down because of sudden changes in either the structure of the model or the
parameter settings.

3. DEMONs, EXPERTs and self structuring

DEMONs and EXPERTs, here introduced in diverse contexts, are closely related entities. They both are autonomous entities, generated by the system, with own local
knowledge and concerns, which take over control and restructure the system.
DEMONs proved to be a very helpful construct to augment the individual's information processing without violating local constraints.
DEMONs serve as links between potential interaction partners. They are used by an
individual to handle 'expected' interrupts, i.e. situations which it knows may
happen and which it knows to handle. In our implementation they in fact take the
role of 'search images'. DEMONs may however outlive the situation which generated
them.
EXPERTs likewise handle expected situations and they act as DEMONs on incoming
REQUESTs. They differ from the prototype DEMON in building up their own expectations, and in containing more knowledge. Moreover, contrary to the prototype DEMON,
they live in state-space rather than space-space. These differences are not pro-

found, however, and many such expansions are conceivable (and useful). Opposite to the MIRROR system, EXPERT (DEMON) control tends to be the norm in MICMAC. Therefore one is inclined to say that MICDYN, i.e. the basic continuous simulation, handles unexpected interrupts, rather than to say that EXPERTs handle the expected interrupts.

References

Bobrow, G.B. and Winograd, T. (1977), "An overview of KRL, a knowledge representation language", Cognitive Science, Vol 1, pp3-46.
Bulirsch, R. and Stoer, J. (1966), "Numerical treatment of ordinary differential equations by extrapolation methods", Numerische Mathematik 8, pp1-13.
Charniak, E. (1972), Toward a model of children's story comprehension, AI-TR-266, MIT.
Dahl, O.J., Myhrhaug, B., Nygaard, K. (1970), Simula information, Common Base Language, Norwegian Computing Center.
Heistad, E. et al (1975), NDRE SIMULA implementation user's manual, FFI-Mat Teknisk Notat S-370, Reference: Job 271/17-, Kjeller, Norway.
Herman, G.T., Arbib, M.A. and Schneider, R.E. (1974), Biologically motivated automata theory, IEEE, New York
Herman, G.T. and Rozenberg, G. (1975), Developmental systems and languages, North Holland/American Elsevier, Amsterdam.
Hewitt, C. (1973), Oral presentation at the Fourth International Joint Conference on Artificial Intelligence.
Hogeweg, P. (1977a), "Locally synchronised developmental systems, conceptual advantages of discrete event formalism", in Zeigler, B.P. (ed.), Frontiers in systems modelling, in press.
Hogeweg, P. (1978), "Simulation of the growth of cellular forms", Simulation.
Lindenmayer, A. (1968), "Mathematical models for cellular interactions in development. I: Filaments with one-sided input.
    II: Simple and branching filaments with two-sided inputs".
Journal of Theoretical Biology, Vol 18, pp280-312.
Lindenmayer, A. and Rozenberg, G. (1975), Formal languages, automata and development, University of Utrecht, Netherlands.
Misner, Ch.W., Thorne, K.S. and Wheeler, J.A. (1975), Gravitation, Freeman and Co, San Francisco.
Neumann, J. von (1960), Theory of self reproducing automata, University of Illinois Press, Urbana.
Simon, H. (1969), The sciences of the artificial, MIT press.
Ulam, S.M. (1962), On some mathematical problems connected with patterns of growth of figures, reprinted in A.W. Burks (ed.), Essays on cellular automata, University of Illinois Press (1970).
Zeigler, B.P. (1977), "System theoretic description of models: a vehicle for reconciling diverse modelling concepts", in: Proceedings of the NATO conference on trends in applied general systems research, ed. G.J. Klir, Plenum Press.
Zeigler, B.P. (1978), "Multi-level multi-formalism modelling - an ecosystem example", in: Theoretical ecological systems, ed. E. Halfon, Academic Press.

METHODOLOGY IN SYSTEMS MODELLING AND SIMULATION
B.P. Zeigler, M.S. Elzas, G.J. Klir, T.I. Ören (eds.)
© North-Holland Publishing Company, 1979

DESIGN OF INTERACTIVE SIMULATION SYSTEMS
FOR BIOLOGICAL MODELLING

Jeffrey R. Sampson and Mireille Dubreuil
Department of Computing Science
The University of Alberta
Edmonton, Canada

Interactive simulation systems can allow biologists
unfamiliar with computing to implement and test models in
a conversational environment. Design of such systems
raises a number of methodological considerations. This
paper looks at the impact on system data structures and
algorithms of interactive facilities for model
specification, simulation monitoring, and data collection.
Examples are drawn from three interactive biological
simulation systems, with emphasis on a recently completed
system for modelling spatial dynamics in population
ecology. A section is devoted to the data structures in
that system.

1. INTRODUCTION

Use of computer simulation to study models should not be limited to
investigators conversant with computers and programming. Interactive
simulation systems can allow modellers from other disciplines, like
biology, to implement, exercise, and modify their models in a
flexible, conversational environment. Such interactive systems
cannot pretend to the scope of general purpose simulation languages,
but must nevertheless serve a broad class of models to justify the
extensive software support involved.

Design of interactive simulation systems raises a number of
methodological considerations, many of which are nicely posed in the
context of modelling biological phenomena. This paper focuses on
ways in which design of system data structures and algorithms is
affected by interactive facilities for model definition and editing,
for simulation monitoring, and for data review and collection. A
special section is devoted to the design of data structures in one
particular system. Finally, there is a brief look at some
implementation considerations.

Illustrations are drawn from three interactive biological simulation
systems recently developed in the University of Alberta Computing
Science Department. An earlier report (Sohnle, Tartar, & Sampson,
1973) treated some of the issues discussed here, in the context of
an interactive system for cell-space models.

One terminological caveat is appropriate. The authors recognize the
hazards associated with the viewpoint which identifies a model of a
system with one of the (potentially very many) programs which
implements that model for purposes of computer simulation.
Nevertheless, for terminological brevity, the term "model" is used

throughout this paper to refer both to the abstract entity and to its realization as some specific program.

2. THREE SIMULATION SYSTEMS

This section presents brief sketches of the systems which are used to illustrate the methodological aspects discussed in subsequent sections. All three systems are concerned with biological modelling, but at successively higher levels of biological organization. Different simulation paradigms are also employed. the first system being continuous and the latter two of the discrete time type. The systems are here described as they are seen by the user who has a model to implement and is not concerned with hardware or software.

For each system, there is a suggestion of the range of models which can be usefully implemented, as well as mention of some which have already been studied. There is also brief comparison with one or two related simulation systems, intended for the reader with deeper interest in the particular domain of biological modelling.

2.1. IBSS

The Interactive Biochemical Simulation System (Huneycutt, 1976) has been developed to provide a facility for simulating biochemical reaction systems. A model is defined by a set of chemical flux equations (e.g., E + S <-> ES), associated numerical data such as reaction rates and initial concentrations of reactants, and (optional) macro commands. The macros cause the system to monitor the state of the simulation and automatically perform specified alterations to the model state or structure if specified conditions are met.

A model defined in IBSS may be stored and later retrieved; it may be modified by deletion of equations, alteration of reactants, and addition or deletion of equation terms. Two sets of equations and associated data can be merged to form a composite model. Any current model component can be retrieved for inspection at any time.

To run a simulation the user specifies the duration of the run, the minimum interval to be used in numerical integration, the maximum tolerable error resulting from such integration, and the number of data points to be saved. A halted simulation can be continued from the point of interruption, after any required modifications to the model.

IBSS has been used to study simple enzyme-substrate reaction kinetics as well as a more complex system of allosteric inhibition, involving some 10 equations and 9 reactants. It is possible for the experienced user to employ the macro features as a kind of simple programming language. This higher level logic is especially useful in modelling of genetic control mechanisms, like those involved in the regulation of expression of bacterial operons.

There have been other simulation systems developed for models of biochemical pathways. Cassano's (1977) system has many points of similarity with IBSS, including an interactive graphics facility. Cassano has used the system to study purine metabolism. In a recent paper in _Simulation_, Garfinkel (1977) has provided an introduction to biochemical simulation methodologies and a survey of his own extensive work in the area.

2.2. NETN

Covington, Sampson, & Peddicord (1978) have developed a system for interactive simulation of neural networks (with synaptic interconnections of a type previously investigated). To define a model the user responds to prompts for characteristics of neurons, synapses, and external inputs. Neurons have individual resting potentials, firing thresholds, and threshold-decay time constants. Synapses are characterized by type (excitatory or inhibitory) and source (another neuron or an external input). The two parameters which govern the behavior of a synapse are its gain, measured as the number of transmitter molecules released per presynaptic pulse, and its persistence, which is an inverse function of the time constant for the decay of transmitter molecule effectiveness. External inputs to a model network may be periodic or stochastic; in the latter case the distribution may be uniform, exponential, or normal, with parameters supplied by the user. There is no inherent limit to the number of model components.

A model may be stored and subsequently retrieved for simulation. In an editor environment, the user can add or delete or change the characteristics of any type of model component. He can specify accumulation and reduction of various types of data concerning the firing behavior of specified individual neurons, to be output in a trace environment.

Covington, Sampson, & Peddicord (1978) report experiments with NETN, including model networks which function as bandpass filters, as pacemaker neurons and nets, and as a good approximation to the physiological behavior of a cerebellar circuit.

A neural network simulation system with similarities to NETN has been described by Perkel (1976). While it would appear to handle a somewhat larger class of models, Perkel's system is not interactive and lacks nearly all of the user oriented features of NETN. Further comparison of the two systems is available in the closing section of the paper by Covington, Sampson, & Peddicord.

2.3. ECOSIM

ECOSIM (Sampson & Dubreuil, 1978) allows a user to model the behavior of an ecological community of many species in space and time. The environment is represented as an unbounded 2-dimensional grid of hexagonal cells. Any number of species can be incorporated in the model, with any number of individuals in each species. After specifying the initial number of individuals in a species, the user either enters their locations (as coordinates) or selects parameters for a stochastic distribution (uniform, normal, exponential, or poisson).

Individuals of mobile species move by random walk, each cell around and including the occupied cell having equal probability unless the user specifies otherwise. A species mobility factor determines the probability of each individual moving in any time step. A competition factor determines the probability of eliminating either of two individuals of the same species which attempt to occupy the same cell.

For each prey species, the user assigns to the predator(s) a probability that predator will kill prey, as well as the maximum number of prey the predator can kill per unit time. By means of a

"satiety level" parameter, a predator can starve to death and have
its mobility depend on how hungry it is. For any species which
produces offspring, the user sets a breeding interval. a probability
of breeding, and the mean and variance of offspring number. The
offspring are distributed in space near the parent.

At the end of each time step, a "global" function defined by the
user can be applied to change the number of individuals of a given
species. Deletions are made at random, while additions of new
individuals are made according to the initial distribution. This
feature can be used to simulate migration and seasonal variations.

Barriers can be established either to create refuges for certain
species or to impede or prevent mobility. For any barrier type, the
user provides probabilities of crossing for any number of species.
Barriers can be distributed either at random or by specifying cell
coordinates.

After defining his model the user may modify it, store it for later
retrieval, and/or initiate a simulation to be run for a specified
interval. Current work with ECOSIM is concerned with stability in
predator-prey systems having various types of species mobility and
spatial distribution of refuges.

Auslander (1976) has defined a syntax and provided implementation
guidelines for a simulation language for population ecology.
Although it is intended for use by biologists wishing to study a
range of models, the differences between Auslander's language and
ECOSIM are more striking than the similarities. In the Auslander
syntax, a model is defined as a set of equations in which spatial
factors are not relevant, while age structures are of primary
importance. (ECOSIM does not support models having explicit
age-structure variables.)

Another population ecology simulation system with an exclusively
temporal orientation is Hogeweg & Hesper's (1978) TRICLE system.
Intended primarily for student use, TRICLE makes available several
models in the form of differential-difference equation systems with
parameters set to "default" values. The user selects a model and
enters any desired modifications to the parameters before running a
simulation. The main similarities between TRICLE and ECOSIM are
dialog with the user and dynamic graphical display of model
behavior. ECOSIM might be improved by providing the user with more
default options for parameters, along the lines of TRICLE.
(Elsewhere in this volume, Hogeweg and Hesper discuss MIRROR, a
non-interactive system for simulation of local dynamics of
organisims interacting in two dimensions.)

3. MODEL DEFINITION

3.1. Model Description Language

To define a model in an interactive mode, the user has to
participate in a dialog with the system. The system designer must
provide a description language that is (1) easy to understand, (2)
reasonably efficient in terms of the time required to define a
model, and (3) effective in the sense that all necessary model
components and parameters are dealt with.

The objectives set out above are not always mutually consistent.
Ease of understanding, for example, is facilitated by a natural
dialog with the user. Such a conversation can be quite costly in
terms of computer time and space resources. And it may be difficult
to devise a dialog that is both natural and necessitates user
specification of all model components. Model design in terms of
equations, on the other hand, accomplishes the latter objective at
the cost of some clarity, especially for users who are not
mathematically inclined.

Shown below are portions of a dialog with ECOSIM which illustrate a
number of important features of a model description language. Notice
that the user needs to remember very little about the system or the
description language syntax, since all relevant information is
requested in a natural order and responses are very simple. The user
can determine from punctuation whether the required information is
YES/NO (the system output ends with a ?) or numerical data (the
output ends with a :). YES and NO can be shortened to their initial
letters for somewhat faster input (as can other response words not
shown in the example). A null response serves to terminate the
specification of unbounded lists, such as the species a predator
preys upon or the species whose mobilities are affected by a
barrier.

When the user supplies an inappropriate response (as to the request
below for INITIAL NUMBER of species 1), the prompting question is
repeated. This simple type of diagnostic is adequate because the set
of possible correct responses is readily identifiable. In other
situations, as when the data type of a supplied number is incorrect,
a more explicit error message is given. Design of simple but
informative diagnostics is an important aspect of the construction
of model description languages.

SYSTEM	USER
CREATE?	YES
SPECIES 1?	YES
DISPLAY SYMBOL:	*
MOBILITY(0-1):	0
PREDATOR?	NO
INITIAL NUMBER:	GHI
INITIAL NUMBER:	400
RANDOM DISTRIBUTION?	YES
U, N, E, or P:	U
XMIN:	-20
XMAX:	20
YMIN:	-10
YMAX:	30
GLOBAL MODIFICATIONS?	YES
FUNCTION:	X**2+5
SPECIES 2?	Y
DISPLAY SYMBOL:	+
MOBILITY(0-1):	.7
COMPETITION FACTOR(0-1):	.2
UNIFORM RANDOM WALK?	Y
PREDATOR?	Y
PREY SPECIES:	1
PROBABILITY OF KILLING:	1.5
PROBABILITY OF KILLING:	.666
MAX KILLS / UNIT TIME:	1
PREY SPECIES:	

```
BREEDING INTERVAL:  ...............  10
BREEDING PROBABILITY:  .............  .85
OFFSPRING MEAN:  ...................  2
OFFSPRING VARIANCE:  ...............  1
INITIAL NUMBER:  ...................  2
RANDOM DISTRIBUTION?  ..............  N
           2 X-Y COORDINATE PAIRS:       5  3
                                         2  3
PRINT INITIAL DISTRIBUTION?  .......  N
GLOBAL MODIFICATIONS?  .............  N
SPECIES 3?  ........................  N
BARRIER 1?  ........................  YES
DISPLAY SYMBOL:  ...................  -
SPECIES:  ..........................  2
CROSSING PROBABILITY:  .............  0
SPECIES:  ..........................
NUMBER OF CELLS:  ..................  4
RANDOM DISTRIBUTION?  ..............  N
           4 X-Y COORDINATE PAIRS:       1  1
                                         1  2
                                         2  1
                                         2  2
           BARRIER 2?  .................  NO
```

The system designer has the option of actually building model data structures during the definition dialog or after it is finished. The former alternative is usually preferable, since the computer is not totally idle during the user's "thinking time" and there is no need to save model data in some temporary internal format. But it must be emphasized that dynamic model creation influences significantly the design of data structures. The kind of data structure flexibility required for building a model "on the fly" is similar to that needed for effective editing of previously defined models, the subject of the next subsection.

3.2. Editing

Any workable interactive simulation system must permit the user to correct or alter a model before simulation or when simulation has been interrupted. The editing facility in ECOSIM is again a dialog in which the user is prompted for the type of model component to be edited, for whether the change is addition, deletion, or modification, and in the last case for which elements are to be m-dified. The user who has developed some familiarity with a simulation system may begin to find such dialog-based editing quite verbose and tedious.

The alternative is to place a somewhat larger demand on the user's memory and supply a simple command language for editing. This is the approach of NETN, where the command EDIT prepares the system for inputs like "6,8dn" (delete neurons numbered 6 through 8), "3ae" (add external input 3), "1cs" (change parameters for synapse 1 of the current neuron), and "1,3pn" (print out the current characteristics of neurons 1 through 3). When a model element is to be added or changed, the user is subsequently prompted for its (new) characteristics and shown the current values (if they exist). A tradeoff between efficiency and ease of understanding has been revealed in this comparison of the editing features of ECOSIM and NETN. Similar tradeoffs occur in the model definition and simulation monitoring phases of interactive simulation.

Regardless of the system designer's decision on the above issue, the mere presence of an editing capability demands highly flexible data structures. Since model components may appear and disappear on user whim, the data structures must be easy to contract and expand. This means that linked allocation of storage is usually much preferable to an array format, since few widely available programming languages have dynamic array facilities. Even in languages like FORTRAN, in which IBSS was written, a system of pointers maintained by the programmer is essential for traversal and modification of data structures. ECOSIM was written in a version of ALGOL, and NETN in "C" (a PASCAL-like language available under the UNIX operating system on a PDP-11/45), both of which have pointers as data types.

Another convenience of using languages like ALGOL and C is the availability of structures or records which can contain the mixed data types frequently associated with components of complex models. Thus in NETN the neuron data structure contains 9 "real" data items, 4 integers, and 6 pointers. The record containing information common to one species in ECOSIM contains 29 reals, 4 integers, 2 pointers, and 3 character strings.

While it is possible to "program around" the limited data structuring features of FORTRAN-like languages, choice of implementation language is important in the design of interactive simulation systems. The role of data structures in such systems is treated further in Section 6 below, in the context of the details of the ECOSIM design.

3.3. Model Storage and Retrieval

All three systems under discussion provide the user with the option of storing a model upon definition or after simulation and/or editing. IBSS and ECOSIM also have facilities for storing some types of output data accumulated during the run.

The ability to retrieve a stored model has at least two important uses in interactive simulation. First, a simulation can be carried out in pieces, with intermediate stages of the model stored, whenever there are limitations on resources like computer availability or the user's time. Second, and even more important, successive simulation studies can be carried out from the same starting or intermediate point, allowing the user to determine the effects of small changes in model parameters or to recover from a hardware or software failure.

Effective model storage and retrieval facilities influence system design in the areas of compactness and bookkeeping. It may be necessary to devise a compressed form of model representation if the run-time version is so large that storage of many models would tax the available mass storage devices. It is of course simpler if the model retrieved from storage has the same format as one just defined by the user.

Since the user will want to recover a model that is identical in all respects to the one saved, the system designer must be careful not to overlook any state variables when saving models. Of special importance are the current seeds of any pseudorandom number generators employed in the system. It is nice to give the user the option of resetting such seeds, enabling him to replicate simulation experiments with a model under controlled stochastic conditions.

4. SIMULATION MONITORING

At the core of an interactive simulation system are the algorithms that actually execute simulation runs. These algorithms frequently play a relatively minor role in system design considerations. As already suggested, much of the data structure is determined on the basis of flexible communication with the user. Procedures must then be designed to manipulate these structures in accordance with the chosen simulation paradigm, which may be discrete event, discrete time (as in NETN and ECOSIM), or continuous (as in IBSS).

Even in the domain of simulation execution procedures there remains the omnipresent concern of interacting with the user. While the simulation is running, the user needs to know as much as possible about what is happening. He must retain a measure of control over the simulation. This section considers how such requirements can be met.

4.1. Duration, Interruption, and Continuation of Runs

At the onset of simulation, the user should have a variety of options as to how long and under what conditions the run is to proceed. Of first importance is the intended length of the simulation in terms of model time. Since computer usage can be expensive, it is also useful if there is a way to avoid runs of unanticipated cost by specifying maximum allowable CPU time.

Beyond these nominal termination conditions, a simulation may halt for other reasons. It is usually not possible to recover gracefully from problems created by bugs in the simulation system, and even less so from operating system crashes. In such cases, however, a record of simulation activity to the point of interruption can often be useful. Accumulation of such a record is done automatically in ECOSIM; but some system crashes will destroy it.

Of much greater relevance to the designer of a simulation system is the controlled premature termination of a run. Conditions for such termination can be implicit or determined by the user. Implicit termination occurs in IBSS if a reactant concentration becomes negative. In ECOSIM, the user may set maximum and minimum bounds on the population size of any species. During every time step, the system verifies that the number of individuals remains within these limits.

The most comprehensive set of user monitoring and interrupt facilities among the systems considered here is found in IBSS. The user may establish one or more Reaction Monitors which look for a specified relation between the values of two model variables (or a variable and a constant) during each integration step. Whenever the relation becomes true, control is transferred to a sequence of one or more macro statements which can actually alter the structure of the model in addition to (or instead of) halting simulation. An interrupt can be generated, for example, whenever the concentration of one reactant exceeds that of another. Or the values of derivatives of reactant concentrations can be monitored to detect local maxima or minima.

When a simulation run undergoes a controlled premature termination, for whatever reason, the issue of continuing the run arises. Both NETN and ECOSIM have "continue" commands which prompt the user for the number of (additional) time steps and resume the simulation from

the point of interruption. Continuation is also possible after
normal termination of a run. Before continuing, the user may invoke
the editor to modify the model. Any number of continuation commands
may be used in succession to create a multi-phase simulation run.
This can be a convenient option when the user wants to simulate an
event which is not part of the system's repertoire, such as an
ecosystem disaster in which several species suddenly disappear.

To make a continuation facility effective, the system designer must
save all state variables and pseudorandom number generator seeds at
any normal termination or controlled interrupt point. He thus
virtually eliminates the need for any reinitiation or
"recompilation" steps in the process of getting things started
again; such steps could both annoy the user and consume costly
computer resources.

4.2. Dynamic State Display

Although the results of a simulation experiment are often most
effectively presented in summary form after run termination (by
methods discussed in the next section), the user may wish to observe
the changes in one or more aspects of the model state as they
change.

Of the three systems discussed here, only ECOSIM offers convenient
display of intermediate states. No such facility is available in
NETN. IBSS can display the current condition of any model component,
but only after simulation has been interrupted. So the user desiring
frequent updates must frequently stop and continue the simulation.

ECOSIM asks the user at the beginning of a run how often he would
like to see a display of the ecosystem being simulated. The response
can range from every time step to a single display at the end of the
run. The model is presented at these specified intervals in the form
of a 2-dimensional array of cells, with locations of barriers and
members of species shown by their respective display symbols. This
type of output facility has important implications for the design of
the system. Since the internal representation of a model in ECOSIM
is not 2-dimensional, the designer had to find an efficient means of
constructing a view of the ecosystem from the linked storage format.

If the occupied area of the ECOSIM grid is too large to be viewed
all at once, the user selectively displays windows of 12 by 12
cells. Figure 1 (see next page) shows how one such window appears on
the screen. The superimposed hexagonal grid may be eliminated from
the display at user option. P and # are display symbols for
organisms; X denotes the cells of an enclosing barrier. Different
combinations of windows (including none at all) may be displayed at
different output points.

J.R. SAMPSON and M. DUBREUIL

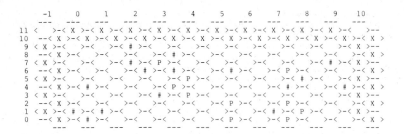

FIGURE 1. Dynamic state display in ECOSIM

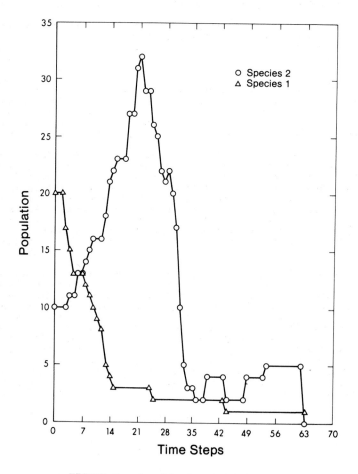

FIGURE 2. Graphical output from ECOSIM

5. DATA COLLECTION

5.1. Tabular Data

In the early stages of model check out, a user may require a rather
detailed summary of model behavior, following a simulation run.
ECOSIM keeps a complete record of the dynamic outputs at user
specified intervals, for purposes of future reference. At the
completion of the run, this record disappears unless the user
chooses to make it permanent. In this way, the masses of data which
are usually not required for validated models do not interfere with
the user's assessment of more condensed data presentations. These
summaries may be both graphical (see below) and tabular, the latter
being a synopsis of the state of the model at the termination of the
run.

At the user's option NETN can summarize information about number of
firings and average firing interval for any neuron(s) in a model.
Both NETN and IBSS provide their most informative and usable
summaries of model behavior in graphical form, as discussed next.

5.2. Graphical Data

Adages concerning the relative value of pictures and words are
particularly applicable to summaries of the simulated behavior of
complex models. Despite the fact that graphical output is frequently
dependent on specialized hardware (see Section 7), some sort of
non-numerical presentation, even a line-printer graph, is virtually
essential in most simulation systems.

IBSS' primary mode of output is a graph of reactant concentrations
against simulated time. The user controls which reactants are
plotted together, as well as the scaling of the axes. Hard copy of a
graph presented on a CRT screen is available on request. In a
similar fashion, ECOSIM can arrange for offline plots of the
population sizes over time, for arbitrary combinations of species.
An example of this type of output is given in Figure 2 (see previous
page). To provide graphical output of the sort just described, the
system designer must arrange to save values of the appropriate model
variables at the required points in time.

More sophisticated processing is required for the elaborate
graphical output features of NETN. In addition to plots against
time, of neuron potentials and firings, this system can summarize
the entire run in the form of histograms showing the distribution of
firing intervals of any neuron and/or the intervals between the
firing of one neuron ("stimulus") and any other neuron ("response").
Interval and stimulus-response histograms can be especially useful
in comparing model behavior with data reported in the
neurophysiological literature (see Covington, Sampson, & Peddicord,
1978). Finally, NETN has a special feature which produces a graph
comparing a number of simulation runs; this can be used to examine
the pass band characteristics of a network designed as a filter.

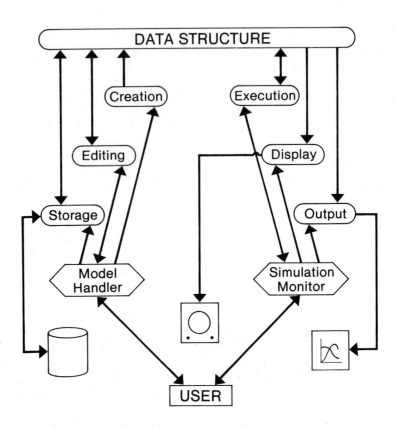

FIGURE 3. Elements of interactive simulation systems

6. DATA STRUCTURES IN ECOSIM

The logical interrelations among the various components of an interactive simulation system, as described in the previous three Sections, are suggested in the block diagram organization presented in Figure 3 (previous page). Note that virtually all of the other components can be viewed as a sort of vast interface between the user and the system data structure.

Since design of data structures is often the most critical phase in the development of an interactive simulation system, the data structures of ECOSIM are here described in some detail. ECOSIM has been chosen because, as the most recently developed of the three systems, its design has benefited most from prior experience.

ECOSIM data structures have been designed in accordance with the following constraints: species can move and grow without bound; editing can insert or delete whole components of a model in the middle of a simulation run; all members of a species must be found and processed at the same time; all organisms and objects in a given cell must be identified to determine if an organism can move there. Since only the last constraint suggests an array type organization, linked allocation of storage is the method of choice.

The primary data structures of ECOSIM are diagrammed in Figure 4 (next page). Species are represented as doubly linked circular lists, with head nodes (shown along the left edge of Figure 4) specifying parameter values applicable to all individuals; an individual node carries the organism's location coordinates, satiety level, breeding clock, and number of kills. The double circular linking allows control of systematic effects which might arise from always processing the individuals in the same order. Each time step, a random individual is chosen and the list is traversed in a randomly chosen direction, so that on the average each individual is processed before and after each other individual equally often. This "sequential" processing strategy is an economical alternative to the "parallel" approach, in which a review of the next state of the world would be required, to find and resolve all conflicts resulting from simultaneous independent activity of model components.

Barriers are represented as location nodes with pointers to head nodes (along the right edge of Figure 4) containing parametric barrier characteristics. A top node in the system contains pointers to the first nodes in lists which link the species head nodes and the barrier head nodes. Species-barrier interactions, interspecies exclusion factors, and predation relations are kept in separate lists.

The last data structure element is a hash table (at the bottom of Figure 4) with entries heading lists of all organism and barrier nodes whose locations hash to that table entry. Thus the last constraint mentioned above is satisfied. The hashing function is used each time an attempt is made to move an individual, to see if there is a barrier affecting its mobility, or if exclusion or competition are relevant. Note that there are two types of "collisions" in this hashing scheme: (1) actual concurrent occupancy of a cell by two or more model components; and (2) collisions in the usual sense of the term, when two or more cells hash to the same table entry. The "collision" detection algorithm must therefore check actual coordinates at individual and barrier nodes to see if a collision is relevant (type (1) above).

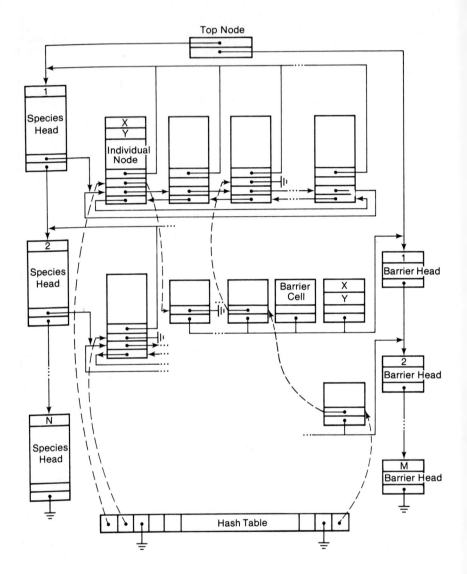

FIGURE 4.   Data structures in ECOSIM

7. IMPLEMENTATION CONSIDERATIONS

The discussion to this point has been relatively independent of the machine and operating system environment in which an interactive simulation system is implemented. This environment may limit the designer's options and make some of the goals described above unattainable. Of nearly equal concern is the extent to which machine, operating system, and programming language make the system more or less capable of easy transfer to other installations. This last section of the paper offers a few remarks on these issues.

7.1. Software and Hardware Requirements

Interactive simulation cannot be carried out in a batch processing environment. Clearly an interactive system, preferably a time-sharing system with a variety of terminal facilities, is required. Two flexible and powerful time sharing systems were used for the systems described here. IBSS and ECOSIM were implemented under the Michigan Terminal System (MTS), while NETN was implemented under the UNIX system developed at the Bell Telephone Laboratories. Current advances in small processors make it possible that interactive simulation will eventually also be carried out in a single user environment on a microprocessor based system. Also on the horizon is the attractive possibility of configuring a microprogrammable machine to fit the particular domain of application of an interactive simulation system.

The impact of hardware on system design is much more significant in the area of input-output facilities than in simulation monitoring. Model definition through extensive dialog, which can proceed quite quickly at a CRT terminal, may drag on unbearably slowly at a teletype. And the types of graphical data summaries which may be presented depend wholly on the terminal and associated support facilities available. NETN, for example, can be run on any terminal connected to the PDP-11/45 on which UNIX runs, but graphical output can be obtained only at a Tektronix terminal. The graphs produced by ECOSIM are generated offline on a Calcomp plotter. The implemented version of IBSS, on the other hand, will produce approximate plots on a teletype terminal.

7.2. System Portability

It is unfortunate that languages ideally suited to the implementation of interactive simulation systems are not widely available. Of the three systems discussed here, only IBSS is genuinely portable, being written in FORTRAN. Even so, some reprogramming would undoubtedly be required to handle input-output in an environment other than the MTS operating system.

Written in ALGOLW, ECOSIM could be moved with some difficulty to installations supporting other versions of ALGOL. Again there is input-output device dependence, since the system was designed to run at a fast CRT terminal, under MTS. NETN will run only under UNIX on a PDP-11 computer, or on a small number of other systems having implementations of C.

Both for these reasons, and because the development of the three systems was in part a learning experience, this paper is not intended as an advertisement for the simulation systems surveyed. Rather it is hoped that this review of interactive simulation methodology will benefit others who may implement similar systems.

8. REFERENCES

Auslander, D.M. Specifications for a simulation language for population ecology. In L. Dekker (ed.), Simulation of Systems, pp. 421-427, North Holland, 1976.

Cassano, W.F. A biochemical simulation system. Computers and Biomedical Research, 10, 383-392, 1977.

Covington, A.R., J.R. Sampson, & R.G. Peddicord. A frequency response neuron model: Interactive implementation and further simulation experiments. Kybernetes, 7, 45-60, 1978.

Garfinkel, D. Simulating biochemical activity in physiological systems. Simulation, 28, 193-196, 1977.

Hogeweg, P. & B. Hesper. Interactive instruction on population interactions. Computers in Biology and Medicine, in press, 1978.

Huneycutt, C.W. An interactive biochemical simulation system. M.Sc. Thesis, Department of Computing Science, University of Alberta, 1976.

Perkel, D.H. A computer program for simulating a network of interacting neurons. Computers and Biomedical Research, 9, 31-73, 1976.

Sampson, J.R. & M. Dubreuil. A computer simulation system for modelling spatial dynamics in community ecology. Fourth International Congress of Cybernetics and Systems, Amsterdam, August, 1978.

Sohnle, R.C., J. Tartar, & J.R. Sampson. Requirements for interactive simulation systems. Simulation, 20, 145-152, 1973.

METHODOLOGY IN SYSTEMS MODELLING AND SIMULATION
B.P. Zeigler, M.S. Elzas, G.J. Klir, T.I. Ören (eds.)
© North-Holland Publishing Company, 1979

THE FORSIM VI DISTRIBUTED SYSTEM SIMULATION PACKAGE

M.B. Carver
Atomic Energy of Canada Limited
Chalk River Nuclear Laboratories
Chalk River, Ontario K0J 1J0, Canada

The FORSIM VI package is a package for simulation of systems described by ordinary differential
and partial differential equations. Partial differential equations are transformed into coupled sets of
ordinary differential equations in time by the method
of lines, using piecewise approximation functions in
the spatial coordinates. The resulting set of equations is integrated by variable order, variable step,
error controlled algorithms. The FORSIM user may
describe his equation system in a single FORTRAN subroutine, or may build large simulations incorporating a
number of user or library supplied routines.

The presentation discusses some of the technology involved in the program modules, such as the
basis functions, sparse matrix assessment of the
system Jacobian, the handling of temporal discontinuities or events, and the solution of initial value and
boundary value problems. Some illustrative examples of
the use of FORSIM are provided, and a brief discussion
of current use in simulation applications is included.

INTRODUCTION

Ordinary differential equations (ODEs), and partial differential

equations (PDEs), may occur in mathematical models arising from a variety of

scientific disciplines. The numerical solution of physically realistic systems

of differential equations, particularly PDEs, is an intricate process which

usually requires the scientist to undertake the development of numerical tech-

niques specialized for the problem in question; expensive, time-consuming work

which is difficult to validate and may have little relevance to subsequent

problems to be solved.

This has caused considerable interest in generally applicable simula-

tion software designed to automate the solution of PDEs. Such software is the

product of the cumulative experience of experts in numerical analysis, simula-

tion, and programming techniques and makes use of established, well tested,

robust algorithms. The software requires of the user, that he present his prob-

lem in a well-posed clearly defined manner, and then relieves him of the further

responsibilities of choosing appropriate algorithms and ensuring accuracy and

validity; as these duties are built into the system.

The positive response of users of such systems provides a continuing

stimulus for their development and improvement. Applications are encountered,

not only in the traditionally mathematical branches of physics and engineering,

but also in fields such as the biological sciences, which are increasingly attracted to the possibilities of computerized modelling.

Depending on the main field of application foreseen, simulation packages may be designed for modelling continuous systems, discrete event sequences,
or a combination of both. The combined simulator is naturally more versatile in
application but to keep a package down to a manageable size, this versatility may
have to be gained at the expense of efficiency in specialized areas. For this
reason partial differential equation packages have normally been designed for
continuous system simulation, but this does not necessarily preclude simulation
of event sequences as we shall see later.

A number of excellent PDE software packages exist, and are cited chron-
ologically in references [1] to [14]. The most versatile of these use some form
of the method of lines to convert the PDE's to ODE's, and hence are also suitable
for use in ODE or mixed ODE/PDE simulation. Of this group, the earlier version
of FORSIM[6] has been in use for several years, at Atomic Energy of Canada
Limited, and at institutions in North America, Europe and Australia. Experience
in using this package for a variety of applied simulations led to the reassess-
ment of the design to improve efficiency, versatility of application and ease of
use, particularly for two and three dimensional PDE problems. The resulting
FORSIM VI package contains a number of improvements in technology and versatility
but retains many of the same features which made its predecessor easy to use.

DESIGN PHILOSOPHY

A package such as FORSIM is used by modellers from a variety of disciplines and must be designed to provide maximum versatility without sacrificing
efficiency. Projects to be solved also may range from the exploration of the
behaviour of a small set of equations, to the implementation of large simulations
containing several hundred differential equations. FORSIM is designed not only
to permit a novice or student to quickly obtain a solution to a simple equation,
but also to act as an efficient numerical analysis base on which a scientist or
engineer may build a simulation involving complex continuous systems with implicitly embedded discrete events.

For these reasons, and also to aid portability, the package is written
in FORTRAN. The user is required to write only one subroutine to specify his
simulation, but if the project demands, this routine may call any number of
auxiliary routines which the user may also supply or access from a library.

Although boundary value problems, and PDE problems may be solved with
FORSIM, both are reduced to initial value ODE problems, so the primary function
of the package is to integrate the equations specified in the user routine, which
relates to the control routines and available libraries as shown in Figure 1.
The integration algorithms are the heart of the package, and as such, must be
fine tuned for efficiency.

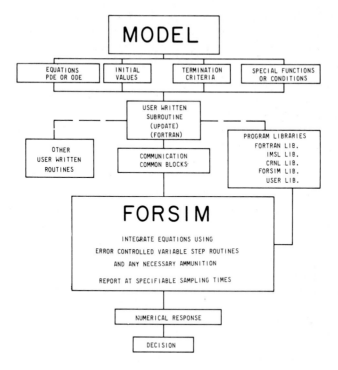

Figure 1: Modelling with FORSIM

ORDINARY DIFFERENTIAL EQUATION SOLUTION

The model is reduced by the package into the initial value problem of solving

$$\overline{\dot{Y}} = \overline{f}[\overline{Y}(t)], \quad \overline{Y}(0) = \overline{Y}_0 \tag{1}$$

where $\overline{Y}$ and $\overline{\dot{Y}}$ are vectors of dependent variables and their derivatives, t is the independent variable, and $\overline{f}$ is the functional relationship defining the equations.

A large number of integration algorithms to solve such a system have been published, and many simulation packages avoid the question of which method is most suitable by providing a large number of algorithms as options from which the user may choose. However, the package designer is better equipped to make such a choice, and it should be part of his duty to make the rational decision of including or excluding algorithms. The integration algorithms in FORSIM have been chosen after exhaustive comparative testing, the results of which appear in [15]. Two algorithms only were selected as the most robust and efficient available, but before discussing these specifically, some of the requisite fundamental features of an effective algorithm should be established.

Requisite Components of an Integration Algorithm

i)   Integration Over a Step, and the Associated Error Estimation

The fundamental feature of an integration algorithm is the combination of the method used to advance the integration from time t to t+h, with that used to estimate the probable accuracy of the result. Although this classifies the algorithm, efficient implementation requires further considerations.

ii) Tolerance and Error Base Specification

The error estimation E_j associated with each dependent variable Y_j is returned from (1) and compared with a specifiable tolerance τ. The step is acceptable providing

$$E_j < \tau * \phi(Y_j). \tag{2}$$

The choice of the error base function $\phi(Y_j)$ has been subject to considerable controversy, but the only generally applicable definition is

$$\phi(Y_j) = \max(|Y_j|, Y_{sj}) \tag{3}$$

Y_{sj} is the value below which Y_j may be considered insignificant. Note that this normally gives a relative error control but relaxes to absolute error control when Y_j is not significant. This not only permits integration to proceed through zero accurately, but also efficiently leave or approach zero in asymptotic fashion[15].

iii) Control and Variation of Step Size

Having established acceptable and probable errors, it is simple to extend this information to compute optimum step size h_2 from current step h:

$$h_2 = h\psi(\tau[\phi/E]_m), \ E\neq 0 \tag{4}$$

where the function ψ is related to the method and m denotes the minimum ratio j.

iv) Discontinuities or Events

Equation set 1 is more generally written as

$$\overline{Y} = f_i[\overline{Y}(t)], \ \overline{Y}(0)=\overline{Y}_0, \ i=\varepsilon(\overline{Y}(t),t). \tag{5}$$

Here the equation definition f has several possible states i which change according to some integer event function ε. An event causes a change in i and a discontinuity in the equation definition at a time which is implicitly dependent on the evolution of the integration.

In fixed time step this causes no stability problem, but the variable time step algorithms necessary for efficient integration will break down on encountering a discontinuity. Any algorithm with good step control will recognize the discontinuity but will waste a great deal of time attempting to negotiate it as the exact time of the event is not known in advance. It is a simple matter to avoid this waste by building in a predictor feature which permits the point of discontinuity to be detected before the natural progression of integration is disturbed. Integration may then proceed exactly to this point and the new state is treated as a new initial value problem. Savings of up to 90% can be realized using such a feature with simple equations, and large equation sets with discontinuities are insoluble without it[16].

vi)   Stiffness and Stability

        An integration algorithm of order k approximates the true solution of an ODE, but also computes k-1 spurious solutions. These spurious solutions are always present, and must remain insignificant in comparison with the principal solution if the algorithm is to be numerically stable. For systems of equations, the significant parameters affecting this behaviour are the associated time constants, which are functions of the eigenvalues of the Jacobian matrix. The Jacobian is defined

$$\overline{J} = \partial \overline{f} / \partial \overline{Y} \tag{6}$$

for the system of equations (1). If the range of magnitude of these eigenvalues is large, the system is said to be stiff, and will cause most algorithms to behave unstably for anything but prohibitively small step sizes.

The FORSIM Integration Algorithms

i)   Runge Kutta Fehlberg (RKFINT)

        This algorithm combines a fourth order Runge Kutta formula with a fifth order error estimate in six function evaluations as published by Fehlberg[17]. The implementation is described in reference [15]. It is an efficient algorithm with optimum step size control. As it requires low auxiliary storage, it is used as default in FORSIM.

ii)   Gear-Sparse (GEARZ)

        This algorithm is used for large and/or stiff equation systems, and is a product of incorporating the Curtis and Reid sparse matrix routines[18] for use with the Hindmarsh[19] implementation of the Gear integration algorithm[20]. It contains backward difference predictor corrector formulae for stiff equations, and the Adams formulae for non stiff equations.

        The efficiency of this algorithm is achieved by using the Jacobian matrix to make the predictor-corrector formulation pseudo implicit. The effectiveness therefore depends heavily on the manner of evaluating and handling the Jacobian, and a drawback of the Gear routines has been that a large fraction of computing time can be spent manipulating this matrix. The FORSIM version has several options for handling the Jacobian, the most efficient of which automatically and dynamically evaluates the sparsity structure of the matrix and optimizes its evaluation. FORSIM also monitors the handling of the Jacobian, and may automatically change options, if excess storage is being used.

THE USE OF FORSIM FOR ORDINARY DIFFERENTIAL EQUATIONS

The Function of the User Routine

The UPDATE routine is written by the user to define the problem to be solved by FORSIM. The routine is so named, as it is called by the integrator at each new time to update the current values of all derivatives.

Whether dealing with partial or ordinary differential equations, the UPDATE routine must specify the form of the equations and store the variables in such a way that the system can get at them. This is done by utilizing a small number of labelled common blocks, a method which has two advantages; it is computationally efficient, and permits all controlling variables to have default values preset by the system, thus requiring the user to specify only those with which he is concerned. It also permits him to use whatever variable names he chooses for block elements.

In ordinary differential equations, the dependent variables are functions of one independent variable, normally time. ODE's are specified, therefore, by writing an explicit expression for the time derivative of each dependent variable.

For example, two ODE's involving variables F and G and time T are specified by defining the derivatives FT (= dF/dt) and GT (= dG/dt):

$$FT = \phi_1(F,G,T)$$
$$GT = \phi_2(F,G,T)$$

(7)

and the initial values F_0, G_0 at T=0.

The coordination of the user's routine to the system has been designed to provide an optimum combination of execution speed, user convenience and versatility, whether used for ODE's or PDE's.

A Simple ODE Case

For illustrative purposes, a small system can be presented with clarity, so for this reason a well-known example of two ODE's has been chosen. The ecological predators and prey model, exhibits cyclical behaviour. In this problem, species A has an infinite food supply, and species B preys entirely on A. The population equations are

$$\frac{dA}{dt} = 2A - \alpha AB$$

$$\frac{dB}{dt} = \alpha AB - B$$

(8)

where α is an encounter frequency parameter.

Although somewhat trite in its import, this example was chosen
to strongly emphasize the importance of quality software in even the simplest
problems. A well-known manufacturer of programmable calculators published an
erroneous analysis of this problem, simply by using a fixed step integration
algorithm with no controlling estimation of truncation error[21]. Neither the
calculator, nor the algorithm was at fault, merely the application, as a correct
solution could have been obtained by using a fixed step size an order of magnitude lower.

The UPDATE routine for all FORSIM simulations is in four functional
sections. The first contains labelled common block storage for communicating to
FORSIM the values A, B of the dependent variables, their derivatives DA,DB, the
independent variable T and a print control INOUT. The second section establishes
initial conditions, the third poses the equations by defining the derivatives,
and the final section prints results at specifiable times. Sections may call
other routines if necessary.

The routine to solve this problem would be simply:

```
         C          SUBROUTINE UPDATE
         C          SECTION 1  COMMUNICATION COMMON BLOCKS
         C
                    COMMON/INTEGT/A,B/DERIVT/DA,DB
                    COMMON/RESERV/T/CNTROL/INOUT
         C
         C          SECTION 2  INITIAL CONDITIONS
         C
                    IF(T.NE.O)GO TO 50
                       A=300.
                       B=150.
                       ALF=.01
         C
         C          SECTION 3  EQUATION STATEMENT
         C
            50      DA=2*A-ALF*B*A
                    DB=ALF*B*A-B
         C
         C          SECTION 4  PRINTOUT
         C
                    IF(INOUT.NE.1)RETURN
                       WRITE(6,100)T,A,B
           100      FORMAT(3F10.5)
                    RETURN
                    END
```

No data cards would be required, as all necessary data is included in
the UPDATE routine. FORSIM will use default system parameters. It is also not
necessary to specify the number of equations. FORSIM determines this by checking
the number of initial conditions imposed, and then reports this. If the number
reported is less than the number of equations, one or more initial conditions
have been omitted and answers will be incorrect. This is a common mistake, and
the built-in diagnosis-check is invaluable.

The FORSIM solution, together with the abortive result from reference [21] is shown in Figure 2.

Practical Cases

FORSIM contains a large number of utility routines for simulation. These normally reside in the FORSIM library, and are selectively loaded only if called by the user routines. They include tabular interpolators, delays, discontinuity handling routines, plotting routines and a number of routines specialized for the handling of partial differential equations. Attention will here be confined to these PDE routines.

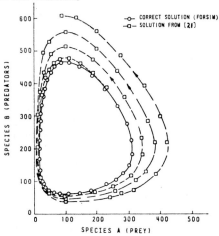

Figure 2:   Solution of the Predator-Prey Problem

PARTIAL DIFFERENTIAL EQUATION SOLUTION

Method of Lines Formulation

There exists a multiplicity of schemes for solving partial differential equations, and in general, a different method can be developed to most efficiently solve each PDE set. However, one class of methods, the method of lines, has been shown to be very general in application. In this method, approximating functions are used to convert the PDEs to a coupled set of ODEs which may then be integrated with controlled accuracy. The method is very flexible, and permits ODEs and algebraic relationships appearing in the simulation to be solved simultaneously with the PDEs.

Although the method applies equally well to arbitrarily defined PDEs in several dependent variables, and independent variables of time and three spatial dimensions, it is conceptually simpler to illustrate with respect to a single PDE defining $u(x,t)$. This may be written as

$$\phi(x,t,u,u_t,u_x,u_{tt},u_{xx},u_{xt},\ldots \text{ etc.}) = 0 \tag{9}$$

Conversion to ODE form is accomplished by representing the spatial variation of u in terms of a spatial basis function B(x) and discrete values of u at nx points:

$$u_a(x,t) = \sum_{i=m}^{n} B_i(x)u_i(t) \tag{10}$$

where 1<m<n<nx, and (n-m) is the order of coupling in space. The approximation (10) is then substituted in (9) and algebraic manipulation produces a set of nx coupled ODEs:

$$[A][U_t] = [B][U] + [C] \tag{11}$$

where A, B and C are nx square matrices, and U_t, U are nx column vectors containing the values of the spatial variable and its time derivative at each point.

In practice, two variations of (11) may be derived, depending on how the approximation function (10) is used. In the finite difference approach, (10) is used merely to derive expressions for the spatial derivatives U_x, U_{xx} etc. in (9). Substitution then normally gives an explicit definition of the time derivatives

$$[U_t] = [B][U] + [C] \tag{12}$$

or in other words the A matrix is a unit diagonal.

In the finite element approach, the approximation function is also used to derive formulae for the time derivatives, and a suitable weighting function is used, integrating over space to minimize the residuals arising from the substitution in (9). This gives rise to an implicit definition of the time derivatives in which the A matrix of (11) is normally banded. In order to integrate, this must then be rendered explicit:

$$[U_t] = [A]^{-1}\{[B][U] + [C]\} \tag{13}$$

Usually the A matrix members are invariant, so the inversion or associated decomposition need only be done once.

The Approximation Functions

Any suitable interpolation formulae may be used in the functions, so the possibilities are diverse. Lagrangian, Hermite and Legendre polynomials, chapeau functions and splines have been implemented in FORSIM using finite difference, and finite element techniques. These are discussed in detail in reference [16]. However, to minimize overhead, the automatic PDE feature in FORSIM permits only two choices of approximation function, Lagrangian polynomials or cubic spline, both implemented in the finite difference sense, although the latter is equivalent to a linear finite element implementation for most equations[22].

The order of the Lagrangian formulae may be chosen by the user, the higher orders increasing accuracy, second order is default. Up to sixth order (seven point coupling in (10)) is practical for equal spatial divisions; for more accuracy with unequal spatial divisions the cubic spline is recommended and gives fourth order accuracy.

Coupled Equations and Multi Dimensional Space

Application of the method to arbitrarily coupled equations in more than one dependent variable causes no particular problem; expressions (10) are merely derived and used for each variable.

In two and three dimensions again the application is straightforward, as to avoid the complexities of deriving three dimensional analogs of (10), the LOD (locally one dimensional) technique is used to apply a version of (10) successively to each dimension. Implementation is then reduced to a logistics problem as the arrays representing U, U_t are now three dimensional.

In two or more dimensions the problem may give rise to singularities or conflicts of definition at common boundary points. They may frequently be eliminated or resolved by application of L'Hopital's rule. As the boundary nodes in two or more dimensions are numerous, significant savings are realized by integrating only internal nodes.

Boundary Value Problems

Boundary value problems in ODEs are solved by forming the associated PDE transient problem and integrating an initial estimate of distribution to steady state. Elliptic PDEs are handled in the same manner. Convergence accelerates well when the Gear algorithm is used.

THE USE OF FORSIM FOR PARTIAL DIFFERENTIAL EQUATIONS

The method of lines may be used within the subroutine which an integration algorithm calls repeatedly to evaluate derivatives. The logic required is shown in Figure 3. In FORSIM, the user needs only the automatic PDE routines for most applications. These routines PARSET and PARFIN perform the duties shown in Figure 3 for one-dimensional problems, and analagous routines are available for two and three dimensions.

Again for clarity, the example used is a set of two simple coupled equations, but the principles apply to all one-dimensional sets. The equations are those used by Collatz[23] to illustrate the method of characteristics. The current i and potential v in an electric cable satisfy the equations:

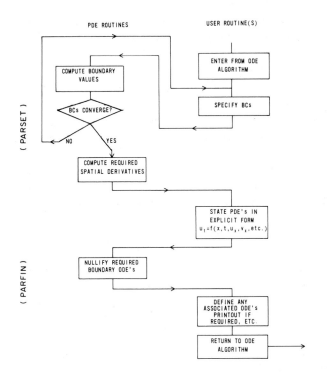

Figure 3:   Method of Lines Logic within an
Integration Algorithm

$$C \frac{\partial v}{\partial t} = - \frac{\partial i}{\partial x} \qquad (14)$$

$$L \frac{\partial i}{\partial t} + Ri = - \frac{\partial v}{\partial x}$$

where C, L and R have the usual electrical connotations. Taking v = -Cv, u = i,
LC = α^2, and RC = β, these equations reduce to

$$\frac{\partial v}{\partial t} = \frac{\partial u}{\partial x} \qquad (15)$$

$$\frac{\partial u}{\partial t} = \frac{1}{\alpha^2} \{\frac{\partial v}{\partial x} - \beta u\}$$

Collatz gives the results of methods of characteristics solutions of (15), with
$0 < x < 2$, $\alpha = 2$, $\beta = 1$ and compares them to the analytical solution

$$u(x,t) = e^{-1/8t} \sin \frac{\pi x}{2} \{\cos \frac{\gamma t}{8} - \frac{1}{\gamma} \sin \frac{\gamma t}{8}\} \qquad (16)$$

where $\gamma = \sqrt{4\pi^2 - 1}$.

Figure 4 illustrates the solution on the FORSIM system. The routine is set up for 11 spatial points, and introduces new common blocks. DERVX and DERVXX contain storage for spatial derivatives, PARTX contains the order of spatial coupling NCUP, the number of spatial points NPOINT, and optional controls concerning the spatial variable, x. PARTB contains boundary condition controls. Additional arrays UA and ER are to store the analytical value of u and the calculated error. The routine RITER prints spatial arrays in a suitable format, and the routine PLOTS produces Calcomp or printer plots of the requested variables. Note that the equation statement is identical in form to equations (15).

```
      SUBROUTINE UPDATE
C
C     FORSIM EXAMPLE 8.11
C     COLLATZ EQUATIONS WITH ANALYTICAL SOLUTION
C
C     EQUATIONS ARE   1 _ DU/DT = .25*(DV/DX-U)   2 _ DV/DT=DU/DX
C
C     SECTION 1    STORAGE
C
      COMMON/RESERV/T,H,DTOUT,EMAX,TFIN
     ,      /CNTROL/INOUT
     ,      /INTEGT/  U(33),   V(33) /DERIVT/UT(33),VT(33)
     ,      /DERVXX/UXX(33),VXX(33) /DERVX /UX(33),VX(33)
     ,      /PARTX/NCUP,NPOINT,XL,XU,AUX,AUXX,NEQDX,DX,X(33)
     ,      /PARTB/BC(4,2,2)
     ,      /CONS/PT1,PT2
      REAL UA(33),ER(33)
C
C     SECTION 2  SET INITIAL CONDITIONS AND CONSTANTS
C
      IF(T.NE.0.)GOTO200
      NSET=2
      NMAX=33
      XU=2.
      PIO2=ASIN(1.)
      GAM=SQRT(16.0*PIO2*PIO2-1)
      IP1=PT1
      IP2=PT2
C
C     BC(4,1,2)=BC(4,2,2)=-2
C
      DO 100 I=1,NPOINT
         V(I)=0.
100   U(I)=SIN(PIO2*(I-1)*XU/(NPOINT-1))
C
C     SECTION 3  DYNAMIC SECTION  CALL PARSET AND SET EQUATIONS
C
 *200 CALL PARSET(NSET,NMAX,U,UT,UX,UXX)
      DO 250 I=1,NPOINT
         UT(I) = .25*(VX(I)-U(I))
 *250 VT(I)  = UX(I)
C
      CALL PARFIN(NSET,NMAX,U,UT,UX,UXX)
C
C     SECTION 4  PRINTOUT AND ANALYTICAL SOLUTION
C
      IF(INOUT.NE.1)RETURN
      TO8=.125*T
      GAMTO8=GAM*TO8
C
      DO 310 I=1,NPOINT
         UA(I)=EXP(-TO8)*SIN(PIO2*X(I))*(COS(GAMTO8)-SIN(GAMTO8)/GAM)
310   ER(I)=UA(I)-U(I)
C
      CALL RITER(X,10HCOORDINATE)
      CALL RITER(UA,10HU EXACT   )
      CALL RITER(U ,10HU CALC    )
      CALL RITER(ER,5HERROR)
C
      CALL PLOTS(1,4,U(IP1),U(IP2),UA(IP1),UA(IP2))
C
      END
```

Figure 4: User Routine for the
Collatz Problem, Equations (15)

Implementation

Although FORSIM requires the user to write his own routines, very few job control cards are required to run a typical case. At Chalk River Nuclear Laboratories, the program is kept on a permanent disc file for common access on the CDC 6600/175 computers.

Data in FORSIM are block oriented, key words highlighted *KEYWORD* are recognized as block titles, and appropriate action is taken. Cards not recognized as a key word card or part of an associated block are reproduced as textual comment. System variables all have default values but may be changed via data; such variables preceded by an X are recognized automatically within certain data blocks. For a job consisting of one case only, all data may be contained in the user routine as in the first example. No data cards are then required. Normal jobs consist of several cases in which certain parameters are changed. Figure 5 illustrates all the job control cards and data required to run two cases of example (b) using different orders of spatial coupling NCUP. Partial results from the first case are shown in Figure 6.

```
MCOL,B235-MBC,REHOVT.          ACCOUNT CARD
FTN.                           COMPILE USER ROUTINE
ATTACH,FORSIM.                 ACCESS FORSIM
COPYL,FORSIM,LGO,EXEC.         MERGE USER ROUTINE WITH FORSIM
ATTACH,FORLIB,FORSIM,CY=1.     READY FORSIM LIBRARY
LIBRARY,FORLIB.                DECLARE LIBRARY
EXEC.                          LOAD AND EXECUTE
  _                    END OF RECORD

      SUBROUTINE UPDATE
         *
         *
         *
      END

                         END OF RECORD
  _
           - - - D A T A  - - -
        FORSIM EXAMPLE 8.11   COLLATZ EQUATIONS OUTPUT

*CONS*            INPUT CONSTANTS
  PT1      5.        PT2     15.     XDTOUT    0.25     XTFIN    2.0
  XJPLOT     -2.

*PLOTS*            [ PLOT LABLES ]
   U(PT1)    U(PT2)      UA(PT1)    UA(PT2)

      NOTE ANY TEXTUAL COMMENT MAY BE INTERSPERSED WITH *KEYWORD* BLOCKS

*PARS*            [ CASE 1  17 POINTS 5 COUPLED ]
 XNPOINT     17.    XNCUP     5.

*PARS*            [ CASE 2  17 POINTS 7 COUPLED ]
 XNCUP       7.

*TGRID*               [ NON LINEAR OUTPUT GRID FOR CASE 3 ]
  .1             0.5     2.5

*PARS*            [ CASE 3  33 POINTS 7 COUPLED NONLINEAR TGRID  NO PLOT ]
 XNPOINT     33.    XJPLOT    0.

*FINIT*           [   NO MORE CASES  ]


                       END OF FILE
```

Figure 5:   Job Control Cards and Data to
            Run Two Cases of Equations (15)

```
*********************************************************
*               F O R S I M   V I                       *
*  FORTRAN ORIENTED DISTRIBUTED SYSTEM SIMULATION PACKAGE *
*      FOR PARTIAL AND OR ORDINARY DIFFERENTIAL EQUATIONS *
* USERS MANUAL - ATOMIC ENERGY OF CANADA REPORT AECL 5821 *
* AUTHORS  M B CARVER AND D G STEWART , C.R.N.L./A.E.C.L. *
*********************************************************

   PRINTOUT TIMES ARE LISTED BELOW
0.      .100E+00  .500E+00  .250E+01
   PARAMETERS NAMES AND VALUES

      XNPOINT=  33.000        XJPLOT =  0.
PARTIAL DIFFERENTIAL EQUATION  1 WILL BE DISCRETISED USING  7 POINT FORMULAE AND 33   EQUAL   SPATIAL DIVISIONS
PARTIAL DIFFERENTIAL EQUATION  2 WILL BE DISCRETISED USING  7 POINT FORMULAE AND 33   EQUAL   SPATIAL DIVISIONS

   CASE NUMBER  3 USING RUNGE-KUTTA RKFEHL (VARIABLE  STEP) INTEGRATION ALGORITHM
                                                    INTEGRATING   66 EQUATIONS


     TIME  0.           STEP  .2071      CPTIME  2.277
VARIABLE   --------  S P A T I A L    V A R I A T I O N    --------
COORDINATE  0.    .1875     .3750     .5625     .7500     .9375    1.125    1.313    1.500    1.688    1.875
U EXACT     0.    .2903     .5556     .7730     .9239     .9952    .9808    .8819    .7071    .4714    .1951
U CALC      0.    .2903     .5556     .7730     .9239     .9952    .9808    .8819    .7071    .4714    .1951
ERROR       0.    0.        0.        0.        0.        0.       0.       0.       0.       0.       0.


     TIME  .1000         STEP  .2071      CPTIME  2.297
VARIABLE   --------  S P A T I A L    V A R I A T I O N    --------
COORDINATE  0.    .1875     .3750     .5625     .7500     .9375    1.125    1.313    1.500    1.688    1.875
U EXACT     0.    .2822     .5402     .7516     .8983     .9676    .9536    .8575    .6875    .4583    .1897
U CALC      0.    .2822     .5402     .7516     .8983     .9676    .9536    .8575    .6875    .4583    .1897
ERROR       0.   .1284E-08 .1658E-08 .2317E-08 .2769E-08 .2983E-08 .2948E-08 .2644E-08 .2119E-08 .1430E-08 .4979E-09


     TIME  .5000         STEP  .2540      CPTIME  2.320
VARIABLE   --------  S P A T I A L    V A R I A T I O N    --------
COORDINATE  0.    .1875     .3750     .5625     .7500     .9375    1.125    1.313    1.500    1.688    1.875
U EXACT     0.    .2358     .4514     .6280     .7506     .8085    .7968    .7165    .5745    .3830    .1585
U CALC      0.    .2358     .4514     .6280     .7506     .8085    .7968    .7165    .5745    .3830    .1585
ERROR       0.   .6354E-07 .1217E-06 .1699E-06 .2027E-06 .2184E-06 .2152E-06 .1936E-06 .1548E-06 .1043E-06 .4474E-07


     TIME  2.500         STEP  .2824      CPTIME  2.420
VARIABLE   --------  S P A T I A L    V A R I A T I O N    --------
COORDINATE  0.    .1875     .3750     .5625     .7500     .9375    1.125    1.313    1.500    1.688    1.875
U EXACT     0.   -.1083    -.2072    -.2884    -.3446    -.3712    -.3659   -.3290   -.2638   -.1758   -.7278E-01
U CALC      0.   -.1083    -.2072    -.2884    -.3446    -.3712    -.3659   -.3290   -.2638   -.1758   -.7278E-01
ERROR       0.   .4489E-06 .8362E-06 .1162E-05 .1389E-05 .1499E-05 .1477E-05 .1326E-05 .1064E-05 .7032E-06 .2925E-06
   ///  THIS RUN TERMINATED BECAUSE THE CURRENT VALUE OF TIME    IS   2.5000


   ///  TERMINAL COMMENTS   ///

//   NO CONVERGENCE PROBLEMS DETECTED DURING THIS RUN
     EXECUTION CPTIME FOR THIS RUN -      .164
     FIELD LENGTH REQUIRED  31616 (075600B)
```

Figure 6: Results from Case 1, Equations (14)

Two-Dimensional Example

The following example was taken from a recent study on flow-induced dipole molecule orientation done in collaboration with the University of New Brunswick, Chemical Engineering Department. We are concerned here only with the equation and its solution by FORSIM. It is

$$d \left\{ \frac{\partial^2 f}{\partial \theta^2} \sin^2\theta + \frac{\partial^2 f}{\partial \phi^2} \right\} = -af \cos\theta \, \cos\phi \, \sin^3\theta$$

$$+ \frac{\partial f}{\partial \theta} \sin^2\theta \, [a \cos^2\theta \, \cos\phi$$

$$- \dot{d}\{\cot\theta + b(\cos\phi \, \cos\theta \, \cos\varepsilon - \sin\theta \, \cos\gamma)\}]$$

$$+ \frac{\partial f}{\partial \phi} \, (-a \cos\theta \, \sin\theta \, \sin\phi + d \, b \, \sin\phi \, \sin\theta \, \cos\varepsilon)$$

$$\frac{\partial f}{\partial \phi} = 0 \text{ at } \theta = 0, \pi \text{ (i.e. } f=c) \tag{17}$$

$$\frac{\partial f}{\partial \phi} = 0 \text{ at } \phi = 0, \pi$$

$$\epsilon = \gamma = \pi/4.$$

This steady-state equation is readily solved using the notation Z_1 = RHS, Z_2 = LHS, and writing a new transient equation

$$\frac{\partial f}{\partial t} = Z_2 - Z_1 \tag{18}$$

which maintains the form of the diffusion equation. We now use the notation $x=\theta$, $y=\phi$, and solve the equation for 31 points in the x axis, and 21 in the y axis.

This form of solution, where $\theta=c$ is a boundary condition, results in a mode shape which may be normalized by any additional criterion. In this case the criterion to be used was

$$\int_0^\pi \int_0^\pi u(\theta, \phi) d\theta d\phi = 1.$$

We merely chose c to be 1.0. The integration can be done in FORTRAN subsequent to solution completion, but is omitted from this example as it is irrelevant to the FORSIM solution.

The solution is set up as shown in Figure 7 to read in the critical variables a,b and d as parameters. We start off with a distorted initial profile assumption of $U(I,J) = I+J$ to avoid the possibility that the initial assumption is itself a trivial solution. Values a=100, b=1.0, d=200 were used with XMETHOD=13, the GEARZ sparse option, and integration converges to steady state in approximately 20 seconds problem time.

OTHER APPLICATIONS

In a general context such as this, it would be superfluous to describe a number of complex examples, so some current areas of application are merely mentioned here, along with the appropriate references.

Three-dimensional PDE's are handled in FORSIM merely by defining variables in three-dimensional arrays and using a 3D utility routine PARTRI, which is again analogous to PARSET. A set of routines for handling fourth order equations is also provided, and has been used to solve the beam vibration equation

$$\frac{\partial^2 y}{\partial t^2} = c \frac{\partial^4 y}{\partial x^4} \tag{19}$$

for multi-span beams, and non-linear constraints[14].

```
C
C
C       ******      FORSIM EXAMPLE   12.1
C                   UNB  FLOW INDUCED MOLECULAR ORIENTATION STUDY   *****
C
C       SUBROUTINE UPDATE
C
C       COMMUNICATION COMMON BLOCKS AND STORGAE
C
        COMMON/RESERV/T,DT      /CNTROL/INOUT
     ,       /INTEGT/F(11,11)/DERIVT/DF(11,11)
     ,       /PARTA/NCUPX,NPOINX,XL,XU,AUX,AUXX,NEQDX,DX,X(11)
     ,       /PARTY/NCUPY,NPOINY,YL,YU,AUY,AUYY,NEQDY,DY,Y(9)
     ,       /PARS/A,B,D
        REAL    BCX(4),BCY(4),BX(4,2,11),BY(4,2,11)
     ,,        STORE(11,11),DUX(11,11),DUXX(11,11),DUY(11,11),DUYY(11,11)
C
C       SET UP INITIAL CONDITIONS AND CONTROLS FOR X,Y AXES
C
        DATA BCX/0.,1.,10.,0./,BCY/1.,0.,0.,0./
        IF(T.NE.0.) GO TO 300
        NPOINY=9
        NPOINX=11
        NPX1=NPOINX-1
        NPY1=NPOINY-1
        PI=2*ASIN(1.)
        XU=YU=PI
        C45=COS(PI/4.)
C
        DO 100 I=1,NPOINX
        DO 100 J=1,NPOINY
100     F(I,J)=FLOAT(I) + FLOAT(J)
C
C       SET BOUNDARY CONDITIONS X=0 _ F=10.   ,   Y=0 _ DF/DY=0
C
        DO 200 J=1,4
        DO 200 I=1,NPOINX
        BX(J,1,I)=BCX(J)
        BX(J,2,I)=BCX(J)
        BY(J,1,I)=BCY(J)
200     BY(J,2,I)=BCY(J)
C
C       DYNAMIC SECTION  CALL PARTWO/PARTOF  DEFINE EQUATIONS IN X,Y,T
C
300     CALL PARTWO(NPOINX,NPOINY,  F,DF,DUX,DUXX,DUY,DUYY,BX,BY,STORE)
        DO 400 I=2,NPX1
        DO 400 J=2,NPY1
        COSX=COS(X(I))
        SINX=SIN(X(I))
        COSY=COS(Y(J))
        SINY=SIN(Y(J))
        COTX=COSX/SINX
C
        Z1 =  -A*COSX*COSY*SINX**3*F(I,J)
     ,        +DUX(I,J)*SINX**2
     ,        +B*COSX**2*COSY-D*(COTX+B*(COSY*COSX*C45-SINX*C45)))
     ,        +DUY(I,J)*(-A*COSX*SINX*SINY + D*B*SINY*SINX*C45)
        Z2 = D*(DUXX(I,J)*SINX**2+DUYY(I,J))
        DF(I,J)=Z2-Z1
400     CONTINUE
C
        CALL PARTOF(NPOINX,NPOINY,F,DF)
C
C       PRINTOUT SECTION
C
        IF(INOUT.NE.1)RETURN
        CALL SECOND(CPTIME)
        WRITE(6,1000)T,DT,CPTIME
        PRINT *, F
        WRITE(6,1000) ((F(I,J),I=1,NPOINX ),J=1,NPOINY)
        PRINT *, DF
        WRITE(6,1000) ((DF(I,J),I=1,NPOINX ),J=1,NPOINY)
C
C       TERMINATE RUN AND DO 3D PLOT
C
        IF(T.NE.0)CALL FINISH(0.1,ABS(DF(5,5)),4HDF55)
        FIN=FINISH(T,0.999,4HTIME)
C
        IF(INOUT.NE.5)RETURN
        CALL PLOTZ(F,NPOINTX,NPOINTY)
1000    FORMAT(1H0,(11GI2.5))
        RETURN
        END
```

Figure 7: User Routine for the Two-Dimensional Problem, Equations (17)

Projects have also been completed in such diverse fields as chemical kinetics[24], neutron kinetics[25], compressible flow analysis[26] two-phase flow transients[27], biomechanics[28], corrosion analysis[29], impact stress analysis[30], high temperature creep[31] and the formulation of transfer functions for subsequent use in an analog computer[32]. At the Swiss Federal University, the well-known discrete event simulation language GASP-IV[33] has been merged with the PDE/ODE capabilities of FORSIM producing what authors describe as a universal simulation package[34].

The development of a general program cannot be successfully done without advice and complaints from users. The FORSIM system has grown progressively, the latest developments being instigated by the current most active user. Recently, a special version of FORSIM has been written[35,36] to obtain long term solution of chemical reaction kinetics equations, as the power of the Gear algorithm can be used effectively here only with the correct application of tolerance criteria as given in equation (3).

REFERENCES

1) Young, D.M. and Juncosa, M.D., "SPADE: A Set of Subroutines for Solving
    Elliptic and Parabolic PDE's", Rand Corporation report P-1709, Santa Monica,
    California, 1959.

2) Morris, S.M. and Schiesser, W.E., "SALEM": A Programming System for the
    Simulation of Systems Described by PDE's", Proceedings, 1968 Fall Joint
    Computer Conference, 33, Part I, p. 353-357, Thompson Book Co., Washington,
    D.C.

3) Zellner, M.G., "DSS: Distributed System Simulator", PhD. Dissertation,
    Lehigh University, June 1970.

4) Cardenas, A.F. and Karplus, W.J., "PDEL: A Language for Partial Differ-
    ential Equations", Comm. of the ACM, 13, No. 3, 1970.

5) Schiesser, W.E., "A Digital Simulation System for Mixed Ordinary/Partial
    Differential Equation Modes", Proceedings, IFAC Symposium on Digital Sim-
    ulation of Continuous Systems, 2, p. S2-1 to S2-9, September 1971, Gyor,
    Hungary.

6) Carver, M.B., "FORSIM: A FORTRAN-Oriented Simulation Program with Particular
    Application to the Solution of Stiff Equation Systems", Proceedings, 1972
    Summer Computer Simulation Conference, SCI, La Jolla, p. 73-80.

7) Csendes, Z.J., "DECL: A Computer Language for the Solution of Arbitrarily
    Partial Differential Equations", Advances in Computer Methods for Partial
    Differential Equations, p. 159-166, R. Vichnevetsky, Editor, AICA Press,
    Rutgers University, New Brunswick, New Jersey, 1975.

8) Carver, M.B., "The FORSIM System for Automated Solution of Sets of Implic-
    itly Coupled Partial Differential Equations", AAICA, 3, 1975, pp. 195-202.

9) Madsen, N.K. and Sincovec, R.F., "Generalized Software for Partial Differ-
    ential Equations", Paper presented at Symposium on Recent Developments in
    Numerical Algorithms for Ordinary and Partial Differential Equations, AIChE
    80th National Meeting, Boston, 1975, published by Academic Press.

10) Schiesser, W.E., DSS/2, "An Introduction to the Numerical Lines Integration
    of Partial Differential Equations", Lehigh University Report, March 1977.

11) Hyman, J.M., "Method of Lines Solution of Partial Differential Equations",
    Courant Institute of Mathematical Sciences, report C00-3077-139, October
    1976.

12) Easen, E.D. and Mote, C.D., "Solution of Nonlinear Differential Equations by
    Discrete Least Squares", Sandia report SAND77-8205, January 1977.

13) Byrne, G.D., "Numerical Techniques for Differential Equations", Proceedings,
    IMACS Conference on Differential Equations, VPI and SU, Blacksburg, Va.,
    March 1977.

14) Carver, M.B., Stewart, D.G., Blair, J.M., and Selander, W.N., "The FORSIM VI Package for the Automated Solution of Arbitrarily Defined Partial and/or Ordinary Differential Equation Systems", Atomic Energy of Canada Limited, report AECL-5821, February 1978.

15) Carver, M.B. and Liu, J., "An Evaluation of Available Integration Algorithms and Selection for the AECL FORTRAN Mathematical Library", Atomic Energy of Canada Limited report to be published.

16) Carver, M.B., "Efficient Integration over Discontinuities in Ordinary Differential Equations", Proceedings, IMACS International Symposium on Simulation Software and Numerical Methods for Differential Equations, Virginia Polytechnic Institute & State University, Blacksburg, Virginia, March 1977.

17) Fehlberg, E., "Lower Order Classical Runge-Kutta Formulas with Step-size Control and their Application to Heat Transfer Problems", NASATR-R315, 1969.

18) Curtis, A.B. and Reid, J.K., "FORTRAN Routines for the Solution of Sparse Sets of Linear Equations", UKEA report AERE-R-6844, 1971.

19) Hindmarsh, A.C., "GEAR: Ordinary Differential Equation System Solver", Lawrence Livermore Laboratory report UCID-30001, Rev. 2, 1972.

20) Gear, C.W., "The Automatic Integration of Ordinary Differential Equations", Comm. of the ACM, March 1971, Vol. 14, No. 3, pp. 176-180, 185-190.

21) "An Example of HP25 Programming", Hewlett Packard Journal, November 1975, p. 6.

22) Carver, M.B. and Hinds, H.W., "The Method of Lines and the Advective Equation", Simulation, August 1978, pp. 59-70.

23) Collatz, L., "The Numerical Treatment of Differential Equations", Springer-Verlag, Berlin, 1960, p. 323-329.

24) Rosinger, E.L.J. and Dixon, R.S., "Mathematical Modelling of Water Radiolysis, a Discussion of Various Methods", Atomic Energy of Canada Limited report AECL-5958, November 1977.

25) Carver, M.B. and Baudouin, A.P., "Solution of Reactor Kinetics Problems using Sparse Matrix Techniques in an ODE Integrator for Stiff Equations", R. Vichnevetsky, Ed., AICA Press, June 1975, pp. 377-381.

26) Carver, M.B., "The Choice of Algorithms in Automated Method of Lines Solution of Partial Differential Equations", Paper presented at Symposium on Recent Developments in Numerical Algorithms for Ordinary and Partial Differential Equations, AIChE, Boston, 1975, published by Academic Press.

27) Ahmad, S.Y. and Carver, M.B., "Automatic Treatment of Shocks in Two-Phase Flow", Fifth Canadian Congress of Applied Mechanics, p. 570-572, Frederickton, May 1975.

28) Johnson, J.R., "Compartment Models of Radioiodine in Man", Atomic Energy of Canada Limited report AECL-5244, October 1975.

29) Burrill, K.A., "Corrosion Production Transients in Water Cooled Nuclear
Reactors", Canadian Journal of Chemical Engineering, 1976.

30) Carver, M.B. and MacEwan, S.R., "Numerical Analysis of a System Described by
Implicitly Defined Ordinary Differential Equations containing Numerous Discontinuities", Applied Mathematical Modelling, 2, p.280-286, 1978.

31) MacEwan, S.R. and Fidleris, V., "Verification of a Model for In-Reactor
Creep Transients in Zirconium", Phil. Mag., 31, p. 1149-1157, 1975.

32) Hinds, H.W., "Application of the PRBS/FFT Technique to Digital Simulations",
Paper presented at the Canadian Conference on Automatic Control, UBC,
Vancouver, June 1975.

33) Pritsker, A.A.B., "The GASP-IV Simulation Language", Wiley, 1974.

34) Cellier, F.E. and Blitz, A.E., "GASP-V: A Universal Simulation Package"
Paper presented at the 1976 AICA International Congress on Simulation of
Systems, Delft, Netherlands, August 1976.

35) Carver, M.B. and Boyd, A.W., "A Program Package Using Stiff, Sparse Ingegra-
tion Methods for the Automatic Solution of Mass Action Chemical Kinetics",
Submitted to Int. J. Chem. Kinet., 1978.

36) Carver, M.B. and Hanley, D.H., "MAKSIM: A Program for Mass Action Kinetics
Simulation by Automatic Chemical Equation Manipulation and Integration Using
Stiff Techniques", Atomic Energy of Canada Limited report AECL-6413, Feb-
ruary 1979.

SECTION IV
STRUCTURE IDENTIFICATION
AND MODEL CALIBRATION

METHODOLOGY IN SYSTEMS MODELLING AND SIMULATION
B.P. Zeigler, M.S. Elzas, G.J. Klir, T.I. Ören (eds.)
© North-Holland Publishing Company, 1979

MODELLING AND PARAMETER IDENTIFICATION
OF INSULIN ACTION ON GLUCONEOGENESIS:
AN INTEGRATED APPLICATION OF ADVANCED MODELLING METHODS

Jaime Milstein*
Department of Mathematics
University of Southern California
Los Angeles, California   90007

Key Words: Parameter Identification, Nonlinear Optimization,
Stiff Differential Equations, Gluconeogenesis,
Insulin Action.

INTRODUCTION

In recent years, the use of mathematical models describing biochemical systems has been widely advocated. Garfinkel (1974) and Chance (1972) have extensively inves-tigated the kinetics of biochemical pathways by means of computer simulation. Henker and Hess (1972) have discussed the difficulties encountered in the appli-cation of numerical techniques to complex enzymatic reactions. Roth and Roth (1969) have used Bellman's Quasilinearization technique for parameter identifica-tion of an inducible enzyme system. Milstein (1975,1978) showed that an itera-tive scheme offers promise for parameter identification of biochemical systems compared with existing techniques (including the quasilinearization technique) by employing multiple trajectories. In this paper we present an integrated applica-tion of advanced modelling methods to describe the effect of insulin action on gluconeogenesis.

Carbohydrates or sugars appearing in nature as starch, glucose, sucrose, fructose etc., are a major source of energy for living organisms. The process of glucose synthesis from non-glucose molecules (such as proteins and glycerols) is called gluconeogenesis. The hormone insulin has a profound effect on this process by affecting regulatory sites in the pathways of protein and fat metabolism. Although the clinical condition of insulin deficiency (diabetes) and its symptoms have been known since last century, there is no adequate explanation of the mechanism of the action of this hormone. Samuel Bessman (1974), proposed a mechanism of insulin action. His theory states that the hormone insulin signi-ficantly effects the process of gluconeogenesis by stimulating protein synthesis.

This approach claims that the anabolic action (stimulation of protein synthesis) of insulin causes the gluconeogenic amino acids to be reincorporated into protein before they can be converted to glucose. This contradicts the generally accepted proposition of Weber, Krebs and others, that insulin blocks one or both of the reactions in the shunt pathway from pyruvate to phosphoenol pyruvate (PEP). This difference in views permits clear formulation of alternate models of insulin action on gluconeogenesis, subject to mathematical analysis and experimental validation.

*This work was partially supported by the National Science Foundation Research Grant MCS 76-09172,and U.S. Government Grants GM 23826 and MB 00146.

Modelling and simulation of gluconeogenesis involves at least two important steps:
(a) a mathematical formulation which captures the natural phenomenon,
(b) correct numerical solution of the dynamics of such a model.

The lack of an accurate mathematical representation of the dynamics, and the lack
of powerful and reliable simulation algorithms for such systems are the main obstacles in both of these steps. To simulate the process of gluconeogenesis, two mathematical models will be presented:

   (1) A linear compartmental model which describes the qualitative nature of
        the process.
   (2) A nonlinear model which describes the kinetics in more detail.

In both of the above models we assume that the underlying structure of the enzyme
kinetics remains at steady state, and does not interfere with the kinetics of the
substrates. To test Bessman's theory it is necessary to calibrate the models, that
is to identify their parameters by comparison with experimental data. Thus, experimental data is being obtained for diabetic rabbits with and without insulin injection. The parameter identification techniques of this paper will be applied to
assess whether there are significant differences in parameters in the direction
predicted by Bessman's theory.

ABBREVIATIONS. The following are the abbreviations of the substrates that will be
                used in this article.

ACETYL-COA.	ACETYL-COENZYME
ADP	ADENOSINE-5-DIPHOSPHATE
ATP	ADENOSINE-5-TRIPHOSPHATE
ALA	ALANINE
DHAP	DIHYDROXY ACETONE PHOSPHATE
DPG	3-PHOSPHOGLYCEROYL PHOSPHATE
FGP	FRUCTOSE 6-PHOSPHATE
F6P	FRUCTOSE 1, 6-DIPHOSPHATE
GAP	3-PHOSPHOGLYCERALDEHYDE
G1P	GLUCOSE 1-PHOSPHATE
G6P	GLUCOSE 6-PHOSPHATE
GLU	FREE GLUCOSE
GLYCE	GLYCEROL
GLYCO	GLYCOGEN
LACT	LACTATE
NAD	NICOTINAMIDE ADENINE DINUCLEOTIDE
NADH	NICOTINAMIDE ADENINE DINUCLEOTIDE REDUCED
PEP	PHOSPHOENOLPYRUVATE
3PGC	3-PHOSPHOGLYCERATE
2PGC	2-PHOSPHOGLYCERATE
PROT	PROTEIN
PYR	PYRUVATE

THE COMPARTMENTAL MODEL

A compartment is a subdivision of the total system in which the concentration of
the relevant substance is uniformly distributed. In gluconeogenesis six compartments are considered. Figure (1) represents the interaction between compartments.

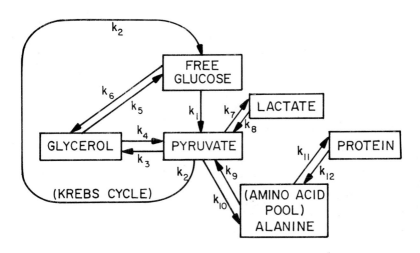

FIGURE 1- Compartmental model of gluconeogenesis. The arrows represent direc-
          tions of flux; k_i, i=1, . . .12 are the rate constants which represent
          the rate of change in concentration from one compartment to another.

DERIVATION OF THE MODEL

To facilitate communication of biochemical models to the computer, a differential
equation generator was written in the PL/C language [4]. This generator takes as
input a schematic representation of reactions as in the following representation
of the compartmental model:

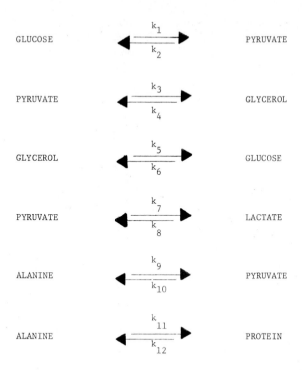

Using the law of mass action, it produces as output a system of differential equations in symbolic form. For example the compartmental model is translated into:

(1) $\dfrac{d}{dt}[GLU]$ = $- K(1) (GLU) + K(5) (GLYCE) + K(2) (PYR) -$
$-K(6) (GLU)$

(2) $\dfrac{d}{dt}[GLYCE]$ = $K(6) (GLU) - K(5) (GLYCE) - K(4) (GLYCE) +$
$+K(3) (PYR)$

(3) $\dfrac{d}{dt}[PYR]$ = $K(4) (GLYCE) - K(3) (PYR) - K(2) (PYR) - K(1) (GLU) +$
$+K(8) (LACT) - K(7) (PYR) + K(9) (ALA) - K(10) (PYR)$

(4) $\dfrac{d}{dt}[LACT]$ = $-K(8) (LACT) + K(7) (PYR)$

(5) $\dfrac{d}{dt}[ALA]$ = $-K(9) (ALA) + K(10) (PYR) - K(11) (ALA) + K(12) (PROT)$

(6) $\dfrac{d}{dt}[PROT]$ = $K(11) (ALA) - K(12) (PROT)$

The advantage of this automatic derivation procedure is that it elimnates the probability of human error in equation writing, in particular when a large number of reactions are considered, Milstein [10] uses this approach to derive a large photosynthesis model. The differential equation generator is more convenient to use than the corresponding portion of the package developed by Garfinkel [5].

THE NONLINEAR MODEL. The process of gluconeogenesis in muscle, from substrate
like protein and amino acids can be described by 18 steps, as in Figure 2.

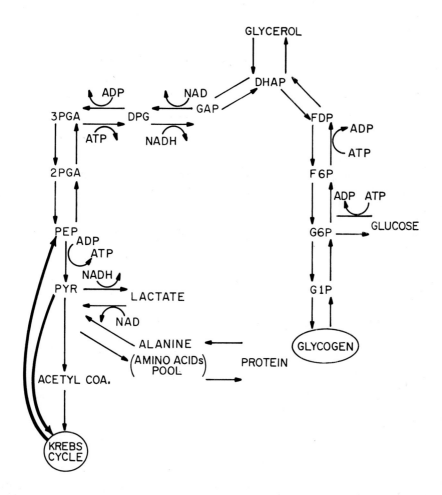

FIGURE 2 - The pathways of gluconeogenesis from protein in muscle. The dark lines
            indicate one way reaction.

In what follows the arrows denote the direction of the reaction and k_i, i=1,...,31 are the rate constants (kinetic parameters).

<u>Figure 3</u>: Schematic representation of the reaction steps.

STEP 1   GLYCO $\xrightarrow{k_1}\xleftarrow{}$ $\underset{k_2}{}$ G-1-P

STEP 2   G-1-P $\xrightleftharpoons[k_4]{k_3}$ G6P

STEP 3   GLU + ATP $\xrightarrow{k_5}$ G-6-P + ADP

STEP 4   G-6-P $\xrightleftharpoons[k_7]{k_6}$ F-6-P

STEP 5   F-6-P $\xrightarrow{k_8}$ FDP + ADP

STEP 6   FDP $\xrightarrow{k_9}$ F-6-P

STEP 7   F-6-P $\xrightleftharpoons[k_{11}]{k_{10}}$ DHAP + GAP

STEP 8   DHAP $\xrightleftharpoons[k_{13}]{k_{12}}$ GAP

STEP 9   GAP + NAD $\xrightleftharpoons[k_{15}]{k_{14}}$ DPGA + NADH

STEP 10   DPGA + ADP $\xrightleftharpoons[k_{17}]{k_{16}}$ 3PGC + ATP

STEP 11   DHAP $\xleftarrow{}\underset{k_{19}}{\overset{k_{18}}{\xrightarrow{}}}$ GLYCE

STEP 12   3PGC $\xleftarrow{}\underset{k_{21}}{\overset{k_{20}}{\xrightarrow{}}}$ 2PGC

STEP 13   2PGC $\xleftarrow{}\underset{k_{23}}{\overset{k_{22}}{\xrightarrow{}}}$ PEP

STEP 14   PEP + ADP $\xrightarrow{k_{24}}$ PYR + ATP

STEP 15   PYR + NADH $\xleftarrow{}\underset{k_{26}}{\overset{k_{25}}{\xrightarrow{}}}$ LACT + NAD

STEP 16   PYR $\xleftarrow{}\underset{k_{28}}{\overset{k_{27}}{\xrightarrow{}}}$ ALA*

STEP 17   ALA $\xleftarrow{}\underset{k_{30}}{\overset{k_{29}}{\xrightarrow{}}}$ PROT

STEP 18   PYR $\xrightarrow{k_{31}}$ ACETYL COA.

*In this reaction pyruvate is converted into amino acids, and in the amino acid
 pool, alanine is the dominant amino acid in the gluconeogenesis process.

Using the schematic representations of the reactions in Figure 3, we automatically
generate the nonlinear mathematical model of Figure 4 using our differential equation generator. The model consist of 20 nonlinear ordinary differential equations
(one for each biochemical pool) having 31 kinetic parameters.

Figure 4: The nonlinear mathematical model of gluconeogenesis.

(1) $\dfrac{d}{dt}$ [GLYCO] = − K(1) (GLYCO) + K(2) (G1P)

(2) $\dfrac{d}{dt}$ [G1P] = + K(1) (GLYCO) − K(2) (G1P)
                          − K(3) (G1P) + K(4) (G6P)

(3) $\dfrac{d}{dt}$ [G6P] $= + K(3) (G1P) - K(4) (G6P)$
$+ K(5) (ATP) (GLU)$
$- K(6) (G6P) + K(7) (F6P)$

(4) $\dfrac{d}{dt}$ [ATP] $= - K(5) (ATP) (GLU)$
$- K(8) (F6P) (ATP)$
$+ K(16) (DPG) (ADP)$
$- K(17) (3PG) (ATP)$
$+ K(24) (PEP) (ADP)$

(5) $\dfrac{d}{dt}$ [GLU] $= - K(5) (ATP) (GLU)$

(6) $\dfrac{d}{dt}$ [ADP] $= + K(5) (ATP) (GLU)$
$+ K(8) (F6P) (ATP)$
$- K(16) (DPG) (ADP)$
$+ K(17) (3PG) (ATP)$
$- K(24) (PEP) (ADP)$

(7) $\dfrac{d}{dt}$ [F6P] $= + K(6) (G6P) - K(7) (F6P)$
$- K(8) (F6P) (ATP)$
$+ K(9) (FDP)$
$- K(10) (F6P) + K(11) (DHAP) (GAP)$

(8) $\dfrac{d}{dt}$ [FDP] $= + K(8) (F6P) (ATP)$
$- K(9) (FDP)$

(9) $\dfrac{d}{dt}$ [DHAP] $= + K(10) (F6P) - K(11) (DHAP) (GAP)$
$- K(12) (DHAP) + K(13) (GAP)$
$- K(18) (DHAP) + K(19) (GLYCE)$

(10) $\dfrac{d}{dt}$ [GAP] $= + K(10) (F6P) - K(11) (DHAP) (GAP)$
$+ K(12) (DHAP) - K(13) (GAP)$
$- K(14) (GAP) (NAD) + K(15) (DPG) (NADH)$

(11) $\dfrac{d}{dt}$ [NAD] $= - K(14) (GAP) (NAD) + K(15) (DPG) (NADH)$
$+ K(25) (PYR) (NADH) - K(26) (LACT) (NAD)$

(12) $\dfrac{d}{dt}$ [DPG] $= + K(14) (GAP) (NAD) - K(15) (DPG) (NADH)$
$- K(16) (DPG) (ADP) + K(17) (3PGC) (ATP)$

(13) $\dfrac{d}{dt}$ [NADH] $= + K(14) (GAP) (NAD) - K(15) (DPG) (NADH)$
$- K(25) (PYR) (NADH) + K(26) (LACT) (NAD)$

(14) $\dfrac{d}{dt}$ [3PGC] $= + K(16) (DPG) (ADP) - K(17) (3PGC) (ATP)$
$- K(20) (3PGC) + K(21) (2PGC)$

(15) $\dfrac{d}{dt}$ [GLYCE] $= + K(18) (DHAP) - K(19) (GLYCE)$

(16) $\dfrac{d}{dt}$ [2PGC] $= + K(20)\ (3PGC) - K(21)\ (2PGC)$
$- K(22)\ (2PGC) + K(23)\ (PEP)$

(17) $\dfrac{d}{dt}$ [PEP] $= + K(22)\ (2PGC) - K(23)\ (PEP)$
$- K(24)\ (PEP)\ (ADP)$

(18) $\dfrac{d}{dt}$ [PYR] $= + K(24)\ (PEP)\ (ADP)$
$- K(25)\ (PYR)\ (NADH) + K(26)\ (LACT)\ (NAD)$
$- K(27)\ (PYR) + K(28)\ (ALA)$
$- K(31)\ (PYR)$

(19) $\dfrac{d}{dt}$ [LACT] $= + K(25)\ (PYR)\ (NADH) - K(26)\ (LACT)\ (NAD)$

(20) $\dfrac{d}{dt}$ [ALA] $= + K(27)\ (PYR) - K(28)\ (ALA)$
$- K(29)\ (ALA) + K(30)\ (PROT)$

(21) $\dfrac{d}{dt}$ [PROT] $= + K(29)\ (ALA) - K(30)\ (PROT)$

THE SIMULATION SCHEME

In general a biochemical model having m pools and p parameters can be represented in the form

$$(1)\quad \dot{\bar{x}} = f(\bar{x}, \bar{k}), \quad \bar{x}(0) = \bar{c}_i, \quad i=1,\ldots, \ell$$

where $\bar{x} \in \mathbb{R}^n_+$ is the vector of state variables; $\bar{k} \in \mathbb{R}^p_+$ is the vector of kinetic parameters; $\bar{c}_i \in \mathbb{R}^n_+$ are distinct vectors of initial conditions; and f is a vector function such as Figure 4 describing the kinetics of the system. Model trajectories are generated from a selected initial state $\bar{c}_i$ with fixed parameters $\bar{k}$ and by numerical solution of (1). Model calibration involves the inverse problem, that is, the determination of the parameter $\bar{k}$ from the solution $\bar{x}$, where $\bar{x}$ is given as observation points at times t, and is subject to experimental error. An algorithmic procedure for parameter identification was developed. To test its efficacy, data was generated by choosing an arbitrary set of values for the kinetic parameters, and integrating the system to obtain its trajectories from states $\bar{c}_1, \bar{c}_2, \ldots \bar{c}_\ell$. An independent uniformly distributed random number generator was used to introduce a noise level of between 5 to 10% noise into the data. We believe this closely approximates realistic data collecting conditions.

ALGORITHMIC PROCEDURE

Milstein (1978), showed how multiple trajectories can be used successfully for parameter identification. An outline of the method will now be given.

As measure of goodness of fit we choose a function $F[\tilde{k}]$ which measures the square of the discrepancies between the given data y and the trajectories of Eq. (1).

From initial states $\bar{c}_i$, $i=1,\ldots,\ell$, employing a parameter $\tilde{k}$:

$$(2) \quad F[\tilde{k}] = \sum_{s=1}^{\ell} \sum_{r=1}^{M} \quad \{y^s(t_r)-x^s(\tilde{k},t_r)\}^T \quad W_r \quad \{y^s(t_r)-x^s(\tilde{k},t_r)\}$$

where $y^s(t_r)$, $x^s(\tilde{k},t_r)$ are the given and simulated value respectively, at time t_r for the sth initial condition; M is the number of a sample points chosen. We wish to minimize $F[\tilde{k}]$ over $\mathbb{R}^P$, i.e. determine a value $\hat{k}$ such that

$$F[\hat{k}] \leq F[k], \forall\ k \in \mathbb{R}^P.$$

To obtain such a $\hat{k}$ we implement the following algorithmic procedure:

STEP 1 Using the initial guess $\tilde{k} = (k_1,\ldots\ k_p)$, of $\hat{k}$ and the first vector of initial conditions $\bar{x}(0) = \bar{c}_o$, integrate the system of differential equations.

STEP 2 Minimize the function (2) for s=1. Denote by $\bar{k}^0 = (k_1^0,\ldots k_p^0)$ the parameter values obtained after the optimization.

STEP 3 Using $\bar{k}^0$, $\bar{x}(0)=\bar{c}_o$, and additional initial conditions $\bar{x}(0)=c_1$, integrate the system as in Step 1 and form $F[\bar{k}^0]$ where we sum over the index, s=2.

STEP 4 Minimize the function (2) for s=2 and obtain a vector of parameter values $\bar{k}^1 = (k_1^1,\ldots k_p^1)$ which is optimal for both trajectories.

STEP 5 Proceed in this way, using the last vector of parameters $\bar{k}^{s-1}$ found and an additional initial state s, we integrate the system from each of the initial states and minimize f over all trajectories simultaneously. To obtain a next estimate $\bar{k}^s$. The final value $\bar{k}^\ell$ is the estimate produced for the optimal vector $\hat{k}$.

STEP 6 Determine the accuracy of the parameters in relation to an assumed noise in the data.

THE ALGORITHMS USED

Three difficult algorithmic tasks are required to estimate the best parameter value $\hat{k}$.

     (1) Numerical integration
     (2) Nonlinear optimization
     (3) Error Analysis

(1) Numerical Integration. Many mathematical models in biochemistry are described by stiff systems of differential equations. Since, in general no closed form solution can be obtained for this type of equation (Rosenlicht 1968) we used a combination of two methods: the Adams-Moulton predictor corrector method and Gear's (1971) method up to 6th order.

(2) Nonlinear Optimization. Bremermann (1970) introduced a global optimization algorithm which minimizes a function of many variables without using derivatives, and using random directions. Milstein (1975) extensively tested this algorithm against certain leading optimization techniques, and found that for higher dimensional problems Bremermann's algorithm performed better. This method, however, can get trapped "close" to the global minima (if it exist, as for quadratic functions, Bremermann 1970) and converge slowly to it. Motivated by its good features and interested in overcoming its bad ones, Milstein

(in preparation) developed an optimization algorithm which uses interpolating
schemes together with weighted random directions. A full description of the
algorithm, and a detailed theoretical analysis of it will be presented elsewhere. The algorithm works as follows:

(1) An initial guess $\bar{k}^o$ of the parameters value is given.

(2) A random direction $\bar{\omega}$ is generated by using the linear transformation
$A\bar{q} = \bar{\omega}$

where

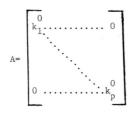

is a pxp matrix whose diagonal entries are the components of $\tilde{k}^o$, and
$\bar{q} = (q_1, \ldots q_p)^T$ is a pxl random vector whose component are independently
generated from a Gaussian distribution.

(3) A set of functional values $\{F_j\}_{j=1}^{s}$ is obtained by evaluating F at
equidistant collinear points $\alpha_1 = \tilde{k}^o + 2\bar{w},\ \alpha_2 = \tilde{k}^o + \bar{w},\ \alpha_3 = \tilde{k}^o,$
$\alpha_4 = \tilde{k}^o - \bar{w},$ and $\alpha_5 = \tilde{k}^o - 2\bar{w}.$

(4) The cubic spline $S(\lambda) \in C^2[\alpha_i,\ \alpha_{i+1}],\ 1 \le i \le 4.$ This cubic approximates
the function F restricted to the Line $\tilde{k}^o + \lambda\bar{w}.$

(5) The roots of $S^1(\lambda) = 0$ are found. There are from 0 to 8 possible real
roots, $\lambda_\ell,\ \ell = 1, \ldots 8.$

(6) Determine the minimum functional value among $F[\tilde{k}^o + \lambda_\ell w],\ \ell = 1, \ldots, 8$
and denote by $\bar{k}_1 = \tilde{k}^o + \lambda_m \bar{w}$ the vector which corresponds to this minimum.
This vector becomes the new approximation for the parameter sought.

(7) If in this step we failed to find a significantly better approximation
to $\tilde{k}^o$, we repeat Steps 1, 2 and 3 and replace Step 4,5,6 of the spline
approximation by a four degree Lagrangian interpolation such as the one
used in Bremermann's optimizer. However, if the step is successful then

(8) Improve the present parameter vector $\bar{k}_1$ by performing a pseudo Newton
Raphson step i.e. $\bar{k}_n = \bar{k}_{n-1} - \dfrac{F[\bar{k}_{n-1}]}{S'(\lambda_m)}.$
if successful retain the present random direction and repeat from Step 1,
substituting $\bar{k}_n$ for $\tilde{k}^o$.

(9) Terminate the search if:

(1) $F \le \epsilon_1$, and $||\bar{k}^j - \bar{k}^{j-1}|| \le \epsilon_2$

where $\bar{k}^j$ is the vector obtained after j times from Step 1 to 9, and ε_1 and ε_2 are preassigned tolerances.

OR

(2) A predetermine number of iterations have been executed.

The technique does not require formal or numerical derivatives. Moreover, there is no need for "close" initial estimates. The algorithm has performed satisfactorily in optimizing functions of many variables (up to 50).

(3) Error Analysis. When a set of parameters value is obtained it is important to investigate their reliability and sensitivity to data perturbed with noise. Milstein (1978) showed that the error analysis technique of Rosenbrook and Storey (1966) is a reliable mathematical procedure for estimating the expected error in parameter values obtained using multiple trajectories. The main result of such analysis follows: let the matrix $D(t_q)$ be defined for each measurement at time t_q by

$$D_q = [D_{ij}(t_q)] = [\frac{\partial \bar{x}_i}{\partial k_j} \; (\bar{k}, t_q)]$$

and define the pxp matrix

$$H = \sum_{s=1}^{\ell} \sum_{q=1}^{M} \left[D_q^T W_q D_q \right]_s$$

and $P = H^{-1}$ respectively, where W_q is a weighting function for each data point; M is the number of data points; and ℓ is the number of initial conditions used.

The expected variance for the ith parameter was shown by Milstein (1978) to be given by

$$\sigma_i^2 = P_{ii}.$$

To compute the P_{ii} entries it is necessary to obtain the matrix D. This is accomplished by considering the variational system derived from Equation (1).

$$[\dot{D}(t)]_s = [A(t)]_s \; [D(t)]_s + [B(t)]_s \; , \; [D(0)]_s = 0$$

where

$$[A(t)]_s = \left[\frac{\partial f_i(t)}{\partial x_j}\right]_s \; , \; [B(t)]_s = \left[\frac{\partial f_i(t)}{\partial k_i}\right]_s \; , \; [D(t)] = \left[\frac{\partial x_i(t)}{\partial k_i}\right]_s, \; \begin{array}{l} i=1,\ldots,n \\ j=1,\ldots,p \\ s=1,\ldots,\ell \end{array}$$

to minimize the probability of making a mistake during the derivation of $[\dot{D}(t)]_s$, we used automated symbol manipulation to formally obtain the matrices of partial derivatives, $\left[\frac{\partial f_i}{\partial k_i}\right]$, $\left[\frac{\partial f_i}{\partial x_i}\right]$, and the necessary sum and product of such matrices. The nonzero elements of $D(t)$ are nonlinear differential equations given in symbolic form. These entries have to be integrated numerically to form the matrix

$$H = \sum_{s=1}^{\ell} \sum_{i=1}^{M} \left[D_i^T W_i D_i \right]_s .$$

The algebraic manipulation of symbolic strings was implemented with the aid of the Altran (algebraic translator) language [3]. The matrix $\dot{D}(t)$ in the cases of our linear and nonlinear models has 31 and 101 nonzero entries respectively. After integrating each entry, the matrix $P = \left[\sum_{s=1}^{\ell} \sum_{i=1}^{M} [D_i W_i D_i]_s^T \right]^{-1}$ is computed and the variance σ_{ii} for the ith parameter is thus obtained.

NUMERICAL RESULTS

The aforementioned algorithmic procedure was applied to simulate the behavior of the gluconeogenesis models. Five vectors of initial conditions c_i, i=1, 2,3,4,5, and four samples of each state trajectory $y(t_r)$, r=1, 2, 3, 4 were generated. The number of data points reflects a feasible number of points that can be obtained in an experimental set up. The data points were generated using a designated vector of parameter values (Table (1), (2), column a) for linear and nonlinear models respectively. A 9% noise is introduced to each data point using a random number generator uniformly distributed in (0,1).

Table (1) summarizes the simulation of the parameter identification task for the linear compartmental model. It is of interest to notice that when using 1 initial condition we only recovered 2 parameter values while when using 5 initial conditions we recovered six (column c and d respectively). Our criterion for judging recovery is that the identified value lies within 25% of the parameter's value of the model which generated the trajectories.

T A B L E   1

SIMULATION OF THE PARAMETER IDENTIFICATION TASK FOR THE LINEAR
COMPARTMENTAL MODEL USING A 9% RANDOM NOISE IN THE DATA

NO.	PARAMETER VALUE A	INITIAL GUESS B	FOUND WITH 1 INITIAL CONDITION C	FOUND WITH 5 INITIAL CONDITIONS SIMULTANEOUSLY D
1.	.5	18.28	.941	.46*
2.	.001	.0013	.00094*	.0008*
3.	.12	82.30	31.29	22.14
4.	.1	.30	.058	.11*
5.	.0012	.0054	.000038	.0049
6.	6.	22.13	876.60	79.34
7.	13.	124.2	27.97	14.98*
8.	450.	526.4	9.94	491.20*
9.	1.	4.31	1.04*	11.98
10.	345.	922.07	1045889.	5119.6
11.	1234.	460.46	78916	4413.4
12.	17.1	80.34	10.14	17.33*

*Parameter Judge Recovered

The results obtained for the nonlinear model are recorded in Table 2. In this
table we recorded the parameter values which were recovered using 5 trajectories
simultaneously. When 1 trajectory was used the 31 parameters value converge to
values different than the ones designated as being the "correct values" (column
c). This is not surprising since the inverse problem will, in general, generate a
non unique solution. However, when 5 trajectories were used 9 parameters converged
within reasonable error bounds, (column d.) Figure (5) shows the goodness of fit
obtained for G-6-P which is one of the worst cases. Figure (6) is a 3 dimensional
representation of Figure (5) where the x-axis represents the data curve, the y-axis
the fitted curve, and the z-axis is the time.

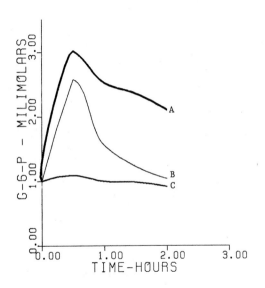

FIGURE 5:

A -   FITTED CURVE

B -   DATA CURVE

C -   PERTURBED CURVE

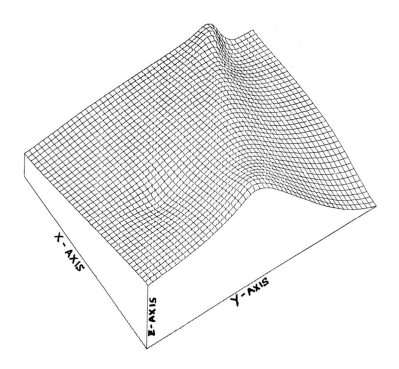

FIGURE 6:   A 3-Dimensional representation of the fitted curves in Figure 5.
This is a qualitative representation of the fitting using symmetry.
The symmetry of the surface about the z-axis indicates the goodness
of fit between the data curve and the fitted curve.

T A B L E   2

PARAMETER IDENTIFICATION RESULTS FOR THE NONLINEAR MODEL
USING DATA HAVING 9% RANDOM NOISE (PARTIAL LIST)

NO.	PARAMETER VALUE A	INITIAL GUESS B	VALUE FOUND WITH INITIAL CONDITION C	VALUE FOUND WITH 5 INITIAL CONDITIONS D
1.	10.	32.84	145.95	9.15
3.	30.	81.20	241.90	32.26
5.	95801.	192161.	820223.	114910.
10.	157.	493.92	1349.96	148.03
11.	.05	.14	.37	.06
12.	78.	129.92	351.63	85.17
16.	.09	.22	.61	.074
17.	1.2	2.99	7.71	1.37
29.	12.	46.52	145.24	9.34

*THE REMAINDER OF THE PARAMETERS VALUES WERE NOT RECOVERED TO WITHIN A 9 to 25
PERCENT ERROR OF THE ACTUAL VALUE (COLUMN A).

The goodness of fit is only a first indication that the parameter extimates are
reliable. A second index of reliability is the sensitivity of the parameter value
to noisy data. This sensitivity is measured by the variance or expected error
$\sigma_{ii}^2 = [H_i]^{-1}$. Since the matrix H represents a 1-1 mapping of $\mathbb{R}^p$ onto $\mathbb{R}^p$, it has
a unique inverse, however, if H is ill conditioned then the variance σ_{ii}^2 will

indicate that some of the parameters are not unique, and cannot be determined (with

the present information) within any range of certainty. Morever, the ill condi-

tioned property of H will indicate that the linear assumption in the error analysis

is not accurate (the analysis is performed under the assumption that $\bar{k} = \hat{k} + \varepsilon$,

where ε is a small perturbation, $\bar{k}$ is the vector parameter sought and $\hat{k}$ is the

vector parameter found). Thus, if a good fitting of the data to the model is

obtained, we can always determine when a vector parameter is reliable or not by
performing the Error Analysis. When the expected errors are within some reasonable
bounds, the parameters can be trusted, otherwise additional analysis of the model
and the data is required. The final validation lies on the ability of the model
to predict new experimental outcomes.

In our case, both the linear and nonlinear model (when 5 trajectories were used) generated a matrix H which is ill conditioned, with condition number close to 10^{74}, and having variances significantly larger than their actual parameters value.

CONCLUSIONS

Linear and nonlinear models describing the process of gluconeogenesis were constructed. The linear model consist of 6 linear ordinary differential equations having 12 unknown parameters. The nonlinear model is represented by 20 nonlinear first order equations having 31 unknown parameters. An integrated algorithmic approach for simulating and determining the kinetic of such models was implemented. These approaches consist of the proper interaction of a new nonlinear optimization technique which uses spline functions and lagrangian interpolating schemes, together with Gears (1970) numerical integration method for stiff differential equations. An extensive numerical simulation of Bessman's theory of insulin action was performed using both of the aforementioned models.

Preliminary results for the linear model shows that reliable parameters value can be obtained when the noise in the data does not exceed 9% and when as many as 9 distinct trajectories are used simultaneously in the parameter identification process. For the nonlinear model the results are still inconclusive and further analysis of the model and the numerical sensitivity of the methods is required before attempting to determine the kinetics, and any meaningful conclusion of Bessman's theory can be obtained.

The results of the simulation are being used in designing experiments in order to obtain the necessary experimental data. This data will be used for estimating optimal values for the rate constants of the gluconeogenesis system, and thus to test Bessman's theory of insulin action.

ACKNOWLEDGMENTS

*I thank Drs. Samuel Basseman and J. Mohan of the pharmacology and nutrition Department at the University of Southern California School of Medicine for their valuable suggestions and for providing the data upon which this paper is based.

REFERENCES

(1)   Bessman, Samuel P. "The hexokimase-mitochondrial binding theory of
       insulin action". Lipmann Symposium, Walter de Gruyter-Berlin, New
       York, 1974.

(2)   Bremermann, J.H., "A method of unconstrained global optimization".
       Math. Biosi. 9, 1(1970).

(3)   Brown, W.S., Altran User's Manual. Bell Laboratories (1973).

(4)   Conway, R., Wilcox, T., Bodenstein, A., Worona, S. "Users Guide to
       PL/C the Cornell compiler for PL/I". Dep. of Comp. Science Cornell
       University (1972).

(5)   Garfinkel, D., Comput. Biomed. Res.,2, 31(1968).

(6)   Garfinkel, D., "Computer applications to biochemical kinetics",
       A Rev. Biochem. 39, 473(1970).

(7)   Gear, C.W., "Numerical initial value problems in ordinary differential
       equations". Prentice-Hall, Inc. (1971).

(8)   Hemker, H. and Hess, B., "Analysis and simulation of biochemical systems".
       Febs, North Holland, American Elsevier (1972).

(9)   Lehninger, A. "Biochemistry". Worth Publishers, Inc. Second Edition
       (1977).

(10)   Milstein, J., Bremermann, H., "Parameter Identification of the Calvin
       Photosynthesis cycle". To appear in the Journal of Mathematical Biology
       (1978).

(11)   Milstein, J. "Error estimates for rate constants of non-linear inverse
       problem", the Siam Journal of Applied Mathematics, Vol. 35, No. 3, Nov.
       (1978).

(12)   Rosenbrook, H., Storey, C., "Computational techniques for chemical
       Engineers", Oxford, Pergamon Press (1966).

(13)   Rosenlicht M., "Liouville's theorem on functions with elementary integrals",
       Pacific Journal of Mathematics, Vol. 24, No. 1, 1968.

METHODOLOGY IN SYSTEMS MODELLING AND SIMULATION
B.P. Zeigler, M.S. Elzas, G.J. Klir, T.I. Ören (eds.)
© North-Holland Publishing Company, 1979

STRUCTURE CHARACTERIZATION FOR SYSTEM MODELING IN

UNCERTAIN ENVIRONMENTS

G.C. Vansteenkiste
University of Ghent
Belgium

J. Spriet and J. Bens
Belgian National Science
Research Foundation
Belgium

1. INTRODUCTION

All modeling practitioners will agree that in system modeling three fundamental stages are present. One is the selection of the type of model. A decision
must be made about what kind of equations will be used, what non-linearities if
any are present, of what order the differential equations are, etc... Here the
researcher is lead by his experience with the problem and by his knowledge of the
fundamental laws that lie behind it.

If nothing is known at all, this stage may present a serious problem, and
trial and error is about the only thing one may attempt to do. This kind of
"black box" problem is very difficult and will not be discussed here. In many
cases however some information about basic physical laws, expressing conservation
of mass, energy and momentum as state equations, is available. Even if the process
dynamics can be described reasonably well, a characterization of the disturbances
is often missing. When the desired models cannot be determined from à priori knowledge alone, experimentation becomes a necessity. In the experiments an input
signal is applied to the process and the process outputs are observed. A description of the process dynamics and the disturbances can then sometimes be obtained
through appropriate processing of the experimental data. The attention in this
paper is focussed on modeling based on a combination of experimentation and physical laws, the so-called "gray box" problems [9] describing uncertain environments. For "white-box" problems the choice of the model structure presents no
problem, since so much is known about how the system behaves that it is easy to
write this knowledge down in computational form.

The second stage of model building is one of parameter identification. This
is a very old problem, with both old and new solutions. Classical techniques include curve fitting and least squares parameter optimization. Modern filter theory
has improved parameter estimation quite a bit, and Kalman filtering techniques
are so powerful that one is now able to extract the model parameters with great
accuracy from noisy measurements. This is not the place to discuss the problem
of parameter identification any further.

The third important step in model building is the validation of the model.
This step is often neglected. Very few criteria exist to say whether a model is
a valid representation for a real world phenomenon or not. It is possible to say
that data obtained from simulation are a close fit to the measured data, but this
does not prove too much. Increasing the number of parameters in the model will
always decrease the deviation (e.g. in the least squares sense) at the cost of
the naturalness of the model : criterion of "best-fit".

When developing a model for some new phenomenon, the procedure of Fig. 1 is
usually followed. One starts with that model structure that is expected to be the
best suited to the problem. Using data from the real world experiments, the parameters of this model are identified. The output of the model is then a posteriori

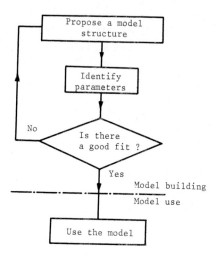

Fig. 1. Model development

compared to the data. If correspondence is not good, the model is rejected. One tries another model structure, perhaps based on another theory, and the whole process is repeated. Finally, one is satisfied about the performance of one model and this structure will be used for the purpose for which the model has been developed. This can be economic planning, chemical reactor steering, or any other kind of control action based on the predicted behavior of the system.

If the model was not accurately chosen, for example, due to some special conditions during the identification phase, no indications will be given. It is only afterwards when the results will be poor, that one will notice that something went wrong. Then the whole procedure for building the model has to be restarted, which can only be done at great expense of time and resources.

In this paper, we shall present another approach to the model building problem, where the emphasis lies on structure characterization instead of parameter identification. The structure of a model is indeed more fundamental to the understanding of what is happening rather than the parameter values.

2. SYSTEMS APPROACH IN UNCERTAIN ENVIRONMENTS

Systems from uncertain environments are characterized by uncertainties as to their topology, their field parameters and their internal characteristics. The phenomena being modeled take place in a medium whose distributed properties are only imprecisely known. The characteristics may even be apt to change in time in an unpredictable manner. The system borders are more fuzzy and the important inputs and outputs are less easily recognized. A badly defined problem is faced.

Mechanical and electrical systems are almost always built up of a number of subunits, for which macroscopic theory yields well-known and accurate descriptions, while the interconnections between these subunits are straightforward. In uncertain environments, subunits are difficult to isolate, as exchanges are intricate and complex. Microscopic theory might be sufficiently developed, but experts in the field are cautious in formulating their opinions about overall-dynamic behaviour and its roots. Isolation of a portion of the system in order

to separate input and output information, may destroy its natural function and
lead to identification of a purely hypothetical process. Many complete systems
however, are too complicated for any reasonable model to include all the factors
that might be important. An optimum choice has to be made between incompleteness
and complexity. No criteria of completedness can be stated and the model does not
meet the "best-fit" criterion.

An often aggravating condition is a paucity of data of relative low quality.
In contrast, in the case of more physical systems, noise is not excessive or
measurements are abundant. It is the acquisition of truly representative field
data that sets the identification of systems in uncertain environments somewhat
apart from the "white-box" simulation problems. Quite often only natural excita-
tion of the process dynamics can be observed. Normal operational records exhibit
just one particular mode of process behaviour, deviations from the steady state
may even be insignificant, blocking all information about the dynamics. This entails a low signal to noise ratio. Measurements are rarely abundant and often
very limited : there is never enough data of the right kind.

A large portion of the ideal process state vector is moreover unmeasurable,
and much of the observed process dynamics is a function of unmeasurable stochastic
process disturbances and of random measurement errors. Only limited information
is available about the statistics of measurement and process noise. A more detailed discussion about these problems can be found in [12],[20] and [23].

3. EXTENDED MODELING APPROACH

Faced with these pecularities of the systems under study, research is conducted to find a more general and more powerful methodology. At the present time
there is not yet a general agreement on the optimal procedure to follow. A sufficiently general scheme is presented on Fig. 2.

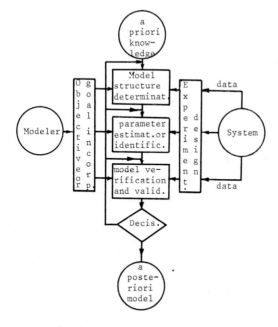

Fig. 2. Extended modeling approach

Parameter estimation and model validation remain key parts of the process, but an additional block : model structure determination proceeds the classic iterative loop. It is believed,that methods should be developed that help formulating the structure of the mathematical equations in a systematic way. In that manner, they should alleviate the problems arising from ill-definition and fuzzy boundaries.

A second modification is the addition of two "interfaces". The first one regulates and conditions the data flow from the process to the model building algorithms. Formerly, data collection was often done in a rather heuristic way, but the data insufficiencies, that almost always occur when "soft" systems are studied, require careful experimental design. Experimental design techniques exist, but they should be adapted to the specific problems of the different modeling stages. The major problem is that a good experimental design depends partly upon a good a priori knowledge (model) of the system under study. The modeling effort provides a feedback in the development of a more rational data collection scheme [18] .

The second interface is between the modeler and the existing options for the modeling methods. It is known that objective and goal considerations should guide the systems-analyst, but little knowledge is available about systematic procedures to incorporate these considerations within the modeling process itself.

This study elaborates on new techniques for model structure determination. The methodology is still in its infancy. The problem itself can be seen as a hierarchical decision tree ([1] and[12]). In a first stage, called "model formulation" important factors, input output quantities and state variables have to be selected. In a second stage, the transformations between these variables have to be chosen, except for a number of unknown parameters that remain to be determined afterwards. There exist guidelines, that can help the model builder during that phase of the modeling effort. The structure should be as parsimonous as possible, identifiable and objective oriented ; the more physical, the better. Unfortunately these principles are quite general and may be conflicting, so they are of little direct help.

If the system is known to be linear and the important variables have been determined from a priori knowledge, the structure characterization boils down to be a problem of determination of the order of the system. Some theories are yet available for this case. Akaike [2] and Rissanen [14] adopt an information-theoretic point of view, while Kashyap [10] stresses a Bayesian approach. For non-linear structures there is a distinct lack of techniques. However a few promising ideas seem to stimulate research.

Before discussing these ideas, a better understanding of the issues can be gained by reconsidering the classic modeling process in Fig. 1. Structure characterization is implied in the overall scheme, but is not explicitly mentioned. In fact, a great part of the structure follows from deductive analysis of a priori knowledge. The remaining ambiguities are resolved after some iterative parameter estimation sessions.

Each parameter identification yields residuals, which were statistically analysed. Whenever, these residuals are sufficiently white, the structure is supposed to be correct. If trends are detected a structure modification is thought necessary. An elegant feature of this approach is that estimation and characterization are performed in one single effort. Drawbacks are evident : bad initial structure will cause long iterations and optimal estimation strategies are likely to be bad structure analyzers. However an important conclusion remains that parameter estimation in a more general sense can be a tool for structure investigations.

A first approach to non-linear structure identification is the use of recursive parameter estimation schemes [Young,22] . The method is a direct extension of the classical approach. One starts with a linear structure and uses the recursive estimations of the parameters as indicators for the structure. Because the recursivity adds an extra-dimension to the estimates, it is believed that additional information is available about structure characteristics. A pleasing feature is the fact that one only introduces non-linearity when it is really necessary, so the principles of parsimony and identifiability are taken into consideration. A doubt, however, remains about the possibility to get from linear configurations to complex non-linear ones which may be more physical in nature.

A second approach relying heavily on probability theory and Bayesian methods has been introduced by Kashyap [11]. A set of candidate structures has to be chosen. Given the data, one looks for the structure class, with the highest a posteriori probability. To some degree, parsimony and identifiability as well as physicality and objectivity can be accounted for, but there is a restriction on the non-linearities. Furthermore, to obtain the a posteriori probabilities, quite a number of assumptions have to be made and the probability densities have to be cooperative. Investigations to test the robustness of the technique when certain assumptions are invalid is certainly necessary.

A third method is known as the "Group Method of Data Handling" introduced by Ivakhnenko [6]. It is based on the principles of heuristic self-organization and relies on bioengineering concepts. As the brain is a neural network, capable of learning and performing complex control and recognition tasks, it is stipulated that an intricate, non-linear, layered network should be able to simulate a system structure if it is appropriately trained by data. The method seems to negate the principle of parsimony and physicality is found less important than objectivity. It may however be that, in such a network, information is processed in a more optimal way.

Another approach sets the problem in a pattern recognition framework. Structures can be considered as different patterns and characterization may be seen as a pattern recognition task using experience and present information. These main ideas have been set forth by Karplus [8] and Saridis [15]. Simundich [16] and Hofstadter [5] have worked out certain details of the procedure. The present work is an extension of Simundich's results for ordinary differential equations. The framework is quite general. Parametric as well as non-parametric methods can be included and the importance of a learning phase is stressed. After proper formulation, the other approaches, discussed above, may most probably be fitted into the pattern recognition methodology, so that pattern recognition can become the unifying framework for structure characterization.

4. PATTERN RECOGNITION FOR STRUCTURE IDENTIFICATION

An ideal situation would be one where the only input to a computer is the data measured from the phenomenon to be modeled and from this the machine would produce a complete model. This is however utopic. A solution has been achieved where one more input is required : an input considered as a library of models. The machine will search among these candidates in order to find a model whose structure is best adapted to the data. The situation that no reasonably good model can be found in the "library" is also possible. Then the operator has to reconsider the problem in order to add new possibilities to the library. The designer has to specify several distinct classes in such a way that later the attribution for the process to a class answers his questions about the characteristics of the model he is looking for.

4.1. Feature extraction

The most crucial part of the method is one of feature extraction. The choice among the different models will be done on the basis of characteristic expressions. The most absolute freedom exists about how to select these expressions. As always, freedom is a nice thing but very difficult to handle. Some experience is needed to be able to develop good features. They must be rather insensitive to noise and also have good discriminating properties. The number of these features is also free. Very often one feature per proposed model is used, but sometimes a better discrimination may be obtained with more. In other cases not all features are really important. For computational reasons the number of features should not be chosen too high.

In the example that follows, some possibilities will be discussed. It is very hard to give a general rule. Usually the features bear some relationship to the variability of certain parameters in the model, but that is definitely not a necessary relationship.

The different features form a so-called feature space. It is very important that one has a good feeling of what the feature space stands for, since it is in this space that most of the characterization operations are performed.

Computation of the features can be considered as an operation on the set of data coming from a candidate model during the initial phase and from the experiment later on. It should be noted that in this stage, during the initial phase, one learns the operations to perform later on the data, in order to answer specific questions about the character of the tentative model. We assume that the data are available in discrete form ; we thus have a set of couples (t_i, x_i) where t_i stands for the independent variable (time instant on which the measurement was taken), and x_i (which may be vector-valued) are the data obtained from the process under study. The experiment is equivalent to a set of such couples

$$D = \{(t_i, x_i) \mid i = 1,2,3,\ldots n\}$$

called the data set. Obtaining the features can be considered as a mapping from the set D of all possible data sets to the feature space

$$\Theta : D \to R^k : D \to f$$

where k is the number of features and f is a point in the feature space. Often, the mapping Θ is multiple valued, that is, more than one point f may be the image of a set D.

This mapping, that is, the procedure for computing points in the feature space from data, is used as a building block for the structure identification algorithm.

From a pragmatic point of view, a given structure results in a set of properties shared by the data generated by that structure. Certain properties are parameter invariant, others are not. Most properties are common to many structures. For a discriminating study, however, it is sufficient to choose for each structure a property, which is sufficiently parameter invariant, and which is not shared, to the same extent, with the other candidate structures. The feature extractor is a set of algorithms that test for each property so that a proper choice can be made, afterwards. There are quite a number of ways to construct feature extractors, the procedure followed in [17] will be explained in general and followed by an example in section 5.

It is assumed that a priori knowledge and prior modeling effort have yielded two sets :
- a set of structures V_c, that are to be tested, with their parametric representations $\underline{F}^i$

$$V_c \equiv \{V_i\} \quad i \in I_c \quad I_c \equiv \{1,2,\ldots,c\}$$

$$\underline{F}^i(\underline{\dot{y}}^i, \underline{y}^i, \underline{u}^i, \underline{p}^i, t) = 0 \tag{1}$$

$\quad \underline{y}^i$ (n x 1) observation vector

$\quad \underline{u}^i$ (m x 1) input vector

$\quad \underline{p}^i$ (r x 1) parameter vector

$\quad \underline{\dot{y}}^i$ (n x 1) derivatives of $\underline{y}^i$

- a set of data V_D

$$V_D \equiv \{(\underline{y}_k, \underline{\dot{y}}_k, \underline{u}_k, t_k)\} \quad k \in I_D \quad I_D \equiv \{1,2,\ldots,d\}$$

For noiseless data, p^i may be computed out of (1) using a few data points.
Under the hypothesis of data coming from that structure, the scatter of many parameter evaluations should be zero and different from zero for data coming from
other structures. Direct feature extraction uses that property. So it is easily
understood that an implementation can have the block diagram shown on Fig. 3.
The different parts are :
- a data selector : the algorithm generates a set W_D of subsets from V_D, each subset capable of parameter extraction

$$W_D = \{W_1\} \quad W_1 = \{(\underline{y}_j, \underline{\dot{y}}_j, \underline{u}_j, t_j)\}$$

$$1 \in I_W \quad j \in I_L \quad I_L \subset I_D$$

- a parameter extractor : these algorithms solve the non-linear algebraic equations (1) for each class separately. So after extraction c sets of parameter
evaluations are obtained:

$$P^i = \{p_1^i\} \quad 1 \in I_W$$

In fact, the extractor may be, for convenience slightly modified : each linear
or non linear operation $q = \underline{h}(\underline{p})$, yields quantities $\underline{q}$ that are also suited for
the present purpose. A good choice may facilitate parameter extraction or give
more noise-insensitive invariants.

- a data concentrato: : this unit performs an operation on P^i and should measure
the variability within each set. The result is a pattern vector $\underline{\alpha}$ of dimension
c.

One realizes that for noiseless data, coming from the ith structure, the ith
entry in the $\underline{\alpha}$ vector will be zero, while positive and hopefully large for the
other entries. At that point, choosing the structure boils down to finding the
smallest entry in the $\underline{\alpha}$ vector. There are however a number of factors that will
obliterate the performance of the scheme :
- computational errors will be present
- derivatives are usually not measured and have to be estimated from the data
- noise, that always corrupts the measurements, will influence the scheme.
As long as the pattern recognizer is able to discriminate variation due to those
causes, from these resulting from structure differences, the scheme will be of
valuable help. Reference [17] elaborates on different choices of operations for
the three blocks in Fig. 3.

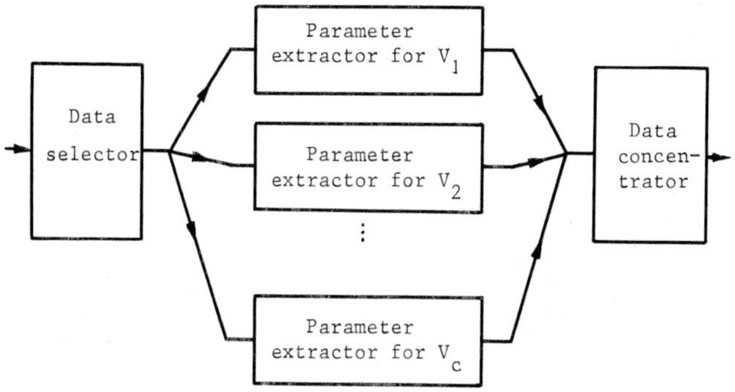

Fig. 3. Feature extractor for structure characterization

4.2. Structure discrimination

Fig. 4 illustrates the different steps of the method. Two stages of operation must be distinguished : training of the classification algorithm using the
candidate models (switches in position I) and using the classifier on the experimental data (switches in position II).

The classifier building block is a standard procedure in the theory of pattern recognition. Excellent handbooks cover this topic [3],[4] and [13] ; a very
good survey of many different methods can be found in [7]. The problem of choosing a classifier algorithm that satisfies the requirements of this context is
discussed in rather great detail in [21].

The classifier splits up the feature space in a partition ; each subset corresponds to a cluster of points and hence to a model from the library. Training
consists of choosing the boundaries in such a way that the partition corresponds
to the different models in an optimal way. It is thus necessary to make some simulation runs with each of the proposed models. The data sets from these runs are
processed by the feature extractor and the classifier output is compared to the
model from which the data were generated. The parameters of the classifier are
adjusted in such a way that maximal correspondence between input and output is
obtained. The more runs that are used to train the classifier, the better. It is
also important to make the different runs of one model as diverse as possible :
that is, combinations of parameter values should be used that cover a wide region,
so that the set of values that may be expected from real life data, falls inside
this region.

4.3. Some additional remarks

The user has to keep a close eye on what is happening in the feature space
during training. Here, his choice of the computation rules for the features is
subject to a very severe test. It may appear that some features are almost useless, or that some kind of modification would entail a considerable improvement
in the classification.

It is also during the training phase that the influence of noise on the data
must be studied. Often a feature has a very specific behaviour for data sets
generated with one particular model (e.g. is a constant). When noise is present,
deviations will occur, and if the feature computation algorithm is ill-conditioned, these deviations will become so important as to make that feature useless.

Noise can originate from different sources. Not only measurement noise can
be present, but noise-like errors may occur in the computation itself. Indeed,
often the values of the derivative of a system variable are required. If a derivative is to be computed from sampled data, the exact value will not be found.
The situation gets much worse for higher order derivatives ; use of these is not
recommended.

Training the system requires quite a lot of computation time. This must however be done only once, before the actual experiments start. Batch processing is
possible. The computations during the experiment itself are minimal, so on-line
computations can easily be performed. An added advantage is that one has permanent
control over the model structure. No time will be lost in identifying the parameters of a model whose structure does not correspond to reality. The scheme of
Fig. 5 is obtained, so the feedback loop of Fig. 1 is eliminated.

Even when the model is in use, it is always possible to keep checking that
the model structure is still adequate for the incoming data. When deviations
occur, it is easy to switch to another model in the library. In this way one will
always use that model that is best adapted to the data.

5. AN EXAMPLE AS ILLUSTRATION OF THE TECHNIQUE

What may look rather complicated is in fact very simple as will be shown.
Let us assume the trajectory of a missile has to be modeled. The object is

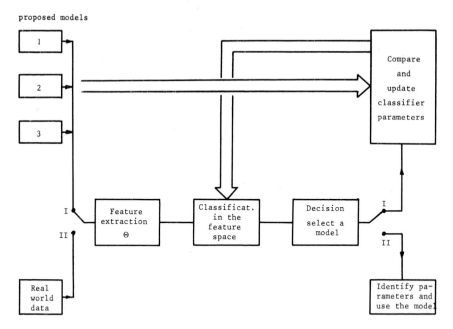

Fig. 4. Structure determination

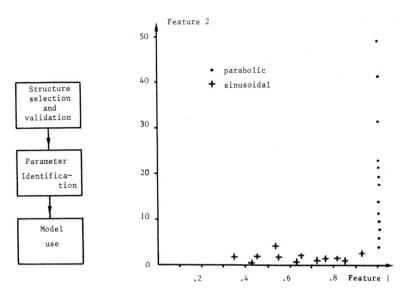

Fig. 5. Proposed modeling approach

Fig. 6. Feature space for the ballistic
trajectory example

launched with a vertical initial velocity and is subject to gravity. One does not need to be a good scientist to know that the altitude x varies parabolically with time, but assume that we have our doubts about this. We say that two equations could model x in function of t, namely

$$x = -\frac{1}{2} g t^2 + v_o t \tag{2}$$

and $x = A \sin (b\ t)$ (3)

We want to use our method to define which equation is a valid model for the ballistic trajectory.

5.1. *Feature selection*

What characteristic expressions may be used to distinguish (2) from (3) on the basis of a data set$\{(t_i, x_i)\}$? A first idea is to use the third derivative of x with respect to time : this should be zero for (2) and assume non-zero values for (3). Although in theory this idea is good, in practice it is worthless. Computation of a third derivative from sampled data is an extremely noise-prone operation ; the resulting errors are much too important. No distinction at all can be made between the two models on the basis of this feature.

A technique that is rather popular is to focus attention on the parameters of the equation. For different time instants, compute one or more parameters of the equation. If that model is the correct one, their value should be about the same at all time instants. Hence, the variability of the computed values is an indication of the correctness of the model. Simundich [16] computes all parameters, at all time instants, and uses the trace of the scatter matrix as a feature. We believe an improvement to this is to define the mapping as a multiple valued one : more than one point in the feature space is assigned to a single data set. This has the advantage that an erroneous measurement at a certain time will not destroy most of the information present in the other measurements. In a previous paper [19] as many points in the feature space were used as there are couples in the data set. Here, we shall limit ourselves to a few points per data set.

The following mapping is used : select two points in the data set : (t_A, x_A) and (t_B, x_B). Compute the value of the derivative of x by the usual difference formula. An approximation for the parameter g of equation (2) is given by

$$g_i = \frac{2x_i}{t_i^2} - \frac{2\dot{x}_i}{t_i} \qquad i = A,\ B$$

The first component of the point in the feature space will express the variability of this parameter ; a way of doing so is to compute the ratio of g_A to g_B.

The second component is inspired by equation (3). The parameter b satisfies the equation

$$\frac{1}{b} \operatorname{tg}(b\ t) = \frac{x_i}{\dot{x}_i}$$

Solve this equation numerically for b in the two points A and B. Then, compute the percent variation of this parameter

$$200\ \frac{|b_A - b_B|}{b_A + b_B}$$

5.2. *Training*

Several runs are now made with both models, for different parameter values. Two elements are randomly picked in the data set, and the feature is computed according to the rules stated above. The result is a point in the feature space.

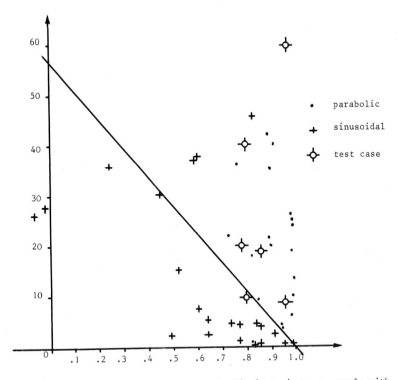

Fig. 7. Feature space for the ballistic trajectory example with added noise on the data

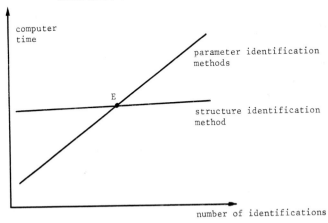

Fig. 8. Computation time of the proposed identification method

For each run, this is repeated a few times. Fig. 6 shows the resulting feature space. The two clusters, corresponding to each model are clearly separated.

In order to make the training more realistic, the data were contaminated with 2% Gaussian noise. The results are shown in Fig. 7. The scatter on the feature points now is much more important. Separation of the two clusters is still possible, but there is some overlapping! It is clear that it is important to use many different points for training : adding new results will only improve the performance of the classifier.

5.3. Use of the classifier

In Table I a possible outcome of measurements is given : an object was launched vertically with an initial velocity. Every second its position is measured. This results in a data set. In order to find out which model best represents the data, two methods have been used.

5.3.1. Least squares curve fitting

A least squares fit was done based on Equations (2) and (3) respectively. Both have two parameters. Table II gives the resulting parameter values, as well as the sum of squared deviations.

These deviations are almost equal. One might be tempted to decide that the sinusoidal law is the better model, but a wise conclusion is to say that we cannot decide which model to withhold.

t	0	1	2	3	4	5	6	7	8	9
x	0	44	78	106	122	130	122	106	78	44

Table I : Trajectory measurements : vehicle location versus time

	A	b	g	v_o	$\Sigma(x_i-x_i^{(m)})^2$
parabola	-	-	9.96	50.335	27.32
sinusoid	128.4	0.326	-	-	26.53

Table II : Least squares curve fitting of equations (2) and (3)

5.3.2. Structure identification

The feature extraction mapping has been performed on this data set, resulting in six new points in the feature space. These points are also represented in Fig. 7. Here, it is clear that most of the feature points associated with the data set are in the region corresponding to the parabolic model. The algorithm will set this model forth as being the better one. The next step is then the identification of the parameters. One way of doing so is by means of a least squares fit.

6. IMPLEMENTATION OF THE TECHNIQUE TO GROWTH-MODEL STRUCTURES

To test the developed methodology, a process was chosen where intensive modeling efforts provides different valid mathematical structures which are all quite similar. The growth of chemo-organotrophs under a limiting substrate was chosen, displaying a wide variety of organisms and consequently of growth patterns.

From a priori information, it is known that for a simple macroscopic model, biomass x, and limiting substrate S are the important quantities. The mass balance for substrate exchange between micro-organism and environment is given by equation (4)

$$\frac{dS}{dt} = - \frac{1}{y} \frac{dx}{dt} \qquad S(t_o) = S_o \tag{4}$$

The biomass itself changes according to equation (5)

$$\frac{dx}{dt} = f(x,S) = \mu(x,S) \cdot x \qquad x(t_o) = x_o \tag{5}$$

The candidate models differ in the form of the growth rate $\mu(x,S)$. Five forms are distinguished, and listed in Table III. The name of the scientist, who first proposed the model, is mentioned. The proposed models entail different physical assumptions.

In contrast with this process, in practice the structure can often not be analytically analyzed, in which case the pattern recognition technique permits a simulation approach to structure analysis. Hence the process allowed the simulation results to be verified.

In reference [17] alternatives are investigated for the processing stages in the different building blocks of a pattern recognizer.

Monod	$\mu(x,S) = \mu_M \cdot S / (k_M + S)$
Contois	$\mu(x,S) = \mu_C \cdot S / (k_C \cdot x + S)$
Verhulst	$\mu(x,S) = \mu_V \cdot (1 - x/x_M), x_M = x_o + y_V S_o$
Piece-wise linear	$\mu(x,S) = \mu_P \qquad S > k_P$
	$\mu_P \cdot S / k_P \qquad S \leqslant k_P$
Tessier	$\mu(x,S) = \mu_T \cdot (1 - e^{-S/k_T})$

Table III : Rate equations

7. CONCLUSION

A new and very promising method for structure characterization has been explained. Although very heavy computations are required before one can start with actual experiments, the computation load per data-set to be processed is very low. The initial computations can be performed off-line, and hence are less costly than those for the usual real time parameter identification methods. It is our belief that the economic point E (Fig. 8) is situated quite far to the left, so that as soon as a few runs have to be identified, this method will be advantageous.

The goal of simulation is also an important factor. If a teacher wants to illustrate Newton's law, it is sufficient to show that the ballistic trajectory is a parabola. The value of the parameters is quite irrelevant. The emphasis lies on structure identification. The problem is somewhat different for an artillerist at his gun. He has no interest whatsoever in the model that simulates the trajectory of his missiles. All he wants to know are the parameters, so he can determine the correct elevation to hit the target. The method presented here is of great help to the teacher, but only of indirect use to the artillerist.

REFERENCES

[1] AKAIKE, H., (1974), Stochastic Theory of Minimal Realization, IEEE T-AC, AC-19, N°6, pp. 667-674.

[2] AKAIKE, H., (1976), Canonical Correlation Analysis of Time Series and the use of an Information Criterion, 'System Identification, Advances and Case Studies', Acad. Press, New York.

[3] CALVERT and YOUNG, (1974), Classification, Estimation and Pattern Recognition, American Elsevier, New York.

[4] DUDA, R.O. and HART, P.E., (1973), Pattern Classification and Scene Analysis, Wiley, New York.

[5] HOFSTADTER, R.F., (1974), A Pattern Recognition approach to the classification of stochastic non-linear systems, Ph.D. dissertation, Purdue, Lafayette, Indiana.

[6] IVAKHNENKO, A.G., (1966), The Group Method of Data Handling, Soviet Automatic Control B-3, 43/55.

[7] KANAL, L., (1974), Patterns in Pattern Recognition 1968-1974, IEEE Trans. on Information Theory, November, IT 20.

[8] KARPLUS, W.J., (1972), System Identification and Simulation, A Pattern Recognition Approach, Proc. Fall Joint Comp. Conf., pp. 385-392.

[9] KARPLUS, W.J., (1976), The future of mathematical models of water resources systems, in System Simulation in Water Resources, Proceedings of the IFIP working conference, Bruges 1975, North-Holland, Amsterdam.

[10] KASHYAP, R.L. and RAO, A.P., (1976), Dynamic Stochastic Models from Empirical Data, Acad. Press, Vol. 122, New York.

[11] KASHYAP, R.L., (1977), A Bayesian Comparison of Different Classes of Dynamic Models using Empirical Data, IEEE T-AC, AC-22, N°5, pp. 715-727.

[12] MEHRA, R., (1978), A Summary of Time Series Modeling and Forecasting Methodology, Proceedings of the IFIP Working Conference on Modeling, Identification and Control in Environmental Systems, Ghent 1977, North Holland, Amsterdam.

[13] NILSSON, N.J., (1965), Learning Machines, 'Foundations of Trainable Pattern Classification Systems', McGraw Hill, New York.

[14] RISSANEN, J., (1976), Minimax Entropy Estimation of Models for Vector Processes, 'System Identification, Advances and Case Studies', Acad. Press, New York.

[15] SARIDIS, G.N. and HOFSTADTER, R.F., (1974), A Pattern Recognition Approach to the Classification of non-linear systems, IEEE Trans. on Systems, Man and Cybernetics, Vol. SMC-4, pp. 362-370.

[16] SIMUNDICH, T.M., (1975), System Characterization : a Pattern Recognition Approach, Ph.D. thesis, UCLA, School of Engineering and Applied Sciences, Los Angeles.

[17] SPRIET, J. and VANSTEENKISTE, G.C., (1978), Design of a preprocessor for a pattern recognizer used for the identification of soft systems, Proceedings of the Summer Computer Simulation Conference, Newport Beach.

[18] VANSTEENKISTE, G.C., (1976), System simulation in the testing role, in System Simulation in Water Resources, Proceedings of the IFIP working conference, Bruges 1975, North Holland, Amsterdam.

[19] VANSTEENKISTE, G.C., BENS, J. and SPRIET, J., (1977), System characterization using a pattern recognition approach - a challenge for the simulation of biosystems ?, Proceedings of the Summer Computer Simulation Conference, Chicago.

[20] VANSTEENKISTE, G.C., (1977), Data collection in large field problems, in MECO 77, Zürich, Acta Press.

[21] VANSTEENKISTE, G.C., BENS, J. and SPRIET, J., (1978), Design of a linear classifier using simulation techniques, Proceedings of the United Kingdom Simulation Conference.

[22] YOUNG, P. and KALDOR, J.M., (1977), Recursive Estimation : A methodological tool for investigating Climatic Change, CRES Report N° AS/R14.

[23] YOUNG, P., (1978), A General Theory of Modeling for badly defined Systems, Proceedings of the IFIP Working Conference on Modeling, Identification and Control in Environmental Systems, Ghent 1977, North Holland, Amsterdam.

METHODOLOGY IN SYSTEMS MODELLING AND SIMULATION
B.P. Zeigler, M.S. Elzas, G.J. Klir, T.I. Ören (eds.)
© North-Holland Publishing Company, 1979

PROBABILISTIC CONSTRAINT ANALYSIS FOR
STRUCTURE IDENTIFICATION: AN OVERVIEW
AND SOME SOCIAL SCIENCE APPLICATIONS

Gerrit Broekstra

Graduate School of Management
Delft, The Netherlands.

A normative theory of structure-model discovery has three principal properties: (i) demarcation of an admissible subspace of
models, (ii) decomposition of both the subspace and its consti-
tuting elements, and the search process conducted within that
space, and (iii) quality criteria for evaluating model adequacy. Constraint analysis, or briefly C-analysis, is particularly well-suited to guide and evaluate the decomposition processes in the subspace of probabilistic models. This paper is
a comprehensive overview of some of the results of C-analysis.
It is also motivated by the observation that the receptivity
of the scientific community to systems problem solving may be
greatly enhanced by comparative studies. To illustrate the
techniques proposed, we shall therefore reanalyze two different
sets of social science data on qualitative variables:
a) a two-attribute turnover table on student attitudes analyzed
earlier by Coleman (1964) and Goodman (1973); and b) a four-way
classification pertaining to the "contact hypothesis" analyzed
earlier by Wilner et.al. (1955), and more recently by Goodman
(1973).

INDEX TERMS: Structure identification, decomposition,
               constraint (analysis), structure system,
               approximation, complexity.

1 NORMIZING, DECOMPOSITION, QUALITY

The systems approach, while emphasizing the lack of disciplinary boundaries [1],
paradoxically enough demarcates the boundary problem as one of its central issues
to study. This may be illustrated in the domain of systems methodology, and, in
particular, in the light of the solution to the problem of structure identification or modelling. [2-6]

Searching for a model structure representing in some optimal way a given set
of data, in a virtually unbounded space of possible models, is a rather ineffi-
cacious approach. From a philosophical point of view, as a matter of fact, such
an endeavour has no solution at all as Hume has shown convincingly. The usual
approach is to demarcate (stake out, normize) some admissible subspace of models
in advance, i.e., before knowledge can be acquired. This kind of observation
induced Gaines' remark of the "precedence of ontology over epistemology."[2]
Similarly, it leads one to ascertain that postulation should precede discovery.[4,6]

On the one hand, in staking out an admissible domain, the demarcation
implies a selection or restriction. On the other hand, it generates freedom
within this restraint by facilitating effective search within the domain. Some
models cannot be reached, since they are outside the admissible domain, but
others can be reached more readily and efficiently.

Closer examination of the demarcation activity reveals the following features. First, it is a purposeful activity. Demarcation is done for a purpose
(prediction, improved functioning of a real-world system, etc.).[2] Second, staking
out a domain of models involves by implication the definition of boundaries.
But, third, and more importantly, such a definition is not entirely arbitrary;
the boundaries should indicate a direction in, or ordering of, the subspace by
means of which the search to be conducted is regulated. And, fourth, the demarcation should allow for the evaluation of the discovery process in terms of,
possibly, conflicting preference orderings due to the investigator. In my opinion,
these features of the demarcation activity indicate that it is neither an arbitrary nor a passive occurrence; it is a normative activity in the sense of
Simon's[7] connotation. We propose to use the term _normizing_ to refer to the
normative activity of staking out and ordering an admissible subspace of models.

It is evident that the efficient and controlled enumeration of models is a
crucial matter in solving the problem of structure identification. I am referring
here to the problem of searching for a structure system in an admissible subspace
of structure systems as it is defined by Klir.[4]

As to the normizing activity, similar remarks can be made for finding a
generative system in a set of data (see also Gaines[2,3] for the analogous problem

of behaviour/structure transformation).

Klir,[4] Klir and Uyttenhove,[5] and Cavallo and Klir[8] have studied extensively the problem of normizing admissible spaces of candidate structures. We
will use Klir's original results on a four-variable structure-model space in
this paper.

The second feature of the structure discovery process is that of decomposition. One can make a distinction in terms of ontological and epistemological
decomposition. The reader may wish to consult Gaines[2] for some philosophical
background as to problems of ontology and epistemology. Klir's hierarchy of
epistemological levels is a good example of an epistemological decomposition of
the process of knowledge acquisition. A hierarchy or lattice of admissible structure candidates is an example of an ontological decomposition. Each structure
candidate, in turn, is ontologically decomposed into a number of coupled elements.

It is the purpose of this paper to develop a formal language for the latter
type of decomposition. It will be shown that the theory of probabilistic
C-analysis as developed by this author is particularly well-suited to take into
account the higher-order interactions. These, in practical research, are quite
often neglected due to the lack of an adequate conceptual framework. Furthermore,
the representation of structure systems is rather naturally expressed in terms
of C-analysis.[9-16]

In modelling probabilistic structure systems to fit the observed behaviour
of a system, the hypothesized structure can only approximate the behaviour. In
general, the degree of approximation will improve with increased complexity
(however defined) of the structure system.[2] An investigator will usually prefer
the less complex models. So, complexity as a preference ordering placed on the
admissible space of structure systems will generally be traded off against the
degree of approximation, which is also a preference ordering. We will use the
generic term quality of a structure system, which may be defined as an ordered
pair consisting of both the degree of approximation and the degree of complexity.
It is noted that quality is a relative concept, i.e., it is defined likewise as
another preference ordering relative to the admissible space of structure systems. Although the solution of the problem of structure identification may lack
uniqueness, the drawback of an intrinsically subjective evaluation of hypothesized
structure candidates is partly overcome by adequate measures of quality, such as
expressed here in terms of the formal language of C-analysis.

In Section 2 we will give a comprehensive overview of the theory of
C-analysis, starting with the basic concepts of variability, constraint, and
structure in subsection 2.1. The next subsection 2.2 deals with the hierarchical
decomposition of a given structure. An important decomposition theorem, previously derived by this author,[10] is introduced. Subsection 2.3 deals with another
type of decomposition, viz., of structural complexity. Section 3 briefly reviews
the necessary concepts inherent to the problem of structure identification, such
as structure system representation (3.1), and measures of approximation and complexity (3.2). Section 4 is motivated by our concern for the acceptance of systems methodological tools as presented here. To this end, two more or less famous
examples from the social science field are reanalyzed here by our techniques. It
will be shown that, without neglecting higher-order interactions, our results are
in remarkable agreement with Goodman's,[17] thus increasing, in my opinion, the
faith in and the susceptibility to these conceptually powerful, simple, and precise systems methodological tools.

2   CONSTRAINT ANALYSIS

2.1   Variability, Constraint, Structure

Consider for a given probability space a discrete random variable X with finite
value or state set $X = \{x_1, x_2, \ldots, x_N\}$, for which a discrete density or probability function is defined, denoted by

$$f(x_i) \quad (f(x_i) \geqslant 0; \ i = 1,2,\ldots,N; \ \sum_{i=1}^{N} f(x_i) = 1)$$

Axiomatically, we define the concept of _variability_ of the probability function of X, denoted H(X), by the Shannon entropy

$$H(X) = - \sum_{i=1}^{N} f(x_i) \log_2 f(x_i) \tag{1}$$

measured in bits, where $0 \log_2 0 := 0$. Variability is a nonnegative real number which
is assigned to a finite probability function as a measure of dispersion (indeter-
minacy, spread) of the distribution. The algebraic properties of (1), such as, for
example, those of symmetry, normality, additivity, and recursivity are well
known,[19] and will not be discussed here. In the axiomatic approach it can be shown
that only (1) has these convenient "natural" properties. For the sake of brevity
we will often speak of the variability of the variable X where the existence of
a probability function is understood, hence the notation H(X).

For a discrete uniform distribution, i.e., $f(x_i) = \frac{1}{N}$ ($i = 1,2,\ldots,N$), the variability attains its maximum value, $H_{max}(X) = \log_2 N$ (property of maximality). The variability is equal to zero if the sample space contains only one mass point.

According to Ashby, the concept of constraint is described as "a relation between two sets, and occurs when the variety that exists under one condition is less than the variety that exists under another."[20] In transposing this definition from the original set theoretic to the probabilistic domain,[9] we define the concept of distributional constraint (of the probability function) of variable X by

$$DC(X) = \log_2 N - H(X) \tag{2}$$

or a normalized version, denoted by DCN(X),

$$DCN(X) = \frac{DC(X)}{\log_2 N} = 1 - \frac{H(X)}{\log_2 N} \tag{3}$$

If the variability is minimal, that is equal to zero, the normalized distributional constraint is maximal, and equal to unity. Similarly, for maximal variability $H(X) = \log_2 N$, the constraint is minimal, and equal to zero. Taking into account the analytic properties of the entropy function,[19] it is evident that $0 \leqslant DCN(X) \leqslant 1$. From these properties the dual character of the variability and constraint concepts are clearly demonstrated. Both concepts may be considered as fundamental to and play a complementary role in the analysis of systems.

Returning to Ashby's definition given above, it may be clear that the two sets he referred to are here associated with two sets of probabilities, one representing the uniform, and the other the actual probability distribution.

Extending the definition of variability to n joint discrete random variables, $(X_1,X_2,\ldots,X_n)$, or briefly $(X_i)_{i \in I_n}$, $I_n = \{1,2,\ldots,n\}$, defined on the same probability space with joint discrete density function

$$f(x_1,x_2,\ldots,x_n) \equiv f(x_i)_{i \in I_n}, \quad x_i \epsilon X_i, \quad \Sigma f(x_i)_{i \in I_n} = 1$$

the joint variability of the discrete random variable $S \equiv (X_i)_{i \in I_n}$ is defined by

$$H(S) \equiv H(X_i)_{i \in I_n} = -\Sigma f(x_i)_{i \in I_n} \log_2 f(x_i)_{i \in I_n} \tag{4}$$

where the summation is performed for all possible values $(x_i)_{i \in I_n}$ in the product

set $\prod_{i \varepsilon I_n} X_i$. It is important to note that the joint variability has the property of symmetry, i.e.,

$$H(X_1,X_2,\ldots,X_n) = H(X_{k(1)},X_{k(2)},\ldots,X_{k(n)})$$

where k is an arbitrary permutation on the index set I_n. We will therefore often
replace the parentheses in S = $(X_1,X_2,\ldots,X_n)$, which may suggest an ordered set
by braces : S = $\{X_1,X_2,\ldots,X_n\}$, where no order is assumed.

The joint or total distributional constraint of S is given by

$$DC(S) = \log_2 N_S - H(S) \qquad (5)$$

where N_S is the cardinality of the product set $\prod_{i \varepsilon I_n} X_i$.
Likewise, a normalized version of (5) may be defined.
It is a well known result that, in general,[19,21]

$$H(S) \leqslant \sum_{i \varepsilon I_n} H(X_i) \qquad (6)$$

where $H(X_i)$ may be referred to as a marginal variability. The equality sign is
valid iff the random variables are jointly independent, i.e., $f(x_i)_{i \varepsilon I_n} = \prod_{i \varepsilon I_n} f(x_i)$ for all possible values x_i.

The degree of deviation of probabilistic independence is expressed by the
difference between the joint variability and the sum of the marginals, and will
be called the total structural constraint of the system of n variables, denoted
by $T(S) \equiv T(X_1:X_2:\ldots X_n)$, and defined as

$$T(S) \equiv T(X_1:X_2:\ldots:X_n) = \sum_{i \varepsilon I_n} H(X_i) - H(S) \qquad (7)$$

Denoting the cardinality of the individual value sets X_i by N_i ($i \varepsilon I_n$), and
noting that $N_S = \prod_{i \varepsilon I_n} N_i$, we may replace (7) by

$$\log_2 N_S - H(S) = \sum_{i \varepsilon I_n} (\log_2 N_i - H(X_i)) + T(S)$$

or, by (5),

$$DC(S) = \sum_{i \varepsilon I_n} DC(X_i) + T(S) \qquad (8)$$

This equation expresses the fact that the total distributional constraint of a
system, considered as a whole, equals the sum of the marginal constraints of

the constituent variables plus the structural constraint holding over the system
of variables. In my opinion, this equation captures precisely the evasive concept of structure in an operational, albeit probabilistic, definition. Eq. (8)
expresses operationally that the whole -DC(S)- is more than the sum of its parts
-DC(X_i)-, and that, what it is more, is precisely the structure of the whole.
Only in the case of probabilistic independence is the whole equal to the sum
of its parts. It is suggested that neither a whole nor a part is "recognized"
as such when the distributional constraints are equal to zero.

We feel that it is rather useless to engage in philosophical discourse by
asking what is a "whole", or what is "structure". This will only result in conceptual ambiguities. The only meaningful way by which science can progress is
in defining structure in the context of an axiomatically developed theory, which
may furnish a measure of it. Other well known examples pertain to the energy
measure and the axiomatized concept of probability itself. Clearly, there can
be no confusion about the underlying assumptions and logical conclusions in the
context of an axiomatic theory of structure. And, of course, one is not compelled
to accept the assumptions of the theory.

For further reference, it may easily be shown that the total n-dimensional
structural constraint can be expressed in terms of joint and marginal probabilities as follows:

$$T(S) = \Sigma \; f(x_i)_{i \varepsilon I_n} \; \log_2 \frac{f(x_i)_{i \varepsilon I_n}}{\prod_{i \varepsilon I_n} f(x_i)} \qquad (9)$$

where the summation is performed for all possible joint probabilities, where
$f(x_i) > 0$, $i \varepsilon I_n$.

We will conclude this subsection with the introduction of the concepts of
conditional variability and conditional structural constraint. It may be of
interest to know the extent to which the variability of a set of variables S
cannot be ascribed to or "explained" from a subset S_1 of S. This residual variability is defined as the conditional variability of $S_2 = S - S_1$, denoted $H_{S_1}(S_2)$,

$$H_{S_1}(S_2) = H(S) - H(S_1) \qquad (10)$$

In other words, the variability of S may be considered as consisting of two
parts; one contribution of the subset S_1 plus another of the complement S_2, given
or controlling for S_1.

With the aid of (10), we are now ready for the introduction of McGill and

Garner's Rule,[22] which states that an identity in information theory will
remain valid when all terms in it are provided with the same index or indices.
To give an example, if the two-dimensional structural constraint $T(X_1:X_2)$ is
conditioned on the variables X_3 and X_4, this yields according to (7)

$$T_{X_3 X_4} (X_1:X_2) = H_{X_3 X_4} (X_1) + H_{X_3 X_4} (X_2) - H_{X_3 X_4} (X_1,X_2)$$

Applying (10), we have in terms of unconditional variabilities

$$T_{X_3 X_4} (X_1:X_2) = H(X_1,X_3,X_4) + H(X_2,X_3,X_4)$$
$$- H(X_1,X_2,X_3,X_4) - H(X_3,X_4)$$

2.2 Hierarchical Decomposition of Structure

By a hierarchical decomposition we refer to a decomposition of a system of
variables into interrelated subsystems, each of the latter being, in turn,
decomposable into subsubsystems, and so on. This terminology is in accordance
with Simon's, who meant by hierarchy the partitioning of systems in successive
subsystems "in conjunction with the relations that hold among its parts."[23]

A partition refers, however, to a decomposition of the set of variables
into disjoint subsets. For example, partitioning $S = \{X_1,X_2,X_3\}$ into two
disjoint subsets $\{X_1,X_2\}$ and $\{X_3\}$ yields with (7) the following decomposition
of structure constraint of S.

$$
\begin{aligned}
T(1:2:3) &= H(1) + H(2) - H(1,2) \\
&\quad + H(1,2) + H(3) - H(1,2,3) \\
&= T(1:2) + T(1,2:3)
\end{aligned}
\tag{11}
$$

where, in order to simplify the notation, the variables are replaced by their
corresponding indices. In words, the total 3-dimensional structural constraint
is decomposed into additive components pertaining to (i) the 2-dimensional
structural constraint within the subsystem $\{X_1,X_2\}$, (ii) the 1-dimensional
structural constraint within $\{X_3\}$, which should, of course, be equal to zero,
i.e., $T(3:\emptyset) = 0$, and (iii) the 2-dimensional structural constraint between the
two subsystems $\{X_1,X_2\}$ and $\{X_3\}$, each taken as a whole, which accounts for the
latter term of (11).

Generalization of (11) to the n-dimensional structural constraint yields
likewise

$$T(1:2:\ldots:(n-1):n) = T(1:2:\ldots:(n-1))$$
$$+ T(1,2,\ldots,(n-1):n) \qquad (12)$$

Repeated application of (12) gives an example of one type of hierarchical decomposition of $T(S)$:

$$T(S) = T(1:2) + T(1,2:3) + T(1,2,3:4) + \ldots + T(1,2,\ldots(n-1):n) \qquad (13)$$

The hierarchical decomposition of systems of variables into disjoint subsets has been widely discussed in the literature.[21,24] It is now of interest to notice that certain developments in systems research, viz., the definition of a structure system by Klir,[4] required the extension of this way of decomposing a system to one in terms of non-disjoint or overlapping subsystems.[10,11] I feel that this is a typical example of the development of a mathematical tool which was generated and motivated by systems problem solving, rather than an intrinsic development within the field of mathematics itself.

Without further proof we will present the following decomposition theorem. Details of proof have been given by this author elsewhere.[10]

Notation

Let $\{E_i\}$, $i \in I_m = \{1,2,\ldots,m\}$, be a family of subsets (not necessarily disjoint) of $S = \{X_j\}_{j \in I_n}$, $I_n = \{1,2,\ldots,n\}$, $m \leqslant n$, such that $\bigcup_{i \in I_m} E_i = S$.

The family $\{E_i\}$ is called a cover of S.
Let

$$C_i = E_i \cap (\bigcup_{k<i} E_k), \quad S_i = E_i - C_i, \quad \text{and } \tilde{C}_i = \bigcup_{k<i} E_k - C_i$$

for all $i \geqslant 2$, and $i \in I_m$. That is, C_i is the overlap of E_i with the preceding E_k's; S_i is the part of E_i not overlapping with the preceding E_k's, and $\tilde{C}_i$ is the union of the preceding E_k's with C_i removed.

(Decomposition) Theorem 1

Let $\{E_i\}_{i \in I_m}$ be a cover of $S = \{X_j\}_{j \in I_n}$, $(m \leqslant n)$.

Then
$$T(S) = \sum_{i \in I_m} T(E_i) - \sum_{i \in I_m-\{1\}} T(C_i) + \sum_{i \in I_m-\{1\}} T_{C_i}(\tilde{C}_i : S_i) \qquad (14)$$

Let us give a few simple examples to illustrate the use of (14).

Example 1

Given $S = \{X_1, X_2, X_3, X_4\}$ with $E_1 = \{X_1, X_2, X_3\}$, and $E_2 = \{X_2, X_3, X_4\}$.
We will use the notation 123/234 to denote the two elements. Hence,
$C_2 = E_2 \cap E_1 = \{X_2, X_3\}$, $S_2 = E_2 - C_2 = \{X_4\}$, $\bar{C}_2 = E_1 - C_2 = \{X_1\}$, which yields
with (14)

$$T(S) = T(1:2:3) + T(2:3:4) - T(2:3) + T_{23}(1:4)$$

In words, the total structural constraint is decomposed into the additive constraints pertaining to elements, minus a correction term pertaining to the intersection of the elements plus a conditional term. The third (correction) term
appeals to intuition, since the constraint between X_2 and X_3 has been accounted
for twice in the structural constraints of the two elements. The conditional term
will usually by hypothesized to be equal to zero when used for the corresponding
structure system, as we shall see.

Example 2

Let $S = \{X_1, X_2, X_3, X_4\}$, with three subsets E_1, E_2, and E_3 given by 12/23/24. We
then find,
$C_2 = \{X_2\}, S_2 = \{X_3\}, \bar{C}_2 = \{X_1\}, C_3 = \{X_2\}, S_3 = \{X_4\}, \bar{C}_3 = \{X_1, X_3\}$.
Insertion in (14) yields with the help of (12)

$$T(S) = T(1:2) + T(2:3) + T(2:4) + T_2(1:3) + T_2(1,3:4)$$

$$= T(1:2) + T(2:3) + T(2:4) + T_2(1:3:4)$$

Again, when applying this identity to structure system identification, it can be
shown that the latter term is hypothesized to vanish. Thus, for this particular
system, the structural constraint equals the sum of the constraints of the elements involved.

Example 3

Let $S = \{X_1, X_2, X_3, X_4, X_5\}$, while the three subsets E_1, E_2, and E_3 are disjoint,
and given by 1/23/45.
Consequently, we have $C_2 = \emptyset$, $\bar{C}_2 = \{X_1\}$, $S_2 = \{X_2, X_3\}$, $C_3 = \emptyset$, $\bar{C}_3 = \{X_1, X_2, X_3\}$,
$S_3 = \{X_4, X_5\}$. Thus,

$$T(S) = T(2:3) + T(4:5) + T(1:2,3) + T(1,2,3:4,5)$$
$$= T(2:3) + T(4:5) + T(1:2,3:4,5)$$

where the latter identity is obtained by using (12). In words, the total constraint equals the sum of constraints within subsets plus the constraint between
the subsets. This is precisely the result of the classical decomposition of a
system into disjoint subsystems, which clarifies that (14) is indeed a generalized decomposition theorem for both disjoint and non-disjoint subsystems.
Further examples may be found in previous publications.[9-12]

2.3 Decomposition of Structural Complexity

Unlike many treatments in the empirical disciplines, the systems approach is
under an obligation to emphasize the existence of interactions among variables
of a higher order than those obtained by treating the system, for instance, as
a set of interactions among pairs. In particular, the conceptualization and
unambiguous development of such higher-order interactions is of special concern
to the system scientist.

It is relatively easy to define the different interactional constraints or
briefly interactions, axiomatically. We will proceed by doing so. However, it
is also very instructive to express the interactions in terms of probabilities.
The more so, as the latter approach will provide a basis for demonstrating that
the theory of C-analysis provides a synthesis of other methods.

Let us commence with the axiomatic definitions of 1-order (2-factor),
2-order (3-factor), etc. interactional constraints in terms of concepts already
defined.

1-order: $\quad Q(1,2) \equiv T(1:2)$

2-order: $\quad Q(1,2,3) = Q_k(i,j) - Q(i,j),$
$\qquad\quad i<j;\ k \neq i,j;\ i,j,k = 1,2,3$

3-order: $\quad Q(1,2,3,4) = Q_l(i,j,k) - Q(i,j,k),$
$\qquad\quad i<j<k;\ l \neq i,j,k;\ i,j,k,l = 1,2,3,4$
$\qquad\quad$ etc.

n-order: $\quad Q(1,2,\ldots,m-1,\ m,\ m+1,\ldots,n) =$
$\qquad\quad Q_m(i,j,\ldots,m-1,m+1,\ldots,k) - Q(i,j,\ldots,m-1,m+1,\ldots,k),$
$\qquad\quad i<j<\ldots<k;\ m \neq i,j,\ldots,k;\ i,j,\ldots = 1,2,\ldots,n,$

where variables are replaced by their corresponding indices, e.g.,
$Q(1,2,3) \equiv Q(X_1,X_2,X_3)$, and McGill and Garner's Rule is employed to define the
conditional interactions. The second-order interaction is a unique quantity

which may be defined in three alternative ways by permuting the indices. Likewise, there are n+1 different definitions which may be used alternatively to
define the nth-order interaction, indicating that it is symmetrical with respect
to the variables involved.

It may be shown that the total structural constraint can be decomposed into
additive combinations of increasing degrees of complexity of the system. The
presence of k-factor positive interactions between k variables in an n-dimensional
system measures the degree to which the system is irreducably complex, i.e., not
to be treated by examination of subsets of (k-1) or less variables at one time.[22]
It may be noted that, in contrast with the nonnegative structural constraints,
second-and higher-order interactions may attain also negative values. Their
meaning is somewhat different from their positive counterparts. A negative interaction may be regarded as a correction on the total constraint, which is "over-
determined" by the respective lower-order interactions.[13,15] The mentioned decomposition of an n-dimensional structural constraint of a system S is as follows

$$T(S) = \sum_{i<j} Q(i,j) + \sum_{i<j<k} Q(i,j,k) + ...+ Q(S) \tag{15}$$

where $Q(S) = Q(1,2,...,n)$, and $i,j,k,...\epsilon I_n$.

The total number of terms on the right-hand side of (15) is equal to

$\sum_{p=2}^{n} \binom{n}{p}$, where p stands for the number of variables involved in the various
interaction terms.

We will now proceed in a somewhat different manner. Previous results have
shown the usefulness of this approach for providing a sound and consistent foundation for C-analysis.[25,26] The present study is an extension and generalization
of these previous studies. We adhere to the following notational conventions.

Notation

Let $f(x_i)$, $f(x_i|x_j)$, etc. be written as $f(i)$, $f(i|j)$, etc. respectively, where
as before, $x_i \epsilon X_i$, $x_j \epsilon X_j$, $i \neq j$, $i,j \epsilon I_n$.

Let $C(i) = f(i)$, while conditioning of $C(i)$ is considered an admissible operation,
i.e., $C(i|j) = f(i|j)$, $C(i|j,k) = f(i|j,k)$, etc.
Let us define next

$$C(i,j,...,l,m,n) = \frac{C(i,j,...,l,m|n)}{C(i,j,...,l,m)} \tag{16}$$

where, again, conditioning is an admissible operation. Because of the recursive character of this definition, we may also write

$$C(i,j,\ldots,l,m,n) = \frac{\dfrac{C(i,j,\ldots,l,m|n)}{C(i,j,\ldots,l|m)}}{C(i,j,\ldots,l)}$$

and so forth.

Let us now derive some of the consequences. By definition, we have

$$C(i) = f(i) \tag{17}$$

From this we find

$$C(i,j) = \frac{C(i|j)}{C(i)} = \frac{f(i|j)}{f(i)} = \frac{f(i,j)}{f(i)f(j)} \tag{18}$$

which measures the conditional probability of $x_i \varepsilon X_i$, given $x_j \varepsilon X_j$, as a proportion of the unconditional probability $f(i) > 0$. If $C(i,j) = 1$ for all values x_i and x_j, the two variables X_i and X_j are probabilistically independent. Taking the $\log_2$ of both sides of (18), and taking the expectation, we find with (9)

$$Q(i,j) = \Sigma\ f(i,j)\ \log_2\ C(i,j)$$

where the summation is performed for all values $x_i \varepsilon X_i$ and $x_j \varepsilon X_j$.
(Remark: even if some marginals were equal to zero, the summation may be performed for all values, owing to $0 \log_2 0 := 0$).[10]

The definition of the absence of first-order interaction has long been agreed upon as far as events are concerned. It may be shown that there is no first-order interaction between the two variables X_i and X_j, i.e., there is independence in the unconditional distribution of X_i and X_j, if and only if $Q(i,j) = 0$, or $C(i,j) = 1$ for all values x_i and x_j.

The agreement on the measurement of the strength of the first-order interaction is however an entirely different matter. As defined here, the first-order interaction provides a natural measure for the strength of association between two variables, rather than events. It equals zero iff X_i and X_j are probabilistically independent, while it attains an upper bound equal to the minimum of $\{H(X_i),H(X_j)\}$, if X_i is entirely dependent on X_j and/or vice versa.

Let us proceed with $C(i,j,k)$, which may be written as follows for those marginal probabilities unequal to zero

$$C(i,j,k) = \frac{C(i,j|k)}{C(i,j)} = \frac{\frac{f(i,j|k)}{f(i|k)f(j|k)}}{\frac{f(i,j)}{f(i)f(j)}} = \frac{\frac{f(i,j,k)}{f(i,j)f(i,k)f(j,k)}}{\frac{f(i)f(j)f(k)}}$$ (19)

where the second from the last term indicates that $C(i,j,k)$ measures the conditional first-order interaction between events $x_i \in X_i$ and $x_j \in X_j$, given $x_k \in X_k$,
as a proportion of the unconditional dependence between x_i and x_j.
Consequently, $C(i,j,k)$ is a measure of the influence of x_k on the dependence
between x_i and x_j, or, since it is a symmetrical term (see the latter one),
of any one variable on the dependence between the other two. Note the compatability of this formulation with that pertaining to $C(i,j)$.

Taking logarithms and the expectation, we find (or define) the second-order
interaction

$$\begin{aligned} Q(i,j,k) &= \Sigma\, f(i,j,k)\, \log_2 C(i,j,k) \\ &= \Sigma\, f(i,j,k)\, \log_2 C(i,j|k) \\ &\quad - \Sigma\, f(i,j)\, \log_2 C(i,j) \\ &= Q_k(i,j) - Q(i,j) \end{aligned}$$

the latter expressing, in a formalized way, what was stated above in words.

Although there seems to be no agreement on definitions of second- (and
higher-) order interactions, we suggest that the ones given here answer as
closely as possible to our intuitive notions of the concept.
Moreover, because of the way of defining higher-order interactions as a "natural"
extension of lower-order ones, they constitute a coherent framework.

Let us continue with the four-variate concepts. Application of (16), yields
for admissible, i.e., unequal to zero, probabilities

$$C(i,j,k,l) = \frac{C(i,j,k|l)}{C(i,j,k)} = \frac{\frac{C(i,j|kl)}{C(i,j|l)}}{\frac{C(i,j|k)}{C(i,j)}} =$$

$$\frac{\frac{\frac{f(i,j|kl)}{f(i|kl)f(j|kl)}}{f(i,j|l)}}{\frac{\frac{f(i|l)f(j|l)}{f(i,j|k)}}{\frac{f(i|k)f(j|k)}{f(i,j)}}} = \frac{\frac{f(i,j,k,l)}{f(i,j,k)f(i,j,l)f(i,k,l)f(j,k,l)}}{\frac{f(i,j)f(i,k)f(i,l)f(j,k)f(j,l)f(k,l)}{f(i)f(j)f(k)f(l)}}$$

The second from the last term shows that $C(i,j,k,l)$ measures the conditional

2-order interaction among events $x_i \varepsilon X_i$, $x_j \varepsilon X_j$, and $x_k \varepsilon X_k$, given $x_l \varepsilon X_l$, as a proportion of the unconditional 2-order interaction between (x_i, x_j, x_k). It is thus a measure of the influence of x_l on the 2-order interaction among (x_i, x_j, x_k). It is also a symmetrical concept. Taking logarithms and expectations, we find (or define) the third-order interaction among X_1, X_2, X_3, and X_4

$$
\begin{aligned}
Q(i,j,k,l) &= \Sigma\ f(i,j,k,l)\ \log_2 C(i,j,k,l) \\
&= \Sigma\ f(i,j,k,l)\ \log_2 C(i,j,k|l) \\
&\quad - \Sigma\ f(i,j,k)\ \log_2 C(i,j,k) \\
&= Q_l\ (i,j,k) - Q(i,j,k)
\end{aligned}
$$

Finally, generalizing to n variables we may write in abbreviated form

$$
C(i)_{i\varepsilon I_n} = \cfrac{f_n}{\cfrac{\P_{n-1}^{f_{n-1}}}{\cfrac{\P_{n-2}^{f_{n-2}}}{\cfrac{\vdots}{\P_1^{f_1}}}}}
\tag{21}
$$

where $f_n = f(i)_{i\varepsilon I_n}$, $\P_{n-1}^{f_{n-1}}$ stands for the product of all $(n-1)$-dimensional probabilities, etc. The $(n-1)$th-order interaction is then given by

$$
Q(i,j,\ldots,n) = \Sigma f(i)_{i\varepsilon I_n} \cdot \log_2 C(i)_{i\varepsilon I_n}
$$

With these concepts and notations, it is easily shown that the joint n-dimensional probability $f(x_1, x_2, \ldots, x_n)$, or briefly $f(i)_{i\varepsilon I_n}$, divided by $\P_{i\varepsilon I_n} C(i)$ may be expanded as follows

$$
\frac{f(i)_{i\varepsilon I_n}}{\P_i C(i)} = \P_{i<j} C(i,j) \cdot \P_{i<j<k} C(i,j,k)\ldots
$$

$$
\cdot \P_{i<j<\ldots<n-1} C(i,j,\ldots,n-1) \cdot C(1,2,\ldots,n)
\tag{22}
$$

where $i,j,\ldots \varepsilon I_n$. By taking logarithms and expectations, and using the above results, Eq.(15) emerges.

3   STRUCTURE IDENTIFICATION

3.1 Structure System Representation

We will briefly review some of the material related to the concept of a structure system as developed by Klir[4] in the context of an epistemological hierarchy
of systems.

Somewhat loosely, a __structure system__ is defined as a set of __elements__, and
a set of __couplings__ between them. An element is defined as a nonempty subset
E_j of $S = \{X_i\}_{i \in I_n}$, such that the variables are all __directly related__,[4] while
$\bigcup_j E_j = S$. Not all structure systems are admissible subsystems. For further details
one is referred to Klir,[14] Klir and Uyttenhove,[5] and Cavallo and Klir.[8]

We have suggested a probabilistic definition of the concept of a direct
relation between pairs of variables.[10,13,15]

__Definition 1__ There exists no direct relation between variables X_i and X_j in a
system of n variables if and only if

$$C(i,j \mid (x_k)_{k \in I_n'}) = \frac{f(x_i,x_j \mid (x_k)_{k \in I_n'})}{f(x_i \mid (x_k)_{k \in I_n'}) f(x_j \mid (x_k)_{k \in I_n'})} = 1 \qquad (23)$$

where $I_n' = I_n - \{i,j\}$, for all admissible values of $x_i \in X_i$, $x_j \in X_j$,
$(x_k)_{k \in I_n'} \quad \varepsilon \quad \prod_{k \in I_n'} X_k$. If the equality does not hold, X_i and X_j are defined to be
directly related.

It is easily shown by taking logarithms and expectations that (23) is
equivalent to

$$T_K (X_i : X_j) = 0$$

where $K = S - \{X_i, X_j\}$. The quantity $T_K(X_i : X_j)$ may thus be regarded as a
measure of the strength of the direct relation between X_i and X_j, being zero
iff the relation is absent, and positive otherwise. The higher its value, the
higher the intensity of the relation.

Variables may be indirectly related through a sequence of couplings
between elements. If E_i is the set of variables associated with one element and
E_j that of another element, a neutral coupling, denoted by $C_{i,j}$, is defined by

$$C_{i,j} = E_i \cap E_j \qquad (i \neq j)$$

while $C_{i,j} = C_{j,i}$, and $C_{i,j} = \emptyset$ if $i = j$.

In order to be able to couple the findings of C-analysis to the representation of a structure system, the following important theorem is needed. Its
proof may be found elsewhere.[10,11] Some examples will be given in the sequel.

Theorem 2

Let, in a system of n variables $\{X_i\}_{i \varepsilon I_n}$, $T_{jL}(i:k) = T_{iL} (j:k) = 0$, where
$L = I_n - \{i,j,k\}$, $i \neq j \neq k$. Then, $T_L(i:k) \overset{n}{=} T_L(j:k) = 0$. (L may include the empty
set).

Example 4 (Ex.1 cont'd)

A structure system is defined by two elements 123/124. This, by definition,
implies that $T_{23}(1:4) = 0$. Consequently, the structural constraint of this
system is given by (see (14), and Ex.1)

$$T(S) = T(1:2:3) + T(2:3:4) - T(2:3)$$

Example 5 (Ex.2 cont'd)

We have the structure system 12/23/24. Hence, by definition, $T_{24} (1:3) =$
$T_{23}(1:4) = T_{12}(3:4) = 0$. The former two equalities yield with Theorem 2: $T_2(1:3) =$
$T_2(1:4) = 0$. Since,

$$T_2(1:3:4) = T_2(1:3) + T_2(1:4) + T_{12}(3:4) = 0$$

the structural constraint of this structure system (see Ex.2)

$$T(S) = T(1:2) + T(2:3) + T(2:4)$$

Structure candidates, denoted as C_i, have been ordered in a lattice by Klir.[4]
A four-dimensional lattice is shown for further reference in Figure 1.
Block diagrams of types of structure systems are shown in 1B. The notations
below the block diagrams serve both as a description of the type of structure
(disregarding the part between parentheses below type 1), and (taking into
account the parenthesized part) as the denotation for the decomposition of the
total constraint. This may be clarified by recognizing that Example 4 is covered
by type 1, and Example 5 by type 4. The reader is referred to Klir,[4] and Klir
and Uyttenhove,[5] for further details concerning lattices of structure candidates
and the structure identification algorithm based thereupon.

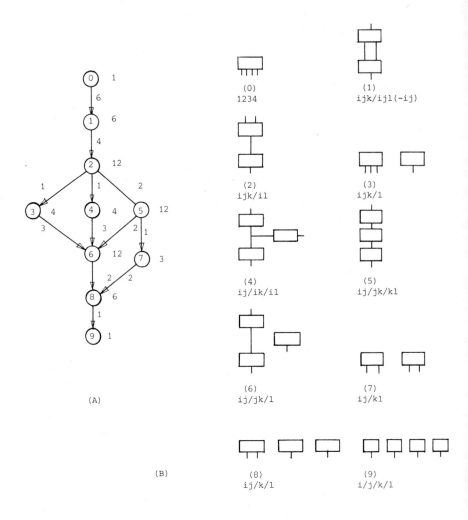

(A)

(B)

(0)
1234

(1)
ijk/ijl(-ij)

(2)
ijk/il

(3)
ijk/l

(4)
ij/ik/il

(5)
ij/jk/kl

(6)
ij/jk/l

(7)
ij/kl

(8)
ij/k/l

(9)
i/j/k/l

Figure 1 (A) : Hierarchy of structure candidates of a four-variable system; numbers within circles refer to the type (see B); numbers next to circles refer to the number of candidates of that type; numbers next to connections between circles refer to the number of immediate successors, (B) block diagrams of types of structure candidates (from Klir and Uyttenhove[5])

3.2 Approximation and Complexity

In the sequel we will briefly deal with the quality criteria: degree of approximation and degree of complexity.[3]

In order to test whether an hypothesized structure candidate is congruent with the data and to what extent, we proposed a simple measure of the degree of approximation called T-distance, defined by

$$\Delta_i^0 = T_0(S) - T_i(S) \tag{24}$$

where $T_0(S)$ is the actual or observed total structural constraint of the candidate C_0, which is also the greatest element of the lattice, and $T_i(S)$ is the estimated total constraint under the hypothesis C_i. It can be shown that the T-distance has a range $0 \leqslant \Delta_i^0 \leqslant T_0(S)$, attaining the value 0 if $C_i \equiv C_0$, while, in passing successive levels of the lattice, it decreases monotonically till the least candidate of the lattice is reached, where $\Delta_i^0 = T_0(S)$.

A normalized version of the T-distance may thus be defined by dividing (24) by $T_0(S)$. An additional advantage of definition (24) is, that a partial T-distance can be defined, which measures the difference of the constraints between two immediate successors of a lattice.[11] It can be shown that the overall T-distance, as defined by (24), is composed of additive combinations of all intermediate partial T-distances.[10,11]

We have suggested elsewhere[11] that the quantity $1.3863\ N\ \Delta_i^0$, where N is the sample size, is approximately distributed like χ^2 with degrees of freedom equal to $df(T_0) - df(T_i)$. The degrees of freedom $df(T)$ may be calculated by using definition (7), and replacing the H-terms by $df(H)$ terms; $df(H(X_i))$ is equal to $-(N_i-1)$, where N_i is the number of categories on X_i. Likewise, $df(H(X_i)_{i \in I})$ equals $-(\P_{i \in I}\ N_i - 1)$, and so on. To illustrate (see Ex.4), in order to test the hypothesis 123/124, the appropriate degrees of freedom (dichotomous variables), briefly denoted as,

$$df(\Delta) = df(T_0) - \left[df(T(1:2:3)) + df(T(2:3:4)) - df(T(2:3)) \right]$$

$$= -4 + 15 - \left[-3 - (-7) + (-3) - (-7) - (-2 - (-3)) \right] = 4$$

Note that in this example the actual testing is tantamount to testing the null hypothesis that $T_{23}(1:4)$ is equal to zero. Examination of (14) and (24) reveals that the test is actually concerned with the simultaneous testing of the null hypothesis that the $T(.)$ terms under the last summation in (14) are equal to zero. It is known that, when the null hypothesis is true, $1.3863\ N\ T(.)$ is asymptotically distributed like χ^2. This equivalent test thus supports the

utility of (24) in testing the null hypothesis that Δ_i^0, as a measure of the
degree of approximation, is equal to zero.

In applying Klir's algorithm of structure generation, the search for the
candidate of the highest quality starts at the top of the lattice (see Figure 1
and Section 4) with C_0, and gradually descends in a selective manner downwards.
In this process the T-distance will in general increase, i.e., the degree of
approximation deteriorates gradually. Conventional criteria of statistical significance may be used to estimate the degree of goodness-of-fit, i.e., the χ^2
values can be assessed by comparing them with the percentiles of the tabulated
χ^2 distribution.

As to the degree of complexity, a simple measure such as the number of
direct relations in a structure system might be used. This does at least discriminate between the different levels of a lattice as to the degree of complexity.
Descending the lattice implies a decrease of complexity. A further discrimination in complexity might be useful within a specific level of the lattice,
whenever more than one structure type occurs. A somewhat more sophisticated
measure could be based on the results of Subsection 2.3. A complexity measure
can be defined in terms of the highest-order interaction associated with the
different elements of the structure. Such a measure is closely related to the
concept of cylindrance.[9]

We will not pursue this matter any further here. It should be clear that
approximation and complexity are conflicting measures of quality of fit. It may
thus occur, that a more complex structure system provides a statistically significant improvement of fit compared with a more simple structure, although the
latter is still acceptably congruent with the data. An example may be found in
Ref. 11. Such considerations strengthen the arguments for an interactive discourse between the computerized algorithms and the investigator, in order to
enable the latter to assess the quality of a structure candidate in view of his
research objectives.

4 APPLICATIONS

For illustrative purposes we will reanalyze data pertaining to an analysis of a
panel and a survey study. Previously we analyzed the famous two-attribute turnover table on voting intention, which was examined repeatedly in the social
science literature over a period of 25 years.[11] It was shown that the results
of a structure identification process based on C-analysis were in remarkable
agreement with those based on the so-called log-linear model.[17,18]

We will now examine two examples which can be found in Goodman.[17]
We will make a brief comparison of his results and ours. The first example is

reproduced in Table 1. It cross-classifies the responses of 3398 schoolboys, each interviewed at two successive points in time with respect to two attributes. The second example, reproduced as Table 2, is a four-way cross-classification of 608 white women in public housing. The two tables can both be viewed as four-way tables in which four dichotomous variables are cross-classified.

Goodman presents a method for determining which of the models of a prescribed class are congruent with the given set of data. In short, Goodman's log-linear model approach is based on the decomposition of the logarithm of the joint probability of any event into a number of additive terms representing main and higher-order interaction effects. Although the method appears rather artificial, the justification is argued in terms of an analogy with the analysis of variance.[18] Hypotheses about the main and interaction effects are tested by conventional chi-square methods.

The general idea is that from the four-way tables pertaining to the variables X_1, X_2, X_3, and X_4 we can obtain two-way and three-way marginal tables, which may be considered as projections from the overall relation. Replacing the original four-way table by some combination of two-and/or three-way tables which covers the four-dimensional relation, will result in a loss of information about the latter. The methods focus on the question whether the information can be ignored by reconstructing the original relation (four-way table) from marginal relations (tables) of smaller dimensions. The latter relations will be referred to as the "fitted marginals" in Tables 1 and 2. The lost information is, of course, measured by the degree of approximation, which in our case is measured by the T-distance and the corresponding assessment of the chi-square value, denoted as χ^2_T in tables 3 and 4. Goodman's computation of the chi-square value based upon the likelihood-ratio statistic is likewise included as χ^2_G.

Tables 3 and 4 list the results for some structure models. The first columns contain the index numbers i of candidates C_i. Within parentheses are included the numbers of the hypotheses as found in Goodman's[17] Tables 11 and 14. The second columns contain the descriptions in terms of fitted marginals of the candidates. The estimated total structural constraints under the model hypotheses are included in the next column, followed by the degrees of freedom computed by the methods outlined in the previous subsection.

For both tables it is noted that Goodman[17] assumed that all second-order or three-factor interactions were equal to zero. In our study, on the contrary, we did take them into account, wherever these occurred, which is the case in practically all models except 33,34 and 35 in Table 3. Since it can easily be shown that the degrees of freedom pertaining to the estimated second-order interaction df$(Q(i,j,k)) = 1$, the degrees of freedom listed in our tables will have to be increased by 1 or 2 in order to obtain the ones listed in Goodman's table,

TABLE 1

Observed cross-classification of 3,398 schoolboys, in interviews at two successive points in time, with respect to the variables X_1 and X_3: self-perceived
membership in leading crowd in the first and second interview, resp. (being in
(out) denoted by +(-)), and X_2 and X_4: favorableness of attitude concerning the
leading crowd in the first and second interview, resp. ((un)favorable attitude
denoted by +(-)).

		Second Interview			
Membership		+	+	−	−
Attitude		+	−	+	−
First Interview					
Membership	Attitude				
+	+	458	140	110	49
+	−	171	182	56	87
−	+	184	75	531	281
−	−	85	97	338	554

Source: Coleman (1964) in Goodman (1973)[17]

TABLE 2

Observed cross-classification of 608 white women living in public housing projects, with respect to the variables X_1 : proximity to a Negro family, X_2 : favorableness of local norms toward Negroes, X_3 : frequency of contact with Negroes,
and X_4 : favorableness of respondent's attitudes (sentiments) toward Negroes in
general.

Proximity	Norms	Contact	Sentiment +	-
+	+	+	77	32
+	+	-	14	19
+	-	+	30	36
+	-	-	15	27
-	+	+	43	20
-	+	-	27	36
-	-	+	36	37
-	-	-	41	118

Source: Wilner et al (1955), Davis(1971) in Goodman (1973)[17]

whenever, respectively, one or two second-order interactions are involved.

Apart from this difference, the listed χ^2 values obtained in testing these models are notably in agreement. All models H_i, except one (see below), published by Goodman have been included in the Tables 3 and 4. It is of interest to notice that the listed candidates C_i were generated by the identification algorithm. In Table 3, candidates 14,15,17 and 18 are immediate successors of the best hypothesis 6 at the preceding level, while 8,9,16 and 17 are the immediate successors of the next best one, candidate 3. Immediate successors of the best fit of the former set is candidate 17. Its immediate successors are 22,33, 34 and 35. In comparing the χ^2 values with the percentiles of the tabulated χ^2 distribution we found that the latter cannot be considered as good fits. In summary, the simplest model, which still provides an acceptable fit in Table 3 is structure system 17 : 234/13.

Comparison of 17 with 3 and 6 indicates that, contrary to Goodman's findings, insertion of the direct relation between X_1 and X_2, or between X_1 and X_4 in model 17, does not contribute in a statistically significant way (p>.05). This is due to the fact that the appropriate test for statistically significant improvement (the difference of the corresponding χ^2 values is assessed with the difference between the degrees of freedom[11,17]) is in our case based on 2 degrees of freedom, while Goodman's test is based on 1 df. Finally, in Table 4, in agreement with Goodman's results, it is shown that candidate 3 provides a rather good fit, while its immediate successors 8,9,16 and 17 are no improvements. For reasons of limited space, we leave further conclusions and implications to the interested reader.

A final remark pertains to the hypothesis 12/13/24/34 (H_{10} in Goodman's[17] Tables 11 and 14). This type of "ring" structure is not included in the four-dimensional lattice defined by Klir (see Figure 1). Nevertheless, there is no problem in testing this hypothesis. By applying (14) it is easily shown that under the hypothesis 12/13/24/34

$$T(S) = T(1:2) + T(1:3) + T(2:4) + T(3:4) - Q(1,2,3,4)$$

or, is equal to the sum of the constraints of the four elements corrected by the third-order interaction of the relation. From a calculation by hand we find for the first example (Table 1) $\chi^2_T = 19.31$ (8 df), while Goodman's calculations yield $\chi^2_G = 17.91$ (7 df; the difference in df is due to df (Q(1,2,3,4)) = 1). For the second example (Table 2) we have $\chi^2_T = 13.74$ (8 df) and $\chi^2_G = 12.37$. Despite the rather different definitions of higher-order interactions, the results are still of the same order of magnitude.

TABLE 3

Total structural constraints T(S), χ_T^2 values based upon the T-distance Δ with associated degrees of freedom, χ_G^2 values based upon the likelihood-ratio statistic as computed by Goodman[17] on the basis of a log-linear model study, for some structure models pertaining to Table 1.

$C_i(H_i)$	Fitted marginals	T(S)	df	χ_T^2	χ_G^2
0	1234	.302			
1(2)	123/124	.299	4	15.55	15.71
2(3)	123/134	.246	4	262.38	262.54
3(5)	123/234	.301	4	4.24	4.06
4(4)	124/134	.298	4	16.02	16.70
5(6)	124/234	.092	4	989.24	989.68
6(7)	134/234	.301	4	3.77	3.84
14(14)	134/23	.245	6	267.09	267.26
15(15)	134/24	.294	6	35.80	36.24
17(12)	234/13	.300	6	8.95	8.78
18(16)	234/14	.088	6	1009.02	1009.22
8(8)	123/24	.294	6	35.33	35.20
9(9)	123/34	.245	6	267.56	267.48
16(11)	234/12	.088	6	1009.02	1009.16
17(12)	234/13	.300	6	8.95	8.78
22	234/1	.080	7	1044.35	
33	13/23/24	.293	8	40.04	
34	13/23/34	.244	8	272.28	
35	13/24/34	.293	8	40.98	

Goodman's denotations A,B,C,D have been replaced by 1,2,3,4 resp.

G. BROEKSTRA

TABLE 4

Total structural constraints T(S), χ_T^2 values based upon the T-distance Δ with degrees of freedom df, χ_G^2 values based upon the likelihood-ratio statistic as computed by Goodman[17] on the basis of a log-linear model study, for some structure models pertaining to Table 2.

$C_i(H_i)$	Fitted marginals	T(S)	df	χ_T^2	χ_G^2
0	1234	.212			
1(2)	123/124	.176	4	29.92	30.00
2(3)	123/134	.185	4	22.25	22.71
3(5)	123/234	.210	4	1.77	2.53
4(4)	124/134	.198	4	11.80	12.31
5(6)	124/234	.158	4	45.18	45.91
6(7)	134/234	.195	4	13.82	14.97
8(8)	123/24	.171	6	34.22	34.31
9(9)	123/34	.183	6	24.19	24.22
16(11)	234/12	.153	6	49.48	50.21
17(12)	234/13	.193	6	15.76	16.48

Goodman's denotations P,N,C,S have been replaced by 1,2,3,4, resp.

5   CONCLUSIONS

C-analysis (constraint analysis) is emerging as a powerful, rigorous, conceptually consistent, and precise theory of decomposition. In modelling the observed
behaviour of a probabilistic system, C-analysis suggests adequate measures of
approximation and complexity. When applied to actual data such as those from the
social science field, new insights may be obtained, especially in the more
difficult problem areas of qualitative rather than quantitative multivariate
modelling. Elsewhere,[27] for example, it was suggested that the structure identification activity may be considered as one phase in a more comprehensive identification procedure. In short, structure identification as defined here may
be followed by a phase of postulating the nature of the relations (in case of
metric variables: linear, non-linear, etc.), which in turn is followed by a
phase of parameter identification. To illustrate how one might proceed in the
cases studied in this paper, let us reconsider structure candidate C_{17} of the
schoolboys' example (Table 1 and 3). The block diagram of this model is shown
in Fig. 2A. In "opening the black box" formed by the element $\{X_2, X_3, X_4\}$ it is
noted that the three-factor interaction $Q(X_2, X_3, X_4) = -.005$ is negligibly small
compared to the total constraint of the element $T(X_2 : X_3 : X_4) = .08$, and especially
so with respect to the overall structural constraint $T(S) = .30$. Next, in postulating causal directions of influence, the causal diagram of binary relations
of Fig. 2B is obtained. Furthermore, the relative strengths of the direct relations may be calculated from the conditional constraints $T_{kl}(i:j)$ normalized as
to their maximum value. It is thus shown from Fig. 2B that 22 percent of the
variance of membership at the time of the second interview is "explained" by
membership at the first interview. It is also evident that the remaining causal
influences are of negligible strength.

The above elaboration serves as a simple illustration of the contention
that the structure identification activity may be followed by further modelling
activities. It is hoped that the present study contributes to supporting
factually the assertion that systems methodology has some valuable tools to
offer to empirical disciplines.

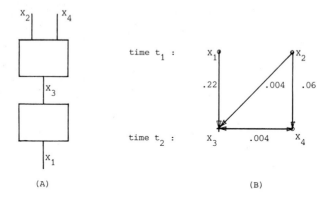

Figure 2   (A): Block diagram of structure system C_{17} (Table 3) of the
            schoolboys' example (Table 1); (B) : Causal diagram of (A)
            indicating relative strengths of the direct relations, and
            postulated direction of influence.

REFERENCES

1. B.R.Gaines, "Progress in General Systems Research." In: Applied General Systems Research, Ed. by G.J.Klir, Plenum Press, New York, 1978.

2. B.R.Gaines, "General System Identification-Fundamentals and Results." Applied General Systems Research, Ed. by G.J.Klir, Plenum Press, New York, 1978.

3. B.R.Gaines, "System Identification, Approximation and Complexity." International Journal of General Systems, 3, 1977, pp.145-174.

4. G.J.Klir, "Identification of Generative Structures in Empirical Data." International Journal of General Systems, 3, No.2, 1976, pp.89-104.

5. G.J.Klir and H.J.J.Uyttenhove, "Computerized Methodology for Structure Modelling." Annals of Systems Research, 5, 1976, pp.29-66.

6. B.P.Zeigler, Theory of Modelling and Simulation, Wiley and Sons, New York, 1976.

7. H.A.Simon, "Does Scientific Discovery have a Logic?" Philosophy of Science, Dec. 1973, pp.471-480.

8. R.Cavallo and G.J.Klir, "The Structure of Reconstructable Relations: A Comprehensive Study". In: Proc.Fourth European Meeting in Cybernetics and Systems Research, Linz, Austria, March 1978.

9. G.Broekstra, "Two Methods of Constraint Analysis." In: Proc. European Meeting on Cybernetics and Systems Research, Linz, Austria, March 1978.

10. G.Broekstra, "On the Representation and Identification of Structure Systems." International Journal of Systems Science, 1978 (to appear).

11. G.Broekstra, "On the Application of Informational Measures to the Structure Identification Problem with an Example of the Multivariate Analysis of Qualitative Data from a Panel Study." In: Recent Developments in Systems Methodology for Social Science Research, Ed. by R.Cavallo, Martinus Nijhoff, Leiden, 1978.

12. G.Broekstra, "Structure Modelling: A Constraint (Information) Analytic Approach." In: Applied General Systems Research, Ed. by G.J.Klir, Plenum Press, New York, 1978.

13. G.Broekstra, "Simplifying Data Systems: An Information Theoretic Analysis." In: Proc.Third European Meeting on Cybernetics and Systems Research, Vienna, April 1976.

14. G.Broekstra, "Constraint Analysis and Structure Identification." Annals of Systems Research, 5, 1976, pp.67-80; 6, 1977, pp. 1-20.

15. G.Broekstra, "Some Comments on the Application of Informational Measures to the Processing of Activity Arrays." International Journal of General Systems, 3, No.1, 1976, pp.43-51.

16. L.A.Goodman, "Causal Analysis of Data from Panel Studies and Other Kinds of Surveys." American Journal of Sociology, 78, No.5, 1973, pp.1135-1191.

17. Y.M.M.Bishop, S.E.Fienberg, and P.W.Holland, Discrete Multivariate Analysis, The MIT Press, Cambridge, 1975.

18. J.Aczél and Z.Daróczy, On Measures of Information and Their Characterization Academic Press, New York, 1975.

19. W.R.Ashby, An Introduction to Cybernetics, Chapman Hall, London, 1956.

20. S.Watanabe, Knowing and Guessing, Wiley, New York, 1969.

21. W.R.Ashby, "Measuring the Internal Informational Exchange in a System." Cybernetica, 8, 1965, pp. 5-22.

22. H.A.Simon, "The Architecture of Complexity." Proc.Am.Phil. Soc., 106, No. 6, 1962, pp.467-482.

23. R.C. Conant, "Detecting Subsystems of a Complex System." IEEE Trans.Syst. Man.Cyber., SMC-2, No. 4, 1972,pp.550-553.

24. A.P.J.Abrahamse, "Constraint Analysis in Structure Modelling: a Probabilistic Approach." In: Applied General Systems Research, Ed. by G.J.Klir, Plenum Press, New York, 1978.

25. A.P.J.Abrahamse, "Constraint Analysis in Structure Modelling." In: Recent Developments in Systems Methodology for Social Science Research, Ed. by R.Cavallo, Martinus Nijhoff, Leiden, 1978.

26. G.Broekstra, "System Identification: On Terminology in Methodology." Paper presented at First Joint Scientific Meeting of the Dutch Soc. Syst. Res. and Ostereichische Stud. Ges. für Kybernetik, April 1977 (Working paper IIB, 1977).

METHODOLOGY IN SYSTEMS MODELLING AND SIMULATION
B.P. Zeigler, M.S. Elzas, G.J. Klir, T.I. Ören (eds.)
© North-Holland Publishing Company, 1979

THE EFFECTS OF DATA VARIABILITY
IN THE DEVELOPMENT AND VALIDATION
OF ECOSYSTEM MODELS

Efraim Halfon
Basin Investigation and Modeling Section
National Water Research Institute
Canada Centre for Inland Waters
Burlington, Ontario
Canada L7R 4A6

INTRODUCTION

One of the purposes of building mathematical models of ecological systems is the
understanding of the system properties. Another important one is prediction
(Fig.s 1 and 2). In this paper some of the problems involved in model development are analyzed, not from a theoretical point of view (see Zeigler, this volume),
but from the availability of information on the real system and the limitations
thereof. Examples are presented mostly from lakes, not because their features are
unique, but because a large amount of research work has taken place on these ecosystems, and much data have been collected over the years. The problem is the
form in which this information is available. Assuming now that all problems
related with uncertainty, and which are discussed below, have been solved, then
some model simulations can be obtained.

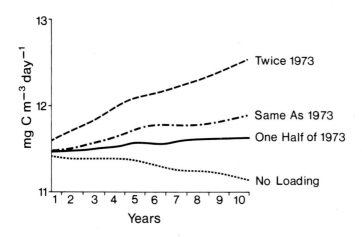

Figure 1: Maximum annual primary production during a ten-year simulation period
of a Lake Superior model. Average whole lake response to different loadings of
soluble reactive phosphorus. Reprinted from Halfon and Lam (1978).

Figures 1 and 2 show outputs of two models, one indicates the expected trend in time of primary production, the amount of carbon fixed by the algae, in Lake Superior if phosphorus loadings are increased or decreased. The other show the expected concentration of chlorophyll a in Lake Huron as a response to total phosphorus loadings. In both figures one feature stands up for its absence: no confidence bounds are presented in the simulations. The simulations are determininstic: to the uninitiated this lack of uncertainty may indicate absolute faith in the model, which is untrue. Let us see now on which source of information these models are based.

INPUT DATA

Input variables to a model are variables over whose values it has no control. These reflect the influence of the rest of the world on the part which has been separated out for modeling. Often input variables are crucial to model behavior. For example, a linear mathematical model must be driven by inputs from the environment, otherwise, when the influence of the initial conditions wears off (transient behavior) all the states would converge to the zero state. In lakes, one of the most important inputs is phosphorus. This is an element which is necessary for life and is limiting algal growth in many lakes. In some instances, another nutrient plays an important role,

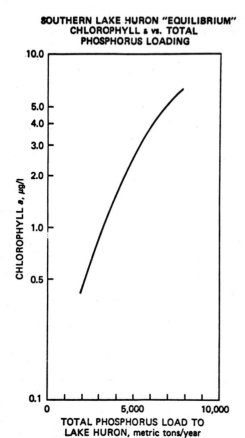

SOUTHERN LAKE HURON "EQUILIBRIUM" CHLOROPHYLL a vs. TOTAL PHOSPHORUS LOADING

Figure 2: Average chlorophyll a concentration as a dynamic response to different phosphorus loadings. (DiToro and Matystik, 1979).

nitrogen. As indicated in Fig. 2, the dynamical relationships between nutrient inputs and algae, in some instances quantified as chlorophyll a concentration, are of interest. Thus, to develop a validated model a fairly good estimate of the inputs should be computed. As indicated in Table I, this is seldom the case. This Table shows some estimated loadings into Lake Ontario, one of the better studied lakes in the world. From the large watershed many big and little rivers take water to the lake. These rivers are not all monitored, only a few are, and not all the time. Therefore, the following sources of uncertainty are present: data are collected only once or twice monthly, floods may take place in between observations and thus large loadings may be missed, concentrations vary with flow conditions. Uncertainty in loading estimates would exist even if all rivers were

Table I. Estimated total phosphorus load into Lake Ontario (metric tonnes)

YEAR	LOAD INTO LAKE	ST.LAWRENCE OUTFLOW	NET LOAD	(data from Simons et al 1978;number in paren- thesis are revised
1967	12977 (11371)	2790 (4694)	10187 (6677)	estimates of Wilson, 1978).
1968	16729 ₍13562)			
1969	17071 (14597)	8833 (6397)	8238 (8200)	
1970	14988 (13774)			
1971	12167 (13374)	5767 (4269)	6400 (9105)	
1972	10906 (13561)	5782 (5394)	5124 (8167)	
1973	11889 (12943)	5876 (4538)	6013 (8405)	
1974	19512 (13231)	5037 (5274)	14475 (7957)	
1975	39047 (10370)			
1976	14749 (9706)			

sampled, which is not the case, since there are too many. Then another source of
uncertainty arises from the fact that statistical methods are used to infer the
loadings from non-sampled rivers. Table I shows that even when a large amount of
data is available from state and federal agencies, research institutes, etc., it
is not always possible to compute the gross or net loadings with some precision.
Therefore, a unique relation between inputs and state of the lake can not be
obtained. Even so, this situation would not be so bad if the state of the system
could be measured accurately, since the input sequances could be reconstructed
(e.g., by using Kalman filters). Unfortunately, as it is shown in the next
section, this is not the case.

STATE OF THE SYSTEM

An estimate of the state of the system is obtained by sampling the lake weekly,
monthly, quaterly, or yearly. This schedule depends on the lake dimensions and
funds available for ships and laboratory analysis. Lake Ontario is about 300 Km
long and 70 Km wide with a maximum depth of 230 meters. Lake Huron and Lake
Superior, two of the Great Lakes of North America, are much larger. This means
that a ship may take one to two weeks just to cruise the lake and collect samples
at 50 to 100 stations and at different depths. This effort can not be made
continuous because of budgeting problems and therefore data are available on a
sparce base, perhaps a few times per year, and not at the same times each year.
Spatial information, however, is usually good, since stations are located uniformly. Not all the information can be collected on routine cruises. Some need
special experiments set up for specific purposes. As a consequence these experiments can not be made at all stations, but perhaps only at two or three locations
in the lake. These efforts require a ship too. When samples are collected (see
Fig. 3 for a detailed map of stations locations in Lakes Huron and Ontario), they
are taken to a laboratory for analysis. Analytical errors may occurr for several
reasons: chemical reagents variability, laboratory technicians inconsistency,
electronic apparatus noise. This source of error is added to the natural variability of the lake waters. Water is not uniformly mixed, there are patches of organisms living a few meters apart and separated by less densly populated waters.
Vertical and horizontal currents move water masses apart. Nearshore waters are
usually different from offshore waters. Bottom depth palys a role in determining
the water temperature, and thus organisms' habitats. All these factors contribute
to let the investigator expect a large standard deviation when a lake average is
computed. When this is not the case (see Fig. 4) systematic errors are likely.
These errors may be due to wrong procedures in the sampling program, or in the

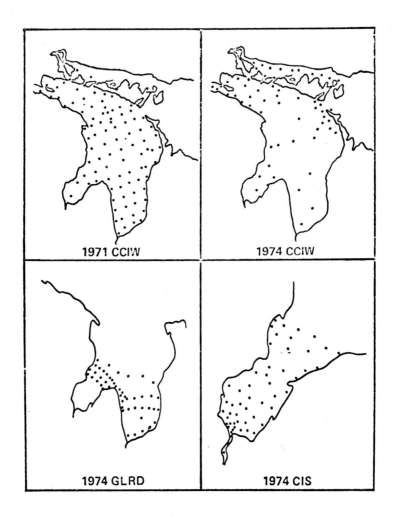

Figure 3a: Map of station locations in Lake Huron and Saginaw Bay.
Reprinted from DiToro and Matystik, 1979.

handling of the samples, or in analytical methods. If large standard deviations
tend to bolster confidence in the data, they are not beneficial when the data are
used for model validation. This topic is discussed in a later section.

CIRCULATION INFLUENCES

As indicated previously, a lake is not a uniformly mixed body of water, it is
therefore interesting to study these differences. This is done with mathematical
models of currents: hydrodynamical models. To make computations easier, the lakes
are usually divided into layers, for example, 0-20 meters, 20-40, and 40 to bot-
tom.When horizontal spatial variation is also needed for special purposes, for
example, nearshore processes, then a three-dimensional model is used. Usually

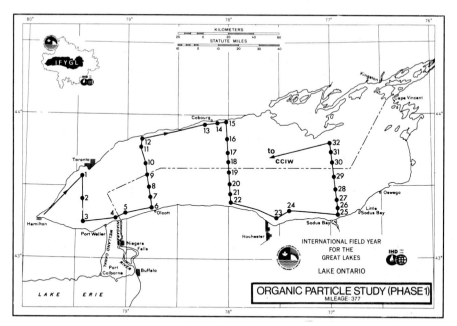

Figure 3b: Map of station location in Lake Ontario in 1972 and 1973. Reprinted from Simons et al. (1978).

however, the lake is considered to be horizontally mixed with only vertical strati-fication, i.e., a one-dimensional model is used for long term prediction. This is due to the increased computation requirements of three-dimensional models, and because of the uncertainties building up when horizontal currents are modelled. Even the vertical dispersion movements can not be modelled perfectly because of observation errors, but this usually considered an acceptable source of error. A uniformly mixed lake model would be too crude an approximation to give reliable prediction in long term prediction. Superficial waters have a marked effect on biological processes, quite different from deep waters, mainly because of the dif-ferent tempearture regime, and influences from the atmosphere (e.g., wind, light, etc.). In Fig. 5 an output of the three-dimensional hydrodynamical model of Lake Ontario is shown. The direction of the arrows indicates the direction of the currents and its length the velocity. When a hydrodynamical model is available, the information from it may be used to reduce the variance of the whole lake average, without giving up the advantage of the variability needed for the data credibility. This effort can help in the validation process.

MODEL VALIDATION

Figure 6 shows a simulation of a model together with some data collected in Lake Huron. As it can be seen, the fit of the model is good but it can not be conside-red extraordinarely good. The standard deviation of the data is such that there may be a very large number of model simulations (different models, or the same model with different combinations of parameters) which could fit the data as well in the least squares sense. The variance of the data influence the development of the model but also its validation. This kind of situation is quite common in ecology and recently Mankin et al. (1977) showed how useless it is to try to validate a model of an ecosystem, given all these uncertainties. Indeed it was this kind of result which led the group at Oak Ridge National Laboratories to

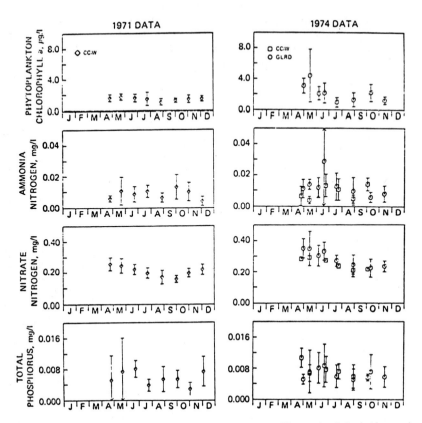

Figure 4: Lake Huron data. The 1971 data show a small standard deviation and a lack of trend in time. This observation suggests that these particular data are not suitable for modeling. Reprinted from DiToro and Matistik (1979).

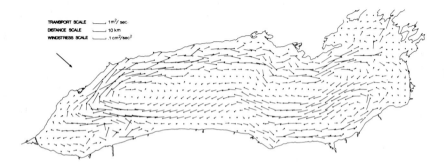

Figure 5: Simulated circulation in Lake Ontario (Simons et al., 1978).

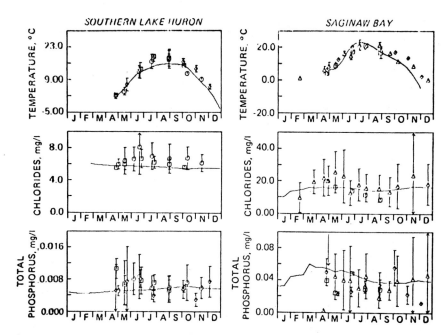

Figure 6: Comparison of model simulations and data in parts of Lake Huron.
Reprinted from DiToro and Matystik (1979).

look for alternative ways to solve this problem. Some of their results are also presented in this volume. Another approach is being developed by Beck and Halfon (in preparation) who do not rely on brute force approach, such as Monte Carlo simulations, but use a more elegant and computationally faster method, i.e., they solve together with the model equations a Kolmogrov's equation which describes the evolution of the transition probability density of the Markov process generated by a stochastic lake model. The model is made stochastic by introducing errors in model structure, parameters, and inputs.

Long time series of data are useful when solving the system identification problem. This assumption implies that a longer time series contains more information, useful for model development. Unfortunately, data have large standard deviations (see previous sections) and if a system is near steady state with the inputs, then the system identification problem becomes more difficult to solve. This is due to the fact that an accurate estimate of the parameter values of the model is obtained when a transient response is observed. To perturb a system with an inpulse or other inputs is a common technique used in system identification. As for the Great Lakes, they are almost in equilibrium with the loadings and no trend is visually apparent. It is not known whether the lakes are getting better, less eutrophicated, or not. Figure 7 shows the scatter of long term data. Large lakes have a very slow turnover time of their waters, ten years for Lake Ontario, and 184 years for Lake Superior, so that transient responses, so useful in system identification, are lacking or not observable with the needed accuracy.

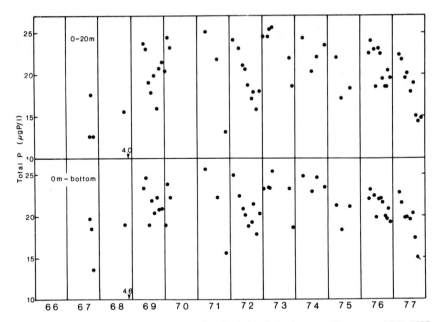

Figure 7: Total Phosphorus concentration in Lake Ontario over the years 1966-1977.
Concentrations are shown for the superficial layer and for the whole lake.
Reprinted from Simons et al. (1978).

SUMMARY

In this paper it has been shown that mathematical models of ecosystems often are
based on information of such nature that the resulting models have questionable
reliability. This kind of result has led to some studies of the effect of uncertainty on the models when these are used for long term prediction. O'Neill (this
volum) gives some preliminary results on parameter uncertainty using Monte Carlo
simulations. These methods, computationally heavy, are only appropriate for models
with a few parameters. More comprehensive results, valid for larger scale models,
are indicated by Beck and Halfon (in preparation) who use the theory of stochastic
differential equations and Kalman filters to solve the problems related to this
uncertainty. Mathematical precision and computer accuracy may be important as
well as the modeller expertise in analyzing an ecosystem, but the properties of
very variable data must always be taken into account.

References

|1| D.M. DiToro and W.S. Matystik: Models of water quality in large lakes, part 1,
    Lake Huron and Saginaw Bay. Manhattan College Unpublished Report (1979).
|2| J.B. Mankin, R.V. O'Neill, H.H. Shugart and B.W. Rust: The importance of
    validation in ecosystem analy is. In G.S. Innis (Ed.) New Directions in the
    analysis of ecological systems. Part 1. The Society for Computer Simulation
    (1977) 63-71.
|3| E. Halfon and D.C.L. Lam: The effects of advection-diffusion processes on the
    eutrophication of large lakes, a hypothetical example:lake Superior.
    Ecological Modelling 4(1978) 119-131.

|4| T.J. Simons, F.M. Boyce, A.S. Fraser, E. Halfon, D. Hyde, D.C.L. Lam, W.M. Schertzer, A.H. El-Shaarawi, K. Willson and D. Warry: Assessment of water quality simulation capability for lake Ontario. Canada Centre for Inland Waters Unpublished Report (1978), 312 pp.

|5| K.E. Wilson: Nutrient loadings to lake Ontario. Canada Centre for Inland Waters Unpublished Report (1978), 104pp.

SECTION V
MODELLING METHODOLOGY
IN DESIGN

METHODOLOGY IN SYSTEMS MODELLING AND SIMULATION
B.P. Zeigler, M.S. Elzas, G.J. Klir, T.I. Ören (eds.)
© North-Holland Publishing Company, 1979

MODELING METHODOLOGY FOR COMPUTER SYSTEM
PERFORMANCE-ORIENTED DESIGN

Gain Wong
General Electric Company
Sunnyvale, California (USA)

Computer systems are highly complex systems involving computer
hardware and computer software functioning together within a
well-defined external data environment. Simulation models of
computer systems must represent the nature of the data environment, the performance characteristics of the computer hardware,
and the operating characteristics of the computer software. A
high-level modeling language, which provides a process and resource orientation, as well as data structuring facilities,
would be invaluable in the timely and accurate implementation
of a model of a computer system.

The process of developing a suitable model of a computer system
is extremely difficult. An appropriate level of model detail
must be selected and validated for each component of the total
model. This process is aided by a methodology for computer
system decomposition which defines the various levels of detail
and provides techniques for selecting the appropriate level.

Once the model of the computer system is implemented using the
high-level modeling language, the verification and validation
of the simulation would be facilitated by a simulator development testbed, a medium where simulation model components can be
exercised and statistics collected on the behavior of the model
component.

INTRODUCTION

Computer systems are highly complex systems involving computer hardware and computer software functioning together within a well-defined external data environment. Figure 1 illustrates the typical representation of a computer system.
Computer hardware includes central processors, central memory, input/output processors, input/output channels, mass storage devices such as magnetic disks,
drums, and tapes, unit record devices such as line printers, card readers, and
card punches, and communications devices. Computer software includes systems
software-operating systems and data base management systems which control the
allocation and scheduling of the system's resources, including hardware, software, and information resources. Computer software also includes applications
software - jobs which compete with each other for the system's resources as they
perform the work of the system. The external data environment of a computer
system includes the commands and data entered by an operator at a terminal, the
stream of batch jobs submitted by cards, and the workload of real time data generated by a physical process which is interfaced to the computer. The volume
and frequency of arrival of the incoming data is what determines the nature of
the external data environment.[5]

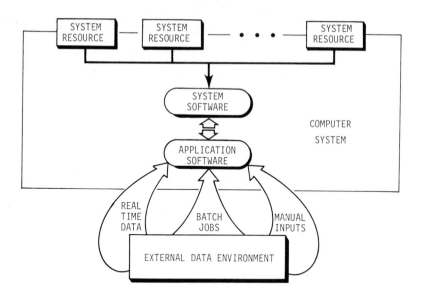

Figure 1 Typical Computer System Representation

One way of approaching the task of creating a model of computer system is as follows. Define all the relevant attributes of the computer system including all the
characteristics of the computer hardware, computer software, and external data
environment. Then, specify how the values of all these system attributes are
related to each other and how changes in the external data environment over time
will affect the values of the attributes for the computer hardware and software.
The first problem we face when we attempt to list the relevant system attributes
is determining the appropriate level of detail in system description.

For example, at a low level, the central processing unit (CPU) can be represented
by:

(1) A set of constant numbers specifying the instruction speed of each
     machine instruction.
(2) Some variables which record whether the CPU is free or busy and if
     busy, what job and what instruction is currently being processed.
(3) A waiting list which records all the jobs that are currently waiting
     to utilize the CPU resource.

With this "instruction level" model of the CPU, there are three events which can
change the state of the CPU:

● A job becomes ready to utilize the CPU.
● A job utilizing the CPU finishes executing an instruction.
● A job utilizing the CPU no longer needs to utilize the CPU.

The CPU can also be represented in a higher level as:

(1) One constant which specifies the average speed of the CPU when
     executing instructions from a specified "mix" of instructions.
(2) Some variables which record what percentage of the CPU resource
     is currently being utilized.

(3) A waiting list which records all the jobs that are currently
     waiting to utilize the CPU resource.

With this "load level" model of the CPU, the CPU is not allocated on an exclusive
basis for the execution of one instruction, but rather on a shared basis for a
specified percentage of the total CPU capability. We are thus modeling the load
that is being placed on the CPU, and not the execution of each instruction.[2]

Computer systems are extremely difficult to model because of the wide range in the
possible levels of model detail. Time granularity is one characteristic of model
detail that must be determined by the modeler. In the actual computer system, the
time between events may range from nanoseconds to minutes. Hardware complexity
is another characteristic of model detail that must be determined by the modeler.
For example, the disk subsystem of a computer system may be represented as a single resource with a certain data rate bandwidth or it may be represented as a
sequence of three resources - an I/O channel with a certain data rate bandwidth
a disk moving arm with a certain access time, and a disk platter with a certain
rotational latency. Software complexity is a third characteristic of model detail
that must be determined by the modeler. For example, the strategy for virtual
memory implemented by the operating system may be represented by a constant system
overhead function or it may be represented by a conditional sequence of operations
for page faulting, task switching, and memory management.

It is clear that there are many levels of detail at which each component of a model
of a computer system can be represented. However, it is not economically feasible
to represent the entire system at the lowest level of detail. Thus, we must select the appropriate level of model detail for each component of the model such
that it is high enough to be economically reasonable and low enough to be a valid
model of the system being modeled. This problem of computer system decomposition
is the primary problem facing the computer system modeler.[8]

A methodology for computer system decomposition which defines the various levels
of detail possible and describes techniques for selecting the appropriate level
will be presented in this paper. Also, a simulator development testbed will be
described where special tools are provided for verifying that an implementation
of a computer system model component is operating correctly, and validating that
the behavior of the model component within the model agrees with the modeler's
expectations.

HIGH-LEVEL MODELING LANGUAGE

Modeling of a complex system such as a computer system by enumerating all the system's attributes and defining how these attributes are affected by changes in the
external data environment is not very easy. The primary difficulty lies in the
lack of structure and organization in a model developed by this approach. What is
needed is a unified approach which not only brings the modeler's attention to all
the relevant aspects of the system being modeled, but provides a natural means of
expressing the characteristics of each component of the model.[4] [6] [7]

The process/resource view of a computer system is a powerful and flexible, as well
as natural, approach to modeling a computer system. With the process view, a computer system is considered to be a set of processes which cooperate with each
other to process data and which compete with each other for the system's resources.
[1] [3] Processes may represent the external data environment, modeling the manner in which the system workload is created; they may represent the system software, modeling the manner in which the system's resources are allocated; or they
may represent the application software, modeling the manner in which the system
work is performed and system resources utilized in the process.

With the resource view, a computer system is considered to be a set of resources,
both real resources such as computer hardware units, and virtual resources such

as a record of information from a data base management system. A resource is thus
anything which might cause a process to be temporarily suspended (unable to continue). For example, when modeling the synchronization of two processes, a "message" resource may be used such that the second process is not activated until the
first process signifies that it is done by passing a message to the second process.

A high-level modeling language is the ideal means of providing the computer system
modeler with a process/resource view. [12] Table 1 shows the desired capabilities of a high-level modeling language. The process view can be realized simply
by providing a mechanism for writing process descriptions which may be executed
concurrently with other process descriptions, and a set of process management
primitives for initiating, controlling the execution of, and termination a process. The resource view can be realized by providing a means of defining a resource with its associated resource handler, and a set of resource management
primitives for requesting, releasing, and obtaining status information regarding
a resource. A library of standard resource handlers can be made available to
model the more common resource allocation policies (e.g., simple facility or
message type resources) but the modeler should be free to develop his own resource
handler if desired.

Table 1   Desired Capabilities of the High-Level
Modeling Language

PROCESS VIEW

- STATIC PROCESS DESCRIPTION
- DYNAMIC PROCESS MANAGEMENT
  - INITIATE PROCESS
  - DELAY (TIME)
  - SUSPEND
  - ACTIVATE PROCESS
  - TERMINATE

RESOURCE VIEW

- STANDARD OR TAILORED RESOURCE HANDLER
- RESOURCE MANAGEMENT
  - CREATE RESOURCE
  - REQUEST RESOURCE
  - RELEASE RESOURCE
  - STATUS RESOURCE
  - DESTROY RESOURCE

DATA STRUCTURING

- LIST PROCESSING
- COMPLEX DATA STRUCTURES
  - TABLES
  - EXTENDED DATA TYPES

In addition to providing a process/resource view, a high-level modeling language
should provide extensive data structuring capabilities. This includes the ability
to define and manipulate linked lists such as stacks and queues. It also includes
the ability to define complex data structures such as table structures and
extended data types. [11]

The purpose of a high-level modeling language is primarily twofold:

First, it imposes a process/resource view which gives the modeler a unified
and structured approach to decomposing the computer system into its model
components. A high-level modeling language with significant "expressive"
power will greatly aid the process of developing a model which accurately
models the computer system being modeled.

Second, the extensive data structuring facility will greatly speed up the
process of developing and testing the implementation of the computer system

model. This is because the simulation programmer will not have to develop
and test his own data structuring mechanisms.

COMPUTER SYSTEM DECOMPOSITION METHODOLOGY

Any computer system can be decomposed into three components:

- Workload characteristics
- Process descriptions
- System resource management policies

Table 2 describes the characteristics of each of these three components. The
workload characteristics basically describe the external data environment in
which the computer system is to operate. The process descriptions describe the
pattern of system resource utilization of each process in the system as well as
the interprocess interactions, both in terms of sequencing and synchronization.
Finally, the system resource management policies describe the manner in which
each system resource is to be allocated - including how requests for the resource
are serviced and how unsatisfied requests are scheduled. These three components
of computer systems will now be examined in more detail. In particular, various
levels of model detail for each component will be discussed.

Table 2   Computer System Decomposition
Methodology

WORKLOAD DESCRIPTIONS

- DESCRIBES EXTERNAL DATA ENVIRONMENT
- CAN BE TRACE-DRIVEN OR MODELED BY STOCHASTIC ARRIVAL PATTERN
- NORMAL AND STRESS CONDITIONS ARE IMPORTANT

PROCESS DESCRIPTIONS

- PATTERN OF RESOURCE UTILIZATION AND INTERPROCESS INTERACTIONS
- INSTRUCTION VS. LOAD LEVEL OF MODELING
- BOTTLENECK TECHNIQUE

RESOURCE DESCRIPTIONS

- PHYSICAL VS. LOGICAL LEVEL OF RESOURCE MODEL
- ACCURACY OF REPRESENTATION OF RESOURCE ALLOCATION POLICY
- LOAD OF RESOURCE ALLOCATION POLICY EXECUTION

WORKLOAD CHARACTERISTICS

The workload of a computer system comprises the external events which trigger the
execution of applications processes. The workload characteristics specify the
times at which applications processes are to execute and also the frequency at
which they execute. Let us consider the case of a computer system where batch
jobs are submitted via cards. If this is an existing system, we can gather data
pertaining to the number of types of batch jobs that are submitted, and the arrival pattern of these jobs. At the lowest level of detail, we can record the type
and arrival time of each job that is submitted to the actual system on a typical
day; then, a trace-driven simulation can be performed in which this record of job
arrivals is used to drive the simulation. [9] ·

There are several potential problems with this technique. First of all, if the
system being modeled does not currently exist, there is no way of obtaining the
trace of events. In this case, a means of approximating the trace of events must
be found. Second of all, it is not clear how to select a "typical" day. A trace

of events of several days can be obtained and combined. Third of all, it may be
desired to exercise the model being developed not only under "typical" conditions,
but also under contrived "stress" conditions.

At a higher level of detail, a random distribution for interarrival times for each
type of application job can be constructed such that the trace of events measured
or anticipated is approximated. Random distributions are often used because they
are easy to generate and more accurate representations of the workload are not
available.

The model of the system workload should ultimately be motivated by the purpose of
the simulation study. If the purpose is to study the resource utilization of the
system under normal conditions, then a random distribution representing the
arrival pattern will be sufficient. If the purpose is to evaluate the system
resource margin under stress conditions, then it is necessary to drive the system
with an input stream which will bring it into a stressed condition. In general,
the combination of these techniques is useful to investigate the workload backlog
created by stress conditions and the subsequent workoff under normal conditions.

PROCESS DESCRIPTIONS

The process descriptions describe the applications processes which actually perform the work of the system. There are two aspects of process descriptions that
we are concerned with:

- The pattern of resource utilization of each process. This includes an
  itemization of each resource required by each process, how many times
  the resource is utilized each time it is acquired, and how long between
  requirements for each resource.
- The sequence of interprocess interactions. This includes what processes,
  if any, to initiate at the termination of each process, what events to
  trigger during the process execution, and what events (that are triggered
  by other processes) to wait on during process execution.

Process descriptions at the "instruction" level of detail represent each access
to a system resource as a separate request. If the number of requests is small,
then this level of detail is both manageable and necessary. However, if the
number of requests is large, then this level of detail is not manageable. The
CPU is an example of a resource which is subjected to a large number of requests.
The CPU resource is allocated by the operating system to user jobs. This allocation is continually being altered because an I/O operation has interrupted the
CPU, or because the current user job needs to perform I/O. Fortunately, in
these cases it is seldom necessary to model at the "instruction" level. If over
a period of time, a large number of accesses is made for roughly the same
amount of resource utilization, then the "load" level of model detail may be used.

A typical process description represents the utilization of the CPU resource in
conjunction with other resources (for example, the input/output channel). A
process may have the characteristics of being compute-bound, I/O bound, or something in between. This can be modeled using the "load" level of detail by
requesting say 75 percent of the CPU resource over a period of time for the
compute-bound process, 25 percent of the CPU resource over a period of time for
the I/O-bound process, and 50 percent of the CPU resource over a period of time
for the in between process. Of course, a process may change its complexion
from one step to the next. In this case, we can subdivide the process into a
number of activities and model the resource utilization differently for each
activity.

The sequence of interprocess interactions of each process must normally be explicitly modeled. Processes which are activated at the termination of a process must
be activated at that time. Process which synchronize with other processes by

waiting for event flags must suspended until the event flag is set. A number of sequential jobs in the actual system may be represented as a sequence of activities in the modeled process as long as the execution of these jobs is strictly sequential. If the jobs may operate concurrently, thus competing among each other for the system's resources, then they must be represented as individual processes and in this case the interprocess interactions must be modeled.

The level of model detail for the process descriptions of applications jobs should ultimately depend on what accurately describes the system behavior, and not only on what is practicable. The question of whether a certain level of model detail can accurately model the system behavior requires considerable analysis. Normally, the "load" level of model detail can be utilized providing the process description is subdivided sufficiently so that each process description activity is accurately described by the average behavior modeled by a "load" level description.

The bottleneck technique of determing the level of model detail is sometimes useful. A bottleneck is defined as any part of the system where contention for resources is so great that throughput of processing is greatly impaired. With this technique, the entire model is developed at the "load" level and the simulator executed to determine where the bottlenecks are. Then, those particular model components are refined until no significant changes in system behavior result from further refinements in model detail. The assumption that is made with the bottleneck technique is that refinements made to model components of low leverage (non-bottlenecks) will not significantly affect the overall system behavior.

RESOURCE DESCRIPTIONS

The system resource mangement policies determine the allocation of the system's resources and the scheduling of the processes which require these resources in order to do their processing. We define a system resource as anything that can cause a process to be suspended until the resource can be allocated. This includes serially reusable resources such as computer hardware devices, and it includes comsumable resources such as messages used to synchronize the execution of two processes. There are three aspects to the question of level of detail as applied to system resource management policies.

First, there is the question of physical versus logical level of detail for the system's resources. For example, when modeling the data base of a computer, does one model only the logical contention for data base records or also the physical contention of the I/O channel, the motion of the moving arm of the disk, and rotational latency of the disk platter?

Second, there is the question of accuracy of representation of the system's resource allocation policy. For example, in a virtual memory operating system, the strategy for assigning memory and the central processor to jobs can be quite complex and it may not be necessary or possible to model this strategy at the lowest level of detail.

Third, there is the question of how to model the load on system resources by the execution of the system's resource allocation policies. For example, some portion of the system's CPU capacity will be used up by the execution of the data base management system. [10]

One objective of resource modeling is to measure the utilization of each system resource. This includes recording the number of times the resource is used, the mean busy period, as well as the percent utilization (proportion of the time the resource is busy as opposed to free). A second objective of resource modeling is to gather statistics on resource contention. This includes how many times a process was placed in the queue to wait for the availability of each resource, the maximum and mean queue lengths, and the maximum and mean waiting times.

The level of detail to which system resources are modeled should ultimately be
that level of detail which is sufficient to accurately answer the questions which
the simulation study is attempting to answer. In one simulation study, the disk
subsystem may have to be modeled down to the rotational latency while in another
it may suffice to model only at the logical disk record level. The problem now
becomes how to determine the appropriate level of detail. This determination
must be made separately for each simulation study.

Two heuristic guidelines are:

    (1)   The coarser the questions of interest, the coarser the model needs to be.
    (2)   There should not be too great a difference in the level of detail
           between model components.

The bottleneck technique of determining which model components could profitably
be modeled in a more detailed manner applies here.

The level of detail to which system resource management policies are modeled can
vary from a simple first-come first-served policy to a priority system to a
complex strategy based on optimizing system throughput. The more complex the
actual system, the more complex the model can become. A typical technique for
determining the appropriate level of detail is to start with a basic simple
resource policy model and then iteratively make refinements until the refinements
no longer significantly affect the general model behavior of interest. [13]

The load on system resources caused by the execution of system software can be
modeled in a number of ways. At a low level of detail, each time the system is
invoked to perform some function, a system overhead process can be initiated to
place some predetermined load on the system's resources. Which resources are
affected and by how much of course depends on the particular system function.
At a higher level of detail, the average system overhead for each system resource
can be estimated and the model can automatically reserve that portion of each
system resource as unavailable for general use. Except for the coarsest models,
the lower level of detail is preferred because of its flexibility in tailoring
different resource overhead rates to different system functions. [10]

APPLICATION OF DECOMPOSITION METHODOLOGY

An example of the application of the decomposition methodology described in
this paper will now be presented. Suppose that we want to develop a simulation
model of a transaction processing system. Using our decomposition methodology,
we decompose the system into its workload characteristics, process descriptions,
and resource descriptions. The workload characteristics define the rate at
which each type of transaction comes into the system. Examples of transactions
are query transactions which obtain information from the system, and update
transactions which modify information in the system. The process descriptions
describe the behavior of the applications programs which are initiated as a
result of each incoming transaction. The aspects of program behavior which are
modeled are resource utilization and interprocess interaction. For example, a
query transaction may be represented as a program which:

    ● Requests allocation of a certain amount of memory
    ● Sends a request to the operating system to read the data base
    ● Waits for completion of system request
    ● Requests allocation of the CPU
    ● Performs some processing
    ● Releases allocation of the CPU
    ● Releases allocation of memory

The resource descriptions identify those system resources for which utilization statistics are desired and/or contention is anticipated. **Examples of** resources are central processors, central memories, and mass storage.

Having decomposed the subject system into its workload, process, and resource components, we are now ready to implement the model. The Simulator Development Testbed described in the next section provides a vehicle for this implementation.

SIMULATOR DEVELOPMENT TESTBED

Assume that we have a computer system decomposition methodology which aids the modeler in confidently constructing a valid computer system model, and a high level modeling language whose expressive power aids the computer system modeler in quickly and accurately implementing the model. It remains to provide the facilities to easily verify and validate the operation of the simulator. The The Simulator Development Testbed (SDTB) to be described in this section provides these facilities. Figure 2 is a block diagram of a typical SDTB.

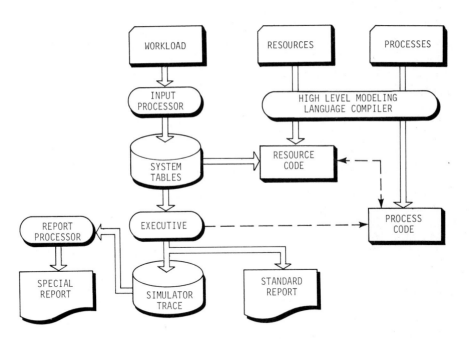

Figure 2   Simulator Development Testbed

Verification and validation of a simulation model are two different processes. On the one hand, verification is the process of establishing that the implementation of the model is correct. This means that the execution of the simulator, while it may differ from the behavior of the actual system nevertheless behaves according to the abstract model which it is based on. On the other hand, validation is the process of establishing that the behavior of the simulation model

sufficiently approximates the behavior of the actual system for the purposes of
the simulation study. [9]

The primary components of the Simulator Development Testbed are:

- The input processor which translates model description data prepared
  according to predetermined syntax rules into system tables for incorporation in the simulator executive.
- The simulator executive which drives the simulation according to the
  workload specifications of the model description data and controls the
  output of the standard simulation report as well as a trace of all the
  primitive operations performed by the simulator.
- The report processor which processes the simulation trace output by the
  simulator executive in order to prepare a special simulation report,
  tailored to the requirements of each particular simulation study.

The model description data processed by the input processor component includes:

- Simulator options which control the operation of the simulator. This
  includes the beginning and end time of the simulation run, and reporting
  options for both the standard report and the simulator trace.
- Resource description data which define all the system resources and
  associate standard or special resource allocation policies of each
  resource.
- Process description data which specify the parameters to be associated
  with each process modeled in the system and also the resource assignments
  to each process. The actual resource allocation policies and process
  descriptions will be coded in the high-level modeling language. The
  purpose of the resource description and process description data is to
  isolate those parameters and correlations which can conveniently be
  isolated and whose values will be experimented with during the course
  of the simulation study. Resource assignments to processes are important
  in simulation studies where alternative computer architecture configurations are being considered and a means is desired for evaluating the effect
  of placing processes in different parts of a distributed configuration.
- Workload descriptions which define the jobs to be processed by the system.
  A trace-driven approach may be taken in which each job to be started is
  represented by a separate workload card which specifies a time for
  initiation along with other parameters such as priority of job. Alternatively, a process may be written whose execution models the arrival of
  jobs of each type according to some arrival frequency distribution. In
  this case, only one workload card is necessary to start the workload
  process. Note that these two methods may be combined in one simulation
  study; that is, some jobs may be trace-driven while others are started
  by a workload process.
- Simulation parameters which specify the values of global simulation variables which affect the behavior of several simulation model components.
  A parameter which will always have the same value for all process
  descriptions can be represented as a simulation parameter.

The simulator executive component basically starts those processes requested by
workload cards in model description data and monitors the operation of the model
to produce the standard simulation report and the simulation trace. One valuable
feature of the simulator executive is the ability to execute the simulation model
for a specified period of simulation time, generating all the normal output
reports and then saving the model state so that the simulation can be continued
at a later time. This feature is especially valuable during the process of determining the appropriate level of detail for some model component. We can implement a new level of model detail and starting from the saved state, simulate
for a short period of time to evaluate the results. We can pickup the

simulation from the saved state as many times as we wish without having to rerun from the beginning. This feature also helps solve the startup versus steady state problem. We can simulate for a period of time sufficient to eliminate the transient effects of system startup and always test new model developments from that point in time.

The report processor component is designed to provide both the convenience of standard reports on resource utilization and contention and process throughput, as well as the flexibility of tailoring special reports. The simulator trace feature allows a simulation run to be made only once, and then a post processor report processor to gather the desired statistics and reports without having to rerun an expensive simulation run. The report processor provides the capability not only of collecting summary statistics on specified conditions (that is, recording values, minimum and maximum values, and counting observations), but also providing graphical capabilities in the form of histograms and plots.

The Simulator Development Testbed provides a vehicle for implementing and testing individual components of a simulation model, and for combining these components into one integrated model. Taking a top down approach, the SDTB user first identifies the primary processing which performs the work of the system under study. This processing can be functionally organized into a number of distinct processes. For example, a transaction processing system can be represented with a query process, an update process, and a database process.

Once the primary processes have been identified, and the interprocess relationships have been defined, the SDTB user can proceed to implement the simulation model at this very coarse level. After the coarse model has been implemented, a sequence of more detailed models can be developed by functionally decomposing each primary process in the same manner that the total system was decomposed. It is the process structuring provided by the high level modeling language which permits the SDTB user to easily express the functional decomposition, including the fact that the decomposition is carried to different levels of detail in different parts of the model. Each primary process can be decomposed into a collection of secondary processes, and each secondary process can be decomposed into a collection of tertiary processes.

Each time a more detailed model is produced by functionally decomposing a previously defined model component, it can be easily tested by substituting it in the previous version of the simulator in place of the decomposed model component. Several model components can be independently decomposed and tested in this way, and later combined. Also, several different decompositions of the same model component can be developed, tested, and evaluated at the same time.

The objective of the Simulator Development Testbed is to provide not only a place where a simulation run can be made and the results efficiently interpreted, but also a place where components of the simulation model can be developed and verified and validated. The SDTB has been designed so that components of a simulation model can easily be run separately and so that the integration of model components is also straightforward. Finally, the SDTB has been designed to maximize visibility to the modeler of the internal behavior of the simulator through the extensive and flexible reporting capabilities.

CONCLUSION

The Computer System Decomposition methodology presented in this paper provides a natural and structured approach to developing a model of a computer system. It decomposes the system into its workload, process and resource components. The critical problem is determining the proper level of detail for each of these components and validating the selection that is made. Aids to the implementation

of a model of a computer system are a high-level modeling language which allows
the modeler to easily express the abstract model developed, and a simulator
development testbed which provides extensive facilities for the verification
and validation of the implementation of the model.

ACKNOWLEDGEMENTS

The approach to modeling computer systems described in this paper has evolved over
the past 4 years at General Electric's Information Systems Programs component in
Sunnyvale, California. Any shortcomings in this paper belong solely to the author,
but many people have contributed to the progress that we have made from the pro-
cess/resource-oriented high-level modeling language SIMTRAN to the Information
and Data System Simulator (IDSS) to the Timing Assessment Model (TAM). In particular, the author is indebted to Jim McCall and Dave Larsen for their many helpful
ideas and dedicated efforts.

REFERENCES

1. G.P. Blunden and H.S. Krasnow, "The Process Concept as a Basis for
    Simulation Modeling," Simulation, Vol. 9, No. 2, August 1967.
2. J.W. Boyse and D.R. Warn, "A Straightforward Model for Computer Per-
    formance Prediction," Computing Surveys, Vol. 7, No. 2, June 1975.
3. O.J. Dahl and K. Nygaard, "SIMULA - An ALGOL-based Simulation
    Language," CACM, Vol. 9, No. 1, September 1966.
4. P.B. Dewan, C.F. Donaghey, and J.B. Wyatt, "OSSL - A Specialized
    Language for Simulating Computer Systems," Proc AFIPS SJCC, Vol. 40,
    1972.
5. M.H. MacDougall, "Computer System Simulation: An Introduction,"
    Computing Surveys, Vol. 2, No. 3, September 1970.
6. M.H. MacDougall and J.S. McAlpine, "Computer System Simulation with
    ASPOL," Symposium on the Simulation of Computer Systems, Gaithersburg,
    Maryland, June 1973.
7. N.R. Neilsen, "ECCS - An Extendable Computer System Simulator," Third
    Conference on Applications of Simulation, New York, December 1969.
8. H.D. Schwetman, "Simulating Computer Systems: An Assessment," Proc
    Simulation '75, Zurich, June 1975.
9. L. Svobodova, Computer Performance Measurement and Evaluation Methods:
    Analysis and Applications, New York, 1976.
10. B.W. Unger, "A Computer Resource Allocation Model with Some Measured
    and Simulation Results," IEEE Transactions on Computers, Vol. C-26,
    No. 3, March 1977.
11. B.W. Unger, "Programming Languages for Computer System Simulation,"
    Simulation, April 1978.
12. G. Wong, "Computer System Simulation with GASP IV," Proc 1975 Winter
    Computer Simulation Conference, Sacramento, California, December 1975.
13. F.W. Zurcher and B. Randell, "Iterative Multilevel Modeling - A
    Methodology for Computer System Design," Proc IFIP Congress, Edinburgh,
    Scotland, August 1968.

METHODOLOGY IN SYSTEMS MODELLING AND SIMULATION
B.P. Zeigler, M.S. Elzas, G.J. Klir, T.I. Ören (eds.)
© North-Holland Publishing Company, 1979

MODELLING AND SIMULATION
IN THE DESIGN
OF COMPLEX SOFTWARE SYSTEMS

William E. Riddle
Department of Computer Science
University of Colorado at Boulder
Boulder, Colorado 80309

John H. Sayler
Computer and Communication Sciences Dept.
University of Michigan
Ann Arbor, Michigan 48109

Abstract: The complexity of large-scale software systems leads to many design problems. The approach to coping with software system complexity discussed in this paper relies upon description schemes permitting the concise, precise and yet abstract description of the connectivity and interactions among system components. The approach allows understanding a complex system by understanding its simpler parts and their interactions so that design problems may be solved in a "divide and conquer" manner. The approach is based upon a software modelling technique allowing designers to explicitly and succinctly describe the structural and behavioral connectivity among a software system's components. The modelling technique allows software design practitioners to more effectively function during the design process because they may make explicit records of their decisions and assess (using either analytic or simulation techniques) the appropriateness of these decisions.

I. INTRODUCTION

There is no single attribute or characteristic which distinguishes a software system as complex — rather, software system complexity may be assessed only in terms of a complicated combination of diverse quantitative and qualitative measurements. Typical quantitative measures are size, number of components and length of the construction period - generally these can be artificially increased by inexperienced or inept designers or programmers. Measures of a more qualitative nature are ones such as understandability and modifiability - these are generally dependent upon the experience and ability of those trying to measure the system's complexity.

One measure does stand out, however, as highly correlated with general, subjective assessments of software system complexity. This measure is *connectivity*, the degree to which interdependencies exist among the system's components. Frequently, the interdependencies manifest themselves in terms of aspects of the system which may be easily measured; an example is the connection established between

two components when one invokes the other through the subroutine-call facilities
of the language in which they are programmed. At least equally frequently, connectivity is *not* manifest in easily measurable aspects of the system; an example
is the typical situation of an initialization component which establishes the environment in which other components operate.

Recent research in the areas of design and maintenance of software systems
has mainly been directed toward discovering ways in which connectivity can be
controlled and reduced. Much of this research has been directed toward small,
sequential programs in the belief that programming in the small must be thoroughly understood before one can appropriately attack large-scale software systems.[1]
But large-scale systems differ qualitatively from small-scale ones, primarily in
the nature of the connectivity among the components. Instead of the simple data
referencing and control flow connections typical of small-scale systems, there
are more complex data sharing and control synchronization connections.

There are two basic approaches to coping with connectivity, and hence complexity, within large-scale software systems. On the one hand, system organizations and operation modes may be sought which lead to clean, clear connections
among the components within the system. Alternatively, description schemes may
be sought which permit the concise, clear description of component connectivity
so that an understanding of the connectivity and its implications concerning the
operation of the system as a whole may be quickly grasped. In either case, the
superstructure and synchronization of the components must be defined clearly and
this is best approached at an appropriately abstract level, where the system's
structure and behavior can be investigated directly without concern for specific
implementation details.

In this paper we discuss an approach to coping with software system complexity of the second type delineated above. Specifically, we review a software
modelling scheme which allows designers to explicitly and succinctly describe
the structural and behavioral connectivity among a software system's components.
We argue that through the explicit description of connectivity and the concomitant ability to reason about its appropriateness, via simulation or analytic
techniques, designers may function more effectively during the design process.

In the next two sections, we discuss software modelling in general, indicating its relationship to other work directed toward the production of correctly
functioning systems and its relationship to programming. Then we review several

---

[1]This viewpoint has been espoused by many, perhaps the most notable being
E. W. Dijkstra ([1],[13]).

software modelling schemes and give an overview of our own, followed by a discussion of the relationships between our scheme and the theory of general systems. We then turn to analysis techniques admitted by our modelling scheme, discussing first the role these techniques play in the general design methodology based upon the modelling scheme and, second, the specific simulation and analytic techniques which we have investigated. In closing, we attempt to give some direction to future work in this area.

Modelling and the Production of Correctly Functioning Systems

Numerous investigators have studied the problems of proving the correctness of programs since Floyd's famous paper [16]. Some understanding of the principles involved is now required of most computer science students and at least two current textbooks ([1],[3]) postulate the necessity of proving programs correct as they are being developed. While the future usage of these techniques is open to debate, it is clear that the majority of working programmers are not now engaged in formally proving the correctness of their programs, and this seems unlikely to change in the near future.

There are two issues here pertinent to the development of large systems:
- it is presently too difficult and time consuming to develop proofs for each small piece of a large system, much less integrate these proofs to achieve a proof of the total system,
- the language of the assertions required in correctness proofs is inimical to the algorithm and data structure constructs used in the abstract description of a large-scale software system.

Another major research area relating to the production of correctly functioning systems is the area of program comprehensibility. Included under this heading are structured programming, code format conventions, and programming language constructs. It is imperative that less effort is required to understand an existing software system than was required to develop the system. Research into program comprehensibility attacks this problem.

A final major area of research concerns the development of software systems - what sorts of environments are best for teams of programmers ([3],[40]), and what sorts of "methodologies" should be applied during development ([17],[43],[46])? While the intuitive notions developed through these investigations are useful, they lack precision and hence function as guidelines only.

There is little doubt that the research mentioned has significantly improved the practice and understanding of the programming process. However, it has not been concerned with the design phase as such, but rather with the program which is

the result of the design phase. In particular, research on the above themes focuses on questions of the following sort (see for example [44]):

What programming language features produce the 'cleanest' code?

What programming language features have good proof rules associated with
them?

What sorts of things can be checked at compile-time and which checks must be
delayed until run-time?

How can one enforce strong typing?

What are good practices for programmers to follow so that the products of
their labor are understandable, correct and modifiable?

The central thesis of this paper is that it is highly profitable to consider
large software systems from a more abstract point of view, that of a general systems theorist. From this viewpoint, some different questions are addressed and
different goals arise, for example:

What are the central characteristics of large software systems? Which are
essential to modelling?

How can these characteristics be quantified, in particular to allow comparison
of different systems or different designs for the same system?

What sorts of mathematical models are appropriate to the study of large software systems?

What homomorphisms can be used to relate different levels of a system design?

Can our mathematical models yield any formally justified design principles?

The ultimate goal of our work is to develop a software system design methodology,
based on answers to these types of questions, which provides a vehicle for employing *program development* guidelines and practices in the *design* of large-scale software systems.

Modelling Versus Programming

Software system modelling and software system programming are superficially
similar but differ qualitatively in three respects. First, during the programming
phase the intent is to produce efficient, well-structured code for each of the system modules. Hence the emphasis is upon the definition of data structures and the
algorithms manipulating them. During the design phase the aims are more global —
one is interested in preparing a complete specification and doing so in an incre-
mental fashion rather than en masse. Second, system properties which are of concern during design may cut across module boundaries whereas during implementation
the system properties of interest are usually local to modules. Finally, during
implementation, behavior is defined implicitly whereas during design one wishes to

give explicit, nonprocedural specifications for behavior.

The constraints in effect during the two phases are quite different also. During implementation one must account for limitations on resource availability, the operation of real synchronization primitives provided by the programming language, and error detection and recovery. During the design phase these concerns are not as important because one is more concerned with global, and partial, descriptions. Thus, during the design phase it is acceptable to assume an infinite supply of resources and idealized synchronization primitives. Further, error correction may be defined implicitly by nonprocedural definitions of the behavior that the system *must* realize, and error detection may be modelled by the inclusion of "error" states as post-conditions of certain actions of the system.

Recent advances in the design of programming languages have tended to raise their level toward the domain of system modelling. For example, abstract data types [23] have recently emerged as an important facility for the specification of sequential programs. While they are convenient for the description of a software system's data storage components, they are not convenient for the succinct description of those system components concerned more with the processing than with the storage of data. This is particularly true when the components operate concurrently.[1] The major problem is that abstract data types are oriented towards describing components as structures of data which are operated upon via procedure calls. Many components (e.g., a text editor in an operating system or a file system in a multiprocessor computing facility) are not naturally described in this manner.

Software Modelling

Thus software system modelling is a task which is distinct from software system programming. Not only are the intents of modelling different from those of programming, but the constructs needed in a modelling language are somewhat different in nature from those of programming languages. While there have been surprisingly few efforts to develop software system modelling languages, some of the essential constructs of these languages have been developed.

The SIMULA language [10] introduced the class concept which has subsequently found a home in the TOPD and the DREAM systems (discussed below), and is central to the definition of abstract data types. Several other features of the SIMULA

---

[1]By "concurrent" we mean parallelism which may be actually achieved by executing the system in a multiprocessing environment or which may be only apparent at abstract levels of system description and never achieved during system execution.

language make it useful for performance modelling of software systems [13].

The Tools for Program Development (TOPD) ([18],[19]), developed at the University of Newcastle upon Tyne, England, introduced the concept of finite state modelling of sequential programs. In TOPD the "values" of data objects are partitioned into "states". Procedures which operate on these objects effect state transitions which may be specified via pre- and post-conditions for the procedure's invocation. One may "run" a TOPD model and receive a listing of the possible states of objects at each statement of the program. In addition, TOPD can perform some checking of the internal consistency of the model description.

Petri nets and equivalent schemes (reviewed in [25]) have been used for the formal modelling of systems with parallelism. These schemes accurately portray the detailed action sequences of processes operating in parallel, but have strong drawbacks as a general modelling scheme for the design of complex systems: first they are not language based; second, they are generally quite cumbersome to use because of the low (essentially machine) level of description; and third, they generally have no explicit behavioral component, i.e., the action sequences may be obtained only through model simulation.

The Program Process Modelling Language (PPML) [26] was developed as a natural, but still formally defined, modelling tool for systems with parallelism. In PPML the component processes of a system are described in a high level modelling language rather than by a graph or a mathematical representation of the potential activity of the system. The component processes of the system are viewed as interacting via message transmission. (This view allows a convenient partitioning of a description in one part which explicitly describes component interdependencies and another part which describes the components' independent activities.) The communication is mediated by link processes which serve to effect all necessary message transmission synchronization. Since overall system coordination is modelled by the transmission of messages, this activity is of paramount importance in understanding PPML-modelled systems. Thus, the PPML scheme allows one to algorithmically derive [27] a closed-form representation of the message transmission behavior of the system, which may be inspected for incorrect functioning of the system. This is obviously not as simple as it sounds, and research continues on ways to extract particular features of interest from this behavioral description. This is discussed more fully in a later section.

Software Modelling in the DREAM System

The considerations outlined above guided us in our development of the Design Realization, Evaluation And Modelling (DREAM) system ([28],[31],[32],[34]) and its

associated modelling language, the DREAM Design Notation (DDN). Our specific aims
were to develop a system which would allow a designer to iteratively develop a
model of an intended system by providing both a modelling language for the description of the system and a data base for retention and extraction of design
fragments. In addition we wished to provide both simulation and analytic techniques which designers could use to incrementally bolster their confidence in
the validity of the evolving design. In this section we describe the basic
modelling constructs provided by DDN; analysis techniques provided by DREAM will
be treated in a later section.

In DDN a system is defined to be a network of hierarchically decomposable
components which execute concurrently and asynchronously. The overall purposes
of the system are achieved via the internal processing of the components as coor-
dinated by communication among them. In DDN, communication is used both as a
modelling construct and for the specification of implementation details. Accordingly, two mechanisms for communication description, message transfer and shared
data, are provided for these two respective purposes. Message transmission is
analogous to control signal processing in the hardware domain, and is primarily a
modelling concept for software. In DDN, message handling models those aspects of
component communication in which components require some knowledge of the rest of
the system since the message pathways must be explicitly defined and fixed and
the message types must be known to the participating processes. The shared data
mode of communication is less of a modelling concept, and nearer to implementation (although it is a modelling concept used in general systems.) This mode of
communication does not require knowledge of how the rest of the system operates
nor how the rest of the system utilizes the data, although constraints may be
imposed within the data definition itself.

The key modelling concepts underlying DDN are:
● the class notion,
● the thorough distinction between the structure and the behavior of a
    system,
● the concept of isolated knowledge ("need to know", "information
    hiding").
The class notion presumes that a system is comprised of many similar units,
hence a model of a system should allow description of templates for the system
parts. In DDN these templates (classes) have associated parameters (or quali-
fiers) which allow slight distinctions to be incorporated within class instances.

The designer of a large system exhibiting parallelism is constantly plagued
by confusion between the structure and the behavior of the target system. The
difficulty arises from the different levels of description the designer wishes

to use, and from the different levels of abstraction inherent in a system description. For example, an operating system may be viewed as an object that
has as its parts a scheduler, an interrupt handler, managers for primary and
secondary storage, an input/output handler, and user jobs. But a description of
any of these sub*objects* is in fact a template of a potential *process*. DDN provides the capability of illuminating these differences via formal specifications
of structure and concomitant descriptions of behavior, either endogenous (associated with the structural specification) or exogenous (not linked to specific
structure).

The concept of information hiding pervades modelling activities in many domains, under an assorted collection of pseudonyms. Basically, the principle
states that component A should be (or is) only knowledgeable about the func-
tion(s) provided by component B, and should be (or is in fact) ignorant of the
internal mechanisms by which component B performs. This concept is also a meta-
modelling one, since designers should not be exposed to the inner workings of
components other than the one currently being designed, else they are likely to
take advantage of this *structure* when they are to be guided only by the behavior of other components. For the same reason, programmers should be guided by
this principle [24], and only be given as much (behavioral) description of the
rest of the system as is necessary to complete their part of the system.

More specifically, in DDN descriptions a software system is decomposed into
components of two types. *Subsystems* are those components which control and
guide the performance of the system's processing. Subsystems operate (conceptually at least) in parallel and asynchronously with respect to other components,
and are individually capable of performing several activities at once.[1] *Monitors*
are those components which also operate concurrently and asynchronously with respect to other components but which serve primarily as repositories of shared information and are individually capable of performing only a single activity at
any point in time.[2] An auxiliary component of DDN descriptions is the *event*
class. This is used for describing part of the system's operation in terms of
a behavioral description which is exogenous to the system. One could, for example, give a description, in terms of several events, of an operating system from
an external user's point of view, thus providing a description which is redundant

---

[1]Those system components which execute concurrently and manipulate shared data
objects are usually considered to be sequential processes, as defined in [21].
A subsystem is a more general object, being essentially a collection of sequential processes.

[2]The monitors of DDN are essentially those defined by Hoare [20]. To the usual
definition of monitors, we have added constructs for behavior specification,
patterned after constructs developed for the TOPD system [19].

with the external description and orthogonal to the internal specification in
the sense that it may establish associations among activities which occur within
physically different parts of the system.

Structural descriptions are given in DDN by specifying the componentry and
the operation of the components. For subsystems this consists of giving fragments of design description, called *textual units*. These describe the ports
through which the subsystem sends and receives messages, the subcomponents (instances of other subsystem classes and of monitor classes) which comprise an
instance of this class, and the algorithms for control processes which control
the flow of messages among the subcomponents and/or ports. As a very simple example[1], consider a data base which we will call a blackboard. Values stored in
the blackboard may be inspected and modified. The blackboard serves as a repository for information and has processing components which "observe" the
changes made to particular entries in the blackboard and notify the outside
world whenever these entries are the subject of a modify operation. The overall
system may be pictured as in Figure 1; the description of this overall system
will be gradually developed throughout the paper.

One possible DDN description of the data base itself is:

[blackboard]: SUBSYSTEM CLASS;

    QUALIFIERS; #_of_entries END QUALIFIERS;

    request: IN PORT;
      BUFFER SUBCOMPONENTS;
        request type OF [bb_request_type]
        END BUFFER SUBCOMPONENTS;
      END PORT;

    answer: OUT PORT;
      BUFFER SUBCOMPONENTS;
        done_signal OF [bb_request_answer]
        END BUFFER SUBCOMPONENTS;
      END PORT;

    signal: OUT PORT;
      BUFFER SUBCOMPONENTS;
        value OF [possible_values]
        END BUFFER SUBCOMPONENTS;
      END PORT;

    SUBCOMPONENTS;
      entries: ARRAY [1::#_of_ entries] OF [data_entry],
      watchers: ARRAY [1::#_of_entries] OF [watcher]
      END SUBCOMPONENTS;

    END SUBSYSTEM CLASS;

---

[1]More extensive examples are given in [8],[9],[31]-[33],[37],[38],[41].

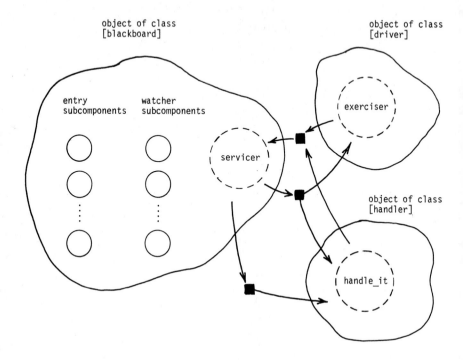

Solid lines indicate actual objects. Dotted lines indicate
models of the operations performed by objects.

Figure 1

The qualifier *#_of_entries* serves to parameterize the class definition to produce a description of a generic class of entities. A [blackboard] subsystem receives the commands requesting the inspection or modification of entries through the *request* port and notifies the outside world that the requested operation has been completed through the *answer* port. The condition checking portion of the blackboard sends notification that the "observed" entry was the subject of a modification operation out through the *signal* port. Each of the ports has a buffer associated with it where messages are stored on the way in and out. The subcomponents are (structures of) instances of other classes which indicate the essential subparts of an instance of this class. The definition could also include the description of a control process which channels messages which come in through the *request* port to the appropriate entry in the *watchers* array as indicated by the content of the request message.

Not all of this definition would normally be visible to a member of the design team other than the designer who prepared it. Port and qualifier definitions would be visible but subcomponent definitions and control process bodies are not visible except when specifically made so by using the attribute "visible" in their declaration. This enforces the desirable result that in defining the use of a subsystem one may normally not rely upon any knowledge of its internal operation.

Monitors are described by defining their subcomponents, their states, and the actions which may be performed on the data objects they model. The states description in DDN contains a great deal of structural information. State variables are defined, and state subsets may then be defined as subsets of the cross product of the values of the state variables. (State subsets are always visible, but state variables are visible only when indicated by the attribute "visible".) In addition one may define an ordering relation on states and an equivalence relation between states defined for the class and sets of states defined for the subcomponents. The definition of the monitor's actions or procedures includes the definition of local subcomponents, sequences of computation steps, the parameters of each procedure, and the transitions of each procedure, i.e., the state transitions that will occur as a result of each procedure's invocation.[1] As a simple example, the following definitions specify the monitor classes referenced in the previous example:

```
[possible values]:  MONITOR CLASS;
   STATE SUBSETS; value1, value2, value3 END STATE SUBSETS;
   END MONITOR CLASS;
```

---

[1]It should be noted that all of the textual units of a DDN description are optional, thus a designer may specify as much or as little detail as desired in any given design step.

```
[bb_request_type]:  MONITOR CLASS;
  STATE SUBSETS; inspect, modify  END STATE SUBSETS:
  END MONITOR CLASS;

[bb_request_answer]:  MONITOR CLASS;
  STATE SUBSETS; value, done  END STATE SUBSETS;
  END MONITOR CLASS;
```

For each, all that has been defined are the externally observable states (which are essentially the same, in this case, as values for an enumerated type in Pascal).

In addition to structural descriptions in terms of templates for the subsystem and monitor classes, designers may additionally define the sharing of system components and how the ports of the components are connected together. Sharing may be described by instantiation control textual units which serve to indicate the "equivalence" of subcomponents which are otherwise (for ease or clarity of description) described as distinct. Connection textual units may be used to describe message communication pathways in terms of "transmission lines" between ports. (More extensive discussion of these constructs may be found in [29] and [30].)

Behavioral aspects of a system may be specified both pseudo-procedurally and non-procedurally. The major pseudo-procedural means of describing a subsystem's behavior is by giving models which define the subsystem's operation, in terms of message flow through the ports, as seen by external subsystems. For example, the operation of [blackboard] subsystems may be modelled as:

```
servicer:  CONTROL PROCESS;
  MODEL;
    ITERATE
      RECEIVE request;
      IF request_type = modify
        THEN value SET TO value1 OR value2 OR value3;
             SEND signal;
             done_signal SET TO done;
        ELSE done_signal SET TO value;
        END IF;
      SEND answer;
      END ITERATE;
    END MODEL;
  END CONTROL PROCESS;
```

Nondeterministic control constructs are also provided in DDN so that models such as these may be even more abstract.

The other means provided by DDN for the behavioral description of a system is the definition of events and sequences of events. In DDN, we distinguish two broad types of events, *endogenous* and *exogenous*. Endogenous events are those occurrences which arise from some activity within the currently DDN-described portions of the software system. Exogenous events are those occurrences which are

relevant to or impinge upon the system's behavior but arise from some activity outside the currently described portions of the software system. Whether an event is
endogenous or exogenous is therefore relative to the extent of the system's description and may change over time - for example, an exogenous event may become an
endogenous event as elaboration of the design leads to the description of the component whose activity gives rise to the event. Some events, however, are inherently exogenous since they pertain to the system's operation but do not stem from the
software portion of the system being designed - examples of such events are activities within some other software system which interacts with the system being designed or operations performed by some physical device controlled by the software
system.

The most elementary method for defining endogenous events is to simply attach
a label, called an *event itentifier*, to some portion of the DDN description - an
example appears later. Exogenous event definitions may not be associated with any
monitor or subsystem definitions and are therefore defined via the <u>event class</u>
textual unit. Once a set of events has been defined, a software system designer
may specify intended behavior by describing the possible sequencing and simultaneity of event occurrences which would be acceptable during system operation. (A
more complete description of DDN constructs for event definition and sequence specification is given in [42].)

Comparison With Formal Modelling Concepts

In this section we will relate the modelling concepts of DDN to the theory of
modelling and simulation developed by Zeigler [45]. Each of the four basic concepts in that theory will be presented in terms of Zeigler's definition, its definition with respect to software systems, and its realization in DDN.

The *real system* is, in Zeigler's scheme, the source of data; moreover the data
or the behavior is all that we can know directly about the system. In particular,
we can have no direct knowledge of the structure of the system. Ignoring the profound epistomological implications of this approach, we note that this allows a
precise distinction to be made between the system and a model of that system. The
situation is somewhat murkier with respect to software. Considering a software
system to be a collection of computer programs, we must infer that a software system is in fact purely a model of some other system since the code has no behavior -
only a prescription of behavior.[1] The appropriate real system is the combination
of the software and some hardware upon which the software instructions are

---

[1]This is a bit pedantic, but in fact there is some confusion in the literature
regarding this point.

executed.[1]

There remains another difference, namely our knowledge of the system. Since we have available the entire structure of the system (the code and the hardware principles of operation), it would appear that our systems are vastly different from those of Zeigler. The complexity of the software systems under discussion vitiates this point. In terms of human understanding, the one million lines of code of the IBM OS/360 operating system are as unknowable as a large ecosystem. In fact, it is because of this complexity that software systems may benefit from application of modelling theory!

In DDN, the real system is the target system being designed, i.e., the real system is a tabula rasa. We know something about the behavior of the system because we generate an exogenous behavior for it, but we do not know the structure of the target system, only a DDN model of it which admits of numerous relizations.

An *experimental frame* is a set of circumstances under which the real system can be observed or can be experimented with. It usually corresponds to some set of questions posed about the real system. Since knowledge about the real system is purely behavioral, an experimental frame may be defined by the entirety of input/output observations from it. In software, since one has the structure of the system to examine (the code), there exist static as well as dynamic experimental frames. Statically, one can examine scope of variables, scope of control, the flow graph of a program or the program schema. Dynamically, one may for example trace variables or resources, or concentrate on the working set or the page movement. In DDN, static experimental frames could include any subset of textual units across the system. Dynamic experimental frames could include any specified exogenous behavioral attributes pertinent to some subset of the subsystems. Note that (as in PPML) one may derive behaviors of the system (in terms of events and event sequences) without the necessity of exercising the system, and these derived behaviors could constitute an experimental frame, i.e., the endogenous behavior description on one level may be viewed as an exogenous behavioral description for the next lower level of detail. The pseudo-procedural models of control processes provide an alternative means in DDN of defining experimental frames since these serve as externally visible definitions of the behavior of a class of subsystems.

The *base model* is a model that accounts for all of the input/output behavior of the system. It is a model that is "valid" (see below) in all permissible

---

[1] We are still faced with a considerable dilemma since to be precise we would have to specify a particular hardware machine and we wish to consider only the software. Hence we assume hereon the existence of a universal order code into which all programs will be translated.

experimental frames. The base model is never fully knowable since the set of possible experimental frames is infinite. In software the base model of the system
is the code (see above) which is in principle fully known. In DDN the base model
is also the code, but is never fully knowable because the DREAM machine is an abstract machine.[1]

A *lumped model* is an abstraction of the base model obtained from the base model by simplification, lumping of variables, or suppression of detail. The lumped
model concept provides a framework for discussing the validity of system models. A
lumped model is tested for validity with respect to an experimental frame, if it
can generate all of the behavior of the experimental frame to within some specified
tolerance. Thus many different lumped models may be valid with respect to the same
experimental frame, and a given lumped model may be valid in some experimental frame
and not in another one. In software, lumped models may be informal prose descriptions of the processes of the system and/or hierarchical organization charts, or
they may be formal predicate assertions defining system purpose. In DDN a lumped
model of the system is a collection of textual units describing the internal operation of subsystems and monitors - its most important constituents are body textual
units for control processes and procedures. Which textual units are chosen depends
on the system properties under investigation, i.e., upon the experimental frame.
Since DDN supports redundant system descriptions it is possible to have different
lumped models which are valid with respect to a selected experimental frame. Note
that this is absolutely necessary for a design system, in order to allow a designer
to investigate alternative structures of the target system. The natural relationship between experimental frames and lumped models developed by Zeigler is of great
value for an understanding of the modelling and design processes.

The modelling scheme advanced in [45] has proven to be very useful in guiding
our thinking about software modelling. There are, however, two major distinctions
between the (natural) systems discussed therein and software systems. The first is
the nature of the *real system*, as already discussed. The other difference is that
DDN is primarily a modelling system for the synthesis of systems rather than for
abstraction from existing systems.[2]

---

[1]The modelling language of DREAM, DDN, assumes the existence of a hardware machine
with capabilities necessary to the modelling process, such as infinite resources,
including the number of processors. As discussed previously, modelling is very
different from programming.

[2]So far, our examples of the usage of DDN ([8],[9],[31]-[33],[38],[41]) have
been in the traditional role of abstraction for understanding rather than for
synthesis.

A Design Methodology

DREAM is a design methodology - a collection of tools and procedures - useful for the design of complex, concurrent software systems. The intents in constructing DREAM have been to provide tools which support a variety of design procedures, as well as to develop a medium for the experimental evaluation of the viability and efficacy of various design procedures. Most of the tools prepared so far, however, tend to make DREAM most useful in conjunction with a top-down design method in which designers iteratively prepare the DDN text describing a system by gradually elaborating the detail of the system's organization and operation. The completed design consists of a set of templates for the components of the system, a description of the behavior exogenous to the system, and an initial configuration of the system: the connectivity and the instantiation control. At each design step, the designer uses DREAM to either modify or augment the existing design retained in a DREAM database by changing, adding, or deleting textual units. It should be noted that the design language admits a hierarchical design, and that DREAM allows a system design to exist at varying levels of detail simultaneously.

DREAM provides a variety of tools to assist in managing this elaboration. Primary among these are an editing facility to assist in text preparation and a data base management facility to assist in the augmentation and modification of the design description. The other tools provided by DREAM allow the designers to obtain information about the characteristics of the system under design that are not explicitly stated in the text of the design description. The intent of these tools is to provide designers with some means of analyzing their design in tandem with its development.

Analysis During Design

A major reason for preparing models of software systems in some formally defined modelling scheme is the concomitant ability to algorithmically assess the system's characteristics. This is all the more important during design, since the analysis provides a preview of the characteristics of the eventual system and allows the designers to incrementally gain confidence in the validity of their decisions. In this section we discuss some approaches to analysis during design which are supported by a modelling scheme of the sort discussed in the previous section.

The analysis schemes that we have in mind are not necessarily fully algorithmic. At any point during design, there exists a partially complete description, D, of the system under design. In addition, the designers have some idea of the system's characteristics along some dimension and we may denote this as C_i. Analysis at this point in the design is then representable as:

$$\rho_i(C_i, \delta_i(D))$$

δ_i derives the system's actual characteristics along the dimension i, to the extent that these are represented by the partial design D. ρ_i compares the system's actual and desired characteristics to determine whether or not they are acceptably related. In the analysis schemes to be discussed here, we require that δ_i be algorithmic but admit ρ_i that cannot be embodied in an algorithm.

The non-algorithmic nature of the analysis may arise from one of two sources. First, the comparison that needs to be done can be theoretically undecidable and thus an algorithm for ρ_i may not exist. Second, it may be impossible for the designers to specify C_i to the level of detail needed to apply the ρ_i comparison algorithm. In either case, the designers must use their intuition, experience and skill at formulating a logical argument in order to prepare a demonstration that the actual characteristics and the desired characteristics are or are not acceptably related.

This type of analysis may be called *feedback analysis* to connote that information concerning the system's characteristics is derived and presented to the designers so that they may assess its implications and subsequently apply any necessary corrective measures. This general class of approaches to analysis may be subdivided according to, first, the characteristic under analysis and, second, the approach used to implement δ_i. This leads to three major types of feedback analysis: feedback analysis of system organization, simulation-based feedback analysis of system behavior, and analytic feedback analysis of system behavior.

Feedback Analysis of System Organization

The simplest form of feedback analysis concerns the connectivity and hierarchical organization of the system. The point of this analysis is to present the designers with a description of the overall system organization, derived from the designer-prepared description which explicitly describes only the components and their local inter-relationships.

The DDN description scheme supports two feedback analysis techniques of this type. The first derives a *connectivity* graph which shows the communication pathways which exist for message transfer among the components. The definition of this graph may be easily derived using the information contained in the port and connection textual units. But actually drawing the graph is not an easy task because of the problem of determining node placement so that a graph with a near-minimal number of arc crossings is produced.

The value of this feedback analysis technique lies in the fact that message communication connections are used in DDN descriptions to model the potential dependency of one component's operation upon the operation of another component. Thus the *physical* connections for message transfer actually reflect the *logical* connections for processing inter-dependencies.

The second feedback analysis technique supported by DDN for the analysis of system organization focuses upon the hierarchical organization of the system's componentry. This technique produces an *instantiation graph* in which arcs represent "part of" relationships among the components - the graph is essentially a map of the elaboration process followed in preparing the design. The definition of an instantiation graph is also easily derived, in this case from the subcomponents and instantiation control textual units. The graph itself is not difficult to prepare since it is a directed acyclic graph[1] rather than a general network.

The value of this feedback analysis technique is that it pictorially represents the component sharing within the designed system. For ease and naturalness of describing individual components, DDN allows the sharing of ports (and therefore buffers) and subcomponents to be described separately from the description of the components themselves. The instantiation graph therefore provides feedback which is helpful in discovering both system organization errors and potential conflicts due to sharing.

Feedback analysis of system organization is primarily of value in checking that the design (at the current level of elaboration) has been correctly described in DDN. The DDN description scheme relies heavily upon relationships among the components such as "sends message to" and "is part of". Graphically presenting these relationships to the designers affords the opportunity to check that the local connectivity with respect to these relationships gives rise to a global system organization which coincides with the desired one.

Simulation-based Feedback Analysis of System Behavior

More critical than an understanding of global system organization is an understanding of the system's overall behavior. The designers have prepared a definition of each component which specifies how it should interact with other components. It is the intent of feedback analysis of system behavior to derive information concerning the overall behavior which results from the local interactions.

One way in which this information may be used is in verifying that the overall behavior is indeed what is required or intended. Another use is in checking the consistency of two descriptions of an interface, one in each of two interacting components. For either of these possible uses, it is necessary to gain information about the dynamic, run-time operation of the system and one way to do this is by simulation.

To effectively perform simulation, designers must supply estimates of the time

---

[1] DDN does not allow recursive subcomponent definition.

consumption characteristics of modelled operations and the behavioral characteristics of modelled operations and the behavioral characteristics of nondeterministic
operations. This can be done in an extended DDN, and this extension[1] is relatively
straightforward. For example the description of the control process within the
blackboard class can be augmented as follows:

```
servicer:  CONTROL PROCESS;
  MODEL;
    ITERATE
      RECEIVE request;
      NULL //100//;
      IF request_type = modify
        THEN value SET TO //3//
              value1 OR value2 OR value3 (.33,.33,.33);
             SEND signal;
             done_signal SET TO //1// done;
        ELSE done_signal SET TO //1// value;
      END IF;
      SEND answer;
    END ITERATE;
  END MODEL;
END CONTROL PROCESS;
```

The notation //n// associates with a modelled operation a time, quoted in some arbitrary time units, for its execution. In general, a time distribution would be specified and techniques developed for simulation languages could be used for this purpose. Also, a more general scheme would allow the specification of timing characteristics of SEND and RECEIVE operations - for simplicity, we assume that this time
is a constant known to the simulator. The notation (p,q,...) indicates the probabilities with which various options in a nondeterministic operation should be chosen.

With descriptions augmented in this manner, the model of the system under design may be exercised by a simulator to obtain an estimate of the distribution of
derived statistics concerning the run-time characteristics of the system. In addition, the timing and probability estimates may be varied to parametrically investigate the sensitivity of these derived statistics to changes in the estimates.

Simulation-based feedback analysis can be used in several ways in addition to
obtaining estimates of the time-related behavioral characteristics of a system.
First, it can be used to investigate the effect of placing a bound upon the number
of messages sent out through a port, such as a [blackboard]'s *signal* port, which
have not been forwarded to a receiver. (This would, of course, require expanding
our example description to reflect this bound.)

Second, simulation can be used to check for the violation of conditions upon
the messages that may flow among the components. For example, DDN would allow the

---

[1]This extension is patterned after one used as the basis for a recent thesis on
performance assessment during design [36].

description of the *signal* port to be augmented as follows:

```
signal:  OUT PORT;
  BUFFER SUBCOMPONENTS;
    value OF [possible_values]
    END BUFFER SUBCOMPONENTS;
  BUFFER CONDITIONS;
    value = value1 OR value = value2
    END BUFFER CONDITIONS;
  END PORT;
```

indicating that only messages with *value* being either *value1* or *value2* may validly flow out through the port. During simulation, observance of this condition could be checked whenever a SEND operation is done involving the *signal* port.

Finally, simulation may be used to investigate the effect of the system's eventual processing environment upon the system's time-related behavioral characteristics. The DDN description scheme is built upon the assumption that resources are infinite in the execution environment - memories are assumed to be unbounded and it is assumed that there is a sufficient number of processes for all the components to run in parallel, each on a dedicated processor. These assumptions are fine for the purposes of modelling and force the designers to conscientiously consider the control needed to operate in a restricted resource environment since no restrictions are levied by the modelling scheme itself. The simulator may be implemented so that it accepts the definition of a run-time environment, in terms of memory bounds and the numbers and types of processors, and takes account of the overhead incurred by sharing within the defined, resource-impoverished, run-time environment.

There are two major problems with a simulation-based approach to feedback analysis of system behavior during design. The most serious is the lack of data upon which to base the timing estimates and, to some extent, the probability estimates. Because the analysis is being done during design, the designers must rely upon their intuition, experience and knowledge to develop reasonable estimates. But, even good estimates of mean values will have large variances and these will cascade to give derived statistics with large variances.

This problem is somewhat alleviated by the fact that as the design progresses, the designers will be able to get better estimates as the design moves to the more detailed level, closer to primitive operations about which the designers have better intuition. The designers may check that these better estimates are consistent with those used previously and revalidate previous designs if there is a significant inconsistency. This iterative approach to behavior assessment is consistent with typical approaches to software design.

The cascading of variances is sometimes tolerable when the designers are only interested in the derived statistics for the purposes of comparing design

alternatives. In these cases, the variances of estimates used in corresponding portions of the alternative structures will typically be roughly the same and it will
be the timing distribution means which differ significantly. Therefore, it may be
possible to relate differences in the derived statistics to differences in the means
of timing distributions and designers may therefore often make correct assessments
of the system's sensitivity to the operation of the individual components and thus
correctly assess the differences among alternative system structures.

The effect of cascading variances can sometimes also be tolerated when absolute
judgments rather than relative comparisons are being made. This can occur, for example, when the system is a real-time one and its specifications indicate some constraints that must be observed. In this case, if the ranges of the derived statistics lie within the constraint, then the magnitude of the variances makes no difference and the designers may validly conclude that the system observes the real-time
constraints. When the ranges fall outside the constraints, it may be possible for
the designers to parametrically assess the relationship of the variances of the individual timing estimates to determine local constraints which lead to satisfaction
of the global constraints.

The second problem with simulation-based approaches to system behavior feedback
analysis is not too severe and may be avoided by careful design of the simulator -
this is the problem of simulating a multiple processor system. During simulation,
operations will generally be sequentialized when in the eventual system they may be
simultaneous. Problems stemming from this sequentialization may be avoided by having the simulator nondeterministically or pseudo-randomly choose among simultaneous
operations. This is effective only if it is permissible to interpret "simultaneous"
as "sequential but unordered" - this is most usually a valid interpretation in the
case of software systems.

Analytic Feedback Analysis of System Behavior

Simulation is an important part of the theory advanced by Zeigler and others.
DDN, however, is *not* a simulation language in the conventional sense for several
reasons. First, current simulation languages do not allow behavioral specification,
thus there can be no static analysis of behavior. That is, simulations produce behavior which is then examined, rather than allowing derivation of behavior from the
structural description, as is possible with DDN. Second, simulations are typically
used to investigate performance characteristics rather than yes/no questions like
system deadlock which whould be answerable from a closed form behavioral description
rather than from a (possibly infinite) number of simulation runs. Thirdly, simulation languages generally do not contain sophisticated synchronization constructs
allowing the direct, explicit representation of solutions to the synchronization

problems found in complex software systems, e.g., operating systems.

The analytic approaches supported by DDN are based upon the ability to derive, from the structural model, a description of behavior in the form of an algebraic expression over the set of event names for the system.[1] To give an example of this, we must augment the [blackboard] class description so that there is a description of that part of the rest of the system which serves to activate [blackboard] subsystems:

```
[driver]:  SUBSYSTEM CLASS;

    ask:  OUT PORT;
      BUFFER SUBCOMPONENTS;
        request OF [bb_request_type]
        END BUFFER SUBCOMPONENTS;
      END PORT;

    listen:  IN PORT;
      BUFFER SUBCOMPONENTS;
        disposition OF [bb_request_answer]
        END BUFFER SUBCOMPONENTS;
      END PORT;

    exerciser:  CONTROL PROCESS;
      MODEL;
        ITERATE
          request SET TO inspect OR modify;
          RECEIVE listen;
          SEND ask;
        END ITERATE;
      END MODEL;
      END CONTROL PROCESS;

    END SUBSYSTEM CLASS;
```

Processes of this class interact with the blackboard in a coroutine fashion, at each interaction requesting either an inspection or a modification operation.

We must also add the definition of some events to the control process within the blackboard definition:

```
          servicer:  CONTROL PROCESS;
            MODEL;
              ITERATE
   hear:      RECEIVE request;
              IF request_type = modify
                THEN value SET TO value1 OR value2 OR value3;
   activate:          SEND signal;
                      done_signal SET TO done;
                ELSE done_signal SET TO value;
                END IF;
   respond:   SEND answer;
              END ITERATE;
            END MODEL
            END CONTROL PROCESS:
```

---

[1] One version of this derivation procedure is given in [27]. It is a variant of the Brzozowski method [6] for deriving a regular expression for the language recognized by a finite-state automaton.

By labelling some of the statements, we have demarcated events by which the behavior
of the system may be described.

Suppose there are two processes of class [driver] and one of class [blackboard]
and they are connected so that the *ask* ports of the drivers are connected to the *request* port of the blackboard and the *listen* ports of the drivers are connected to the
answer port of the blackboard. Then we have one possible configuration for the system and may ask: what is the system's behavior in terms of the sequences of events
which arise from its operation? An answer may be algorithmically obtained once an
initial configuration of messages in the ports has been specified. In our example,
an appropriate initial configuration is that the *listen* port has a single message and
the other ports are empty. Using a notation called event expressions [27], the behavior of the system could be expressed as:

$$(\text{hear}(\text{activate} \cup \underline{\lambda})\text{respond})^*$$

This indicates that the *servicer* loops through the sequence of events: receive a request (hear); possibly send a message out through the *signal* port (activate); respond
to the request (respond). (This particular behavior description is easy to derive by
inspection, but that is because of the simplicity of the example.)

With this information, designers are able to draw inferences about whether the
system is appropriately designed or not. They might, for example, not have realized
that the blackboard is constructed so that it operates as a subroutine (with the undesirable effect that responses may be received by the wrong driver process) and wish
to redefine it as a subsystem with two components, one for each of the drivers. In
general, designers may use the derived information to confirm that the system will behave as they intend. Of course, the level of confidence that they can attain in this
way depends on the completeness of the design, and to the extent to which they have
defined events that capture the behavior they are interested in analyzing (i.e., the
suitability of the chosen experimental frame).

A second use for the behavior expression derivation technique is in the guidance
of the design process itself. In this case, the designers use the derived information
to guide future design decisions. A particularly important use along these lines is
the choice of synchronization mechanisms for the efficient coordination of parallel
processes. A large number of synchronization mechanisms have been developed ([5],
[11],[20]) and they are all essentially equivalent in power. The choice of a particular mechanism is frequently done on the basis of a designer's experience or the seem-
ing appropriateness of the mechanism. But a wrong choice can lead to a much higher
than necessary overhead in the system's operation. A totally appropriate choice can
be made only if the designers know how the mechanism is *actually* used as well as how
it is *intended* to be used.

In our example, the intended usage of the *signal* port would indicate that some bouned message transmission mechanism, such as communication semaphores [35], would be appropriate. But if event expression analysis were to uncover that this port was never used to send other than *value1* messages, then a more efficient solution would be to use a normal semaphore mechanism which merely counts the number of messages rather than actually storing them. This solution may later prove to be an incorrect one and the designers must be careful to periodically check that the solution is valid.[1] But at the point in design represented by our example, there is no reason to choose a more general, less efficient solution merely to rid oneself of the responsibility to carefully assess the effect of subsequent design decisions.

This is an example of a larger class of problems concerning the choice of appropriate strategies, policies and algorithms for controlling the interactions among a system's parallel components. By using DDN in conjunction with a top-down design method, designers will have a much clearer idea of how a facility will actually be used before they design the components which provide the facility. They will therefore be able to tune the implementation of a facility to its use coincident designing its basic organization and operation.

A final way in which the expression derivation technique may be utilized is in performance assessment. To give an example, we must first create a description of a closed system by adding the definition of the class of processes that are activated by the blackboard when an "observed" entity is assigned a new value:

```
[handler]:  SUBSYSTEM CLASS;

   ask:  OUT PORT;
      BUFFER SUBCOMPONENTS;
         request OF [bb_request_type]
      END BUFFER SUBCOMPONENTS;
   END PORT;

   listen:  IN PORT;
      BUFFER SUBCOMPONENTS;
         disposition OF [bb_request_answer]
      END BUFFER SUBCOMPONENTS;
   END PORT;

   start:  IN PORT;
      BUFFER SUBCOMPONENTS;
         next_value OF [possible_values]
      END BUFFER SUBCOMPONENTS;
   END PORT;
```

---

[1]Whenever a design is modified or elaborated, some check must be made to assure that the design description is internally consistent. The DREAM approach is to provide tools by which advisory information may be obtained and used in assessing the implications of a change in the design - these tools are not automatically applied when a change is made but are rather selectively applied by the designers.

```
handle_it:  CONTROL PROCESS;
   MODEL;
      ITERATE
         RECEIVE start;
         ITERATE 0 OR MORE {{5}} TIMES
            request SET TO //1// inspect;
            SEND ask;
            RECEIVE listen;
            END ITERATE;
         END ITERATE;
      END MODEL;
   END CONTROL PROCESS;

   END SUBSYSTEM CLASS;
```

In the nondeterministic ITERATE statement, the notation {{m}} indicates the expected
number of times that repetition will occur.

An event expression could then be derived which contains the timing and probability information [36]. We do not give this expression for two reasons. First, it
is fairly complex and uses some constructs which we do not want to have to explain
here. Second, and more important, in and of itself it does not explicitly give any information about the overall timing characteristics of the system - it is really just
another representation of the system that happens to account more directly for the
interactions among the parts of the system. In DREAM, the derived expression is not
displayed to the user but rather there are facilities for deriving summary statistics,
such as means and variances. Through the interactive use of these facilities the designer could determine, for instance, that if the objects of class [handler] operate
as described and if the timing estimates are an accurate reflection of the real system, then the operation of the object of class [driver] would be slowed down by about
20 percent, due to the interference introduced by the interactions of an instance of
class [handler] with a [blackboard]. That this is true is not immediately obvious
from inspection of the descriptions, but is rather easy to deduce from statistics about
the execution and waiting times of the processes.

A Look to the Future

We have argued throughout that there exists a need to view complex software systems from a modelling as well as traditional viewpoint, and have discussed one
rapprochement. There is some evidence that other research groups are beginning to use
this approach to solving software design problems (e.g., [2],[7],[14],[15]).

We believe that there are many benefits to both the modelling and the software
communities by collaboration. The modelling community will find that software is rich
in problems to be solved by formal means, and that software systems are readily available for study at all levels of system complexity. The software community could

particularly benefit from answers in the following areas:

What is a natural domain for expressing the semantics of software description languages?

Can we describe a hierarchy of software system models à la Zeigler [45], and morphisms between them?

How may we formalize notions of subsystem connectivity and strength?

What sorts of analyses of behavior can we perform?

It should also be apparent that the primitives of DDN (or similar languages) are well suited for the description of systems other than software. The concepts of rigorous specification of subsystems, inter-system communication mechanisms, and behavioral specification are particularly interesting in this regard.

Conclusion

General systems theory has traditionally worked on two fronts, system characterization and unification of models of systems. The DREAM system furnishes a testbed for the investigation of both of these issues with respect to complex software systems. There does not currently exist a formal theory of software system synthesis (or any other type of system for that matter[1]) nor a formal theory of software system organization, nor even a commonly accepted set of definitions or vocabulary for the discussion of such systems. It is hoped that DREAM is a rich enough system to provide the beginnings of answers to these questions (or at least a framework for the discussion of such issues), in addition to its intended practical value.

Bibliography

1. S. Alagic, and M. A. Arbib (1978), The Design of Well-Structured and Correct Programs, Springer-Verlag, New York.

2. A. A. Ambler et al. (1977), Gypsy: A language for Specification and Implementation of Verifiable Programs, Proc. of an ACM Conference on Language Design for Reliable Software, Software Engineering Notes, 2, 1-10.

3. F. T. Baker (1975), Structured Programming in a Production Programming Environment, IEEE Trans. on Software Engineering, SE-1, 241-252.

4. J. W. Baker, D. Chester and R. T. Yeh (1978), Software Development by Step-wise Evaluation and Refinement, SDBEG-2, Dept. of Comp. Sci., Univ. of Texas, Austin.

5. P. Brinch Hansen (1972), A Comparison of Two Synchronization Concepts. Acta Informatica, 1, 190-199.

6. J. A. Brzozowski (1964), Derivatives of Regular Expressions, J.A.C.M., 11, 481-494.

7. I. M. Campos and G. Estrin (1978), SARA Aided Design of Software for Concurrent Systems, Proc. Natn. Computer Conf., Anaheim, Calif., pp. 325-336.

---

[1]This has been recognized as one of the major problems that general systems theorists should attack [22].

8. J. Cuny (1977), A DREAM Model of the RC 4000 Multiprogramming System, RSSM/48, Dept. of Computer and Communication Sciences, Univ. of Michigan, Ann Arbor.

9. J. Cuny (1977), The GM Terminal System, RSSM/63, Dept. of Computer and Communication Sciences, Univ. of Michigan, Ann Arbor.

10. O. J. Dahl and K. Nygaard (1966), SIMULA--an Algol Based Simulation Language, Comm. ACM, 9, 671-678.

11. E. W. Dijkstra (1968), Cooperating Sequential Processes, In Programming Languages, Genuys (ed.), Academic Press, New York.

12. E. W. Dijkstra (1972), Notes on Structured Programming, in Structured Programming by O. J. Dahl, E. W. Dijkstra, and C. A. R. Hoare, Academic Press, New York.

13. E. W. Dijkstra (1976), A Discipline of Programming, Prentice-Hall, New Jersey.

14. G. Estrin and I. Campos (1978), Concurrent Software System Design, Supported by SARA at the Age of One, Proc. Third International Conference on Software Engineering, Atlanta, Georgia.

15. G. Estrin (1978), A Methodology for the Design of Digital Systems - Supported by SARA at the Age of One, Proc. Natn. Computer Conf., Anaheim, Calif. pp. 313-324.

16. R. W. Floyd (1977), Assigning Meanings to Programs, in Mathematical Aspects of Computer Science, J. T. Schwartz (ed.), 19, AMS, Providence, RI.

17. P. Freeman and A. Wasserman (eds.) (1977), Tutorial on Software Design Techniques, IEEE Computer Society, New Jersey.

18. P. Henderson (1975), Finite State Modelling in Program Development, Proc. 1975 International Conf. on Reliable Software, Los Angeles.

19. P. Henderson, et al. (1975), The TOPD System, Tech. Report 77, Computing Laboratory, Univ. of Newcastle upon Tyne, England.

20. C. A. R. Hoare (1974), Monitors: An Operating System Structuring Concept, Comm. ACM, 17, 549-557.

21. J. J. Horning and B. Randell (1973), Process Structuring, Computing Surveys, 5, 5-30.

22. G. J. Klir (1972), Trends in General Systems Theory, Wiley-Interscience, New York.

23. B. H. Liskov and S. N. Zilles (1975), Specification Techniques for Data Abstractions, IEEE Trans. on Software Engineering, SE-1, 7-19.

24. D. L. Parnas (1972), On the Criteria to be Used in Decomposing Systems into Modules, Comm. ACM, 15, 1053-1058.

25. J. L. Peterson and T. H. Bredt (1974), A Comparison of Models of Parallel Computation, Proc. IFIP Congress 74, Stockholm.

26. W. E. Riddle (1972), The Hierarchical Modelling of Operating System Structure and Behavior, Proc. ACM 72 National Conf., Boston.

27. W. E. Riddle (1976), An Approach to Software System Modelling, Behavior Specification and Analysis, RSSM/25, Dept. of Computer and Communication Sciences, University of Michigan, Ann Arbor. (To appear: J. of Computer Languages.)

28. W. E. Riddle, J. H. Sayler, A. R. Segal and J. C. Wileden (1977), An Introduction to the DREAM Software Design System, Software Engineering Notes, 2, 11-23.

29. W. E. Riddle (1977), Hierarchical Description of Software System Organization, RSSM/40, CU-CS-120-77, Dept. of Computer Science, Univ. of Colorado at Boulder.

30. W. E. Riddle (1977), Abstract Process Types. RSSM/42, Dept. of Computer Science, Univ. of Colorado at Boulder.

31. W. E. Riddle, J. H. Sayler, A. R. Segal, A. M. Stavely and J. C. Wileden (1978), A Description Scheme to Aid the Design of Collections of Concurrent Processes, Proc. Natn. Computer Conf., Anaheim, Calif., pp. 549-554.

32. W. E. Riddle, J. C. Wileden, J. H. Sayler, A. R. Segal and A. M. Stavely (1978), Behavior Modelling During Software Design, IEEE Trans. on Software Engineering, SE-4, 283-292.

33. W. E. Riddle (1978), DREAM Design Notation Example: T. H. E. Operating System, RSSM/50, Dept. of Computer Science, Univ. of Colorado at Boulder.

34. W. E. Riddle, J. H. Sayler, A. R. Segal, A. M. Stavely and J. C. Wileden (1978), DREAM--A Software Design Aid System, Proc. 3rd Jerusalem Conf. on Info. Tech., Jerusalem.

35. H. J. Saal and W. E. Riddle (1971), Communicating Semaphores, CS-71-202, Comp. Sci. Dept., Stanford Univ., Stanford, Calif.

36. J. W. Sanguinetti (1977), Performance Prediction in an Operating System Design Methodology, RSSM/32 (Thesis), Dept. of Computer and Comm. Sciences, Univ. of Michigan, Ann Arbor.

37. A. R. Segal (1977), DREAM Design Notation Example: A Multiprocessor Supervisor, RSSM/53, Dept. of Computer and Communication Sciences, Univ. of Michigan, Ann Arbor.

38. A. M. Stavely (1977), DREAM Design Notation Example: An Aircraft Engine Monitoring System, RSSM/49, Dept. of Computer and Communication Sciences, Univ. of Michigan, Ann Arbor.

39. B. Unger (1978), Programming Languages For Computer System Simulation. SIMULATIC

40. G. Weinberg (1971), The Psychology of Computer Programming, Van Nostrand Reinhold New York.

41. J. Wileden (1977), DREAM Design Notation Example: Scheduler for a Multiprocessor System, RSSM/51, Dept. of Computer and Communication Sciences, Univ. of Michigan, Ann Arbor.

42. J. C. Wileden (1978), Behavior Specification in a Software Design System, RSSM/43 Dept. of Computer and Info. Science, Univ. of Massachusetts, Amherst.

43. N. Wirth (1971), Program Development by Step-wise Refinement, Comm. ACM, 14, 221-227.

44. D. B. Wortman (ed.) (1977), Proceedings of an ACM Conference on Language Design For Reliable Software, SIGPLAN Notices, 12.

45. B. P. Zeigler (1976), Theory of Modelling and Simulation, Wiley-interscience, New York.

46. F. W. Zurcher and B. Randell (1968), Iterative Multi-level Modelling - A Methodology for Computer System Design. Proc. IFIP Congress 68, Edinburgh.

METHODOLOGY IN SYSTEMS MODELLING AND SIMULATION
B.P. Zeigler, M.S. Elzas, G.J. Klir, T.I. Ören (eds.)
© North-Holland Publishing Company, 1979

SOME RECENT DIRECTIONS OF RESEARCH
IN MODELLING REAL TIME COMPUTER SYSTEMS

H.G. MENDELBAUM - Département Informatique, Université PARIS V - IUT

143, Avenue de Versailles - Paris FRANCE.

ABSTRACT

In the field of real time process control, the activity of
modelling serves to describe:

The physical world to automatize

The control algorithm

The internal computer system,

To be clear, we use the words physical system when speaking of the
plant to control, and operating system for the programs which control the plant.
In this symposium many papers dealt with techniques and methods which can
also be used to model and describe the physical system to control. We shall
not discuss it here any longer. Let us admit that the physical system and the
control algorithms are known.

In this paper we shall rapidly overview the main methods for the
modelling and description of the controllers and the operating systems
which perform the control algorithms.

MODELLING CONTROL SYSTEMS :

The control of a plant can be seen in a symetrical view as the
interaction between a physical system and an operating system. This interaction
is realized through an exchange of information between the two entities, as
if there were a dialogue between the plant and the driver (see fig. 1).

The operating system receives events and measures from the plant
and the plant receives commands from the operating system. The latter contains
"the controller" which is the description of the control algorithm. It works as
a memory containing the actions to do when something happens in the plant
and how to react.

Fig. 1 :

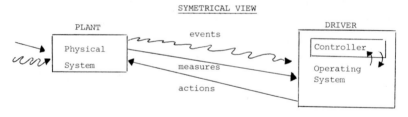

SYMETRICAL VIEW

Interactions between the two systems :

        Measures

        Events (Internal, External)

        Actions

Internal interactions between the controller and the operating system :

        Conditions (Flags)

           Status

The two systems can be considered as two interactive automata.

        Often in control computer systems, the controller and the operating system are not well distinguished, and the problem is to describe them clearly and simply.

GRAPHIC TOOLS :

        Some tools have been proposed for the modelling of such control systems. Roughly, we may classify them in two kinds depending on the types of problem and on the former education of the control engineer.

        A program can be

1) A description of a flow of actions influenced by data

2) A set of black boxes which manipulate the data

        1) Some designers see their applications as a flow of control actions. They can use Petri graphs (1) or Real-time graphs (2) which express the chaining, synchronizations and exclusions between modular actions and events. For example, in the figure 2, the R.T. graph expresses that action B has to be performed only after the end of action A and the occurrence of event i_T ; then action C can be run. When action C is finished, according to a boolean output value, action A may be reactivated, or the parallel actions D-F and E-G will be started.

Fig. 2

PETRI GRAPHS

(1)

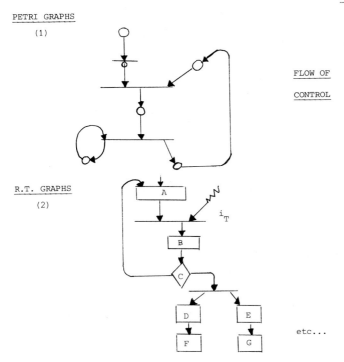

FLOW OF

CONTROL

R.T. GRAPHS

(2)

i_T

A

B

C

D   E   etc...

F   G

        2) Some other designers conceive their application as a flow of
Data. For instance they may use phase graphs (3) which express the various
status (STA, STB etc... on fig 3) of the applications and the conditions
(events, data) which provoke status changes. Estrin (4) and Jackson (5)
use special graphs to express the data interconnections between action modules
(for ex. between actions A, B and C in fig. 3).

Fig. 3

PHASE GRAPHS
    (3)

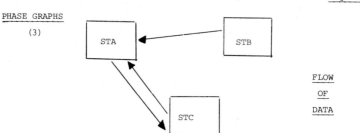

FLOW
OF
DATA

ESTRIN GRAPHS
    (4)

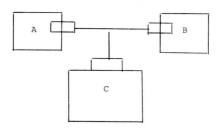

JACKSON GRAPHS
    (5)

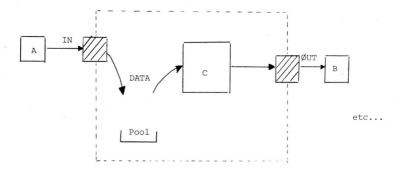

etc...

SYSTEM MODEL OF COOPERATING TASKS :

Control systems can be modelled as a set of task programs manipulating the same data separately. These tasks must cooperate in order to avoid destruction of the common data.

Tools have been designed to allow this cooperation : semaphores (6), Monitors (7) etc... These tools are used in each task. When the controller is working, it is as if each task had its own independent life, at some point some tasks have to synchronize with others (in order to manipulate some common data) and will use cooperating tools.

This kind of modelling is the most wide spread used over the world:

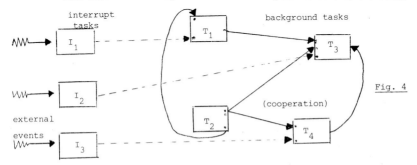

Fig. 4

This model is mainly due to the structure of present computers : based on the two fundamental concepts of sequentiality of execution and of interruptability of programs on a single machine.

A task is executed as a sequential program until a cooperation is necessary, then another task can be started and the first one will be continued later. External events are attached to special interrupt tasks.

The control Algorithm has to be forced in such a model in order to build the corresponding controller. The designer has to split his application in a set of tasks and create the control cooperation between them.

SYSTEM MODEL OF GLOBALLY CONTROLLED SEQUENTIAL MODULES :

A new approach consists in describing directly the control algorithm as a flow of actions. This can be made by separating the description of the flow of control from the descriptions of the sequential modular actions which manipulate the data.

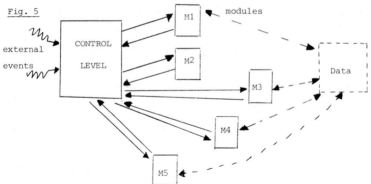

Fig. 5

When the controller is working, the control level is the master
which starts modules depending on the occurred events and the synchronization
conditions. Each time a synchronization decision has to be performed the
sequential modules give back the execution to the control level.

The first advantage of this kind of model is that the controller
is closer to the definition of the problem than to the structure of the computer.
A second advantage is that all the description of the control is gathered in a
unique program (the control level) so that it is easier to build and to verify.

The application is safer and simpler, since we have structured on
one side the synchronization and chaining rules and on another side the
sequential actions on data. On the opposite in the cooperating tasks model,
synchronizations and treatments are mixed. A workshop on this new kind of
methods was held in Paris, last year (8). Three types of approach of this method
could be seen :

1 - TASK-CONTROL DESCRIPTION LANGUAGES :

Some searchers try to find the canonical structures for the
description of the control algorithm in terms of a programming language.

At running time this program will be performed and will control
the flow and the synchronizations of the sequential modules (as procedures).

In this approach simulation languages could be used for the
description of flow of actions and events. More specific descriptions have
been proposed : GAELIC (9), SUPER-PROCOL (10), PATH EXPRESSIONS (11)...

example of canonical structures of a control flow description (9) :

Alphanumerical Algorithms

    SEQUENCE

    IF.. THEN.. ELSE..

    WHILE ... DO

Real-Time Control

    IF EVT ...

    WHEN EVT ...

    M(P...)   ASOON AS EVT ...

    // M1 (P1...) ; .... SYNCHRO L

    // M2 (P2...) ; .... SYNCHRO L

    L : SEQUENCE ...

    EXCL i SEQUENCE .... ENDEXCLi

    EACH Δt TIME   ....

    AFTER Δt TIME   ....

    EXACT t TIME   ....

    BEFOREΔt TIME   ....

These structures are not definitive, and research is going on in this field.

2 - EXTENDED MONITORS FOR TASK CONTROL :

The controller level can be made of a program of predefined structure like a monitor for example. Each time a task has to make a synchronization, it will call special functions in a control monitor instead of activating directly independent cooperating tools (semaphores etc...).

All the synchronization actions will be gathered as functions in a unique program (the control monitor) and access to these functions will be filtered through a set of condition rules.

Fig. 6 :

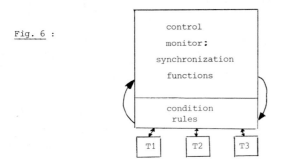

VERJUS (12) has proposed conditions rules of the form :

CONDITION (T2) :   # auth T1 - # Term T3 = O .

VAUDENE (13) uses path expressions access rights (11), and RIDDLE (14)   uses
TOPD rules (15).

3 - AUTOMATA-LIKE CONTROLLERS :

        The control algorithm can also be described directly in an automata
form, and at running time such automata will be directly interpreted by an
operating system.

        The simple structure of such a system is the following :

Fig. 7 :

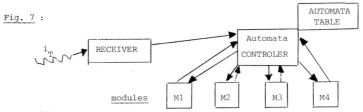

MENDELBAUM et al. has proposed such a structure (2) and HENDERSON has also
presented an automata approach in TOP-D (15).

        For example, in the ABOS system (2) the control programmer will
translate his control algorithm in the following form of direct access rules :

$$e1, \; Q_2, C_3 \;\rightarrow\; M_4 \;(P...), \; Q_3, \; \overline{C}_3$$

$$e_1, \; Q_1, \overline{C}_1 \;\rightarrow\; M_1 \;(\;...), \; Q_4, \; \overline{C}_2$$

$$e_0, \; Q_4, \overline{C}_2 \;\rightarrow\; M_3 \;(\;...), \; Q_1, \; C_2$$

which mean that when an event e arrives in state Q, with boolean conditions C
then the system must activate the module M (parameters ...) and set the new

state Q and new conditions C.

The Petri nets (1), the R.T. graphs (2) and phase graphs (3) quoted above can be directly (and even automatically) translated in such an automata form.

At running time, each time an event occurs, the control is given to the automata nucleus which reaches directly to the corresponding rules and activates the following module. These events can be emitted from plant (external signals) and/or from finishing modules. Boolean conditions can serve to record past events, or to flag mutual exclusion or synchronizations etc...

The correctness of the control can be easily checked either "a priori" or at run-time.

CONCLUSION :

In this paper we have overviewed rapidly the two main ways of modelling process control computer systems : The "cooperating tasks" and the model of "globally controlled sequential modules". This latter new approach is now being developped in several laboratories in different countries. It is only at the beginning and it seems a promising approach.

APPENDIX

EXAMPLE OF A CONTROL SYSTEM DESIGN

1) STUDY OF PHYSICAL SYSTEM

2)   MCDEL

3) STUDY OF A CONTROL THEORY OF THIS MODEL

4)   CONTROL ALGORITHM

396   H.G. MENDELBAUM

DESCRIPTION OF THIS CONTROL :

5)

(R.T. Graphs)

6) TRANSLATION IN THE CONTROLLER LANGUAGE

  for instance (GAELIC) :

  DO A ASOONAS i_t THEN B WHILE M ENDDO

  // D ; F SYNCHRO S

  // E ; G SYNCHRO S

  S : H   ASOONAS i_k THEN E ENDAS

  WHEN i_z THEN L ; A SYNCHRO S ENDWH

7) TRANSLATION IN

NON SEQ-MACHINE LANGUAGE

for instance (ABOS)

e_o, o $\longrightarrow$ A,O i_z, o $\longrightarrow$ L,o

e_A, o, c_1 $\longrightarrow$ wait, o, $\overline{c}_1$ e_1, o $\longrightarrow$ A, 1

e_A, o, c_1 $\longrightarrow$ B, o, c_1

i_t, o, c_1 $\longrightarrow$ wait, o, $\overline{c}_1$ e_A, 1, $c_2 = 3 \longrightarrow$ H, o, $c_2 := o$

i_t, o, c_1 $\longrightarrow$ B, o, c_1 "   "

e_B, o $\longrightarrow$ M, o   "   "

$\downarrow$

8) EXECUTION UNDER

A CONTROLLER MODELLED SYSTEM

REFERENCES :

(1) M. HACK "Petri Nets languages" (MIT, TR 159, March 1976)

(2) H.G MENDELBAUM, F. MADAULE "Automata as structured tools for real time
systems" (IFAC/IFIP workshop on real-time programming, P.D. GRIEM ed,
ISA-Publ, Boston, 1975)

(3) MM. PRUNET, DUMAS "Phase graphs" (ibid)

(4) G. ESTRIN "The SARA system" (symposium on Modelling and simulation
methodology, B.P ZEIGLER Ed, Weizman Institute, Rehovot, 1978)

(5) M. JACKSON "Moral system" (IFAC/IFIP workshop on real time, N. MALAGARDIS
Ed, IRIA, Paris 1976)

(6) EW. DIJKSTRA "The structure of THE multiprogramming system" (CACM 11 (5),
341, 1968)

(7) CA. HOARE "Monitors, an operating system structuring concept" (CACM,
Octobre 1974)

(8) AFCET-IUT Workshop on "Global description methods for synchronizations in
real-time applications" (H.G.MENDELBAUM, G. VERROUST Eds, AFCET, Paris 1977)

(9) F. LE CALVEZ, H.G MENDELBAUM, F. M ADAULE "Compiling GAELIC : a global R-T
Language" (IFAC/IFIP workshop on R-T programming, C.H. SMEDEMA Eds, North
Holland Publ, Eindhoven 1977).

(10) P. DESCHIZEAUX, P. LADET "Programmation en temps réèl des synchronizations
par gestion des liens évènements processus" (AFCET-IUT workshop on global
descriptions ..., H.G MENDELBAUM Ed, Paris, 1977).

(11) P.E. LAUER, R.H. CAMPBELL "Formal semantics of a class of high level primitives (Acta Informatica, 1975).

(12) P. ROBERT, J.P. VERJUS "Toward autonomous description of synchronization
modules" (IFIP Congress, Montreal 1977)

(13) D. VAUDENE, J. VIGNAT "Enoncés de synchronization et moniteurs: deux
entités regroupées"
(AFCET-IUT Workshop on global descriptions ..., Paris 1977)

(14) W.E. RIDDLE, J.H. SAYLER "Modelling and simulation in the design of complex
software systems" (this volume).

(15) P. HENDERSON "Finite state modelling" (Intern. Conf. on Reliable
software, Los Angeles, 1975).

METHODOLOGY IN SYSTEMS MODELLING AND SIMULATION
B.P. Zeigler, M.S. Elzas, G.J. Klir, T.I. Ören (eds.)
© North-Holland Publishing Company, 1979

ENGINEERING MODELLING AND DESIGN SUBJECT TO MODEL
UNCERTAINTIES AND MANUFACTURING TOLERANCES

John W. Bandler
Group on Simulation, Optimization and Control
Faculty of Engineering
McMaster University, Hamilton, Canada L8S 4L7

This paper deals with engineering design problems in which,
for example, either a large volume of production is envisaged
or in which only a few units are to be custom made. Designs
are considered subject to manufacturing tolerances, material
uncertainties, environmental uncertainties and model
uncertainties. The reduction of cost by increasing
tolerances, the determination and optimization of production
yield, the problem of design centering and various aspects of
tuning are discussed. Nonlinear programming approaches are
considered. The important problem of searching for
candidates for worst case solutions is briefly mentioned.
Simplicial approximation, quadratic modelling, linear cuts
and space regionalization are reviewed. A fairly extensive
bibliography to relevant work in the modelling and design of
electrical circuits is provided.

INTRODUCTION

This paper deals principally with those engineering design problems in which
either a large volume of production is envisaged, e.g., integrated circuits, or
in which only a few units are to be custom made, e.g., filters for satellite
applications. In the latter case, reliability and high performance are
essential, whereas in the former, low production cost at the expense of
performance is more typical. All designs are considered to be subjected to
manufacturing tolerances (e.g., on physical dimensions of components),
uncertainties on the materials used in fabrication (e.g., on dielectric constants
of insulators), uncertainties on the environment in which the product is to
operate (e.g., on temperature), uncertainties in the computational models used in
the design process (e.g., on equivalent circuits purporting to represent the
actual circuits) and so on.   Attention is directed towards the relevant
modelling, the design and the manufacture of electrical circuits such as filters,
amplifiers and switching circuits. Hence, certain undesirable effects which
deteriorate performance over and above those already indicated may be due to
electromagnetic coupling between (usually adjacent or closely located) components
and to terminations which, inevitably, are not ideally matched to the circuit
under consideration. The effects are particularly acute at high frequencies of
operation.

The main objective is to discuss some approaches to the reduction of cost as
effected by increasing or optimally assigning tolerances, maximizing production
yield with respect to an assumed probability distribution function, utilizing any
available design margins and the interrelated optimization problem called design
centering which, as the term implies, involves the optimal location of a set of

This work was supported by the Natural Sciences and Engineering Research Council
of Canada under Grant A7239.

nominal design parameter values. Post-production tuning is also highly relevant. Such tuning is customary in engineering production as a means of improving or repairing actual outcomes in an attempt to improve performance. Production yield in the present context may be simply defined by the ratio of number of outcomes which satisfy the specifications to the total number of outcomes. A potential yield may be similarly defined by replacing the number of outcomes by the number of outcomes which satisfy the specifications after tuning if necessary [13].

An extensive bibliography of relevant material is appended [1-48].

REVIEW OF COMPUTER-AIDED DESIGN

In order of general complexity, Fig. 1 attempts to highlight typical problem formulations in modern computer-aided engineering design, in particular using the nomenclature appropriate to the optimal design of electrical circuits and systems [1,10]. Fig. 2 shows some typical design situations. The concept of upper and lower specifications or desired bounds on a response function of an independent variable ψ, e.g., frequency or time, implies a constraint region in the k-dimensional space of designable variables ϕ, viz.,

$$\phi \triangleq \begin{bmatrix} \phi_1 \\ \phi_2 \\ \cdot \\ \cdot \\ \cdot \\ \phi_k \end{bmatrix} . \tag{1}$$

This concept is easily generalized to response functions of a number of independent variables similarly assembled in the vector ψ. Error functions involved in a minimax or Chebyshev approximation problem expressed along a sampled ψ axis can be represented in terms of ϕ by contour diagrams of the maximum, with its distinctive discontinuous derivatives. A family of possible responses, involving, e.g., a whole production line of circuits with independent parameters lying within a tolerance region of a nominal design is shown also. Since all the points in the tolerance region are also in the constraint region the envelope of responses lies within the upper and lower specifications.

The ideas of Figs. 1 and 2 may be amplified as follows. Nominal design: here we seek a single point in the space of designable variables which best meets a given set of performance specifications and design constraints. A suitable measure of deviation such as least squares, least pth or minimax [27] is typically chosen as the objective function to be minimized. Sensitivity minimization: here it is recognized that the solution point is subject to fluctuation or change due to any phenomenon associated with design parameters. An overall measure of this sensitivity, usually involving first-order sensitivities with respect to the parameters, is often included in the objective function [45]. The aim is, of course, to trade off some performance by gaining greater insensitivity of the possible outcome.

When undesirable effects such as model uncertainties and manufacturing tolerances are considered explicitly, two important classes of problems emerge: statistical design and worst-case design. In statistical design it is recognized a priori that a production yield of less than 100% is likely and there are, consequently, two principal aims. We attempt to minimize the cost or, alternatively, maximize the yield. Worst-case design requires that all units meet the performance specifications under all circumstances [7]. Typically, we either attempt to center the design with fixed assumed absolute or relative tolerances (analogous to maximizing yield) or we attempt to optimally assign tolerances to reduce production cost. What distinguishes all these problems from nominal designs or sensitivity minimization is the fact that a single point is no longer of

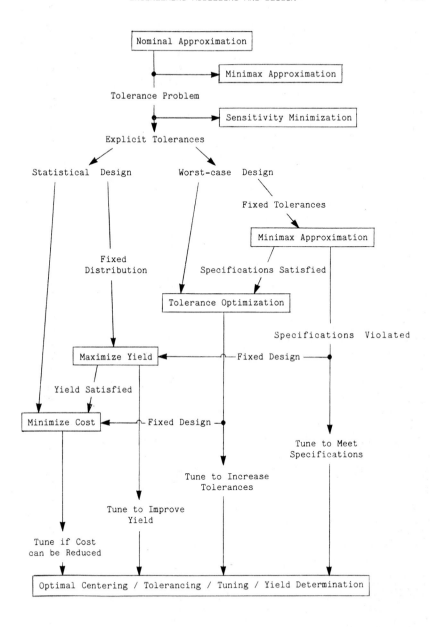

Fig. 1 A typical sequence of computer-aided design problems. As one proceeds
down the diagram the problems tend to increase in complexity [1,10].

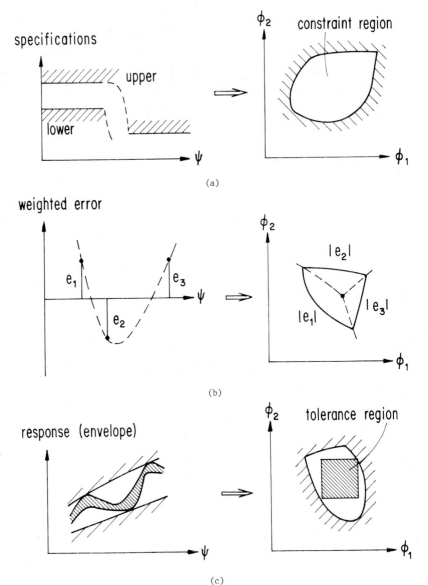

Fig. 2 Some typical design situations.   In (a) we have upper and lower
        specifications on the response function of one independent variable
        implying a constraint region with respect to the designable parameters ϕ.
        In (b) is shown a weighted error function sampled at three points giving
        rise to contours of $\max_{1 \leq i \leq 3} |e_i|$.   In (c) we have a toleranced design
        satisfying performance specifications.

interest: a (tolerance) region of possible outcomes is to be optimally located with respect to the feasible (acceptable, constraint) region.

We can deal with fixed designs or tunable designs. A fixed design or model implies that no subsequent tuning or adjustment is available to correct a posteriori for unacceptable performance. Thus, if a worst-case design is sought, then every unit has to be individually tested and each violating outcome discarded. On the other hand, a tunable design implies that at least one variable can be adjusted after such testing in an effort to meet the specifications [19]. There is no guarantee that the specifications will be met, of course, even after tuning unless it was properly accommodated in the original design.

The presence of tuning tends to increase cost, not only in making the tunable component available but possibly in having to pay a skilled person to carry out the tuning [19,41]. Repairing seems closely related to tuning in this respect. Tuning (repairing) may often be a design feature to permit the customer to alter his unit to meet different specifications. This leads to the concept of tunable constraint regions [11,19], which can be handled readily mathematically albeit computationally at greater expense.

Fig. 3 is a representation, in the space of designable parameters, of the concepts of the manufacturing tolerance over which it is presumed there is no control (by definition), the manufacturer's tuning variable, which is usually designed to be inaccessible to the customer and the tuning variable designed specifically for the customer which is exteriorized and often embodied in a large knob with attendant scale. It may be remarked that Fig. 3 applies equally in a descriptive sense whether the tolerance and tuning effects are all associated with one physical variable or whether they relate to different physical variables.

Setting any or all of the aforementioned problems up as nonlinear programs poses numerous difficulties [7,19]. There is the problem of identifying a suitable objective (cost) function to be minimized. Very little is known, in general, about production cost as a function of the variables entering into a design. This observation probably applies as much to most manufacturers as to outsiders. Hence, highly simplified points of view are usually taken in an attempt to render the problem tractable: functions which force the expansion of a tolerance orthotope within the constraint region accompanied by its optimal location, the optimal location of an expandable sphere within the region and so on. The matter is complicated by generally unknown correlations between variables, empirical assumptions about models and model parameter uncertainties and unreliable or unknown distributions of outcomes of component values between tolerance extremes. The number of constraints and even variables that could be chosen for an otherwise deceptively simple design problem is virtually unlimited.

One of the most challenging problems, therefore, is the development of general rules or methodology at a high level, procedures or algorithms at a lower level, for rapidly determining candidates for active constraints [47,48]. Various obvious mathematical assumptions have been proposed to simplify the selection problem: linearity of the constraints with respect to the variables [41], convexity of the constraint region [29,30], one-dimensional convexity of the constraint region [7,17,25].

The most frequently made assumption (virtually axiomatic in the light of the current state of the art) is that extremes of performance correspond to extremes in designable variables. Hence, most approaches to optimal design subject to tolerances concentrate on extreme points of parameter ranges.

The direct use of the Monte Carlo method of tolerance analysis within the optimization loop is extremely expensive [32]. Many approaches have therefore been suggested, either to avoid repeated use of the Monte Carlo method for

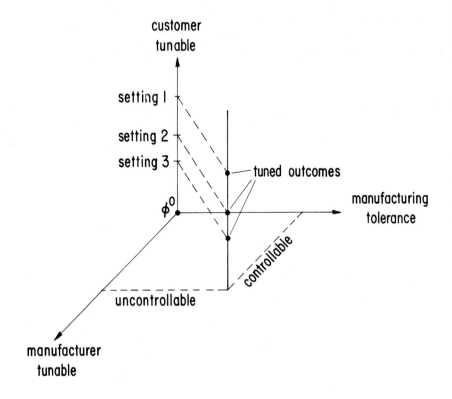

Fig. 3 A representation of the concepts of tolerance and independent tuning both by manufacturer and customer, with respect to a nominal design ϕ^0.

estimating yield, or to employ multidimensional approximations of the design constraints [1-4,9-14,29-31]. However, the need to optimize yield in the context of a huge production has sometimes dictated the computational impetuosity of uniting a Monte Carlo analysis with a general purpose simulator. That such extemporaneousness is extant only serves to underline the importance of the anticipated results.

SOME OPTIMIZATION APPROACHES

Bounding the Constraint Region R_c

The constraint region R_c is the set of all points ϕ for which all performance specifications and design constraints are satisfied. Thus

$$R_c \overset{\Delta}{=} \{ \phi \mid g(\phi) \geq 0 \}, \tag{2}$$

where g is the vector of constraint functions. Upper and lower bounds on each
parameter ϕ_i (the ith component) for which $\phi \in R_c$ provide useful design
information [28]. In a statistical analysis, for example, constraints can be
stacked in order of increasing computational effort, suitable upper and lower
bounds appearing at the top of the stack. If any constraint is violated further
testing of the candidate solutions becomes unnecessary.

Fig. 4 illustrates the bounding of R_c. In general, 2k optimizations, where k is
the dimensionality of ϕ, are required. In practice, however, since common
solutions are likely to exist [8], significantly fewer optimizations would be
expected. Upper bounds on tolerances for fixed designs can, following such an
analysis, be immediately calculated. Similarly, upper bounds on yield can be
calculated for fixed designs.

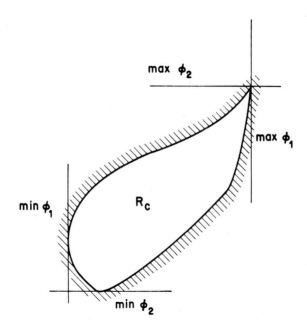

Fig. 4 Bounds on the constraint region obtained by independently maximizing or
minimizing the parameters subject to the constraints [8].

Minimax and Least pth Objectives M and U

Often, in engineering design, we define a function of the form

$$M(\phi) \overset{\Delta}{=} -\min_{i \in I_c} g_i(\phi), \tag{3}$$

where I_c is an index set, each element of which identifies a particular
constraint defining R_c. More generally, the so-called generalized least pth
objective [15,27] can be defined, for example, as

$$U(\phi) \triangleq \begin{cases} 0 \quad \text{if } M(\phi) = 0, \\\\ M(\phi) \left[\displaystyle\sum_{i \in J(\phi)} \left[\dfrac{-g_i(\phi)}{M(\phi)} \right]^q \right]^{1/q} \quad \text{if } M(\phi) \neq 0, \end{cases} \tag{4}$$

where

$$q = p \, \text{sgn} \, M(\phi) \; , \; p \geq 1 \tag{5}$$

and

$$J(\phi) \triangleq \begin{cases} \{i \mid i \in I_c, \, g_i(\phi) < 0\} \text{ for } M > 0, \\\\ I_c \text{ for } M < 0. \end{cases} \tag{6}$$

The important feature of $U(\phi)$ is that it coincides with $M(\phi)$ when $M = 0$, it shares the same sign as $M(\phi)$, yet under mild restrictions on p it is differentiable everywhere except when $M = 0$. The role of the M in the definition of U in (4) is twofold. Firstly, it scales the functions automatically ameliorating ill-conditioning attributable to the exponent q. Secondly, it facilitates the matching of the two otherwise discontinuous least pth objectives, namely, the objective for $M > 0$ and the one for $M < 0$. A discussion of the properties of this function is available in the literature [15,27].

The problem

$$\underset{\phi}{\text{minimize}} \; U(\phi)$$

where $U(\phi)$ is given by (4)-(6), is, consequently, not only a feasibility check but is also a centering process if R_c is not empty, since ϕ then tends to move away from the constraint boundary to the interior of R_c. Tolerances are not explicitly optimized by this formulation. However, this kind of optimization is virtually mandatory prior to introducing explicit tolerances. Active or near active constraints can then, for example, usually be identified.

Interior Approximation R_I and Exterior Approximation R_E

In accordance with the foregoing concepts we can now let

$$R_c \triangleq \{\phi \mid U(\phi) \leq 0\}, \tag{7}$$

where $U(\phi)$ is given by (4)-(6). Under assumptions of convexity of R_c or one-dimensional convexity [7] we can eliminate the corresponding regions indicated in Fig. 5. This allows further refinement, for example, of finding an upper bound to production yield over and above that outlined earlier.

Interior or exterior approximations to R_c can be conceived [8] as illustrated by Fig. 6. A best exterior approximation may be found by deflation of a suitable

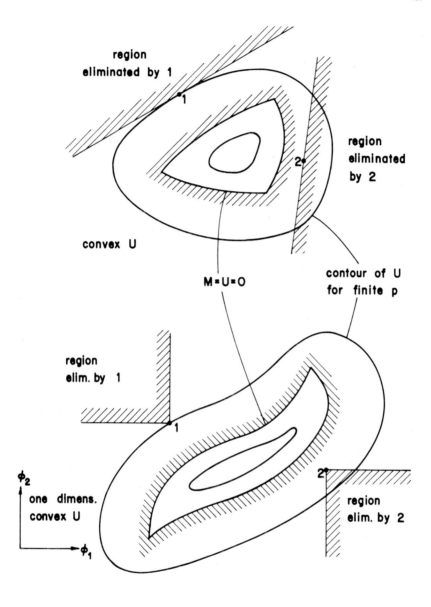

region
eliminated by 1

convex U

region
eliminated
by 2

M=U=0

contour of U
for finite p

region
elim. by 1

ϕ_2

one dimens.
convex U

ϕ_1

region
elim. by 2

Fig. 5 Elimination of regions during an analysis process exploiting assumptions
        of (a) convexity and (b) one-dimensional convexity [8]. The condition U
        > 0 implies specification violated, U < 0 specification satisfied.

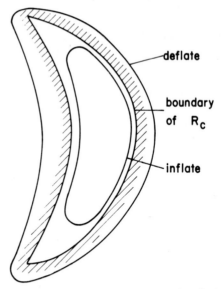

Fig. 6   Interior approximation R_I and exterior approximation R_E to the constraint
region R_c. Optimal approximations are obtained by inflating the interior
approximation and deflating the exterior approximation [8].

region R_E and a best interior approximation by inflation of a suitable region R_I,
while retaining

$$R_I \subset R_c \subset R_E. \tag{8}$$

Under these circumstances, the calculation of the original functions describing
R_c would only be necessary for $\phi \in R_E - R_I$, which is a region of uncertainty as
to the location of the boundary.

While the exterior approximation may be exploited computationally to provide an
upper bound, the interior approximation obviously leads to a lower bound.

The Tolerance Region R_ϵ

The interior or exterior approximation could be used in design centering
procedures. Consider the tolerance region [7]

$$R_\epsilon \overset{\Delta}{=} \{\phi \mid \phi^0 - \epsilon \leq \phi \leq \phi^0 + \epsilon\}, \tag{9}$$

where ϕ^0 is called the nominal point and $\epsilon \geq 0$ is a vector of associated
tolerances. Thus,

$R_c \subset R_\epsilon \Longleftrightarrow R_\epsilon$ is an exterior approximation,

$R_\epsilon \subset R_c \Longleftrightarrow R_\epsilon$ is an interior approximation.

A serious problem, in general, in the above expressions is the implication that
all points $\phi \in R_\epsilon$ for the interior approximation and all $\phi \notin R_\epsilon$ for the exterior
approximation must be accounted for, i.e., we have to deal with an infinite
number of constraints.

The Outcome Region R_μ

Consider, for example, the region

$$R_\mu \overset{\Delta}{=} \{ \underset{\sim}{\mu} \mid -1 \leq \mu_i \leq 1 \}, \tag{10}$$

where $\underset{\sim}{\mu}$ is a k-vector. In this case, we could define an alternative expression for the tolerance region of (9), namely, the orthotope

$$R_\varepsilon = \{ \underset{\sim}{\phi} \mid \underset{\sim}{\phi} = \underset{\sim}{\phi}^0 + \underset{\sim}{E}\, \underset{\sim}{\mu},\ \underset{\sim}{\mu} \in R_\mu \}, \tag{11}$$

where

$$\underset{\sim}{E} \overset{\Delta}{=} \begin{bmatrix} \varepsilon_1 & & & & \\ & \varepsilon_2 & & & \\ & & \cdot & & \\ & & & \cdot & \\ & & & & \cdot \\ & & & & & \varepsilon_k \end{bmatrix}. \tag{12}$$

Suppose also that a discrete set of $\underset{\sim}{\mu}$ is available, say R_{μ_V}, and let

$$R_V \overset{\Delta}{=} \{ \underset{\sim}{\phi} \mid \underset{\sim}{\phi} = \underset{\sim}{\phi}^0 + \underset{\sim}{E}\, \underset{\sim}{\mu},\ \underset{\sim}{\mu} \in R_{\mu_V} \}. \tag{13}$$

In practice, we are forced to consider a discrete set of $\underset{\sim}{\mu}$ out of computational necessity. What would distinguish one algorithm from another is the strategy for discarding elements of the set and/or adding new elements.

Candidates for Worst Case

If

$$R_V \subset R_c \implies R_\varepsilon \subset R_c \tag{14}$$

then R_V or R_{μ_V} are said to provide candidates for a worst-case design. In practice [18-20], it is usual to consider vertices of R_ε or R_μ as candidates so that

$$R_{\mu_V} = \{ \underset{\sim}{\mu} \mid \mu_i \in \{-1,\ 1\} \}. \tag{15}$$

This corresponds to intuition, which suggests that extremes of performance correspond to extremes of designable variables. Mathematics, fortunately, was invented to harness intuition and avoid its pitfalls, consequently the assumption that R_{μ_V} yields candidates for worst case requires justification in particular problems.

Optimization with Fixed Tolerances

The extension of the generalized least pth objective to handle all $\underset{\sim}{\mu} \in R_{\mu_V}$ and the minimization of the resulting function $U(\phi^0)$ with respect to ϕ^0 for fixed absolute or relative tolerances is rather simple. Difficulties occur essentially in housekeeping of arrays. Of course, depending on the sizes of the tolerances and the assumption (14) the solution obtained is not necessarily an acceptable worst case [21]. It may be that violations still occur at vertices and in the neighborhood of violating vertices.

Optimizing the Tolerance Orthotope

If we have a cost function $C(\phi^0, \varepsilon)$ with the properties [7]

$$C(\phi^0, \varepsilon) \to c \text{ as } \varepsilon \to \infty$$

$$\hspace{6cm} (16)$$

$$C(\phi^0, \varepsilon) \to \infty \text{ for any } \varepsilon_i \to \infty$$

then the minimization of $C(\phi^0, \varepsilon)$ with respect to ϕ^0 and ε subject to $R_\varepsilon \subset R_c$ can be used to optimize (maximize) the tolerances. Again, candidates for worst case R_v are considered in practice to reduce computational effort.

Typical objective functions which fit the requirements of (16) and which have some physical justification [7,18,19,35,40,44,45] can take the form

$$C = \sum_{i=1}^{k} \frac{c_i}{\varepsilon_i} \hspace{4cm} (17)$$

or

$$C = \sum_{i=1}^{k} c_i \frac{\phi_i^0}{\varepsilon_i}, \hspace{4cm} (18)$$

where c_i are constants which reflect the cost or importance of the corresponding term.

Tolerancing and Tuning

Tuning is the post-production process which permits a manufacturer to correct for unavoidably large deviations from the specifications or the customer to optimize his unit under operating conditions. We can take into account a simple situation where one or more variables are tunable by considering the slack region

$$R_\rho \triangleq \{\rho \mid -1 \leq \rho_i \leq 1\}, \hspace{3cm} (19)$$

where ρ is a k-vector. A tuning region $R_t(\mu)$ may then be defined as

$$R_t(\mu) \triangleq \{\phi \mid \phi = \phi^0 + E \mu + T \rho, \rho \in R_\rho\}, \hspace{2cm} (20)$$

where

$$T \triangleq \begin{bmatrix} t_1 & & & \\ & t_2 & & \\ & & \ddots & \\ & & & t_k \end{bmatrix} \hspace{3cm} (21)$$

and a suitable cost function $C(\phi^0, \underset{\sim}{\epsilon}, \underset{\sim}{t})$ minimized. Such a cost function should probably have the properties

$$C(\phi^0, \underset{\sim}{\epsilon}, \underset{\sim}{t}) \to c \text{ as } \underset{\sim}{\epsilon} \to \underset{\sim}{\infty}$$

$$C(\phi^0, \underset{\sim}{\epsilon}, \underset{\sim}{t}) \to \infty \text{ for any } \epsilon_i \to \infty$$

$$C(\phi^0, \underset{\sim}{\epsilon}, \underset{\sim}{t}) \to C(\phi^0, \underset{\sim}{\epsilon}) \text{ as } \underset{\sim}{t} \to \underset{\sim}{0} \qquad (22)$$

$$C(\phi^0, \underset{\sim}{\epsilon}, \underset{\sim}{t}) \to \infty \text{ for any } t_i \to \infty.$$

See, for example, Bandler et al. [19]. The cost function might also involve μ [13], since the cost of tuning may well depend on a statistical distribution to outcomes.

The important concept to note in any worst-case optimization program, for example, that is set up to solve a centering, tolerancing and tuning problem is that for <u>all</u> selected sets of μ taken, for example, from R_μ of (15) there must exist <u>some</u> ρ (i.e., one independent ρ vector for each μ vector) which permits a corresponding point ϕ to be in R_c (i.e., the intersection of the $R_t(\mu)$ of (20) and R_c must not be empty).

Vertex Selection

An efficient vertex selection scheme in a tolerance assignment or centering problem would involve finding local or global solutions $\overset{\vee}{\mu}$ to

$$\min_{\underset{\sim}{\mu} \in R_\mu} g_i(\phi^0 + \underset{\sim}{E} \underset{\sim}{\mu}) .$$

It is easily shown that the components of $\overset{\vee}{\mu}$ satisfy [21,46,47]

$$\overset{\vee}{\mu}_j = - \text{ sgn } \frac{\partial g_i(\overset{\vee}{\underset{\sim}{\mu}})}{\partial \mu_j} \qquad (23)$$

for $\overset{\vee}{\mu} \in R_\mu$ of (15). Iterative approaches for solving the associated nonlinear system of equations

$$\underset{\sim}{\mu} = - \text{ sgn } \nabla_{\underset{\sim}{\mu}} g_i(\underset{\sim}{\mu}) \qquad (24)$$

for all $i \in I_c$, where

$$\nabla_{\underset{\sim}{\mu}} \overset{\Delta}{=} \begin{bmatrix} \partial/\partial\mu_1 \\ \partial/\partial\mu_2 \\ \cdot \\ \cdot \\ \cdot \\ \partial/\partial\mu_k \end{bmatrix} . \qquad (25)$$

is the partial derivative operator with respect to the $\underset{\sim}{\mu}$, have been suggested [21,47,48].

The Generalized Tolerance Problem

Tromp [47,48] has carried out extensive research which has permitted the generalization of the tolerance problem to accommodate physical tolerances, model uncertainties, external disturbing effects and dependently toleranced parameters in a completely unified manner. In essence, his approach begins by defining the k_{0i}-dimensional vector ϕ^{0i}, the k_i-dimensional vector ϕ^i and the $k_{\mu i}$-dimensional vector μ^i so that ϕ^i is a function of ϕ^{0i} and μ^i for all $i = 1, 2, \ldots n$, and ϕ^{0i} itself depends on all ϕ^{i-1} for $i = 2, 3, \ldots, n$.

Input parameters, for example, the physical parameters (dimensions, constants of materials used, etc.) available to the manufacturer might be identified as ϕ^1, whereas ϕ^n would be the output vector, for example, the sampled response of a system or the constraint vector g, introduced earlier. The quantities $\phi^2, \ldots,$ ϕ^{n-1} can be identified, for example, as appropriate intermediate or model parameters. The variables μ^i, $i = 1, 2, \ldots, n$, embody unavoidable, undesirable or unknown fluctuations generally. Hence, we may assemble the vectors

$$\phi^0 \triangleq \begin{bmatrix} \phi^{01} \\ \phi^{02} \\ \cdot \\ \cdot \\ \cdot \\ \phi^{0n} \end{bmatrix}, \quad \phi \triangleq \begin{bmatrix} \phi^1 \\ \phi^2 \\ \cdot \\ \cdot \\ \cdot \\ \phi^n \end{bmatrix}, \quad \mu \triangleq \begin{bmatrix} \mu^1 \\ \mu^2 \\ \cdot \\ \cdot \\ \cdot \\ \mu^n \end{bmatrix}. \tag{26}$$

The preceding discussions involving the μ variables and the μ-space generalize in an obvious manner. However, the tolerance region in the ϕ-space need no longer turn out to be an orthotope [47,48].

This kind of analysis permits more than strictly design or manufacturing problems to be simulated and optimized. Planning problems, anticipated system operation, aging, measurement errors and so on may be embodied into the original simulation, and the design solution accordingly optimized to reflect these phenomena. It is obvious that uncertainty can enter into the problem at any stage: at a high (conceptual) level or at a low (computational) level. Distributions of and correlations between parameters exacerbate the situation, but one may even conceive of building desirable correlations into a manufacturing process.

Centering via Large-change Sensitivities and Performance Contours

The design centering approach of Butler involves large-change sensitivities in conjunction with pairwise changes in parameter values with respect to chosen performance contours [26]. A scalar continuous function of design parameters which reflects the goodness of a design is chosen as a performance criterion. A nominal design which satisfies this performance criterion is assumed to exist. The concept of large-change sensitivities is that of finding changes in function values due to significant deviations in designable parameters. This concept is used to draw contours of the performance criterion changing parameters in a pairwise manner for each contour. The design center is obtained by inspection, i.e., by choosing a nominal value which is well centered for all contours. As an example of a performance criterion, we might use $-M(\phi)$ of (3). The method is illustrated in Fig. 7.

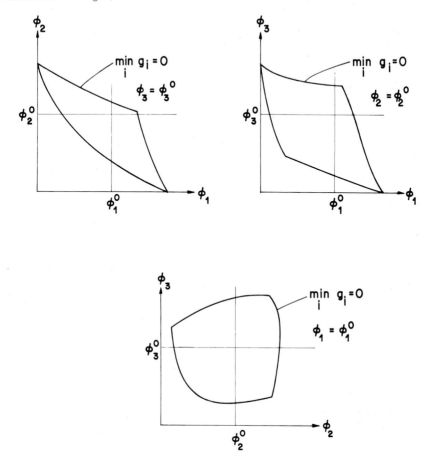

Fig. 7 Performance contours for pairwise changes in parameters [26]. Reducing ϕ_1^0 will result in a better centered nominal design [1].

Centering via Simplicial Approximation

The simplicial approximation approach of Director and Hachtel [29,30] involves linear programming and one-dimensional search techniques. Their approach is to inscribe (inflate) a hypersphere inside the constraint region. During the process of enlarging this hypersphere a polytope which approximates the boundary of the constraint region is constructed. Fig. 8 illustrates the procedure.

The algorithm initially searches for points on the constraint boundary in both positive and negative directions for each parameter from a feasible point (a point within the constraint region). The convex hull described by these boundary

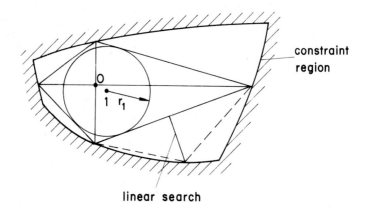

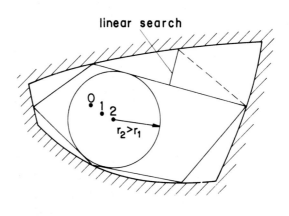

(b)

Fig. 8 An illustration of the simplicial approximation approach due to Director and Hachtel [29]. In (a) we show the results of an initial search for boundary points. In (b) is depicted the polytope approximating the boundary of the constraint region after two iterations [1].

points provides the initial polytope approximating the boundary of the constraint region. This polytope will be an interior approximation only if the constraint region is convex. Using linear programming a hypersphere is to be inflated inside this polytope in a k-dimensional space. The tangent hyperplanes are determined. These hyperplanes, faces of the polytope, are simplices in a space of k-1 dimensions. The largest simplex, i.e., the one which contains the largest hypersphere, is to be broken and replaced by k simplices. This is performed by adding a new vertex to the polytope obtained by searching for a boundary point along the normal direction to the largest simplex from the center of the corresponding hypersphere.

The inflation of an orthotope, as distinct from a sphere, describing the tolerance region is the essence of the work of Bandler et al. [1-21].

Quadratic Modelling of the Constraints

A nonlinear programming approach but employing approximations to the design constraints has been presented [1-5,8-14]. An interpolation region centered at the initial guess to the nominal design is chosen. The simulation program is used to provide the value of the response functions (constraints) at a certain set of base points. The base points are points within the interpolation region and defined in terms of values of the designable parameters. Based upon the corresponding values of the resulting responses, multidimensional quadratic polynomials are constructed. These quadratic polynomials have the general form

$$P(\underset{\sim}{\phi}) = a_0 + \underset{\sim}{a}^T(\underset{\sim}{\phi} - \overline{\underset{\sim}{\phi}}) + \frac{1}{2}(\underset{\sim}{\phi} - \overline{\underset{\sim}{\phi}})^T \underset{\sim}{H}(\underset{\sim}{\phi} - \overline{\underset{\sim}{\phi}}), \qquad (27)$$

where a_0 and $\underset{\sim}{a}$ are, respectively, a constant scalar and a constant vector, H is a constant symmetric Hessian matrix of the quadratic and $\overline{\phi}$ is the center of the chosen interpolation region.

The base points are simply those points where the approximated response function and the quadratic polynomial coincide. A system of simultaneous linear equations has to be solved to obtain the polynomial. The number of base points (exactly equal to the number of simulations required) is the minimum necessary to fully describe the responses and is given by

$$N = (k+1)(k+2)/2, \qquad (28)$$

where k is the number of designable parameters. The number N is the number of the unknown coefficients. An arrangement of the base points is depicted by Fig. 9.

Space Regionalization for Statistical Analysis

Space regionalization was suggested by Scott and Walker [43]. Based upon the probability of having an outcome to fall within a region, a weight is assigned to this region and the center of the region is checked against the nonlinear constraints to determine whether this whole weight will contribute to the yield or not. See Fig. 10. The number of required analyses, however, increases exponentially with the number of variables subject to statistical variations, since the response at the center of each region is to be evaluated.

Regionalization was also used by Leung and Spence [36-38] in conjunction with systematic exploration. The centers of the regions are scanned systematically by changing one parameter at a time and employing efficient matrix inverse modification methods for the (linear) circuit analyses required. Leung and Spence also suggested checking the worst outcome in each region, instead of the center of the region, if a lower bound on yield is required.

Yield Determination and Optimization

Elias applied the Monte Carlo method directly to the constraints [32]. Director and Hachtel suggested applying the Monte Carlo method in conjunction with their simplicial approximation [30]. Their polytope could be updated according to the points which fall within the constraint region but not in the polytope. Pinel and Singhal used importance sampling, concentrating the distribution of sample points at critical regions to reduce computational effort [42].

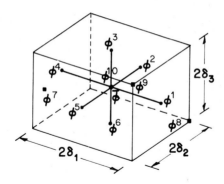

Fig. 9   Arrangement of base points for computing the quadratic interpolating polynomial in 3 dimensions [3].   In order to exploit sparsity in the associated system of linear equations ϕ^7, ϕ^8 and ϕ^9 should be, respectively, placed in the planes containing $\{\bar{\phi}, \phi^1, \phi^2\}$, $\{\bar{\phi}, \phi^1, \phi^3\}$ and $\{\bar{\phi}, \phi^2, \phi^3\}$, where $\bar{\phi}$ is the center of interpolation [3].

Bandler and Abdel-Malek [1-4,9,14] dealt with the mass of calculations involved in determining and optimizing yield as follows.   Multidimensional quadratic polynomials are fitted to the constraints and updated periodically during the optimization process.   An analytical approach is used to calculating yield and its sensitivities with respect to all the variables employing linear cuts of the tolerance region.   The sensitivities permit the use of gradient methods of optimization.

The basic idea is to use weighted hypervolumes.   Evaluating hypervolumes, in general, is expensive because it involves a multidimensional integration.   For the special case of cutting an orthotope by a linear constraint, however, a simple formula can be found [2,14,46].   See, for example, Fig. 11.

The method of Bandler and Abdel-Malek does not assume that these linear cuts are fixed in the parameter space.   It is possible for these linear cuts be continuously updated to follow the generally nonlinear constraints.   This facilitates a good approximation to the boundary of the constraint region as the tolerance region is allowed to move in the parameter space during, for example, an optimization process.   Methods for continuously updating the linear cuts have been given [3,14].

CONCLUSIONS

Most of the approaches in current use for design centering, tolerance assignment and yield optimization employed by electrical circuit designers appear quite general.   While many applications of the techniques are directed towards design problems involving linear systems of equations, nonlinear circuits are of

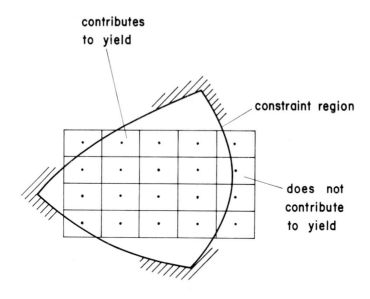

Fig. 10 Regionalization [43] of the parameter space for estimating production yield [1].

particular interest to circuit designers [3,4,11] and it is expected that considerable effort will be devoted to such systems in the future [22].

Some of the formulations described here, if expedited intact for problems involving masses of nonlinear constraints, for example, are formidable beyond our present computational means. With the anticipated breakthrough of computers with massive parallel processing capabilities, however, some of these problems are certain to be dealt with by engineers on an almost routine basis in the near future. Even at this time relevant and enormous computational problems are begging to be solved, not the least of which involve optimal topology and reliability of large systems. Of course, feasible solutions to such problems can be and are currently obtained suboptimally. It is the integration of all the concepts mentioned in the foregoing pages into a unified design methodology that is called for whether or not, for computational expedience, the superproblem thus created is subsequently partitioned or decomposed.

ACKNOWLEDGEMENT

The author would like to thank Dr. H.L. Abdel-Malek, now with the Department of Engineering Physics and Mathematics, Faculty of Engineering, Cairo University, Giza, Egypt, whose excellent Ph.D. thesis is the source of some of the material reported here. Furthermore, the author must acknowledge extensive and rewarding discussions with Dr. H. Tromp, Faculty of Engineering, University of Ghent, Ghent, Belgium, and Dr. Abdel-Malek, which helped clarify some of the finer points of the presentation. N.M. Sine of the Faculty of Engineering, McMaster University, Hamilton, Canada, is thanked for her efforts in the preparation of this manuscript.

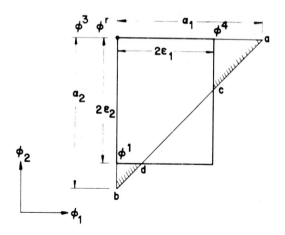

Fig. 11   A two-dimensional example illustrating the calculation of nonfeasible
hypervolumes [1,2]. The nonfeasible area V is given by subtracting the
areas of triangle ϕ^1bd and ϕ^4ac from ϕ^rab where ϕ^r is termed a reference
vertex.   Hence, it is easy to show that

$$V = 0.5 \, \alpha_1 \alpha_2 \left(1 - \left(1 - \frac{2\epsilon_1}{\alpha_1}\right)^2 - \left(1 - \frac{2\epsilon_2}{\alpha_2}\right)^2 \right) .$$

REFERENCES

[1]   H.L. Abdel-Malek: A unified treatment of yield analysis, worst-case design
and yield optimization, Ph.D. Thesis (1977) McMaster University, Hamilton,
Canada.

[2]   H.L. Abdel-Malek and J.W. Bandler: Yield estimation for efficient design
centering assuming arbitrary statistical distributions, Int. J. Circuit
Theory and Applications 6 (1978) 289-303.

[3]   H.L. Abdel-Malek and J.W.   Bandler: Yield optimization for arbitrary
statistical distributions, Part I: theory, Proc. IEEE Int. Symp. Circuits
and Systems (New York, 1978) 664-669.

[4]   H.L. Abdel-Malek and J.W. Bandler: Yield optimization for arbitrary
statistical distributions, Part II: implementation, Proc. IEEE Int. Symp.
Circuits and Systems (New York, 1978) 670-674.

[5]   H.L. Abdel-Malek and J.W. Bandler: Subroutines for implementing quadratic
models of surfaces in optimal design, Report SOC-191 (1978) Faculty of
Engineering, McMaster University, Hamilton, Canada.

[6] H.L. Abdel-Malek and J.W. Bandler: Centering, tolerancing, tuning and minimax design employing biquadratic models, Report SOC-211 (1978) Faculty of Engineering, McMaster University, Hamilton, Canada.

[7] J.W. Bandler: Optimization of design tolerances using nonlinear programming, J. Optimization Theory and Applications 14 (1974) 99-114.

[8] J.W. Bandler, H.L. Abdel-Malek, P.B. Johns and M.R.M. Rizk: Optimal design via modeling and approximation, Proc. IEEE Int. Symp. Circuits and Systems (Munich, 1976) 767-770.

[9] J.W. Bandler and H.L. Abdel-Malek: Optimal centering, tolerancing and yield determination using multidimensional approximations, Proc. IEEE Int. Symp. Circuits and Systems (Phoenix, AZ, 1977) 219-222.

[10] J.W. Bandler and H.L. Abdel-Malek: Modeling and approximation for statistical evaluation and optimization of microwave designs, Proc. 7th European Microwave Conf. (Copenhagen, 1977) 153-157.

[11] J.W. Bandler, H.L. Abdel-Malek, P. Dalsgaard, Z.S. El-Razaz and M.R.M. Rizk: Optimization and design centering of active and nonlinear circuits including component tolerances and model uncertainties, Proc. Int. Symp. Large Engineering Systems (Waterloo, Canada, 1978) 127-132.

[12] J.W. Bandler and H.L. Abdel-Malek: Algorithms for design centering involving yield and its sensitivities, Proc. 21st Midwest Symp. on Circuits and Systems (Ames, Iowa, 1978) 242-248.

[13] J.W. Bandler and H.L. Abdel-Malek: Advances in the mathematical programming approach to design centering, tolerancing and tuning, Joint Automatic Control Conf. (Philadelphia, PA, 1978) 329-344.

[14] J.W. Bandler and H.L. Abdel-Malek: Optimal centering, tolerancing and yield determination via updated approximations and cuts, IEEE Trans. Circuits and Systems CAS-25 (1978) 853-871.

[15] J.W. Bandler and C. Charalambous: Practical least pth optimization of networks, IEEE Trans. Microwave Theory Tech. MTT-20 (1972) 834-840.

[16] J.W. Bandler, J.H.K. Chen, P. Dalsgaard and P.C. Liu: TOLOPT - a program for optimal, continuous or discrete, design centering and tolerancing, Report SOC-105 (1975) Faculty of Engineering, McMaster University, Hamilton, Canada.

[17] J.W. Bandler and P.C. Liu: Some implications of biquadratic functions in the tolerance problem, IEEE Trans. Circuits and Systems CAS-22 (1975) 385-390.

[18] J.W. Bandler, P.C. Liu and J.H.K. Chen: Worst case network tolerance optimization, IEEE Trans. Microwave Theory Tech. MTT-23 (1975) 630-641.

[19] J.W. Bandler, P.C. Liu and H. Tromp: A nonlinear programming approach to optimal design centering, tolerancing and tuning, IEEE Trans. Circuits and Systems CAS-23 (1976) 155-165.

[20] J.W. Bandler, P.C. Liu and H. Tromp: Integrated approach to microwave design, IEEE Trans. Microwave Theory Tech. MTT-24 (1976) 584-591.

[21] J.W. Bandler, P.C. Liu and H. Tromp: Efficient, automated design centering and tolerancing, Proc. IEEE Int. Symp. Circuits and Systems (Munich, 1976) 710-713.

420   J.W. BANDLER

[22] J.W. Bandler and M.R.M. Rizk: Optimization of electrical circuits, Report SOC-183 (1977) Faculty of Engineering, McMaster University, Hamilton, Canada.

[23] P.W. Becker and F. Jensen: Design of Systems and Circuits for Maximum Reliability or Maximum Production Yield (Polyteknisk Forlag, Lyngby, Denmark, 1974).

[24] R.K. Brayton, G.D. Hachtel and S.W. Director: Arbitrary norms for statistical design via linear programming, Proc. IEEE Int. Symp. Circuits and Systems (New York, 1978) 161-164.

[25] R.K. Brayton, A.J. Hoffman and T.R. Scott: A theorem of inverses of convex sets of real matrices with application to the worst-case D.C. problem, IEEE Trans. Circuits and Systems CAS-24 (1977) 409-415.

[26] E.M. Butler: Realistic design using large-change sensitivities and performance contours, IEEE Trans. Circuit Theory CT-18 (1971) 58-66.

[27] C. Charalambous: A unified review of optimization, IEEE Trans. Microwave Theory Tech. MTT-22 (1974) 289-300.

[28] C. Charalambous: Discrete optimization, Int. J. Systems Science 5 (1974) 889-894.

[29] S.W. Director and G.D. Hachtel: The simplicial approximation approach to design centering, IEEE Trans. Circuits and Systems CAS-24 (1977) 363-372.

[30] S.W. Director and G.D. Hachtel: Yield estimation using simplicial approximation, Proc. IEEE Int. Symp. Circuits and Systems (Phoenix, AZ, 1977) 579-582.

[31] S.W. Director, G.D. Hachtel and L.M. Vidigal: Computationally efficient yield estimation procedures based on simplicial approximation, IEEE Trans. Circuits and Systems CAS-25 (1978) 121-130.

[32] N.J. Elias: New statistical methods for assigning device tolerances, Proc. IEEE Int. Symp. Circuits and Systems (Newton, MA, 1975) 329-332.

[33] G.D. Hachtel and S.W. Director: A point basis for statistical design, Proc. IEEE Int. Symp. Circuits and Systems (New York, 1978) 165-169.

[34] B.J. Karafin: The optimum assignment of component tolerances for electrical networks, BSTJ 50 (1971) 1225-1242.

[35] B.J. Karafin: The general component tolerance assignment problem in electrical networks, Ph.D. Thesis (1974) Univ. of Pennsylvania, Philadelphia, PA.

[36] K.H. Leung and R. Spence: Multiparameter large-change sensitivity analysis and systematic exploration, IEEE Trans. Circuits and Systems CAS-22 (1975) 796-804.

[37] K.H. Leung and R. Spence: Efficient frequency domain statistical circuit analysis, Proc. IEEE Int. Symp. Circuits and Systems (Munich, 1976) 197-200.

[38] K.H. Leung and R. Spence: Idealized statistical models for low-cost linear circuit yield analysis, IEEE Trans. Circuits and Systems CAS-24 (1977) 62-66.

[39] K. Madsen and H. Schjaer-Jacobsen: New algorithms for worst case tolerance optimization, Proc. IEEE Int. Symp. Circuits and Systems (New York, 1978) 681-685.

[40] J.F. Pinel and K.A. Roberts: Tolerance assignment in linear networks using nonlinear programming, IEEE Trans. Circuit Theory CT-19 (1972) 475-479.

[41] J.F. Pinel, K.A. Roberts and K. Singhal: Tolerance assignment in network design, Proc. IEEE Int. Symp. Circuits and Systems (Newton, MA, 1975) 317-320.

[42] J.F. Pinel and K. Singhal: Efficient Monte Carlo computation of circuit yield using importance sampling, Proc. IEEE Int. Symp. Circuits and Systems (Phoenix, AZ, 1977) 575-578.

[43] T.R. Scott and T.P. Walker, Jr.: Regionalization: a method for generating joint density estimates, IEEE Trans. Circuits and Systems CAS-23 (1976) 229-234.

[44] A.K. Seth: Electrical network tolerance optimization, Ph.D. Thesis (1972) University of Waterloo, Waterloo, Canada.

[45] M. Styblinski: Sensitivity minimization with an optimal assignment of network element tolerances, Proc. IEEE Int. Symp. Circuits and Systems (Phoenix, AZ, 1977) 223-226.

[46] H. Tromp: unpublished formula (1975) Faculty of Engineering, University of Ghent, Ghent, Belgium.

[47] H. Tromp: The generalized tolerance problem and worst case search, Conf. Computer-aided Design of Electronic and Microwave Circuits and Systems (Hull, England, 1977) 72-77.

[48] H. Tromp: Generalized worst case design, with applications to microwave networks, Doctoral Thesis (in Dutch) (1978) Faculty of Engineering, University of Ghent, Ghent, Belgium.

SECTION VI
THEORY IN MODELLING
AND SIMULATION

METHODOLOGY IN SYSTEMS MODELLING AND SIMULATION
B.P. Zeigler, M.S. Elzas, G.J. Klir, T.I. Ören (eds.)
© North-Holland Publishing Company, 1979

THE ROLE OF STATISTICAL METHODOLOGY IN SIMULATION

Jack P.C. Kleijnen
Department of Business and Economics
Katholieke Hogeschool Tilburg
5000 LE Tilburg, The Netherlands

Statistical methods relevant to both discrete and continuous
simulation are presented, using a minimum of formulas. A strategic
issue is the ad hoc character of simulation. Statistical methods are
surveyed which help to generalize and interpret simulation output
data. Moreover statistical tools can show which system variants
should be simulated, in order to obtain an understanding of the simulated system configurations. For Monte Carlo simulations some tactical problems are discussed: runlength and variance reduction.
More specifically, the ad hoc character of simulation is
mitigated by a formal metamodel (auxiliary model), for which familiar
regression analysis is used. The metamodel may include interactions
among factors in the simulation experiment, and can be tested for its
adequacy. Selecting the values of the input variables is the domain
of experimental design. An example demonstrates that seven factors
can be examined in only sixteen rather than $2^7 = 128$ runs. Situations
with, say, a thousand factors require special screening designs. Requirements for "optimal" designs are briefly discussed.
In stochastic simulation two tactical problems exist: Variance reduction can be achieved through special techniques such as
common random numbers, antithetic variates, control variates (regression sampling), importance sampling (virtual measures). The variance is further affected by the simulation runlength, which leads
to questions such as: are we really interested in steady-state behavior, how can we compute a confidence interval for the simulation
output, how should we initialize the run, etc.
The survey includes seventy references, many just recently
published, and provides references to a number of practical applications of statistical methods.

CONTENTS

Abstract

1. Strategic problems in simulation
2. Tactical problems in stochastic simulation
3. Formal metamodels: regression analysis
4. Experimental design
5. Summary of strategic problem
6. Tactical problems in stochastic simulation: variance reduction
7. Tactical problems: runlength
8. Miscellaneous statistical problems
9. Conclusion

Notes

References

1. STRATEGIC PROBLEMS IN SIMULATION

At this conference both _discrete_ and _continuous_ simulations
are covered. We hope that the present contribution is relevant to the

two simulation areas, though we shall emphasize discrete-event simulation. In practice, simulation is a method very frequently used to
study complex systems (1). However, a major practical drawback of simulation is its <u>ad hoc</u> character: After spending much mental effort
and computer time to develop, program and run a computer simulation
- strictly speaking - the results are valid only for the specific
parameter values and structural relationships of the executed simulation program. Changing a parameter or relationship means that the simulation program has to be run again. Nevertheless, such changes are
necessary to answer "what if" questions, or to find optimal system
configurations, or - not to be forgotten - to establish the sensitivity of the conclusions to specific model assumptions.

Even after a great many simulation runs have been performed,
it is difficult to obtain a general understanding of how the simulated
system works: During the construction of the simulation model and its
program much knowledge about the details of specific system components
is acquired. However, insight into the behavior of the total system
requires execution of the simulation program. These computer runs
yield a mass of data but this mass may turn into a mess. Hence the
output data should be summarized by a limited number of measures such
as averages, peaks, and correlation coefficients (or spectra). In this
way the various system configurations (system variants) are characterized by a few "statistics". We have the problems of how to determine
whether system variants show <u>significantly</u> different outputs, how to
discern "patterns" in output changes as system configurations change,
and so on.

In this paper we shall present a <u>statistical methodology</u> for
generalizing the results of simulation experiments. Because of the
survey character of our paper we shall avoid the use of mathematicalstatistical formulas as much as possible. As we shall see, the accuracy of the resulting generalizations (metamodels) can be made explicit. Moreover, we shall present a systematic <u>and</u> efficient methodology for the exploration of the great many <u>systems</u> that can be simulated. In Kleijnen (1977) we further described how this methodology fits
in the sequence of mental modeling - formal modeling - computer programming - program running - system understanding - optimization - implementation. Before the formal presentation of our methodology in
section 3, we introduce tactical issues in section 2.

Note that to solve the above strategic problems we propose
techniques that have been known for a long time, namely regression
analysis and experimental design. Unfortunately, most simulation users
have ignored these statistical techniques. Statisticians have not been
consulted in simulation experiments, or have been thwarted by communication problems (e.g. an expression like "simulation runs 1 and 2"
has to be translated into a "one observation from population 1 and 2
respectively"). The most practical course seems that simulationists
obtain a basic knowledge of the relevant statistical techniques, which
can be borrowed from the discipline of mathematical statistics except
for some minor adaptations (2). The tactical problems in simulation
do involve specific statistical techniques such as variance reduction
by means of autithetic random numbers.

2. TACTICAL PROBLEMS IN STOCHASTIC SIMULATION

Besides the "strategic" problems in simulation models of any
type, there are some nagging "tactical" issues, if the simulation is
of the <u>Monte Carlo</u> type, i.e., when the program contains stochastic
variables generated by means of pseudo-random numbers. Even if we concentrate on a single system configuration with all parameters fixed,
we have to decide on the simulation runlength. Once we terminate the

run, we wish to know the output's accuracy, specified by a statistical confidence interval. The determination of this accuracy may be complicated by serial correlations (autocorrelations) among successive simulation observations so that simple statistical methods are misleading. However, as we shall see, the majority of practical, non-academic simulation experiments can be analyzed without sophisticated statistical analysis techniques. The reason is that most practical studies do not concern long run, steady-state behavior!

Even with modern high-speed digital computers, Monte Carlo simulation may require runs of such lengths that computing time becomes a bottleneck. We might then try to apply special statistical techniques to reduce the variability of the output, and hence the required runlength. This is the area of Variance Reduction Techniques (VRT's), also known as Monte Carlo Techniques, to be discussed in section 6.

3. FORMAL METAMODELS: REGRESSION ANALYSIS

A real-life system can be modeled by means of a simulation model. The relationship between the inputs and outputs of the simulation program (model) can in turn be modeled by a metamodel or auxiliary model, to meet the strategic demands mentioned in section 1. Several types of metamodels are surveyed in Kleijnen (1977): common sense graphical approach, tables with two or three factors, explicit formal metamodels such as the Meisel & Collins (1973) piecewise linear approximations. In this paper, however, we concentrate on linear regression metamodels. The common-sense graphical approach is related to the regression approach, as follows. We may change one factor, say x; observe the resulting output y; repeat this procedure a number of times; plot the (x,y) combinations; fit a curve by hand; and conclude whether x has an important effect on y. The regression approach formalizes this hand-fitting by applying the least squares algorithm. It extends the procedure into multiple dimensions. It further systematizes the various steps, including tests for the importance (significance) of factors together with their interactions, and tests for the adequacy of the fitted regression metamodel. Linear regression analysis has the great advantage of being a familiar technique for most scientists: Regression models have been extensively applied to interpret and generalize experimental results in agriculture, chemistry, engineering, psychology, etc., where these models are known as Analysis of Variance (ANOVA). Regression metamodels in simulation have been advocated by a few other authors. For instance, Week & Fryer (1977) estimated "main" effects and "interactions" in their job shop simulation study (effects to be defined more precisely below). In their study regression models are further utilized to answer "inverse" or "control" questions, i.e., which input values yield fixed, desired output values; see also Kleijnen & Rens (1978), Kleijnen et al. (1978), Koons & Perlic (1977) and Sherden (1976). The need for some kind of metamodel to assist the more detailed model (simulation, mathematical programming model) has been emphasized by several more authors: Blanning (1974, 1975), Geoffrion (1976), Pegels (1976, p. 205), Rose & Harmsen (1978).

Note that the variables in a model may be partitioned into decision and environmental variables (factors). Decision (control) variables are under the control of management. Environmental variables are not under management control but do influence the outputs (responses, crieria). In the long run some environmental variables may become controllable, for instance, arrival times may be influenced by sales promotion. The criterion variables can be either satisfied or optimized by a correct selection of the decision variables. The sensitivity of this choice to the assumed environmental factors, must also be investigated. Observe that variables are observable, whereas parameters are not.

Let x denote a _factor_ influencing the outputs of the real-world system. The factor may be qualitative or quantitative, continuous or discrete. In Kleijnen (1975, p. 300) it is shown how we can represent a qualitative factor by several dummy variables assuming only the values zero or one. We shall concentrate on qualitative factors (besides quantitative factors) which are studied for only two "levels" or "values". Then this qualitative factor can be represented by a single dummy variable x with the values -1 and +1; see Tables 1 and 2 below. The _response_ (output) of the real-world system is a time-series. We shall concentrate on a single response variable; for multiple outputs we apply our procedure to each variable separately. In order to compare system configurations, we characterize a whole time path by either a single measure or a few measures: average, standard deviation, correlation coefficients (spectra), slope of a fitted linear trend, peak, etc. Let y_R denote such a measure, characterizing a time path of the real-world system. Hence the response variable y_R is a function of the factors x:

$$y_R = f_1(x_1, x_2, \ldots, x_m) \qquad (3.1)$$

The real system is approximated by a _simulation model_, where $\underline{y}$ is a function of k factors x_j (j=1,...,k), plus a vector of random numbers $\vec{\underline{r}}$. (Vectors and matrices are denoted by $\rightarrow$; stochastic variables underlined.) Hence the simulation output $\underline{y}$ can be represented as

$$\underline{y} = f_2(x_1, x_2, \ldots, x_k, \vec{\underline{r}}) \qquad (3.2)$$

where k is much smaller than the unknown m in eq. (3.1), and $\vec{\underline{r}}$ symbolizes the joint effect of all factors x in eq. (3.1) not explicitly represented in eq. (3.2). The simulation model is specified by a computer program denoted by the function f_2. This model may be approximated in turn by a _meta-model_ (within a specific experimental domain E; see below). We propose a metamodel that is _linear_ in its _parameters_ β. This linearity does not mean that the metamodel is linear in its variables x; see eq. (3.5) below. We have not yet been confronted with situations which require non-linear regression analysis to obtain valid metamodels for simulation experiments. Models not linear in their parameters β, lead to non-linear regression analysis, involving more complicated mathematical procedures; see Bard (1974) and Malinvaud (1975). Before we proceed to various specific metamodels we observe that the metamodel approach also applies when no sampling is used so that $\vec{\underline{r}}$ in eq. (3.2) vanishes.

The simplest metamodel to express the effects of the k factors would be:

$$\underline{y}_1 = \beta_0 + \beta_1 x_{i1} + \cdots + \beta_k x_{ik} + \underline{e}_i \qquad (i=1,\ldots,N) \qquad (3.3)$$

where in simulation run i (observation i) factor j has the value x_{ij} (j=1,...,k) and $\underline{e}_i$ represents the noise (disturbance, error) in the metamodel which is assumed to have zero expectation. Such a simple metamodel implies that a _change_ in x_j has a constant effect on the expected response, $\mathcal{E}(\underline{y})$:

$$\frac{\partial[\mathcal{E}(\underline{y})]}{\partial x_j} = \beta_j \qquad (j=1,\ldots,k) \qquad (3.4)$$

Such a (simplistic?) model was used to perform sensitivity analysis of a simulation model for medical services in disaster planning; see

Lawless et al. (1971).
   A more general metamodel postulates that the effect of factor j also depends on the values of the other factors j' (j' $\neq$ j). This can be formalized as in eq. (3.5) where for illustration purposes we take k=3:

$$\underline{y}_i = \beta_0 + (\beta_1 x_{i1} + \beta_2 x_{i2} + \beta_3 x_{i3}) +$$
$$+ (\beta_{12} x_{i1} x_{12} + \beta_{13} x_{i1} x_{13} + \beta_{23} x_{i2} x_{13}) + \underline{e}_i$$

$$(i=1,\ldots,N) \qquad (3.5)$$

Here the parameters (coefficients) β_{12}, β_{13} and β_{23} denote <u>interactions</u> between the factors 1 and 2, 1 and 3, and 2 and 3 respectively. A graphical illustration of interaction in the case of two factors (k=2), is shown in FIG. 1. In case (a) of FIG. 1 the curves are parallel, i.e., the effect of x_2 on $\&(\underline{y})$ does not depend on the level of x_1. In case (b) the interaction coefficient β_{12} is positive. Hence the two factors are <u>complementary</u>, i.e. the increase in $\&(\underline{y})$ is stimulated when the increase of x_2 is accompanied by an increase in x_1. In case (c) the marginal output of x_2 is much smaller when more of x_1 is available which can be substituted for x_2 (3). The need to consider interactions among factors when analyzing simulation results, has been emphasized by Koons & Perlic (1977) in their case-study of a steel plant.
   If all factors are quantitive, continuous variables, then we add "purely quadratic" effects β_{jj} to eq. (3.5). This yields

$$\underline{y}_i = \beta_0 + \sum_{j=1}^{3} \beta_j x_{ij} + \sum_{j<j'}^{2} \sum^{3} \beta_{jj'} x_{ij} x_{ij'} + \sum_{j=1}^{3} \beta_{jj} x_{ij}^2 + \underline{e}_i$$

$$(3.6)$$

which represents the <u>Taylor series</u> expansion of eq. (3.2), cut off after the second-degree terms. For an application of eq. (3.6) we refer to Koons & Perlic (1977, p.7). We conjecture that in practice it is rare that all factors are quantitative, so that we shall concentrate on the metamodel with k main effects β_j , k(k-1)/2 two-factor interactions $\beta_{jj'}$ and the general (overall) mean β_0. In symbols:

$$\underline{y}_i = \beta_0 + \sum_{j=1}^{k} \beta_j x_{ij} + \sum_{j<j'}^{k-1} \sum^{k} \beta_{jj'} x_{ij} x_{ij'} + \underline{e}_i$$

$$(i=1,\ldots,N) \qquad (3.7)$$

   We start by <u>assuming</u> a metamodel, such as eq. (3.7), but next we test statistically whether this assumption was realistic! Two statistical tests can be used:
1) Generate some <u>new</u> observations $\underline{y}$ using the <u>simulation</u> model. Apply the familiar Student t-test to compare these observations $\underline{y}$ to the predicted value $\hat{\underline{y}}$ based on the regression metamodel which was estimated from the old observations (4).
1) A so-called lack-of-fit F-test can be computed which compares the "mean residual sum of squares" to the "pure error".
   For details on both approaches we refer to Kleijnen et al. (1978).

   If the assumed metamodel turns out to be unreasonable, we have several alternatives:
1) Make the metamodel more complicated by adding terms such as <u>three-factor interactions</u>. If $\underline{y}'$ is a shorthand notation for $\underline{y}$ in eq. (3.5), then we may expand (3.5) to

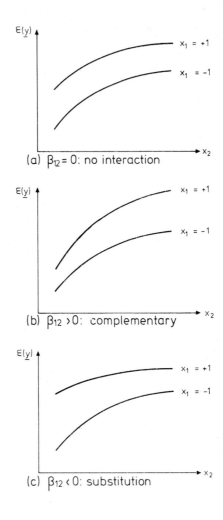

(a) $\beta_{12} = 0$: no interaction

(b) $\beta_{12} > 0$: complementary

(c) $\beta_{12} < 0$: substitution

FIG.1   Interactions

$$\underline{y}_i = \underline{y}'_i + \beta_{123}x_{i1}x_{i2}x_{i3} \qquad (3.8)$$

The intuitive interpretation of three-factor interactions is difficult. Moreover these interactions increase the number of parameters.
Estimating additional parameters creates statistical problems:
noisy, insignificant estimates. It also conflicts with the purpose
of science: explain a phenomenon as parsimonously as possible.
Therefore we prefer the next alternative.

2) Look for <u>transformations</u> of x. For instance, if $\underline{y}$ denotes waiting
time, and x_1 and x_2 denote mean arrival and service rate, then
the transformation $x' = x_1/x_2$ implies that the original metamodel

$$\underline{y} = \beta_0 + \beta_1 x_1 + \beta_2 x_2 + \underline{e} \qquad (3.9)$$

is replaced by a new metamodel:

$$\underline{y} = \gamma_0 + \gamma_1 x' + \underline{e}' \qquad (3.10)$$

We strongly recommended to look for such transformations, from the
very beginning of the study! The transformation can be based on
the relevant theory, for instance, queuing analysis in the above
example. A popular transformation in econometrics is $x' = \log x$,
$y' = \log \underline{y}$, so that the parameters β represent elasticity coefficients. A more complicated example is provided by Yen & Pierskalla
(1977). A simple transformation, namely $x' = 1/x$, is also utilized
in the case study by Kleijnen et al. 1978).

3) Reduce the <u>experimental domain</u> E. This option limits the generality of our conclusions. However, if the only purpose of the metamodel is to find the optimum x-values, then a small domain E can
be used, a metamodel fitted, and the direction of better x-values
determined. See Montgomery & Bettencourt (1977) for details on
this so-called Response Surface Methodology (RSM). For a bibliography we also refer to Kleijnen (1975).

Observe that the specification of the metamodel's form is
based on prior knowledge, intuition, pilot experiments, etc. Bens &
Vansteenkiste (1978) suggest a systematic approach - based on pattern
recognition - for deriving the structure of the model. (They discuss
this approach in the context of modeling systems through differential equations.)

After we have used the metamodel to meet the demands of
sensitivity analysis, optimization, and so on, we can return to the
original simulation model to study the system behavior in detail,
e.g. study its dynamics.

The parameters β in the above equations can be estimated and
tested for their significance, using the familiar technique of linear
regression analysis, applying either Ordinary or Generalized Least
Squares; see e.g. Draper & Smith (1966). Least Squares is also
summarized in Kleijnen et al. (1978).

4. EXPERIMENTAL DESIGN

As we mentioned in section 1 parameters, variables, and
structural relationships in the simulation model are changed in order
to perform sensitivity analysis, answer what-if questions, do optimal-
ization studies, etc. "Infinitely many" system configurations (system variants or briefly systems) may be of potential interest. Even
with modern fast computers simulating all system variants is out of

the question. The problem of selecting a limited number of system
variants for actual simulation evaluation can be solved by means of
statistical methods known as experimental design methodology. Experimental design theory has been developed since the 1920's and has
been widely applied to experiments in agriculture, chemistry, etc.
Unfortunately, its application to the management and social sciences
is still in its infancy. The reason is that in sociotechnical systems
the scientific design of experiments is difficult and expensive (disruption of the organization). However, in a simulation model of such
a social system, the experimental factors are completely under the
scientist's control, so that experimental design methods become relevant. Let us consider some simple examples.

If we study the effects of just a few factors, say three,
then we may start by letting each factor assume only two "values" or
"levels" denoted by $x = +1$ and $x = -1$ respectively, and evaluating
the responses at all combinations, i.e., at $2^3 = 8$ combinations. (Remember that one factor combination specifies one system variant to
be simulated.) But, if we are willing to assume that the "first order"
model of eq. (3.3) is adequate, then we need to estimate only four
effects, namely β_0, β_1, β_2 and β_3. Before reading on, the reader is
challenged to specify his own selection of the x-combinations! Next
he may compare his selection to Table 1. This table is constructed
using a "trick" developed in experimental design theory: The last
column x_3 is obtained by multiplying the corresponding elements in
the x_1 and x_2 columns. Note that all columns in Table 1 are orthogonal so that least squares estimation of the β's becomes very simple.
These tricks are certainly necessary as the number of factors increases, since in that case the number of combinations N grows dramatically. For instance, for seven factors we have $N = 2^7 = 128$. Table 2
displays a design for seven factors. The first eight combinations
in Table 2 are selected as follows: Write down all 2^3 combinations of
x_1, x_2 and x_3. This yields x_{ij} for $i = 1,...,8$ and $j = 1,...,3$. The
column for x_4 is obtained by multiplying the corresponding elements
in the x_1 and x_2 columns (analogous to the x_3 column in Table 1). The
shorthand nation is 4=12. Likewise we use 5=13 and 6=23. Column 7 is
generated by multiplying the elements of columns 1, 2 and 3; in shorthand: 7=123. The last eight rows in Table 2 are obtained by switching
the signs of the first eight rows: $x_{i'j} = -x_{ij}$ (for j=1,...,7 and
i' = i+8 with i=1,...,8). Additional discussion on these "tricks" can
be found in the experimental design literature. If we assume a first
order model, then only the first eight combinations need to be evaluated, in order to estimate the eight effects β_0, $\beta_1,...,\beta_7$. If we
leave open the possibility that such a metamodel is inadequate, then
the next eight combinations should also be evaluated. The first eight
combinations form a so-called 2^{7-4} fractional factorial design. The
sixteen combinations together form a 2^{7-3} fractional factorial, and
can give us an idea of the importance of interactions besides main
effects; see Kleijnen (1975) for more details. Observe that a design
matrix such as Table 1 also specifies the cross-products x_1x_2, x_1x_3
and x_2x_3 in eq. (3.5).

The above phase of the investigation may be preceded by a
pilot or screening phase. In the latter stage a great many, say a
thousand factors may be conceived of, but hopefully only relatively
few are really important. Detecting these important factors can be
based on special experimental designs: group-screening, random designs, etc. Random designs (randomly selected factor combinations)
were applied in a water-resource simulation; Maass et al. (1962).
Group-screening (5) was utilized in the simulation of computer systems - see Schatzoff & Tillman (1975) - but applications of this class
of designs are extremely rare. Nevertheless we imagine that it is
quite common to have simulation models with a great many parameters

Table 1

Experimental Design for Three Factors

Combination	x_1	x_2	x_3 $(= x_1 x_2)$
1	+1	+1	+1
2	-1	+1	-1
3	+1	-1	-1
4	-1	-1	+1

Table 2

Experimental Design for Seven Factors

Combination	x_1	x_2	x_3	x_4	x_5	x_6	x_7
1	-	-	-	+	+	+	-
2	+	-	-	-	-	+	+
3	-	+	-	-	+	-	+
4	+	+	-	+	-	-	-
5	-	-	+	+	-	-	+
6	+	-	+	-	+	-	-
7	-	+	+	-	-	+	-
8	+	+	+	+	+	+	+
9	+	+	+	-	-	-	+
10	-	+	+	+	+	-	-
11	+	-	+	+	-	+	-
12	-	-	+	-	+	+	+
13	+	+	-	-	+	+	-
14	-	+	-	+	-	+	+
15	+	-	-	+	+	-	+
16	-	-	-	-	-	-	-

and variables, which could benefit from group-screening designs. Several types of screening designs are evaluated in Kleijnen (1975 b).

Let us briefly evaluate the designs derived in the statistical design literature. The traditional, "common sense" approach is to change one factor at a time. We proposed to change several factors simultaneously; see Tables 1 and 2. Such factorial designs (full or fractional) are more efficient, i.e., they yield more accurate estimators of the factor effects, and they provide estimates of possible interactions among factors. Some problems, however, remain: Specific designs such as 2^{k-p} designs, and the concomittant regression analysis yield "optimal" results, only under certain statistical assumptions such as constant variances. How robust are these optimality properties and what are the alternatives? Kleijnen et al. (1978) give a survey of ad hoc optimal designs specified by computer, generalized least squares, robust estimation procedures, etc. Designs such as in Tables 1 and 2 may be evaluated against the following requirements:

1. A small number of runs N: Obviously, to estimate q parameters it is necessary that $N \geq q$. However, N may be much smaller than 2^k, for instance, in Table 2 N=16 whereas $2^7 = 128$.

2. Maximum statistical accuracy, given the number of runs: If the classical statistical assumptions hold, then the accuracy requirement is satisfied by choosing an orthogonal design; otherwise the selection of the design matrix poses a problem not yet solved. Note that Tables 1 and 2 do yield orthogonal columns.

3. Providing a measure for the adequacy of the fitted metamodel: If $N > q$ then a lack-of-fit F-statistic exists. If besides the N observations we have one or more runs, not used in the estimation of the parameters, then "validation" of the model is possible, using a t-test.

4. Desirable "confounding" (bias, alias pattern): If not all factor effects can be estimated from N runs only, then main effects should be biased by high order effects, not by other main effects. The designs derived by experimental design theory immediately show how effects are biased by other effects. For instance, Table 2 shows that the main effect β_4 is completely confounded with the two-factor interaction β_{12} (but we hope that β_{12} is unimportant, i.e. is zero). The effect β_4 is not biased by any other main effect.

5. Flexibility of the design: Unfortunately, in many standard designs the number of runs N is restricted to a power of two (2^{k-p} designs) or a multiple of four. Fortunately, it remains possible to start with only a small number of runs, to test the results, and to proceed to a larger design that yields more detailed estimates, so-called sequentialized designs; see Kleijnen (1975, pp. 344-345, 367-370).

6. Numerical inaccuracy caused by an ill-conditioned matrix $\vec{X}$: An orthogonal matrix $\vec{X}$ eliminates such problems. When using normalized variables (between -1 and +1) we should not forget to translate the estimated effects back into the original effects; see Kleijnen et al. (1978).

Note that, if the assumed metamodel turns out to be completely misleading, then the "optimal" properties of the experimental design break down. For instance, if the interaction between the factors 1 and 2 in Table 2 is actually important then we cannot estimate the main effect of factor 4 since it is completely confounded with that interaction. To reduce the possibility of such events, preliminary experimentation and analysis is necessary; see also Kleijnen (1975, pp. 391-393) and Kleijnen et al. (1978).

In conclusion, the literature on experimental design is overwhelming, and still growing! As a sample we mention the recent textbook Daniel (1976). In Kleijnen (1975, pp. 287-450) we have given a

selection from the vast literature, tailored to the needs of the simulation practitioner: The focus is on simple designs such as 2^{k-p}
designs, excluding sophisticated designs such as "partially balanced
incomplete block designs". Excluded are techniques not needed in simulation, e.g., randomization and blocking (remember footnote 2).

5. SUMMARY OF STRATEGIC PROBLEM

Formal metamodels are a useful technique for generalizing
and interpreting simulation output. An important aspect of this interpretation is the concept of factor interaction. Efficient exploration of the simulation space requires an experimental design. Work
on statistical designs is abundant but unfamiliar to the majority of
simulation practitioners. Our experience is that the necessary statistical techniques can be learned without too many problems. Observe,
however, that these techniques alone cannot solve the problem for
the scientist! The models and hypotheses to be evaluated by the statistical techniques, have to be provided by management or by other,
non-statistical specialists. The use of the techniques leaves much
freedom: choice of significance levels α, form of the regression
model, etc. Hence an automated application of techniques is impossible. Moreover, all statistical techniques are based on certain statistical assumptions such as constant variances, which are not satisfied in practice. It remains a challlenge to develop more general
and robust techniques. In the mean time, the practioner must use his
judgement in the selection and use of his statistical tools. Nevertheless we feel that these tools result in a more efficient exploration of the experimental area, and in a better idea of both the limitations and the generalizations of the simulation experiment. In this
way, one important drawback of simulation is mitigated, namely its
ad hoc character. For an elaborated case study using a variety of
statistical techniques, we refer to Kleijnen et al. (1978).

6. TACTICAL PROBLEMS IN STOCHASTIC SIMULATION: VARIANCE REDUCTION

In the following sections we shall focus on problems arising
when simulating one specific system configuration, i.e., a single
factor combination. The problems we shall discuss arise in stochastic
simulation models only: runlength determination in relation to estimation of the variances of the simulation response, and reduction of
that variance through special statistical techniques, so-called Variance Reduction Techniques (VRT's)
Though we devoted a doctoral dissertation to the issue of
VRT's - see Kleijnen (1975, pp. 105-285) - over the years we have
grown very pessimistic as to the practicality of such techniques.
Note that computer time can be saved (and variance reduced through
additional runs) by other devices than VRT's, e.g., more efficient
random number generation, faster sampling procedures, better software.
In Kleijnen (1975, pp. 105-285) we discussed six VRT's in detail including some applications, and provided references to many more techniques. Four VRT's will be briefly discussed in the present contribution.

1) Common random numbers

A system configuration may be simulated using the same random number seed as in the other system variants, so that systems are
compared "under the same circumstances". This is the only VRT often
applied by practitioners. Kleijnen (1975) discussed the practical

omplication of synchronizing random number streams per type of stochastic process. A complication overlooked by most practitioners, is
that the _analysis_ of the simulation results gets more difficult when
the outputs become dependent. For instance, ordinary least squares
assumes independent responses y. The following VRT is (nearly) as
simple as the use of the same random numbers, but does not complicate
the analysis.

2) Antithetic variates

Suppose the first run of a specific system configuration is
generated from the random number stream r_0, r_1, r_2,... and yields the
result y_1. Then the "antithetic" run is generated from the complements
$1-r_0$, $1-r_1$, $1-r_2$,... and yields y_2. The idea is, that when y_1 happens
to undershoot its expected value, then y_2 is expected to overshoot
that value. (Example: In run 1 most r's happen to be small, so that
service times are short, and waiting times are short. In run 2 most
r's are large, etc.) Statistically speaking, $\underline{y}_1$ and $\underline{y}_2$ are conjectured
to be _negatively_ correlated, so that the variances of their average
decreases. The statistical analysis remains simple, since it can be
based on the n/2 averages of the antithetic pairs (y_1,y_2) (y_3,y_4),...,
(y_{n-1},y_n). Kleijnen (1975 a) discussed the surprising fact that it is
not necessarily optimal to combine antithetic variates and common
random numbers when comparing two system variants. Recently Schruben
& Marjolin (1977) investigated the joint application of these two
VRT's when investigating N system variants in an experimental design.
They found that applying either common random numbers only, or a specific combination (6) of common and antithetic random numbers, reduced the estimated variance by 80% compared to independent random number streams.

3) Control variates or regression sampling

During a simulation run we may keep track of the average
value $\hat{\mu}$ of the _input_ variable, say, interarrival time. Note that the
expected value μ is known, since we sample the input from a known
distribution function. If we wish to estimate, say, average waiting
time η for a specific average input value μ, then we may correct our
estimate via the regression model

$$\underline{y}_i = \beta_0 + \beta_1 \cdot \underline{x}_i + \underline{u}_i \qquad (i=1,...,n) \qquad (6.1)$$

where $\underline{y}_i$ is the average waiting time of run i, $\underline{x}_i$ is the average interarrival time $(\underline{x} \equiv \hat{\underline{\mu}})$ of run i, $\underline{u}$ is an error term, and the β's are
regression coefficients. Hence

$$\hat{\underline{y}}_\mu = \hat{\underline{\beta}}_0 + \hat{\underline{\beta}}_1 \cdot \mu$$

$$= \bar{\underline{y}} + \hat{\underline{\beta}}_1 (\mu - \bar{\underline{x}}) \qquad (6.2)$$

Where the last equality follows from the least squares properties of
$\hat{\underline{\beta}}_0$. So $\bar{\underline{y}}$, the "crude" average response of the n simulation responses
$\underline{y}_i$, is corrected for the deviation of the "control variable" $\bar{x}$ from
its theoretical mean. We found a variance reduction factor of 3.84
in a case study, a telephone exchange simulation; however, the esti-
mator (6.2) does involve some statistical complications as shown in
Hopmans & Kleijnen (1977).

4) Importance sampling and virtual measures

There is one class of systems that might benefit very much
from variance reduction, namely, systems where we are interested in
"rare events" such as "excessive" waiting times, inventory stockouts,
etc. During most of the simulation run nothing of interest happens. A

VRT especially relevant for such systems seems importance sampling
(IS), and a closely related technique known as "virtual measures".
These techniques sample more frequently that part of the time path
during which rare events tend to occur more frequently (and correct
for that oversampling). In a recent case study we applied this idea
to a telephone-exchange simulation studying blocking probabilities
(all lines occupied). Unfortunately, the practical results were very
disappointing; see Hopmans & Kleijnen (1978).

7. TACTICAL PROBLEMS: RUNLENGTH

          Under the heading "runlength" we shall discuss a set of re-
lated questions such as:
- How long to continue the simulation run?
- How to start the run (initialization)?
- How often to replicate (repeat) the run with different random number seeds?
- How accurate is the estimated response (confidence intervals)?
An important remark to start with, is that in practical - as opposed
to academic - simulations these questions can often be answered using
only elementary statistical techniques, such as t-statistics.
          In practice simulation models are usually terminating, i.e.,
the simulation run is stopped when a specific event occurs. Simple
examples are:
1) In studying maintenance policies the simulation run may end when
   the equipment (say, a computer) breaks down. A new run starts with
   a "perfect" piece of equipment.
2) A queuing system such as a bank or hospital clinic is closed at
   5 P.M. (critical event). The new run corresponds with a new day,
   and starts in the "empty" state.
3) A corporate simulation model can be utilized to examine a policy's
   effect on profit over the next three months (planning horizon). The
   simulation run of three months is repeated for different policies
   (what-if), starting each run from the most recent "situation"
   (system state). Note that corporate models are often non-stochast-
   ic.
4) Queuing systems that never close down are, e.g., a telephone exchange and a highway crossing. Such systems may be simulated to
   see whether the system configuration can handle peak traffic. As
   soon as the rush hour is over (critical event, defined fuzzily),
   the simulation run is terminated. A next run starts from a pre-
   rush-hour situation. The same or different random numbers (see
   section 6 on variance reduction) may be utilized in that next run,
   to study the sensitivety to the starting conditions (and to the
   random number seed). Example 4 seems to be the most problematic
   example, and may merit additional research.

          In the above examples there is no interest in steady-state
responses! Steady-state (stationary, equilibrium, long-run) behavior
means that the distribution function (probability law) does not change
over time. In the above examples, however, start-up and end effects
are part of the relevant output. Each simulation run yields a single
observation on the output, say, the average waiting time or the total
profit. (The relevance of transient behavior is also emphasized in
Fox (1978), Lam & Pedersen (1977), Law (1978).) If the simulation is
stochastic, more accurate estimates can be obtained by repeating (re-
plicating) the run with different random numbers (possibly antithetic, see section 6). The statistical analysis is straightforward
for terminating systems. For example, if y_i denotes average waiting
time in run i ($i=1,\ldots,n$), then its standard deviation (standard error)

is estimated by

$$\underline{s}_{\bar{y}} = \{ \sum_{i=1}^{n} (\underline{y}_i - \bar{\underline{y}})^2 / (n-1) \}^{\frac{1}{2}} \qquad (7.1)$$

A confidence interval for the expected value η can be based on the
Student t-statistic:

$$P\{\eta \leq \bar{\underline{y}} + t_{n-1}^{\alpha} \cdot \underline{s}_{\bar{y}} / \sqrt{n}\} = 1-\alpha \qquad (7.2)$$

For an application we refer to Kleijnen (1979) where Chapter IX concerns a simulation experiment with an IBM management game, used to
study the financial benefits of accurate information.

If the confidence interval in eq. (7.2) turns out to be too
long, we may improve the accuracy of the average simulation output $\bar{\underline{y}}$
by generating additional runs. The total number of runs for a fixed
length c of the confidence interval, should be

$$\underline{n} = \{t_{n-1}^{\alpha}/c\}^2 \cdot \underline{s}_{\bar{y}}^2 \qquad (7.3)$$

For additional comments (7) on such a sequential approach we refer
to Kleijnen (1975).

In the simulation literature most attention is focussed on
steady-state behavior. Such behavior is primarily of academic interest: Simulation is used by many academics in the study of analytic
models such as queuing models; see the examples in Ignall et al.
(1978). Transient behavior of such models is difficult to analyse
- Kotiah (1978), Liittschwayer & Ames (1975) - so that most academic
studies concentrate on steady-state (limiting) behavior. Simulations
with the (practical!) aim of assisting such theoretical studies, are
confronted with serious problems: Should the simulation be continued
or should replicated runs be used? Replicated runs yield independent
observations (on, say, steady-state average waiting time) but each
run creates an initialization problem (transient behavior). A single
prolonged run consists of many dependent individual responses: autocorrelation or serial correlation problem. For instance, if customer
i has to wait "very" long (longer than average) then the next customer probably has to wait longer too (positive correlation). Elementary statistical techniques assuming independence, are misleading in
that case. The variance of the average of the continued run $\bar{\underline{w}}$ based
on m autocorrelated individual observations $\underline{w}$ is given by:

$$\text{var}(\bar{\underline{w}}) = \{1+2 \sum_{j=1}^{m} (1 - \frac{j}{m}) \rho_j\} \sigma_w^2 / m \qquad (7.4)$$

where σ_j is the autocorrelation between $\underline{w}_t$ and $\underline{w}_{t+j}$, and σ_w^2 is the
variance of an individual observation $\underline{w}$. If the w's were independent
($\rho_j = 0$), then the variance of their average would reduce to the
familiar expression σ_w^2/m. So the autocorrelation ρ inflates the variance of the average, and should not be ignored. Several approaches
are possible.

1) Repeated runs

As mentioned above, replicated runs result in independent
observations so that the analysis becomes simple. However, each run
confronts us with the initialization problem. References on this approach and its alternatives will be given below.

2) Prolonged run with (nearly) independent subruns of fixed length

Assume that the serial correlation among the individual ob-

servations decreases as the observations are father apart. Divide
the total run (usually after removing the initial phase as in approach
1 above), into subruns of fixed length. Though the first "few" observations of a subrun still depend on the last "few" observations of the
preceding subrun, the subrun averages will be independent, practically speaking, provided the subrun length is "long enough". Therefore
a subrun length may be selected intuitively; the autocorrelation
among subrun averages be tested (through the estimated autocorrelation coefficient of lag 1 or through von Neumann's ratio); if the autocorrelation is too high (empirical threshold) then the subrun length
is increased, etc. A recent paper on this approach is Law & Carson
(1977). In practice, approach 2 is often followed, but with the subrun length being selected purely intuitively. Relying on intuition
alone seems dangerous: Analytical results for simple queuing systems
demonstrate that for heavy traffic individual observations remain
correlated over surprisingly long lags. On the other side, over-
estimating the autocorrelation means that the subruns are too long,
so that too few subruns remain (8).

3) Prolonged run without subrun distinction, but with estimated individual autocorrelations

Eq. (7.4) displayed the effects of the autocorrelation coefficients ρ_j (j=1,2,...) among the individual observations. In the
sixties Fishman and Kiviat - at that time with the RAND Corporation -
published several reports in which the variance of the average simulation output was based on the estimation of those coefficients ρ_j (or
their Fourier transformations: spectral analysis). For a recent discussion we refer to Clark (1977). In practice this approach has never
been popular: cumbersome estimation of the ρ's; difficult selection
of m (the number of ρ's to be incorporated).

4) Prolonged run with truly independent subruns based on the renewal
property

Both Iglehart and Fishman have pioneered the application of
the renewal or regenerative property shown by many simulated systems.
Consider a queuing system that has become empty. Then the next history (timepath) is independent of the past history! Consequently, the
total run can be divided into subruns, each subrun starting as soon
as a customer arives into an empty system. In contrast to the subruns
of fixed length (approach 2) the new subrun definition creates subruns of stochastic lengths: when does the system return to the empty
state? These subruns are exactly independent! The estimation of the
total run's average involves some statistical complications: ratio
estimators for point estimation, jackknifing for confidence intervals,
etc. Estimating quantiles (9) such as the 90%-point, involves some
more problems; see Seila (1978) and also Coppus et al. (1977). Percentiles (e.g., the probability of queue-sizes exceeding the waiting
room capacity) are studied by Fishman & Moore (1977). However, it is
our experience (with graduate students in management science and
econometrics) that these statistical complications are easily overcome. In general, any Markov system shows the renewal property. However, practical problems remain: The renewal state may occur so in-
frequently that too few subruns result; see Hopmans & Kleijnen (1978)
for an example. Approximative renewal states may then be formulated;
see Gunther (1975). An excellent textbook on the renewal approach is
Crane & Lemoine (1977); see also Fishman (1977). Since we feel that
this is an important technique for the exact analysis of steady-state
(academic) simulations, we refer to some more applications: Lavenberg
& Slutz (1975) and Schwetman & Bruell (1976). A case study (a simple
time sharing system) comparing this approach to replicated runs (ap-

proach 1) is provided by Sargent (1977).
        The initialization problem was mentioned several times in
the present section. Remember that this problem exists primarily in
steady-state, academic simulations. An example of initialization in
a practical simulation for planning purposes, is provided by Jain
(1975, p. 85): "At the start of a simulation the model represents the
actual state of the machine shop." Note further that in the renewal
approach to steady-state simulations, there is no start-up problem:
observations can be collected immediately when the simulation starts
in the renewal state (e.g. the empty state). In the other three approaches the transient phase does pose a problem: Usually initial observations are thrown away though this is not necessarily optimal.
Recently Wilson & Pritsker (1978) investigated a variety of heurist-
ics; it seems best to start the simulation run in the most likely
stationary state, and to retain all observations (transient and
steady-state).

8. MISCELLANEOUS STATISTICAL PROBLEMS

        In the above sections we concentrated on those statistical
problems we thought to be most relevant in the analysis and design
of simulation experiments. In the present section we briefly discuss
some remaining issues.

1) Multivariate responses

        In practice a number of criteria and measures are of inter-
est, e.g., in a queuing situation we may be interested in both waiting time and server utilization, measured by their means and their
90% quantiles. Though sophisticated statistical tools exist (e.g.,
Multivariate Analysis of Variance or MANOVA), it is practical to use
univariate techniques and to account for the multivariate character
by the choice of an appropriate error rate (Bonferroni inequality);
see Kleijnen (1975) for more details.

2) Multiple comparison and ranking procedures

        In the first few sections we discussed situations where a
number of factors define a great many system variants of potential
interest. Relevant techniques are regression analysis and experimental design. A different situation exists, when there are only a few
system variants, say, 10 or in general k. These systems may correspond to different queuing disciplines, etc. Multiple comparison procedures (MCP) are suited to situations with k system (or populations
in statistical jargon) and a fixed number of simulation runs (observations) per system. MCP give exact statistical results (controlled α
errors) when comparing $k(k-1)/2$ systems with each other, or when selecting a subset containing the "best" system (say, highest mean response), etc. Multiple ranking or selection procedures (MRP) have
been developed for situations where the number of runs is not fixed,
but has to be determined such that the best system can be selected
with a prespecified probability of correct selection. Both MCP and
MRP are discussed at length in Kleijnen (1975); a recent bibiography
is Dudewicz & Koo (1978). At present simulation practitioners have
shown little interest in these procedures; the only applications we
are aware of are Lin (1975) and Vicéns & Schaake (1972).

3) Statistical input: random numbers, etc.

        We have investigated the statistical analysis of the simulat-
ion output. On the input side we have the traditional problems of

random number generation (multiplicative and shift back generators) and sampling from distributions (including multivariate distributions). References can be found in Kleijnen (1975).

4) Model validation

Checking whether the model's output conforms with the real world observations can be based on a variety of statistical techniques: t-tests, goodness-of-fit tests, regression analysis, etc. However, this issue involves many more aspects than just statistics; see, e.g., Zeigler (1976).

9. CONCLUSION

Simulation means experimentation, albeit experimentation using a mathematical model instead of the real world. Any experiment requires a sound design. Without such a design even the most sophisticated analyses fails, e.g., if the factors 1 and 2 are changed simultaneously, their separate effects cannot be estimated. Scientific designs such as 2^{k-p} factorials further make it possible to explore the simulation space much more efficiently. The statistical analysis, given the design, should extract as much information from the experiment as is possible, e.g., estimate interactions. Such an analysis can be done systematically by means of a formal metamodel, i.e., a regression model. Moreover, such an analysis shows the limitations of the conclusions. For instance, if the simulation run is too short, so much stochastic noise may be present that instead of an expensive simulation model, a toss of the coin had better been used.

NOTES

1) In Operations Research, for instance, several surveys have shown that simulation is a most popular technique (together with linear programming and statistical techniques such as regression analysis); see Ledbetter & Cox (1977) for a recent survey and for additional references.

2) For instance, randomization and blocking are important topics in statistics. However, in simulation the experimenter has complete control over the experiment so that these topics become unimportant; see, however, footnote 6.

3) Actually the curves in FIG. 1 are not straight lines so that they represent more general formulations than eq. (3.5).

4) These "new" observations might correspond to the "center" of the design ($x_j = 0$ for all factors), in order to check whether pure quadratic effects are zero.

5 Suppose the individual factors are $x_1, x_2, \ldots$ and we know the signs of the β's, e.g., $\beta_1 \geq 0$ and $\beta_2 \geq 0$. Then x_1 and x_2 can be combined into a single group-factor z_1 with main effect $\gamma_1 = \beta_1 + \beta_2$ (≥ 0). Hence z_1 is +1 (or -1 respectively) if all its component factors are +1 (or $x_1 = x_2 = -1$ respectively). Execute only two runs (instead of $2^2 = 4$ runs):

run 1: $x_1 = -1 \qquad x_2 = -1 \rightarrow z_1 = -1$

run 2: $x_1 = +1 \qquad x_2 = +1 \rightarrow z_2 = +1$

If responses do not differ significantly, we can conclude that
neither x_1 nor x_2 are important, and eliminate x_1 and x_2 from further experimentation. In general, k individual factors x can be
combined into g group-factors z which can be tested in a 2^{g-p}
fractional factorial design; see Kleijnen (1975 b).

6) Split the N design points in two orthogonal blocks, and run one
block with common random numbers, and the other block with the
antithetic numbers.

7) Notice that n in eq. (7.2) is deterministic, whereas $\underline{n}$ is stochastic in eq. (7.3). Nevertheless eq. (7.3) gives satisfactory results.

8) When the number of subruns is M, then the variance of the estimated variance is $2\sigma^4/M$ so that the confidence interval for the mean
η becomes less stable (but has the same expected length).

9) For the definition of quantiles and percentiles consider
$P(x < y) = z$. If $z(0 \leq z < 1)$ is fixed to, say, 90% then we have
to estimate the "quantile" y. If we fix y, then we have to estimate
the "percentile" z.

REFERENCES

BARD, Y., NONLINER PARAMETER ESTIMATION. Academic Press, Inc.,
New York, 1974.

BENS, J. and G.C. VANSTEENKISTE, Structure characterization for
system modeling in uncertain environments (this volume).

BLANNING, R.W., The sources and uses of sensitivity information.
INTERFACES, 4, no. 4, August 1974, pp. 32-38. (See also 5, no. 3,
May 1975, pp. 24-25).

CLARK, G.M., AN IMPROVED PROCEDURE FOR ESTIMATING THE VARIANCE OF A
TIME-SERIES AVERAGE CALCULATED FROM SAMPLE AUTOCOVARIANCES OF SIMUL-
ATION OUTPUT DATA. Department of Industrial and Systems Engineering,
The Ohio State University, Columbus, 1977.

COPPUS, G., M. VAN DONGEN and J.P.C. KLEIJNEN, Quantile estimation
in regenerative simulation: a case study. PERFORMANCE EVALUATION
REVIEW, 5, no. 3, Summer 1977, pp. 5-15. (Reprinted in SIMULETTER, 8,
no. 2, Jan. 1977, pp. 38-47.)

CRANE, A. and J. LEMOIGNE, AN INTRODUCTION TO THE REGENERATIVE METHOD
FOR SIMULATION ANALYSIS. Springer-Verlag, Berlin, 1977.

DANIEL, C., APPLICATIONS OF STATISTICS TO INDUSTRIAL EXPERIMENTATION.
John Wiley & Sons, Inc., New York, 1976.

DRAPER, N.R. and H. SMITH, APPLIED REGRESSION ANALYSIS. John Wiley &
Sons, Inc., New York, 1966.

DUDEWICZ, E.J. and J.O. KOO, A CATEGORIZED BIBLIOGRAPHY ON RANKING
AND SELECTION PROCEDURES. Technical Report No. 163, Draft, Department
of Statistics, The Ohio State University, Columbus, June 1978.

FISHMAN, G.S., Achieving specific accuracy in simulation output analysis. COMMUNICATIONS ACM, 20, no. 5, May 1977, pp. 310-315.

FISHMAN, G.S. and L.R. MOORE, ESTIMATING THE MEAN OF A CORRELATED BINARY SEQUENCE WITH AN APPLICATION TO DISCRETE EVENT SIMULATION. Technical Report No. 77-2, Curriculum on Operations Research and System Analysis, University of North Carolina, Chapel Hill, April 1977.

FOX, B., Estimation and simulation. MANAGEMENT SCIENCE, 24, no. 8, April 1978, pp. 860-861.

GEOFFRION, A.M., The purpose of mathematical programming is insight, not numbers, INTERFACES, 7, no. 1, Nov. 1976, pp. 81-92.

GUNTHER, F.L., THE ALMOST REGENERATIVE METHOD FOR STOCHASTIC SYSTEM SIMULATIONS. Research Report No. 75-21, Operations Research Center, University of California, Berkeley, Dec. 1975.

HOPMANS, A.C.M. and J.P.C. KLEIJNEN, REGRESSION ESTIMATORS IN SIMULAT-ION. Report FEW-70, Department of Economics, Katholieke Hogeschool, Tilburg (Netherlands), Dec. 1977.

HOPMANS, A.C.M. and J.P.C. KLEIJNEN, IMPORTANCE SAMPLING IN SYSTEMS SIMULATION: A PRACTICAL FAILURE? Report FEW-73, Department of Economics, Katholieke Hogeschool, Tilburg (Netherlands), June 1978.

IGNALL, E.J., P. KOLESAR and W.E. WALKER, Using simulation to develop and validate analytic models: some case studies. OPERATIONS RESEARCH, 26, no. 2, March-April 1978, pp. 237-253.

JAIN, S.K., A simulation-based scheduling and management information system for a machine shop. INTERFACES, 6, no. 1, part 2, Nov. 1975, pp. 81-96.

KLEIJNEN, J.P.C., STATISTICAL TECHNIQUES IN SIMULATION. (In two parts.) Marcel Dekker, Inc., New York, 1974/1975.

KLEIJNEN, J.P.C., Antithetic variates, common random numbers and optimum computer time allocation. MANAGEMENT SCIENCE, APPLICATION SERIES, 21, no. 10, June 1975 (a), pp. 1176-1185.

KLEIJNEN, J.P.C., Screening designs for poly-factor experimentation. TECHNOMETRICS, 17, no. 4, Nov, 1975 (b), pp. 487-493.

KLEIJNEN, J.P.C., Discrete simulation: types, applications, and problems. In: SIMULATION OF SYSTEMS, edited by L. DEKKER, North-Holland Publishing Company, Amsterdam, 1976.

KLEIJNEN, J.P.C., GENERALIZING SIMULATION RESULTS THROUGH METAMODELS. Working Paper 77.070, Department of Business and Economics, Katholieke Hogeschool, Tilburg (Netherlands), Dec. 1977. (A summary will be published in IEEE TRANSACTIONS ON SYSTEMS, MAN, AND CYBERNETICS, 1979.)

KLEIJNEN, J.P.C., COMPUTERS AND PROFITS: QUANTIFYING FINANCIAL BENEFITS OF INFORMATION. Addison-Wesley, Reading, 1979.

KLEIJNEN, J.P.C. and P.J. RENS, IMPACT revisited: a critical analysis of IBM's inventory package "IMPACT". PRODUCTION AND INVENTORY MANAGEMENT, 19, no. 1, first quarter, 1978, pp. 71-90.

KLEIJNEN, J.P.C., A.J. VAN DEN BURG en R.T. VAN DER HAM, Generalization of simulation results: practicality of statistical methods.
EUROPEAN JOURNAL OPERATIONAL RESEARCH, 1978 (to appear).

KOONS, G.F. and B. PERLIC, A STUDY OF ROLLING-MILL PRODUCTIVITY
UTILIZING A STATISTICALLY DESIGNED SIMULATION EXPERIMENT. Research
Laboratory, United States Steel Corporation, Monroeville (Penn.),
(1977?).

KOTIAH, T.C.T., Approximate transient analysis of some queuing systems.
OPERATIONS RESEARCH, 26, no. 2, March-April 1978, pp. 333-346.

LAM, C.F. and J. PEDERSEN, Continuous simulation of a complex queuing
system. SIMULATION, 29, no. 2, August 1977, pp. 42-48.

LAVENBERG, S.S. and D.R. SLUTZ, Regenerative simulation of a queuing
model of an automated tape library. IBM JOURNAL OF RESEARCH AND DE-
VELOPMENT, 19, Sept. 1975, pp. 463-475.

LAW, A.M., STATISTICAL ANALYSIS OF THE OUTPUT DATA FROM TERMINATING
SIMULATIONS. Report 78-4, Department of Industrial Engineering,
University of Wisconsin, Madison, April 1978.

LAW, A.M. and J.S. CARSON, A SEQUENTIAL PROCEDURE FOR DETERMINING THE
LENGTH OF A STEADY-STATE SIMULATION. WP 77-12, Department of Industrial Engineering, University of Wisconsin, Madison, April 1977.

LAWLESS, R.W., L.H. WILLIAMS and C.G. RICHIE, A sensitivity analysis
tool for simulation with application to disaster planning. SIMULATION,
17, no. 6, Dec. 1971, pp. 217-223.

LIITTSCHWAGER, J.M. and W.F. AMES, On transient queues: practice and
pedagogy. In: COMPUTER SCIENCE AND STATISTICS, EIGHTH ANNUAL SYMPO-
SIUM ON THE INTERFACE, edited by J.W. FRANE, Health Sciences Computing Facility, University of California, Los Angeles, 1975.

LIN. W.T., MULTIPLE OBJECTIVE BUDGETING MODELS: A SIMULATION. Working
Paper 12-01-75, Department of Accounting, Graduate School of Business
Administration, University of Southern California, Los Angeles, 1975.

LEDBETTER, W.N., and J.F. COX, Operations research in production
management: an investigation of past and present utilization.
PRODUCTION AND INVENTORY MANAGEMENT, 18, no. 3, third quarter, 1977,
pp. 84-92.

MAASS, A., et al., DESIGN OF WATER-RESOURCE SYSTEMS. Harvard University Press, Cambridge (Massachusetts), 1962.

MALINVAUD, E., STATISTICAL METHODS OF ECONOMETRICS. North-Holland
Publishing Company, Amsterdam, second revised edition, 1975.

MEISEL, W.S. and D.C. COLLINS, Repro-modeling: an approach to efficient model utilization and interpretation. IEEE TRANSACTIONS ON
SYSTEMS, MAN, AND CYBERNETICS, vol. SMC-3, no. 4, July 1973, pp. 349-
358.

MONTGOMERY, D.C. and V.M. BETTENCOURT, Multiple response surface
methods in computer simulation. SIMULATION, 29, no. 4, Oct. 1977,
pp. 113-121.

PEGELS, C.C., SYSTEMS ANALYSIS FOR PRODUCTION OPERATIONS. Gordon and Breach Science Publishers, New York, 1976.

ROSE, M.R. and R. HARMSEN, Using sensitivity analysis to simplify ecosystem models: a case study. SIMULATION, 31, no. 1, July 1978, pp. 15-26.

SARGENT, R.G., Statistical analysis of simulation output data. SIMU-LETTER, 8, no. 3, April 1977, pp. 21-31.

SCHATZOFF, M. and C.C. TILLMAN, Design of experiments in simulation validation. IBM JOURNAL OF RESEARCH AND DEVELOPMENT, 19, no. 3, May 1975, p. 252-262.

SCHRUBEN, L.W. and B.H. MARJOLIN, PSEUDO-RANDOM NUMBER ASSIGNMENT IN STATISTICALLY DESIGNED SIMULATION AND DISTRIBUTION SAMPLING EXPERI-MENTS. School of Operations Research and Industrial Engineering, College of Engineering, Cornell University, Ithaca, Nov. 1977.

SCHWETMAN, H.D. and S.C. BRUELL, When to stop a simulation run: a case study. SIMULETTER, 7, no. 4, July 1976, pp. 131-137.

SEILA, A.F., ON THE PERFORMANCE OF TWO METHODS FOR QUANTILE ESTIMAT-ION IN REGENERATIVE PROCESSES. Bell Telephone Laboratories, Inc., Holmdel (N.J. 07733), (1978?).

SHERDEN, W.A., Origin of simultaneity in corporate models. In: WINTER SIMULATION CONFERENCE, edited by H.J. HIGHLAND, T.J. SCHRIBER and R.G. SARGENT, ACM, New York, 1976.

VICÉNS, G.J. and J.C. SCHAAKE, SIMULATION CRITERIA FOR SELECTING WATER RESOURCE SYSTEM ALTERNATIVES. Report no. 154, Ralph M. Parsons Laboratory, Department of Civil Engineering, M.I.T., Cambridge (Mass.), Sept. 1972.

WEEKS, J.K. and J.S. FRYER, A methodology for assigning minimum cost due-dates. MANAGEMENT SCIENCE, 23, no. 8, April 1977, pp. 872-881.

WILSON, J.R. and A.A.B. PRITSKER, Evaluation of startup policies in simulation experiments. SIMULATION, 31, no. 3, Sept. 1978, pp. 79-89.

YEN, H.C. and W. PIERSKALLA, A SIMULATION OF A CENTRALIZED BLOOD BANKING SYSTEM. Michael Reese Hospital, Office of Operations Research, Chicago (Ill. 60616), 1977.

ZEIGLER, B.P., THEORY OF MODELLING AND SIMULATION. John Wiley & Sons, Inc., New York, 1976.

METHODOLOGY IN SYSTEMS MODELLING AND SIMULATION
B.P. Zeigler, M.S. Elzas, G.J. Klir, T.I. Ören (eds.)
© North-Holland Publishing Company, 1979

SOURCES OF UNCERTAINTY IN ECOLOGICAL MODELS[1]

R. V. O'Neill and R. H. Gardner
Environmental Sciences Division
Oak Ridge National Laboratory
Oak Ridge, Tennessee 37830

Three sources of errors in ecosystem models have been
analysed: (1) model bias or errors in model structure, (2)
measurement error or uncertainty in model parameters, and
(3) variability of natural ecosystems.

Bias may result from aggregation of a k component model into
k-n components. For some models, total error can be
segregated into the errors resulting from pair-wise
aggregation of components. Minimal errors result if
components with similar turnover times are aggregated, or
small components are lumped with large ones. Propagation of
measurement error by Monte Carlo methods shows that errors
are affected by model structure, the relationships among
parameters and the position of a component within the
system. Explicit results indicate that natural variability
(genetic, spatial, etc.) is poorly represented by mean
parameter values, but predictability is improved if
approximate probability distribution are considered.

INTRODUCTION

Ecological models are playing an ever-expanding role in research planning, in
presentation of research results, and in application of results to management
problems. In spite of the intense activity in model development and
application, relatively little attention has been given to examining the
magnitude of uncertainty associated with model predictions. O'Neill (1973)
showed that small errors in model parameters could result in significant errors
in model output. Recently, several workers have pointed to the uncertainty
introduced into a model as a result of representing a complex system with a few
state variables (Harrison, in press; Cale and Odell, in press; O'Neill and Rust,
in press). A few studies have applied Monte Carlo simulation to investigate
error propagation (Gardner and Mankin, in press; Garten et al., in press;
Gardner et al., submitted). However, these studies make no attempt at a
systematic analysis and review of the various sources of uncertainty in
ecosystem model predictions.

Variance in model output is the result of a number of interacting causes. For
purposes of the present review, we have grouped the sources under three
headings: model structure, parameter error, and natural variability in

---

[1]Research sponsored by the National Science Foundation under Interagency Agreement 40-700-78 with the U. S. Department of Energy, Oak Ridge National
Laboratory (operated by Union Carbide Corporation under contract W-7405-eng-26
with the U.S. Department of Energy). Publication No. 1249, Environmental
Sciences Division, ORNL.

ecological systems. Uncertainty associated with model structure results from
constructing a simple mathematical model to describe a complex natural system.
It is clear that the model can only represent limited aspects of the system
behavior. We will review results we have obtained (1) for error resulting from
representing a two-component system by a single-variable model; (2) for relative
error propagation in alternative models for the same ecological system, and (3)
for error propagation in a series of models of increasing structural complexity.

Errors in model parameters measured in independent laboratory and field
experiments represent a second major source of uncertainty. For present
purposes, we will consider each parameter as measured in a separate experiment.
We will review our Monte Carlo studies on (1) the sensitivity of variance on
model outputs to variance on individual parameters, (2) the effects of the
distribution of parameter values on the distribution of output, and (3) the
propagation of parameter errors in a time-varying model and its implications for
gathering validation data.

Ecological models at conventional scales of aggregation (e.g., using population
or biomass variables) also contain uncertainty due to (1) environmental (e.g.,
meteorological) variability, (2) genetic variability within populations, and (3)
spatial heterogeneity. These sources of intrinsic variability also lead to
errors if they are ignored in model development. We will review results we have
obtained on all three of these sources of natural variability.

UNCERTAINTY RESULTING FROM MODEL CONSTRUCTION

By definition, a model is a partial representation of a system. As a result of
condensing or aggregating (Zeigler, 1976) a system into a relatively small
number of state variables, an element of uncertainty is introduced (O'Neill,
1973; Zeigler, 1976; Harrison, in press; Cale and Odell, in press). This
uncertainty is a type of bias (Goodall, 1973) which results from the omission or
the inadequate representation of important processes and components.

Model Aggregation

The problem of model aggregation can be seen in an example presented by O'Neill
and Rust (in press) for a two-component linear system:

$$dx_1/dt = I_1 + cx_2 - (a + b) x_1$$
$$dx_2/dt = I_2 + bx_1 - (c + d) x_2 \ ,$$

These equations are a generalized form for many cases of ecological interest.
For example, they might represent a predator-prey system if $c = d = I_2 = 0$.
Let us represent this system by a single-component model:

$$dx_{(1+2)}/dt = I_1 + I_2 - k \ x_{(1+2)} \ ,$$

where

$$k = \frac{(ac + ad + bd)(I_1 + I_2)}{(b+c+d) \ I_1 + (a+b+c) \ I_2}$$

This expression for k is derived by requiring that $x_{(1+2)}(\infty) = x_1(\infty) + x_2(\infty)$. This is the case of greatest ecological interest since parameter values are ordinarily calculated by assuming mass balance at some equilibrium condition.

The error which is introduced by this lumping is:

$$E_t = x_1(t) + x_2(t) - x_{(1+2)}(t) \quad .$$

The differential equations for this problem are simple, and it is possible to derive an explicit closed solution for the error. O'Neill and Rust (in press) present a detailed analysis of the general case, represented at the top of Fig. 1. In the lower portion of the figure, we present solutions for the error term in four special cases.

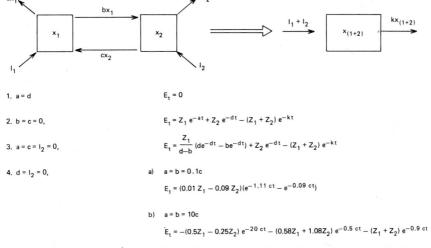

1. $a = d$ $E_t = 0$

2. $b = c = 0$, $E_t = Z_1 e^{-at} + Z_2 e^{-dt} - (Z_1 + Z_2) e^{-kt}$

3. $a = c = l_2 = 0$, $E_t = \dfrac{Z_1}{d-b} (de^{-dt} - be^{-dt}) + Z_2 e^{-dt} - (Z_1 + Z_2) e^{-kt}$

4. $d = l_2 = 0$,   a) $a = b = 0.1c$

                                    $E_t = (0.01 Z_1 - 0.09 Z_2)(e^{-1.11 ct} - e^{-0.09 ct})$

                              b) $a = b = 10c$

                                      $E_t = -(0.5Z_1 - 0.25Z_2) e^{-20 ct} - (0.58Z_1 + 1.08Z_2) e^{-0.5 ct} - (Z_1 + Z_2) e^{-0.9 ct}$

Fig. 1.   Error at time t resulting from representing a two-component system by a single-variable model. The error expression is given for four special cases. The quantity $Z_i = x_i(0) - x_i(\infty)$, $i = 1,2$.

When $a = d$, the system can be represented exactly by a single-component model and no error results from lumping. This is a special case of the conditions for lumpability (Zeigler, 1976). In general, lumpability does not require identical turnover rates but only a weaker "uniformity of influence" (Zeigler, 1976).

When $b = c = 0$, the two components are unconnected and we are simulating the aggregation of two populations in the same trophic level. In this case the error is strongly dependent on the difference between the magnitudes of a and d. If $a = c = l_2 = 0$, we are simulating the aggregation of two members of a food chain. In this case, the error function is more complex, containing an additional negative term, and the magnitude of the error depends strongly on the difference in magnitude of the rates, d and b.

If $d = I_2 = 0$, we are representing a "shunt" system in which compartment 2 has its only input and output through compartment 1. If the shunt (i.e., compartment 2) is large and slow (case a), the error is positive for all t, meaning that the aggregated model underestimates the $x_1 + x_2$ pool. If the shunt is small and fast (case b), the error is everywhere negative and the lumped model overestimates the $x_1 + x_2$ pool.

In all of the special cases given in Fig. 1, as well as in the more complex general case (O'Neill and Rust, in press), the error is strongly dependent on the difference between the output rates (i.e., a and d) of the two-component system. The more similar the parameters, the smaller the error introduced by lumping.

Alternative Models

Another approach to studying model bias is to compare alternative models for the same ecological system. For example, consider the five models of calcium cycling in four different forested ecosystems listed in Table 1. The models are: a four-compartment model parameterized for a tropical and two deciduous forests (Jordan et al., 1972), a four-compartment model for a deciduous forest (Waide et al., 1974), and a 29-compartment deciduous model condensed into seven compartments for this analysis (Shugart et al., 1976).

To determine the effect of model structure on error propagation, we randomly sampled the parameter values for these models from distributions with variances equal to 10% of the mean, generated 250 parameter sets for each model under three conditions (denoted in the columns of Table 1) and obtained the steady state solutions (see Gardner et al., submitted, for details of the Monte Carlo methods). The results were compared by partitioning the sum of squares of the analysis of variance into appropriate orthogonal t tests (Kirk 1968).

All five models were significantly different for both the litter and available nutrient compartments (case 1, Table 1). In this first case, mean parameter values were taken directly from the studies cited above. Repeating the analysis with the parameters normalized so the turnover of each compartment equals 1.0 (case 2, Table 1) tends to deemphasize differences due to the parameters and emphasize differences due to model structure. The litter values are similar for all three Jordan et al. (1972) models, while the available nutrient compartments are still distinctly different. When the forcing functions are also normalized so the sum of all inputs to the model equals 1.0 (case 3, Table 1), the differences in litter and available nutrient compartments are not significant for the first three models, but the differences among model types are still distinct.

The three Jordan et al. (1972) models are similar because the structures are identical. Although the normalized transfer matrices and forcing functions for the three models are not identical, they are similar enough that a 10% variation of the parameters obscures any differences. However, the consistent differences between the first model and the fourth model (Waide et al., 1974) is unexpected because these two models are based on data for the same forest. Even without normalization, these models should be indistinguishable. The similarity in behavior of the available nutrient compartment of the three Jordan et al. (1972) models is also unexpected because it indicates that differences in the inputs to these systems are sufficient to explain the dissimilarities between tropical and deciduous systems. The above conclusions are not ecologically valid and are due to the unintended results of model bias.

Table 1

Comparison of final state litter and available nutrient pools for
five calcium cycling models. The third model is for a tropical
system, the rest represent deciduous forests. Models which share
the same integer could not be distinguished by orthogonal t tests
(Kirk 1968) at $p < 0.05$. For example, in the first column all five
models are different. In the last column, the first three models
are indistinguishable. In all three cases, model parameters were
varied at   10% of their means.

	Case 1		Case 2 [a]		Case 3 [b]	
Model	Litter	Available nutrient	Litter	Available nutrient	Litter	Available nutrient
A [c]	1	1	1	1	1	1
B [c]	2	2	1	2	1	1
C [c]	3	3	1	3	1	1
D [d]	4	4	2	4	2	2
E [e]	5	5	3	5	3	3

[a] Parameters normalized so the sum of the output coefficients of each
compartment is 1.0.

[b] Parameters normalized and forcing functions normalized (sum of all
forcings = 1.0).

[c] Jordan et al. (1972).

[d] Waide et al. (1974).

[e] Shugart et al. (1976).

As another example, the 29-compartment model of Shugart et al. (1976) was solved
to steady state, the fluxes calculated, and four condensations made (Fig. 2).
The four condensations represent different logical ways of looking at the
system. In the first case (Fig. 2), above ground vegetation is aggregated
according to function. Thus, leaves are lumped into compartment one whether
they occur in canopy trees or herbs on the ground. In the second case, the
vegetation is subdivided by a vertical stratification into canopy, subcanopy,
etc. In this instance, leaves and woody tissue are combined for each stratum.
In the third case, the aboveground vegetation is lumped into a single
compartment and the belowground portions of the system are emphasized. In the
fourth case, the emphasis is on representing the route taken by calcium from
soil to root to bole to branch to leaf. In each condenstation, the system is
assumed to be in steady state and compartment values and flux rates between
compartments are summed from the original 29 compartment model. The appropriate
parameters are then calculated from these summed values.

The objective of this exercise was to compare the steady state values and the
variances of the original model with the four condensed versions. The

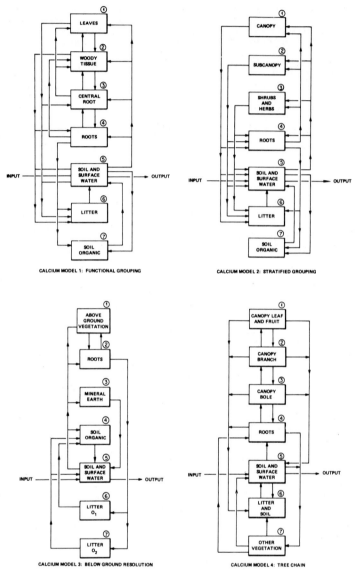

Fig. 2.   Four condensations of the 29-compartment model of Shugart et al.
           (1976). Transfer coefficients were adjusted to ensure equivalent
           total steady state values.

parameters of all models were varied at 10% of their mean, 300 Monte Carlo
solutions were obtained, the steady states and the variances of the original
model were analytically combined (see Snedecor and Cochran, 1967, p. 190), and
results compared between models. The ratios of the original to the condensed
models showed that steady state values differed by 2 to 32%. These results are

within the range expected from Monte Carlo simulations. Thus, the four model
condensations are unbiased, adequate representations of the larger system so
long as we are only interested in mean values and variances at steady state.
The transients of each model may well be distinctly different.

The task of model condensation is more difficult when the system is not
completely defined. The original model represented a completely known and
measured system. But consider the situation in which data on calcium in the
soil and surface water are not available to the modeler. A second series of
condensed models was formed by deleting the soil water compartment from the
original series. The mean steady state values were again within 2% of the
values predicted by the 29-compartment model, but the variances were an average
of 70% less (range 44 to 96%) than the variances of the original models. With
this reduced variance, the condensed models, which completely omitted an
important functional component, appeared to be dramatically improved over the
original.

Inspection of the recycling dynamics of the two series of models shows that a
calcium molecule would pass through the soil water compartment of the first
series of condensations an average of 29 to 30 times before it was lost from the
system. The deletion of this feedback mechanism in the second series of
condensations completely altered the dynamics of these models and, thus, the way
in which the errors were propagated.

Model Structure

The complex effects of model structure are illustrated by the following
example. Consider a series of four-compartment, donor-controlled, linear models
with increasing numbers of connections between the compartments (Fig. 3). All
models have unitary forcings (input) to compartment 1, respiration losses from
all compartments, and a normalized turnover rate (sum of transfer coefficients
for each compartment equal to 1.0). For each model structure (Fig. 3) we define
five parameter sets, each set having a 20% increase in material lost by
respiration (respiration increased from 0.2 to 0.8 in steps of 0.2), with the
proportion passing to succeeding compartments reduced by a complementary amount
so that the sum remains equal to 1.0. For each parameter set we randomly varied
the individual transfer coefficients by ± 10% of their mean, normalized value
and obtained 250 steady state solutions. The normalization was used to obtain
mean values and was not retained for each individual perturbation. The variance
of the steady state values can be partitioned into effects due to model
structure and effects due to parameter values.

Figure 3 shows that the relative magnitude of these effects are dependent on the
position of the compartment within the model. The dynamics of compartment 1 are
dependent on the forcings (held constant for these iterations) so the steady
state values of this compartment are nearly invariant. The second compartment
is most affected by structural differences between models (73% of the variance),
which result from material bypassing compartment 2. However, this effect is
less important to compartment 3 (29% of the variance) and unimportant to
compartment 4 (9% of the variance). Compartments 3 and 4 are most affected by
changes in the parameter values which cause successive increases in the amounts
of material lost from the system (58 and 84% respectively). Thus, the route
that material takes to reach the end of the chain is unimportant to compartments
at the end of that chain, but the rate of loss from the system is important.

Repeating the same analysis for six models with increasing numbers of feedbacks
(Fig. 4) shows that compartments 1 and 2 react to the interaction of parameter
and structure effects. There is no clearly defined final compartment for
feedback models, so material passing through compartment 1 has the chance to
return if the structure of the model provides the necessary linkages. However,
if the rate of loss from the system is relatively high, the effect of any

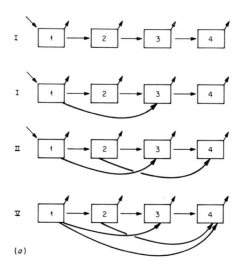

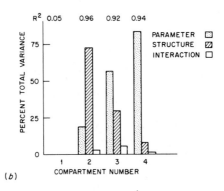

Fig. 3.   Comparison of error propagation in four four-compartment models. The
          relative magnitude of parameter versus structural effects for each of
          the four compartments is shown in the histogram.

feedback is reduced. Structural effects are less important to compartments 3
and 4, which are most affected by parameter changes which result in losses from
the system (Fig. 4).

UNCERTAINTY RESULTING FROM PARAMETERS MEASURED IN INDEPENDENT EXPERIMENTS

A second major source of uncertainty in model predictions is error in the
measurement of parameter values. In this section, we will focus on the
situation in which each parameter is measured in a separate laboratory or field

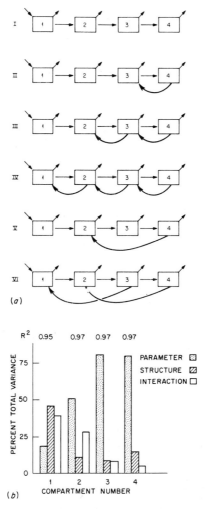

Fig. 4.   Comparison of error propagation in six four-compartment models with
           feedback. The relative magnitude of parameter versus structural
           effects for each of the four compartments is shown in the histogram.

experiment. If the distribution of parameter measurements is known, the error
can be propagated through a model to calculate confidence intervals for
predictions (Furness, 1978; Gardner and Mankin, in press). Such knowledge is of
considerable practical and theoretical importance, permitting quantitative
comparison of model output and validation data, clearly indicating the
resolution or precision of the model prediction, and placing model predictions
in the proper context for management applications.

Parameter Sensitivity

The sensitivity of model output to individual parameters can vary greatly. Thus, small changes in the variance, or small shifts in the expected value of some parameters may have a large impact on model predictions, while even large changes in other parameters will have little effect. There is a considerable literature dealing with parameter sensitivity analysis (e.g., Tomovic, 1963).

Monte Carlo techniques used in our studies have the advantage that they provide the information necessary for a multivariate analysis of the uncertainty in model output. For example, in a recent study (Garten et al., in press), the time to steady state, τ, was taken as a measure of model performance [$\tau = 1/\lambda$, where is the largest (i.e., least negative) eigenvalue of the linear model)]. Varying all 12 parameters in the model at $\pm 10\%$ of their means, the variance of λ was 6.7% of its mean. Variance in τ was most related to the two coefficients (r's = -0.75 and -0.65) which describe input and output of the largest compartment. The variance of τ drops to 0.02% when these two coefficients are held constant.

Another technique which is useful for sensitivity analysis is correlation analysis which is capable of separating into independent factors those parameters which contribute most to the variance of a particular variable. Applying this technique to the Monte Carlo simulations of global carbon models (Gardner et al., submitted) showed that the atmospheric levels of CO_2 were affected most (1) by the big slow components (54% of the variance due to carbon deposition and release from sediments); (2) by the rapid exchange mechanisms between the ocean surface and the atmosphere (14%); and (3) by the carbon turnover in the ocean circulation (10%).

Parameter Distributions

The effect that variances of individual parameters may have on the confidence levels of prediction will depend on the distribution of parameter values (including the likelihood of extreme values), construction of the model (including parameter sensitivities), and the self-limiting or damping of errors by the model. Consider the simple linear model with time invariant parameters,

$$x_t = x_0 e^{-\alpha t} + u .$$

If the exponent, α, is varied normally about its mean value, this model produces steady state values, x_α, which are log-normally distributed. The mean of the steady states is a biased estimate of the expected value, and the estimation of fluxes and transfer coefficients based on the mean will also be biased (Gardner and Mankin, in press).

The situation is more complex for multiple-compartment linear models. If the transfer coefficients of a system with no feedbacks are all normally varied and the forcings held constant, the variance of steady state values of a specific compartment, x_i, will be related to the variance of its turnover rate, $_{ii}$, and the distance (i.e., the number of intervening compartments) from the forcing function (Gardner and Mankin, in press). The more intervening compartments there are, the greater will be the variance of the material transferred to compartment i. Because the variance increases with distance, the highest variance of steady state values will be associated with compartments at higher trophic levels, even though these compartments may be set up to have the smallest variance of turnover rates.

The effect becomes even more complex in a model of material recycling, with the extreme case reached in closed systems. Steady state values for smaller compartments may be skewed left (versus skewed right for the log-normal distribution). This effect is most pronounced in "stiff" systems, where transfers of large quantities occur by slight shifts in some parameter values.

In all the above cases, the expected values for fluxes and transfers are difficult to estimate. Median values (rather than means) are the best estimators of expected compartment values for calculating instantaneous fluxes and transfer coefficients. There is a significant bias in using only the mean values as is common in ecological modeling. Our results suggest that both median and mean values should be estimated and compared to detect potential bias.

Model Validation

In time-varying models, error propagation is a complex function of time, and can be an important consideration in choosing data for model validation. O'Neill et al. (in press) analyzed measurement error (parameters at ± 4% of their means) propagated through an herbivore-carnivore model. Model simulations were made over an annual cycle with parameters expressed as a function of temperature which varied sinusoidally over the year.

An impression of the uncertainty of model output as a function of time can be gained from Fig. 5. The figure shows the number of simulations retained up

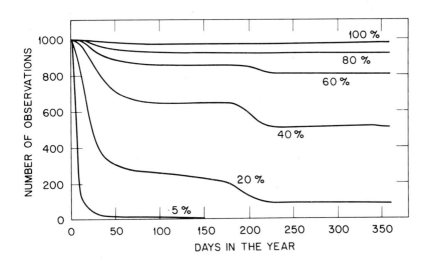

Fig. 5.   Error as a function of time in a complex nonlinear model. The observations are the number of Monte Carlo simulations within a given percentage of the expected value, up to and including any specific day during the year.

until a specific date according to several error criteria. Thus, the number of simulations shown by day 200 has satisfied that error criterion for every date up to and including day 200. The error criterion is expressed as a percentage plus or minus the expected value. If a particular Monte Carlo simulation was

within the error "window" on a given day, it was retained as an observation. If
a simulation exceeded the allowable interval around the expected value, it was
eliminated.

Almost every run satisfies the criterion when the window is set at 100%.
Setting the window at 5% disqualifies almost every run before day 50. For
windows less than 50% a significant number of runs are eliminated before day
50. From day 50 to day 175 very few additional runs are eliminated. This
indicates that model and data are well-behaved in this region and little
discrimination is possible. Immediately following day 175 the curves drop
sharply indicating that this specific time interval would be extremely useful in
detecting uncertainty in model output.

Following the mid-summer minimum, very few additional simulations are
eliminated. This indicates that model error is well-behaved in this region just
as in the interval 50-175. Of course, it also indicates that any simulations
which have satisfied the stringent requirements up to day 200/220 are likely to
be well-behaved for the remainder of the year.

Our analysis indicates that optimal sampling dates for validation would be
during the initial 50 days and then again intensively over the interval
surrounding the mid-summer minimum. Data collected during other times of the
year would add little to the validation process. Although these exact time
intervals are limited to the model used, the present study indicates that Monte
Carlo simulation of a preliminary model would be useful in determining optimal
sampling strategies for validating a model. Current practice is to gather field
data at uniform intervals throughout the year. Our results indicate that a more
efficient approach would be to conduct intensive sampling only during specific
intervals of the year.

UNCERTAINTY RESULTING FROM NATURAL VARIABILITY

The third major source of error in model predictions is the natural variability
of ecological systems. This type of error is introduced through assumptions
adopted during initial model formulation. Of the numerous possibilities, we
will emphasize three in our present discussion: (1) the assumption that
environmental variables can be represented as deterministic, (2) the assumption
that ecological systems can be represented as spatially homogeneous, and (3) the
assumption that population parameters can be represented by mean values,
ignoring genetic variability. We will discuss each of these three assumptions
in turn.

Environmental Variability

Ecosystems are directly influenced by changes in environmental variables (e.g.,
temperature, precipitation, light, etc.). In ecosystem models, rate processes
(e.g., growth, respiration, etc.) are represented as functions of these
environmental variables. The lack of predictability of the environment thus
leads to an uncertainty in the model predictions.

In existing models, environmental uncertainty is usually handled through the
interpretation placed on model output. Thus, if average values for
environmental variables are supplied to the model, the output is considered to
represent ecosystem behavior during some hypothetical average year. This
interpretation seems legitimate, but the fact remains that our ability to
predict behavior in any given future year depends on our ability to predict the
time course of environmental variables influencing system behavior.

To the best of our knowledge, only a single ecological study (O'Neill, 1971) has
dealt explicitly with the contribution of environmental variability to model
uncertainty. In this study of centipede population energetics, hourly

temperatures during May were recorded under the forest litter. The data were fitted by least squares to a sine curve with a 24-hr period and with a standard error of estimate of 1 degree. In the Monte Carlo model, temperature for a given hour was selected from a uniform distribution ranging from -1 to +1 around the value calculated from the sine function. The model contained other random variables dealing with intrinsic biological factors such as the time required for successful capture of a prey organism. By examining the variability in model output (e.g., calories of prey ingested over the month), it was possible to analyze for the contribution due to randomly varying the temperature. In this case, setting temperature to its expected value (i.e., value given by the sine function) results in a 17% reduction in overall variance, which is not a significant reduction given the degrees of freedom of the problem. Thus, for this example, ignoring the intrinsic uncertainty in temperature did not significantly increase uncertainty in the model prediction. However, the variability in temperature was small and it is unlikely that examination of parameters, such as rainfall, in other models would lead to such fortuitous results. In general, environmental variability can be expected to be a significant contributor to uncertainty whenever model objectives require predictions of future behavior of the system.

Spatial Heterogeneity

The current generation of models, for the most part, considers the ecosystem as spatially homogeneous. This assumes that either the system is uniform in space or else some average property can be described as independent of spatial patterning. Since it is well established that spatial heterogeneity occurs (e.g., Pielou, 1969), the important point becomes the degree to which the dynamics of an average system property are dependent on the pattern. The effects of spatial heterogeneity on the dynamics of small population systems has received attention and a review of some aspects of this work can be found in Zeigler (1978).

In a study of stream ecosystems (O'Neill et al., in press), distributed and lumped representations of the same system were compared. If rate processes were distributed uniformly along the length of the stream (i.e., all stream segments equal), the spatially homogeneous, or lumped, representation was fully adequate. However, as a gradient was introduced (i.e., downstream segments processed material at different rates from upstream segments), the ability to represent the system as spatially homogeneous decreased rapidly. The greater the difference between upstream and downstream segments (i.e., the steeper the gradient), the greater the error in representing average properties of the system. Thus, when adjacent segments differed in rate processes by 30%, the spatially homogeneous model overestimated average heterotroph biomass by 20% and underestimated average particle size by almost 40%. In addition, the lumped model was unable to represent behavior in response to perturbation. If the disturbance was not distributed uniformly along the length of the stream, the lumped model could simulate neither initial response nor recovery to normal behavior. Smith (submitted) has given a detailed analysis of the effect of spatial gradients on the stability of ecosystem models. He was able to demonstrate a relationship between time to recovery from a perturbation and the steepness of the spatial gradient. In addition, he has suggested approaches to lumping the spatial system into a homogeneous model that minimizes the error.

Genetic Variability

With few exceptions (O'Neill, in press), ecosystem models are deterministic. Individual parameters (e.g., respiration rate for a population or trophic level) are considered to be constants, ignoring the intrinsic genetic variability within the population or species variations within the trophic level. The assumption that a parameter is constant across a population introduces an error into model predictions.

The potential for this error should be obvious from experience with pest
populations and DDT. Since the pesticide increases average mortality, the
logical prediction is extinction in any model that ignores genetic variability.
In fact, susceptability to DDT is distributed within the population so that some
segments are highly resistant. As a result, the population recovers in spite of
continued application of DDT. This recovery is a result of the genetic
variability of the population and could not be represented in a model that
ignores such variability.

Because of the mathematical complexities of dealing with model parameters as
random variables, little work has been done on this source of error. However,
significant insight can be gained by simple assumptions about the way the
parameter is distributed. We have had some success by assuming that parameters
were distributed within the population according to a triangular distribution.

Consider a single parameter model for a population process:

$$dx/dt = -ax \ ,$$

where x is the average quantity of some material in each individual of a
population and a is a rate constant expressing instantaneous loss. Equations of
this form are useful for describing the loss of a substance, such as a
radioisotope, from the organisms. For our purpose, we will assume that the
number of individuals in the population does not change over the time interval
of interest. A solution for this equation is given by

$$x_t = x_0 \ e^{-at} \ . \qquad\qquad (1)$$

Now let us consider that the parameter a is not constant for every individual in
the population but is distributed according to a triangular distribution, f(a).
We can describe this distribution as ranging from plus to minus some fraction,
k, of the mean value, a (i.e., $a(1-k) \leq a \leq a(1+k)$). For simplicity, we will
assume that initially each individual contains exactly the same quantity of
material, x_0. Calculation of the true value for concentration of the
substance in the population can now be made by integrating over the left and
right halves of the triangular distribution:

$$E(x_t) = x_0 \int_{a(1-k)}^{a} f(\alpha)e^{-\alpha t} \ d\alpha + x_0 \int_{a}^{a(1+k)} g(\alpha)e^{-\alpha t} \ d\alpha \ , \qquad (2)$$

where simple trigonometry yields

$$f(\alpha) = \frac{\alpha - a(1-k)}{(ak)^2} \ , \ \text{and}$$

$$g(\alpha) = \frac{a(1+k) - \alpha}{(ak)^2} \ .$$

Integration of Eq. 2 shows

$$E(x_t) = x_0 \, e^{-at} \left[\frac{e^{akt} + e^{-akt} - 2}{(akt)^2} \right] . \qquad (3)$$

Comparison of Equations 1 and 3 reveals that the term in brackets is a multiplicative error term that expresses the error that results from ignoring the genetic variability in the parameter a. As one might expect, the bracketed term shows that the error grows through time. The greater the variability of the parameter within the population (i.e., the larger the value of k), the greater is the value of the maximum error.

A similar approach can be taken with any simple function. Whenever the function of interest is nonlinear in the distributed parameter, the error is non-zero. Thus, there can be a significant bias in predictions from models that ignore genetic variability in model parameters.

DISCUSSION

It is apparent from the wide variety of examples reviewed here that uncertainties and errors can be introduced into model predictions from many sources. The number of potential sources of error and the continuing discovery of counterintuitive results in our studies make it difficult to synthesize the information at the present stage of development. It is particularly difficult at this time to outline practical approaches to minimizing error when approaching a new modeling problem. Such a synthesis will have to await the results of continuing studies.

Although some interesting analytical results have been achieved with linear and simple nonlinear models, the real challenge lies with complex ecosystem models that have been developed during the past decade (e.g., Shugart et al., 1974; Innis, 1978; Patten, 1975). Investigation of error propagation in these large nonlinear models will have to be inductive, with numerous individual applications required before general patterns begin to emerge.

In addition to applying the approach to a great variety of complex models, it will also be important to investigate interactions among the various sources of error. Changing the structure of a model (e.g., from linear to nonlinear) not only changes the dynamics of the model but also changes error propagation properties. O'Neill (1973) was able to show that, for a wide neighborhood around the equilibrium, a biased linear model was able to provide predictions with less uncertainty than an unbiased nonlinear model. This result indicates that, in some cases, errors from different sources may tend to cancel each other. Undoubtedly, in other cases, the errors will be found to augment each other. These types of interactions will have to be understood before we will be able to suggest ways for minimizing overall error.

Error analysis is just beginning to be applied to ecological models. This is evident from the number of "in press" citations in our references. Hopefully this new effort will result in a systematic and rigorous approach to model building that will help transform ecological modeling from an art to a science.

REFERENCES

Cale, W. G., and P. L. Odell. (In press). Concerning aggregation in ecosystem modeling. IN E. Halfon (ed.), Theoretical Systems Ecology. Academic Press, NY.

Furness, R. W. (1978). Energy requirements of seabird communities: A bioenergetics model. J. Anim. Ecol. 47:39-53.

Gardner, R. H., J. B. Mankin, and W. R. Emanuel. (Submitted). A comparison of three carbon models. Ecol. Model.

Gardner, R. H., and J. B. Mankin. (In press). Analysis of compartment models as representations of woodland ecosystems. IN D. E. Reichle (ed.), Dynamic Properties of Forest Ecosystems, Cambridge University Press, England.

Garten, C. T. Jr., R. H. Gardner, and R. C. Dahlman. (In press). A compartment model of plutonium dynamics in a deciduous forest ecosystem. Health Phys.

Goodall, D. W. (1973). Building and testing ecosystem models. pp. 173-194. IN J. N. R. Jeffers (ed.), Mathematical Models in Ecology. Blackwell Scientific Publications, London. 398 pp.

Harrison, G. W. (In press). Compartmental lumping in mineral cycling models. IN Proceedings of Environmental Chemistry and Cycling Processes Symposium. Savannah River Ecology Laboratory.

Innis, G. S. (ed.). (1978). Grassland Simulation Model. Springer-Verlag, NY. 298 pp.

Jordan, C. F., J. R. Kline, and D. S. Sasscer. (1972). Relative stability of mineral cycles in forest ecosystems. Am. Nat. 106:237-253.

Kirk, R. E. (1968). Experimental Design Procedures for the Behavioral Sciences. Brooks/Cole Publishing Co., Belmont, CA. 577 pp.

O'Neill, R. V. (1971). A stochastic model of energy flow in predator compartments of an ecosystem. pp. 107-121. IN G. P. Patil, E. C. Pielou, W. E. Waters (eds.), Statistical Ecology. Volume III: Many Species Populations, Ecosystems and Systems Analysis. Pennsylvania State University Press, University Park, PA.

O'Neill, R. V. (1973). Error analysis of ecological models. pp. 898-908. IN D. J. Nelson (ed.), Radionuclides in Ecosystems. USAEC-CONF-710501. Technical Information Division, Washington, DC. 1268 pp.

O'Neill, R. V. A review of stochastic modeling in ecology. (In press). IN S. E. Jorgensen (ed.), State of the Art in Ecological Modelling. Copenhagen.

O'Neill, R. V., and B. W. Rust. (In press). Aggregation error in ecological models. Ecol. Model.

O'Neill, R. V., J. W. Elwood, and S. G. Hildebrand. (In press). Theoretical implications of spatial heterogeneity in stream ecosystems. IN D. L. DeAngelis (ed.), Computer Models in Ecology. Springer-Verlag, NY.

O'Neill, R. V., R. H. Gardner, and J. B. Mankin. (Submitted). Propagation of parameter error in a nonlinear model. Ecol. Model.

Patten, B. C. (ed.). (1975). Systems Analysis and Simulation Ecology, Volume III. Academic Press, NY. 601 pp.

Pielou, E. C. (1969). An introduction to mathematical ecology. Wiley-Interscience, New York. 286 pp.

Shugart, H. H., R. A. Goldstein, R. V. O'Neill, and J. B. Mankin. (1974). TEEM: A terrestrial ecosystem energy model for forests. Oecol. Plant. 9:231-264.

Shugart, H. H., D. E. Reichle, N. T. Edwards, and J. R. Kercher. (1976). A model of calcium cycling in an east Tennessee Liriodendron forest: Model structure, parameters and frequency response analysis. Ecology 57:99-109.

Smith, O. L. (Submitted). The influence of environmental gradients on ecosystem stability. Am. Nat.

Snedecor, G. W., and W. G. Cochran. (1967). Statistical Methods. Iowa State University Press, Ames, Iowa. 593 pp.

Tomovic, R. (1963). Sensitivity Analysis of Dynamic Systems. McGraw-Hill, New York. 142 pp.

Waide, J. B., J. E. Krebs, S. P. Clarkson, and E. M. Setzler. (1974). A linear systems analysis of the calcium cycle in a forested ecosystem. Prog. Theor. Biol. 3:261-345.

Zeigler, B. P. (1976). Theory of Modelling and Simulation. Wiley, New York.

Zeigler, B. P. (1978). Persistence and patchiness of predator-prey systems induced by discrete event population exchange mechanisms. J. Theor. Biol. 67:687-713.

METHODOLOGY IN SYSTEMS MODELLING AND SIMULATION
B.P. Zeigler, M.S. Elzas, G.J. Klir, T.I. Ören (eds.)
© North-Holland Publishing Company, 1979

COMPOSITION SPACES AND SIMPLIFICATIONS

Sudhir Aggarwal
Department of Mathematics
University of California
Riverside, California   92521

Any investigation of a complex model often uses
several methods of analysis. One method is to
analyze a complex model by studying a "simpler"
or lumped model. This process is termed simplification. A second method is to decompose the
model into component parts. The formalization
of this aspect is via the concept of a composition space. Given these two methods it is natural to combine them. We show that this is
possible by proving that if lumped model components are simplifications of base model components, then the lumped model is a simplification
of the base model. This is shown to hold for
both the case of deterministic simplification as
well as the case of probabilistic simplification.

INTRODUCTION

An investigation of a complex model---whether a real physical model
or an abstract mathematical one---often uses several methods of analysis. One method is to analyze a complex model by studying an alternate "simpler model." Thus, given the original or base model, we derive from it a lumped model that has performance characteristics of
interest which are similar to those of the base model. This process
is termed simplification. A second method of analysis is to decompose (or redefine) a model in terms of a set of constituent parts.
The performance characteristics of the model are defined in terms of
the component characteristics. For this to be possible, it is apparent that the method of interconnection must also be given. Rather
than discussing decomposition, we develop our results in terms of
composing elements and introduce the concept of a composition space.

Given the two methods of analysis defined above, it is natural to

attempt to combine them. We show that this is possible by precisely
formulating and answering the following question. If components of
a lumped model are simplifications of corresponding components of a
base model, under what conditions is the lumped model a simplification of the base model?

This paper is organized as follows. Section 1 introduces the concept
of a composition space and relates it to homogeneous and heterogeneous algebras. This makes precise the notion of building a model from
more primitive components. Section 2 formalizes the simplification
concept by comparing base and lumped model performance values. These
formalizations show that for both the case of deterministic modeling
(Section 3) and probabilistic modeling (Section 4), our original
question can be answered in the affirmative.

1.   COMPOSITION SPACES

The concept of a composition space that is introduced will be seen to
be compatible with the simplification idea that is considered later.
Our focus of interest is on a set of models S for which we can extract some feature(s) of interest using a performance evaluation map
Φ . The values that are calculated from the models in S lie in the
set V .

Definition A <u>systems space</u> is a 3-tuple $\langle S,\Phi,V \rangle$ where S and V
are sets and $\Phi:S \to V$. The set S is the set of models, V is the
set of performance values, and Φ will be called the performance e-
valuation map.

As a simple example, S could be the set of chairs in the world, V
the set of colors, and Φ the map that defines the color of a chair
under certain standard conditions. In general, a model $s \in S$ often
has a mathematical structure---for example, $S = \{X,Y,Z,\delta,\lambda\}$ for S
a finite state automaton, with the definitions of $X,Y,...$ as appro-
priate---but this will not concern us in the present development.
For our purposes, S and V are just abstract sets and Φ is a
mapping.

We first consider the definition of an n-composition space that will
motivate our more general definition. S^n denotes the n-fold cross
product of S and Φ^n denotes the map that is Φ on each component.

Definition $\langle S',\Phi',V' \rangle$ is an <u>n-composition space</u> of $\langle S,\Phi,V \rangle$ by

$\langle h,k \rangle$ if $h:S^n \rightarrow S'$ is a partial surjective (onto) map and $k:V^n \rightarrow V'$ is a map such that the following diagram commutes.

$$
\begin{array}{ccc}
S^n & \xrightarrow{\ \Phi^n\ } & V^n \\
\downarrow h & & \downarrow k \\
S' & \xrightarrow{\ (\Phi')^n\ } & V'
\end{array}
$$

We call h the <u>system structuring</u> map and k the <u>performance structuring</u> map. Note that by commutes is meant commutes for values of $(s_1,\ldots,s_n) \in S^n$ for which $h(s_1,\ldots,s_n)$ is defined.

The above definition provides the minimal structure we feel is necessary to discuss compositions. The mapping h defines how components are linked together to form a larger model. Since we are interested in the larger composed model, there may be n-tuples $(s_1,\ldots,s_n)$ that cannot be linked up and we wish to exclude these from consideration. Thus, h is a partial function. Usually, h is more usefully defined in terms of simple operations such as those that yield serial connections, but we wish to keep h arbitrary for the present. Any type of interconnectivity can thus be subsumed under the definition---it need not be a "physical wire linkage" and could, in fact, be an information flow link. All that we require is to determine the element of S' that results from an (ordered) n-tuple of S . Notice that the composition of the components is characterized with reference to the space V . Thus, composing of elements is dependent upon the aspects of the components that we are considering. In particular, if the performance of a physical component was its color, it might not be possible to form a valid composition space if the Φ' map was also a map that extracted color as the performance of interest.

<u>Lemma 1.1</u> Let $\langle S',\Phi',V' \rangle$ be an n-composition space of $\langle S,\Phi,V \rangle$ by $\langle h,k \rangle$ and let $\langle S'',\Phi'',V'' \rangle$ be an m-composition space of $\langle S',\Phi',V' \rangle$ by $\langle \hat{h},\hat{k} \rangle$. Then, $\langle S'',\Phi'',V'' \rangle$ is an mn-composition space of $\langle S,\Phi,V \rangle$ by $\langle \hat{h} \cdot h^m, \hat{k} \cdot k^m \rangle$.

<u>Proof</u>: $\hat{h} \cdot h^m$ is a partial surjective map from $S^{mn} \rightarrow S''$ since h and $\hat{h}$ are surjective. The commutativity of the resulting maps is immediate. □

Lemma 1.1 allows us to see the inherent transitivity of the concept of composition spaces. Essentially, by looking at a 3-tuple

$\langle S, \Phi, V \rangle$, we are focusing on a particular level of decomposition. We can go up or down the levels by either going to more detail and looking at another component level or we can aggregate and look at a coarser level of analysis.

Definition If $\mathcal{S}' = \langle S', \Phi', V' \rangle$ is an n-composition space of $\mathcal{S} = \langle S, \Phi, V \rangle$ for some n and mapping pair $\langle h, k \rangle$, then $\mathcal{S}' \leq \mathcal{S}$.

Although the ordering relation defined above is transitive (Lemma 1.1) and obviously reflexive, it can easily be shown that it is not antisymmetric ($\leq$ is antisymmetric if $a \leq b$ and $b \leq a \Rightarrow a = b$). Consequently, we cannot define a partial ordering on such 3-tuples. It might appear that nonantisymmetry is counter-intuitive, but it seems reasonable when we recall that our structuring of components is always dependent on their performance values. Thus, it seems perfectly reasonable to consider situations in which, for example, transistors can be linked into higher level units, which in turn can be linked to perform as transistors. We now present some examples of composition spaces.

Example 1.1 Let S be a set of linear, continuous time, lumped parameter, time invariant models. A typical element will be represented by g and graphically by $\rightarrow \boxed{g} \rightarrow$. The performance characteristic of interest is the corresponding transfer function which we represent by capital G . Notice that, in general, elements of S will have some specific mathematical representation---for example, as a 4-tuple of matrices $\langle A, B, C, D \rangle$ ---and there will usually be a "procedure" to determine G from the representation g . The mathematical representation might also include other features of interest such as the maximum operating temperature of the component g . Let V be the set of rational algebraic fractions (ratios of polynomials with real coefficients over one indeterminate). Thus, $\Phi : S \rightarrow V$ and $\Phi(g) = G$. We wish S' to consist of the series connection of pairs of elements from S . That is, a typical element is $\rightarrow \boxed{g_1} \rightarrow \boxed{g_2} \rightarrow$ and the system structuring map $h : S^2 \rightarrow S'$ is defined by $h(g_1, g_2) = \rightarrow \boxed{g_1} \rightarrow \boxed{g_2} \rightarrow$. We let $V' = V$ and $\Phi' : S' \rightarrow V'$ also extract the transfer function from elements of S' since series composition also yields time invariant, linear, continuous time systems. It is easy enough to see that we obtain a 2-composition space if we define $k : V^2 \rightarrow V'$ by $k(G_1, G_2) = G_1 \cdot G_2$. Notice that if Φ and Φ' were different maps, for example, if the performance of interest was the

maximum operating temperature, then with the same   h , the function
k   might be $k[\phi(g_1),\phi(g_2)] = \min[\phi(g_1),\phi(g_2)]$.

Example 1.2   Suppose   S , V ,   and   V'   are the same as in example 1.1,
and   S'   is some composed space of pairs of components of   S .   Let
ϕ and ϕ' be as in example 1.1---that is, they extract the transfer
function.   If the performance structuring map   k   happens to be
$k(G_1,G_2) = \dfrac{G_1}{1+G_1G_2}$ then one possibility for the map   h   is clearly a
feedback composition.   That is, we can describe

$h(g_1,g_2)$ =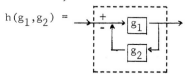

as the system enclosed by the dotted lines.

So far, we have essentially considered combining components by a single operator   h .   The natural extension to many operators is easy.

Definition   An <u>operator domain</u> is a disjoint sequence of sets Ω =
$(\Omega_n:n=0,1,2,\ldots)$. Ω_n is the set of n-ary operator labels of Ω .

Definition $\langle S',\phi',V'\rangle$ is a <u>Ω-composition space</u> of $\langle S,\phi,V\rangle$ by
$\langle\delta,\gamma\rangle$ if for each $\omega\in\Omega_n$, there exists a partial map $\delta_\omega:S^n\to S'$ and
a map $\gamma_\omega:V^n\to V'$ such that the following diagram commutes.

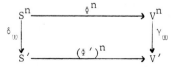

The maps δ_ω and γ_ω are, respectively, <u>system structuring</u> and <u>per-
formance structuring</u> maps.

A close connection exists between the concept of an Ω-composition
space and certain ideas from universal algebra.   In fact, the difference is really just a point of view.   We introduce some terminology
from the theory of homogeneous algebras that we shall need later.

Definition   An <u>Ω-algebra</u> is a pair (X,δ) where   X   is a set, Ω is
an operator domain, and δ assigns to each ω in Ω_n an n-ary operation $\delta_\omega:X^n\to X$.

Definition   Given Ω-algebras (X,δ) and (Y,γ) , an Ω-homomorphism

from (X,δ) to (Y,γ) is a function $f:X \to Y$ which commutes with
the Ω-operations, that is, for all $\omega \in \Omega_n$ and n-tuples $(x_1,\ldots,x_n)$
of X , we have $f \cdot \delta_\omega (x_1,\ldots,x_n) = \gamma_\omega (f(x_1),\ldots,f(x_n))$. Equivalently, we require that the following square be commutative:

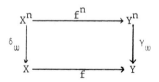

Definition Let (X,δ) be a Ω-algebra. Then, a binary relation θ
on X is called a congruence relation if it is an equivalence relation that satisfies the substitution property (SP) .
(SP) For each $\omega \in \Omega_n$, $a_i \equiv b_i(\theta)$ $a_i,b_i \in X, 1 \le i \le n$, then
       $\delta_\omega (a_1,\ldots,a_n) \equiv \delta_\omega (b_1,\ldots,b_n)(\theta)$.
(Note: $a_i \equiv b_i(\theta)$ means a_i is equivalent to b_i under the equivalence relation θ . Similarly $[a](\theta)$ would be the set of elements
equivalent to a .)

Definition Given an Ω-algebra (X,δ) and a congruence relation θ ,
we can define a new Ω-algebra $(X/\theta,\hat{\delta})$ called the quotient algebra
by:
    (1) $X/\theta = \{[a](\theta)|a \in X\}$
    (2)   For each $\omega \in \Omega_n$, $\hat{\delta}_\omega : (X/\theta)^n \to X/\theta$ by
          $\hat{\delta}_\omega ([a_1](\theta),\ldots,[a_n](\theta)) = \delta_\omega (a_1,\ldots,a_n)(\theta)$.
    (Note: $\hat{\delta}$ is said to be the induced set of operators.)

Definition Let G be an arbitrary set called the generator set.
Let F be the Ω-algebra such that for any Ω-algebra A and mapping
$f:G \to A$, there exists a unique mapping $\bar{f}:F \to A$ such that $\bar{f}$ is an
Ω-homomorphism and the following diagram commutes (Note: i is an
inclusions map of G into F .):

$G \xrightarrow{\quad i \quad} F$

f $\bar{f}$

A

F is said to be the free $\underline{\Omega\text{-algebra}}$ generated by G . It is well
known that F is unique up to isomorphism. The following word alge-
bra can easily be seen to be the free Ω-algebra and for our purposes,
we can think of the free Ω-algebra as the word algebra.

Definition Let G be a set. We first define the set of Ω-terms

over G . (1) g∈G is an Ω-term over G . (2) if ω∈Ω$_n$, and
$p_1, \ldots, p_n$ are Ω-terms over G , then ω$(p_1, \ldots, p_n)$ is an Ω-term
over G . The set of Ω-terms over G , call it Ω(G) , can be easily
seen to be an Ω-algebra. It is called the <u>word</u> <u>algebra</u>.

The word algebra is essentially a rather simple concept. It can be
thought of as simply syntactic strings of symbols that allow one to
reconstruct how a particular word was constructed. Thus, a particular word in the word algebra carries with it the information as to
how it was constructed from the generator set. Furthermore, there is
no identification of strings for the free Ω-algebra as we have defined it.

There is a very close connection between Ω-composition spaces and Ω-
algebras. In fact, the following proposition is obtained.

<u>Proposition 1.1</u> Let ⟨S,Φ,V⟩ be an Ω-composition space of ⟨S,Φ,V⟩
by (δ,γ) , and assume the maps δ$_ω$,γ$_ω$ are total functions. Then,
we can consider (S,δ) and (V,γ) to be Ω-algebras and Φ is an Ω-
algebra homomorphism.

<u>Proof</u>: The proof is immediate from the definitions. □

Although Ω-composition spaces for the homogeneous case are just a-
nother way of viewing Ω-algebras, it is worth noting that the focus
is different. We started by considering the 3-tuple ⟨S,Φ,V⟩ with
no particular structure on S and no restrictions on Φ . Thus, for
Ω-composition spaces, this 3-tuple is the basic unit and the mappings
{δ$_ω$} come later and arise naturally as system structuring maps. On
the other hand, for Ω-algebras the Ω-algebra is the fundamental unit
and the homomorphisms are a natural consequence.

Given sets S,V and a mapping Φ:S → V we can define a binary relation Φ̂ on S by saying a≡b(Φ̂) ⇔ Φ(a) = Φ(b) . For simplicity
when there is no chance of confusion, we shall just use Φ to indicate this relation.

<u>Proposition 1.2</u> Let ⟨S,Φ,V⟩ be an Ω-composition space of ⟨S,Φ,V⟩
by (δ,γ) with all maps total functions. Assume that Φ is surjective. Then, Ω-algebras (S/Φ,δ̂) and (V,γ) are isomorphic (an isomorphism is a bijective homomorphism).

<u>Proof</u>: From the previous proposition, Φ is a homomorphism. It is
easy to check that Φ induces a congruence relation on S and thus

$(S/\Phi, \hat{\delta})$ is a quotient Ω-algebra. The result follows from a well-known isomorphism theorem of universal algebras (c.f. Grätzer, p. 57). □

Example 1.3 Consider the composition space defined as in example 1.1. We can clearly consider $h: S^2 \to S$ since $\to \boxed{g_1} \to \boxed{g_2}$ can be considered to be an element of S. Consequently, it is clear that for a performance structuring map k to exist, if we replace $\to \boxed{g_1} \to$ by $\to \boxed{\hat{g}_1} \to$ with $\Phi(g_1) = \Phi(\hat{g}_1)$, then $h(\hat{g}_1, g_2)$ must be equivalent to $h(g_1, g_2)$. This simply says that only those performance functions are valid that allow $(S/\Phi, \hat{h})$ to be a quotient algebra of (S, h). (Note: $\hat{h}$ is the induced map from h on the equivalence classes.) One other interesting point can be made in this example. We would probably want $h(h(g_1, g_2), g_3) = h(g_1, h(g_2, g_3))$. That is, serial connection should be associative. This can be represented by using the concept of equationally defined classes in universal algebra (c.f. Manes, p. 7).

The previous example touched on the question of inferring the maps $\{\gamma_\omega\}$ from the maps $\{\delta_\omega\}$ and Φ. This can be thought of as a problem of __analysis__. We know the system structuring maps and we wish to discover the performance structuring maps. The converse question---discovering $\{\delta_\omega\}$ from $\{\gamma_\omega\}$ and Φ ---can be termed a question of __synthesis__. The previous propositions show that analysis is a property of the algebra S and its quotient algebras, for the homogeneous case. In particular, given the systems space $\langle S, \Phi, V \rangle$ and a set of system structuring maps $\{\delta_\omega\}$, either S/Φ is a quotient algebra, in which case we can define the corresponding performance structuring maps, or, Φ is a performance evaluation map that is not compatible with a Ω-composition space framework.

Before we conclude our discussion on composition spaces, we discuss a point of further research interest. Notice that the performance evaluation maps Φ are the __questions we can ask of the models__. It would thus be interesting to structure the set S mathematically so that the totality of valid Φ maps is apparent. An example will clarify this point.

Example 1.4 Consider a world modeling situation in political science in which the elements of S are the nation-states. Let h be the system structuring that forms an element of S' from the elements of S. This new element is a world model and two world models are the same if they have identical alliance configurations. Now, let us suppose that we use the performance evaluation map Φ' to ask whether a world model with a particular alliance configuration would be predicted to be involved in global war. A similar question, however, may not make sense at the nation-state level since global war may not be a performance evaluation map that we can apply to the component nation state. Thus, the possible questions Φ will differ from the possible questions Φ' .

Although we have shown the connection of (homogeneous) composition spaces to homogeneous algebras, we have made no mention of heterogeneous algebras. Actually, one can view composition spaces as a special case of heterogeneous algebras just as homogeneous composition spaces were essentially algebras. Since composition spaces arise naturally for the types of questions we ask and are less complex notationally, we do not pursue this connection any further.

2. SIMPLIFICATIONS

We now address the question of when a model s' (the lumped model) can be considered to be a "model" of another element s (the base model). In general, we normally consider lumped models that are coarser or simpler models of a base model and thus s' can be said to <u>simplify</u> s . We shall use the characterization s' <u>models</u> s and s' <u>simplifies</u> s interchangeably, although it must be kept in mind that we are only relating s' to s in a specific way, and we make no attempt to measure the complexities of s' and s .

Let S be a set of base models. These are the objects of interest as previously defined for which we have a performance evaluation map Φ . We wish to "model" an element $s \in S$ by an element $s' \in S'$ where S' is the set of lumped models. There are two issues that arise immediately.

The first point to note is that we do not expect to model every aspect of s . For example, if s happens to be a representation for some electrical component, it is clear that we might not be interested in making sure that the model was the same color as the component modeled. This can be formalized by a performance evaluation function $\Phi : S \to V$ where V is the set of possible values of interest. Φ extracts the features of interest from the base models S . Parallel to this, we have the set of lumped models S' from which we hope to obtain a model $s' \in S'$ and a map $\Phi' : S' \to V'$, that is, the

performance evaluation map on the lumped systems space. The question
of the nature of s′ and s is thus circumvented by recognizing
that we only wish to compare $\Phi(s)$ and $\Phi'(s')$. Thus the modeler
has been explicitly introduced in this framework since the modeler
must determine the performance maps of interest. We shall see later
that the modeler will also play a crucial role in deciding how to
compare $\Phi(s)$ and $\Phi'(s')$.

The second issue that arises is that there are always limitations on
measuring accuracy as well as **desired** accuracy. We may not really be
able to calculate $\Phi(s)$ because of physical limitations. Our measurement might turn out to be $\hat{\Phi}(s)$ where $\hat{\Phi}$ represents the actual
value in V obtained rather than the true value $\Phi(s)$. However,
depending on the structure of the space V , we may wish to say that
we are sure that $\hat{\Phi}(s)$ is "close" enough to $\Phi(s)$. Also, for
modeling purposes, we may in fact be content with only the value
$\hat{\Phi}(s)$, even though we could do better. Both these aspects are formalized by introducing the concept of an **approximand space** $G(V)$.

Definition Given a set V , for v∈V let A(v) be a subset of V
such that $\{v\} \subseteq A(v)$. Then A(v) is the v-approximand of V . An
approximand space $G(V) = G$ is a collection of v-approximands index-
ed by v . That is, $G(V) = \{A(v)|v∈V\}$.

The idea of an approximand space is that for each value v∈V we have
a set of values in V that are acceptable approximations to v .
That is, if $v_2∈A(v_1)$ then v_2 is an acceptable approximation to
v_1 . Before we give the definition of s′ models s , we need only
recognize that the modeler must interpret values in V′ or the lumped model values, as values in V which is the set of base model
values. Essentially, the modeler decides the meaning of a measurement on the lumped systems space relative to the base systems space.

Definition A **system** is a 4-tuple $\langle S,\Phi,V,G\rangle$ where $\langle S,\Phi,V\rangle$ is a
systems space and G is an approximand space on V .

Definition An **interpretation map** H is a map $H:V' \to V$.

The approximand space structure provides us with sufficient structure
to allow us to define what we mean by s′ models s or s′ simplifies s .

Definition Let a base system be represented by a 4-tuple $\langle S,\Phi,V,G\rangle$
where S is a set of base models, Φ a performance map $\Phi:S \to V$

with V the value space. Let G be an approximand space of V .
Similarly let ⟨S′,$̇′,V′,G′⟩ be a corresponding lumped system. Let
there be given an interpretation map H . Let A(v) be a v-approximand with A(v)∈G and A′(v′) be a v′-approximand with A′(v′)∈G′ .
Then for s∈S and s′∈S′

    (1) s′ <u>weakly models</u> s if H[$̇′(s′)]∈A[$̇(s)]

    (2) s′ <u>models</u> s (or <u>strongly models</u>) if H[A′($̇′(s′))] ⊂
       A[$̇(s)] .

(Alternatively we may use the term s′ weakly simplifies s or s′
simplifies s .)

The above definition captures the features of importance for us.
Given a lumped model s′ , its <u>true</u> performance is $̇′(s′) = v′ . If
this value is interpreted within an approximand region of the base
model true value, then for our purposes, this can be used instead of
the base model. This is the case of "weakly models." However, because of tolerated or unavoidable inaccuracies, the actually calculated value of the lumped model may be $̂′(s′) = v̂′ with v̂′∈A′(v′) .
Thus, for the case of "models," we require that v̂′ also be interpreted close enough to the true base model performance value---that
is, we require H(v̂′)∈A(v) .

We shall mainly use the notion of "models," although the concept of
"weakly models" is a useful motivation for the generalization to
"models with confidence-α" that we discuss later.

<u>Notation</u> Since we shall always be concerned with the approximand of
the performance value of a model, we shall use the notation APROX(s)
to mean A($̇(s)) . Similarly, APROX(s′) will mean A′($̇′(s′)) .
Consequently, we can pictorially represent s′ models s as follows:

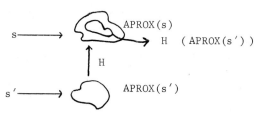

Although we have introduced the idea of an approximand space to capture our modeling notion, we have not related it at all to the concept of composition spaces. It is natural to consider a situation in
which ⟨S′,$̇′,V′,G′⟩ is an Ω-composition space of ⟨S,$̇,V,G⟩ via
(δ,γ) . Everything is as before except that we have added the

approximand structure onto the spaces V and V' . Now, although for $s' \in S'$ and $(s_1, \ldots, s_n) \in S^n$ such that $\delta_\omega(s_1, \ldots, s_n) = s'$ we have $\gamma_\omega \circ \phi^n(s_1, \ldots, s_n)$ by definition, we have not specified how the approximand structure $\{APROX(s_i)\}_{i=1}^n$ is to be related to

$APROX(s')$. We now consider two intuitively plausible ways these approximands can be related.

<u>Definition</u> $\langle S', \phi', V', G' \rangle$ is a <u>permissive Ω-composition space</u> of $\langle S, \phi, V, G \rangle$ if for any $\omega \in \Omega_n$,

$$\gamma_\omega[A(\phi(s_1)), \ldots, A(\phi(s_n))] \subset A[\phi'(\delta_\omega(s_1, \ldots, s_n))] \ .$$

<u>Definition</u> $\langle S', \phi', V', G' \rangle$ is a <u>strict Ω-composition space</u> of $\langle S, \phi, V, G \rangle$ if for any $\omega \in \Omega_n$,

$$\gamma_\omega[A(\phi(s_1)), \ldots, A(\phi(s_n))] \supset A[\phi'(\delta_\omega(s_1, \ldots, s_n))] \ .$$

The idea of a permissive Ω-composition is that if the observed values of a set of components lie in the approximand regions of the true performance values, then the observed performance value of the model composed from the components must also lie in the approximand region of the true performance value of the composite model. The concept of a strict Ω-composition is somewhat stronger. Strict Ω-composition implies that any observed performance value of the composite model can be obtained by an appropriate choice of observed component performance values that lie within the approximand regions of the true component performance values. Although this is rather strong, it will be seen that strict Ω-composition need only apply to the lumped or derived model, and thus is a characterization that can be imposed by the modeler on the models that are being considered.

In the next section, we use the concepts of composition spaces together with the simplification ideas in order to prove our desired theorem for the case of deterministic modeling.

3. DETERMINISTIC MODELING

Deterministic modeling is the situation that we have described so far in which it is possible to say that an element s' models an element s . We thus distinguish this case from that of probabilistic modeling for which such a statement can only be made with some confidence α .

In the previous section, we gave a definition for s' models s given lumped and base systems $\langle S', \phi', V', G' \rangle$ and $\langle S, \phi, V, G \rangle$ and an

arbitrary interpretation map   H .   In many situations, however, the
interpretation map   H   is injective.   Such a situation is natural
since the lumped model performance space is generally coarser than
the base model performance space.   Whenever   H   is injective, we can
simplify matters by considering a new performance evaluation map
$\Phi'' = H \circ \Phi'$, with $\Phi'' : S' \to V$.   Thus, both the base and lumped models
have performance values that are in the same space   V .   It is readi-
ly seen that   H   induces another approximand structure G'' on   V
that is derived from the approximand structure G' .   That is, if
$v'' = H(v')$, then we define $A''(v'') = H[A'(v')]$.   For v'' not in
the image of the map   H , we can define $A''(v'') = \{v''\}$.   This induced approximand structure is associated with S' and Φ'' .   In the
theorem that we prove, we thus consider a situation in which the performance evaluation maps have as range the identical space   V , with
however two different approximand structures on   V ---one associated
with the evaluations of the base models, and the other associated
with the evaluations of the lumped models.

We now show that if components of a lumped model simplify components
of a base model, then the lumped model simplifies the base model,
given suitable hypotheses.

<u>Theorem 3.1 (Deterministic Modeling)</u>   Let a base system $\langle S_B, \Phi_B, \hat{V}, G_B \rangle$
be a permissive Ω-composition space of a base component system
$\langle S, \Phi, V, G \rangle$ via (δ, γ) .   Similarly, let a lumped system $\langle S_L, \Phi_L, \hat{V}, G_L \rangle$
be a strict Ω-composition space of a lumped components system
$\langle S', \Phi', V, G' \rangle$ via (δ', γ) .   For $\omega \in \Omega_n$, let $\delta_\omega(s_1, \ldots, s_n) = s$ and
$\delta'_\omega(s_1', \ldots, s_n') = s'$.   Then, s_i' models s_i for $i = 1, \ldots, n \Rightarrow s'$
models   s   (see the figure below).

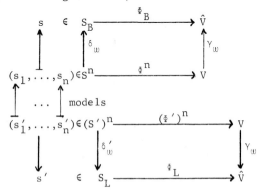

Proof: By definition of s' models s , we must show that
APROX(s') ⊂ APROX(s) . (Note that there is no interpretation map
H ; instead we have approximand structures G_L and G_B on $\hat{V}$ as-
sociated with S' and S respectively.)

Since the lumped system is a strict Ω-composition space of the lumped
component system, we have

$$\gamma_\omega(APROX(s_1'),\ldots,APROX(s_n')) \supset APROX(s') .$$

Because s_i' models s_i , we get

$$APROX(s_i') \subset APROX(s_i) \qquad i=1,\ldots,n$$

and thus $\gamma_\omega(APROX(s_1'),\ldots,APROX(s_n')) \subset \gamma_\omega(APROX(s_1),\ldots,APROX(s_n))$.
Finally, since the base system is a permissive Ω-composition space of
the base components system, we have

$$\gamma_\omega(APROX(s_1),\ldots,APROX(s_n)) \subset APROX(s) .$$

Putting together the sequence of inclusions yields the desired result

$$APROX(s') \subset APROX(s) . \qquad \square$$

The above theorem gives a characterization of how the effects of components simplifying corresponding components carries over to the composed models. Each hypothesis yields the precise information to make
the theorem work. Notice in particular that the performance evaluation maps $\{\gamma_\omega\}$ are the same for the base and lumped composition
spaces. This is essential and simply means that the components were
"hooked together" in parallel fashion, although the actual system
structuring maps $\{\delta_\omega\}$ and $\{\delta_\omega'\}$ are different. Without this,
there could be no equivalent connection pattern for the two spaces,
and it would be impossible to expect any resemblance between the base
and lumped composed models.

The above theorem requires the base and lumped composition spaces to
be permissive and strict respectively. Actually, we can consider
this condition as holding between only the specific elements of interest and get the same result.

Definition Let $\langle S',\Phi',V',G'\rangle$ be an Ω-composition space of
$\langle S,\Phi,V,G\rangle$. Suppose that for $\delta_\omega(s_1,\ldots,s_n) = s'$

$$\gamma_\omega[A(\Phi(s_1)),\ldots,A(\Phi(s_n))] \subset A[\Phi'(s')] .$$

Then s' is a permissive composition of $s_1,\ldots,s_n$. (A similar
definition for strict composition arises by changing ⊂ to ⊃ .)

<u>Corollary 3.1</u> Theorem 3.1 remains true if we require s to be a permissive composition of $s_1,\ldots,s_n$ and s′ to be a strict composition of $s_1',\ldots,s_n'$ and drop the permissive and strict restrictions on the base and lumped Ω-composition spaces respectively.

<u>Proof</u>: Simply particularize the theorem to the specific models and components in question. □

We now present two examples that illustrate the above ideas.

<u>Example 3.1</u> This example relates to reliability theory. Let us consider structural elements that consist of K nodes each of which is working (= 1) or not working (= 0). The internal logic of the box determines the global functioning (0 or 1) of the box depending on the values of the nodes which we label $x_1,\ldots,x_K$. Thus the property of interest of the structural element is a Boolean function f: ${\{0,1\}}^K \to \{0,1\}$. For simplicity, let us imagine the nodes to be on electrical lines and say that parallel nodes on lines imply any one of the nodes functioning is sufficient for global functioning, and series nodes on a line implies all nodes must be functioning. Although the black box may have other characteristics, we ignore these since our performance function ϕ will only extract the global functioning (or Boolean function). We present a couple of 3 nodes and their associated Boolean functions

$$f(x_1,x_2,x_3) = \begin{cases} 0 & \text{for } x_1 = x_2 = x_3 = 0 \\ 1 & \text{otherwise} \end{cases}$$

$$f(x_1,x_2,x_3) = \begin{cases} 0 & \text{if } x_3 = 0 \text{ and } x_1 \text{ or } x_2 = 0 \\ 1 & \text{otherwise} \end{cases}$$

Let s be the space of 4-nodes as previously defined with $\phi : S \to V$ mapping a structural element to the corresponding Boolean function. Consider the following 2-composition space:

where

S = set of 4-nodes

S_B = set of 8-nodes

h is a parallel connection of 4-components, that is, $k(f,g) = f \vee g$:
$\{0,1\}^8 \to \{0,1\}$ such that $f \vee g(x_1,\ldots,x_8) = f(x_1,\ldots,x_4) \vee g(x_5,\ldots,x_8)$
with $1 \vee 1 = 1$, $1 \vee 0 = 0 \vee 1 = 1$ and $0 \vee 0 = 0$.

$V, \hat{V}$ are the obvious spaces of Boolean functions, and
Φ, Φ_B extract the Boolean functions as discussed above.

We can graphically give an example of h as follows:

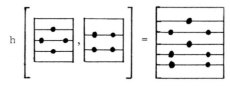

We still need to define an approximand structure on $V, \hat{V}$ and do this
as follows. Let f be a Boolean function. Then a Boolean function
g with the same domain D is k-close to f is $\sum_{x \in D} |f(x) - g(x)| \leq k$.
Then, we let the approximand structure G_B on $\hat{V}$ be the 2-close
functions to any element $f \in \hat{V}$. Similarly, the approximand structure
G on V is the set of 1-close functions. This completes the description of the base composition space. We wish to consider a space
S' of 3-nodes that we will try to use as models of 4-nodes. Consequently, we define a lumped 2-composition space as follows:

$$S' \times S' \xrightarrow{\Phi' \times \Phi'} V' \times V' \xrightarrow{H_1 \times H_1} V \times V$$

with vertical maps h', k; and

$$S_L \xrightarrow{\Phi_L} V_L \xrightarrow{H_2} \hat{V}$$

S' = set of 3-nodes

S_L = set of 6-nodes

Φ', Φ_L extract the Boolean functions as before except that we now
wish to interpret a 3-node as a 4-node and a 6-node as an 8-node.
We do this as follows: $H_1[f'] = f$ with
$$\begin{cases} f(x_1,x_2,x_3,1) = 1 \\ f(x_1,x_2,x_3,0) = f'(x_1,x_2,x_3) \end{cases}$$

and

$H_2[f_L] = f_B$ with
$$\begin{cases} f_B(x_1,x_2,x_3,0,x_5,x_6,x_7,0) = f_L(x_1,\ldots,x_6) \\ f_B(\text{other combinations}) = 1 \end{cases}$$

Let the approximand structure on V',V_L be the 1-close functions, and that on $V,\hat{V}$ be induced by H_1,H_2 . The function h' is a parallel connection and k is the same as before.

Given the base and lumped composition spaces defined above, it can readily be checked that the 3-node s' models s where

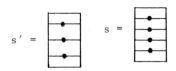

Since (s,s) is a permissive decomposition of $h(s,s)$ and (s',s') is a strict decomposition of $h'(s',s')$, it follows by the corollary that $h'(s',s')$ models $h(s,s)$. Although this seems apparent, the corollary actually says much more. If, for example, s' fails somewhat and has as a performance a Boolean function that is 1-close to $\phi'(s')$, we can still be sure that our modeling statement holds. Thus, we have an indication of the reliability of the composite base model modeled by a composite lumped model that is dependent on the component reliabilities.

It is the case that S_B is <u>not</u> a permissive 2-composition space of S , and consequently, the above statements cannot be generalized to arbitrary elements of S_L and S_B .

<u>Example 3.2</u> We now present an example that relates to neural networks. Similar modeling situations have been discussed by Zeigler [9] and further detail may also be found in Aggarwal [3]. Consider a neuron to be an idealized structure that operates as follows. Operating on a discrete time scale, it receives inputs from other neurons and fires (outputs a 1 pulse) if the total strength of the input is greater than the neuron's firing threshold. Let the firing threshold be a function [THOLD] of the recovery state, i.e., number of time units since the neuron last fired. For simplicity, assume that a neuron has identical firing threshold for all time units ≥ 2 . In addition, assume each neuron is independently perturbed by some level of noise which modifies the original threshold. Let this noise be determined by a cumulative distribution. function F . The neurons may be connected together, in which case they form a neural network.

In our example, we first describe a base system $\langle S_B,\phi_B,\hat{V},G_B \rangle$ that is a permissive 2-composition space of a base component system $\langle S,\phi,V,G \rangle$ via (h,k) . Let an element $s \in S$ be a pool of neurons

with no interconnections. The neurons may be considered to be defined by specifying the recovery state of each neuron. The performance evaluation map extracts the calculated proportion in each recovery state and thus results in an element $v \in V$ that is a 3-tuple,
say $\langle .8,.15,.05 \rangle$. This says that .8 neurons are in recovery
state 0 (firing), .15 are in recovery state 1 and .05 are in
recovery states greater or equal to 2 . Let s_1 and s_2 be two
pools of neurons. We now define the map h that composes these
pools as follows. Each neuron in pool s_2 receives input from every
neuron in s_1 . The performance evaluation map ϕ_B extracts the
calculated proportion in each recovery state in each pool at the next
time step. Thus $\hat{V} = V \times V$. A pictorial view of h is as follows
with 3 and 2 neurons per pool for simplicity.

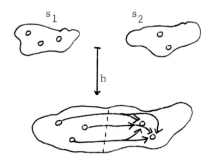

It has been shown by Zeigler [9] that when the number of neurons in
each pool is quite large, we can define the performance structuring
map k that makes this system a composition space. Let the recovery
state vector for pool s_i at time t be $\phi(s_i) =$
$\langle r_0^i(t), r_1^i(t), r_2^i(t) \rangle = r^i(t)$. Then we have $r^2(t+1) = r^2(t) \cdot P(t)$,
where $P(t)$ is the following matrix.

$$
\begin{bmatrix}
P_{00} & 0 & 0 \\
P_{10} & 0 & P_{12} \\
P_{20} & 0 & P_{22}
\end{bmatrix}
$$

where P_{ij} = probability that a neuron goes from recovery state i
to recovery state j . Let M be the maximum input that a neuron in
pool s_2 can receive. Then $P_{i,0} = F[M \cdot r_0^1(t) - \text{THOLD}(i)]$, and $P_{12} =$
$1 - P_{10}$ and $P_{22} = 1 - P_{20}$. In our example we take F and THOLD to
be the following functions.

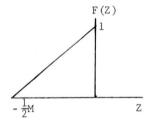

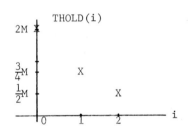

For simplicity, we shall assume that these functions also define the
noise and threshold characteristics of pool one. Actually, since
pool one neurons receive no input, any arbitrary functions could be
used. Fixing the functions, however, we can define a matrix $Q(t)$
such that $r^1(t+1) = r^1(t) \cdot Q(t)$. Thus, the performance evaluation
map is $k(r^1(t), r^2(t)) = (r^1(t) \cdot Q(t), r^2(t) \cdot P(t))$.

It remains to define the approximand structures G and G_B . We
shall assume that the approximand structure on V is such that for
a recovery state vector $v = \langle v_0, v_1, v_2 \rangle$, any vector $w = \langle w_0, w_1, w_2 \rangle$
is in $G(v)$ if $|w_i| \leq 1.01|v_i|$. That is, we allow errors in the
base model evaluation up to 1% on each vector component. Similarly
we shall define G_B on $\hat{V}$ by allowing a 5% error in each vector
component. (Recall that $\hat{V}$ has elements of the form $\langle r_0^1, r_1^1, r_2^1 \rangle$,
$\langle r_0^2, r_1^2, r_2^2 \rangle$.)

The linearity of F can be used to easily establish that this system
is a permissive composition space. The linearity property, of
course, is used only as a convenience. If F were nonlinear, we
would need only to be able to calculate precisely how errors expand
and be certain that a 1% error does not result in more than a 5% error at the next time step. Notice that the figure of 5% was chosen
only to insure that the system could be proved to be a permissive
composition space.

We now turn to a definition of the lumped composition space. Instead
of a pool of neurons, the lumped component is a single block that
aggregates information about the neurons in a pool. The state of a
block is a list of predicted proportions in particular recovery
states (but note that the latter is <u>derived</u> from the finer state of
the base model) and thus a 3-tuple similar to $\langle r_0, r_1, r_2 \rangle$. Thus, an
element $s' \in S'$ is defined by $\langle s_0', s_1', s_2' \rangle$. We now let the map ϕ'
be a map that defines the 3-tuple to within the limits of normal com-
puter accuracy; that is, it represents this vector as a rational vec-
tor $\langle v_0', v_1', v_2' \rangle$. Thus ϕ' is "almost" the identity function. The

composition map h′ is defined by specifying that block 1 influences block 2 with strength M . Φ_L extracts the values of the lumped network at the next time step similar to Φ' . That is, it also is "almost" the identity. It remains to define the approximand structures G′ and G_L . We shall let the approximand region for v′∈G′ be simply a .01% error rather than a 1% error as defined for v∈G . Similarly, G_L is defined also by a .01% error. It can readily be seen that since Φ' and Φ_L only reflect representation accuracy (say to within 6 decimal places), the lumped system is a <u>strict</u> composition space.

What have we shown with this example? With the stringent requirements imposed on the connection pattern of the neurons we can say that the network of blocks models the network of neurons whenever each block models its respective neuron pool. This seems clear from the manner in which we defined k . However, if we <u>relax</u> the assumptions on the connection pattern of neurons by saying that each neuron in pool s_2 only receives input from a "large" set of neurons in pool s_1 , we can argue that the performance evaluation maps of the base systems will still yield values within the approximand regions. Thus, the modeling relationship still holds and says something much stronger than before since the connection pattern is not total, even though we use the same lumped system. We were not able to initially describe the base model with this relaxed assumption because of the difficulty in proving the composition space property. Although the composition space property does not hold for the base system with the relaxed assumptions, it is not really necessary since the theorem really relates to how approximand regions behave. Thus, a version of the theorem could be formulated that does not require the exact composition space property but does require approximand regions to behave appropriately. Similar arguments can be used to consider the case of larger networks, the difficulty being the necessity of checking that the composition space is permissive.

So far, we have considered the case of deterministic modeling. We now turn to the situation in which we are not modeling with complete confidence but rather only with some α-confidence.

4. PROBABILISTIC MODELING

In this section, we confine ourselves to the special case of composition spaces that is equivalent to homogeneous algebras. Thus, we shall only consider the situation in which a system 4-tuple ⟨S,Φ,V,G⟩

is an Ω-composition of itself via (δ,γ) total functions. As noted before, this is equivalent to (S,δ) and (V,γ) being Ω-algebras and Φ being an Ω-homomorphism. For simplicity, whenever we now use the term <u>system</u>, we implicitly assume the associated (homogeneous) composition space structure or, equivalently, the associated homogeneous algebra structure.

We wish to formalize what is meant by modeling with α-<u>confidence</u>. That is, rather than being sure that a lumped element models a base element we wish only to be able to say that with confidence α , the lumped element models the base element.

First, recall the definition of weakly models.

<u>Definition</u> Let a base system 4-tuple be $\langle S,\Phi,V,G\rangle$ and a lumped system 4-tuple be $\langle S',\Phi',V',G'\rangle$. If $H:V'\rightarrow V$ then for $s\in S$ and $s'\in S'$,

        s' weakly models s if $H[\Phi'(s')]\in A[\Phi(s)]$.

Notice that there is no reference to the approximand structure of the lumped model. In this definition, it was assumed that $\Phi'(s')$ was known precisely. Suppose, instead, that we only know a probability distribution of the performance value of s' in V' . That is, corresponding to s' , we have a probability $P_{s'}$ defined on V' . In order for this to make sense, we require V' to be a measure space and we let the set of events in V' be the smallest σ-algebra containing $H^{-1}(A(v))$ for $v\in V$. See reference [5] for more on probability spaces and measures.

<u>Definition</u> Let $\langle S,\Phi,V,G\rangle$ be a base system. Let $\langle S',\Phi',V'\rangle$ be a lumped space and consider V' to be a measurable space $\langle V',G'\rangle$, with G' the smallest σ-algebra containing $H^{-1}(A(v))$ for $v\in V$ where $H:V'\rightarrow V$ is an interpretation map and $A(v)$ is the v-approximand of v . Corresponding to $s'\in S'$, let $P_{s'}$ be a probability measure on $\langle V',G'\rangle$. Then

        s' α-models s if $P_{s'}(H^{-1}(A(\Phi(s)))) = \alpha$.

Pictorially, we have the following:

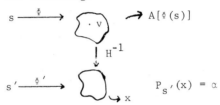

We shall call $\langle S',\Phi',V',G'\rangle$ a lumped measurable system if G' is as above. Alternatively, we say that s' models s with confidence α . Intuitively, the definition arises as follows. We are interested in deciding with what probability s' models s . Since all that is necessary is that the interpretation of s' fall in the approximand of the performance value of s , we check to see with what probability this occurs.

Notice that the lumped space no longer has an approximand structure on it but rather has a σ-algebra associated with it and a set of probability measures for each $s' \in S'$.

<u>Definition</u> An approximand space $G(V)$ <u>refines</u> the approximand space $B(V)$ if for all $v \in V$, $A(v) \subset B(v)$ where $G(V) = \{A(v)|v \in V\}$ and $B(v) = \{B(v)|v \in V\}$.

The following proposition easily follows from our definitions.

<u>Proposition 4.1</u> Let $\langle S,\Phi,V,G\rangle$ and $\langle S,\Phi,V,B\rangle$ be base systems such that $G(V)$ refines $B(V)$. Let $\langle S',\Phi',V',G'\rangle$ be a lumped measurable system and $H:V' \to V$ be the interpretation map. For $s' \in S'$ and $s \in S$, if

$$s' \quad \alpha\text{-models} \quad s \quad \text{(relative to approximand space} \quad G \text{)}$$

and

$$s' \quad \beta\text{-models} \quad s \quad \text{(relative to approximand space} \quad B \text{)}$$

then

$$\alpha \leq \beta \quad .$$

<u>Proof</u>: Since s' α-models s we have $\alpha = P_s, [H^{-1}[A(\Phi(s))]]$. By the definition of <u>refines</u>, $A(\Phi(s)) \subset B(\Phi(s))$. Thus

$$\alpha = P_s, [H^{-1}[A(\Phi(s))]] \leq P_s, [H^{-1}[B(\Phi(s))]] = \beta \quad . \quad \square$$

The proposition illustrates the fact that our α-modeling notion does capture the notion of modeling with a certain confidence probability. We now seek to prove a theorem for the probabilistic case that is analogous to our deterministic modeling theorem which shows that a composed lumped model simplifies a composed base model if corresponding components satisfy the modeling relationship.

The probabilistic modeling notion requires that for a lumped measurable system 4-tuple $\langle T,\bar{\Psi},W,B\rangle$ there be defined a probability measure P_t on the measurable space $\langle W,B\rangle$ for every $t \in T$. However, there have been no consistency conditions on the various measures. (Notice that $\bar{\Psi}$ is the "true" performance map $\bar{\Psi}:T \to W$.) Let us consider T

to be a free algebra generated by a primitive set of components G .
We have already seen that $\bar{\psi}$ is the unique map that extends the performance map $\psi:G \rightarrow W$. Suppose now that corresponding to each component $g \in G$, we have a probability distribution P_g on W . That
is, rather than knowing the exact performance of elements of G , we
only have probabilistic information on the performances. Is there a
"natural" extension to probability distributions for each element
$t \in T$? We see that indeed there is.

<u>Definition</u> Let $t_1,\ldots,t_n \in T$ for the Ω-algebra $T = (T,\gamma)$. Suppose
$t = \gamma_\omega(t_1,\ldots,t_n)$. We assume that there exist probability distributions $P_{t_1},\ldots,P_{t_n}$ on W . Then, the <u>naturally induced distribution</u>
P_t on W is the probability measure defined as follows. Let $P_{t_1} \times$
$P_{t_2} \times \cdots \times P_{t_n}$ be the product measure on W^n . Then, it can easily be
shown that the map $\gamma_\omega:W^n \rightarrow W$ induces a unique measure on W by
$P_t(B) = P_{t_1} \times P_{t_2} \times \cdots \times P_{t_n} [\gamma_\omega^{-1}(B)]$ for any set $B \in \mathcal{B}$ where $\mathcal{B}$ is the
σ-algebra for the measurable space $(W,\mathcal{B})$. This unique measure will
be the naturally induced measure.

<u>Proposition 4.2</u> Let $\langle T,\psi,W,\mathcal{B} \rangle$ be a lumped measurable system and
suppose $T = (T,\gamma)$ is the free algebra generated by G . Assume
that for each $g \in G$, there exists a probability distribution P_g on
W . Then, for any $t_1,\ldots,t_n \in T$ there exists a unique and well defined naturally induced distribution P_t on W for $t = \gamma_\omega(t_1,\ldots,t_n)$.

<u>Proof</u>: We may consider T to be the word algebra over the generator
set G . Thus, any element $t \in T$ has a unique word representation.
The unique measure is thus the naturally induced probability measure
for this word representation since we are assuming there is a probability measure that is well defined for each $g \in G$. Because each
word is uniquely determined from the operators $\{\gamma_\omega | \omega \in \Omega\}$ and the el-
ements of G , it can be seen that P_t is well defined. □

Notice that we have eliminated any possibility of conflicting definitions of distributions for an element $t \in T$ by insisting that T be
a free algebra. If this were not the case, two different ways of
"building t " might conflict. For example, if $t = g_1 \cdot g_2 =$
$r_1 * r_2 * r_3$, a consistent definition might not be possible (assume $\circ, *$
are two binary operations of the Ω-algebra). Our definition of the
induced probability distribution effectively says that if the distribution for one element is unknown, and similarly unknown for another

element, then the uncertainty for the two elements combined is just
the product uncertainties. Thus, all the uncertainty is captured in
the performance distributions of the components. There is no uncertainty due to the interconnections. Also, there is also no interactive effect of connections.

We are now in a position to prove our main result for the probabilistic case.

<u>Theorem 4.1 (Probabilistic Modeling)</u> Let the base system $\langle S, \Phi, V, G \rangle$
be a permissive composition space. Let $\langle T, \psi, W, \beta \rangle$ be a lumped measurable system and let the interpretation map $H : W \to V$ be an algebra
homomorphism. Assume T is the free algebra generated by G with
the probability distributions P_t , $t \in T$ being the naturally induced
distributions given P_g , $g \in G$.

Assume that for $t_i \in T$, $s_i \in S$, $i = 1, \ldots, n$,

$$t_i \quad \alpha_i\text{-models} \quad s_i \qquad i = 1, \ldots, n \; .$$

Then, if $t = \tau_\omega(t_1, \ldots, t_n)$ and $s = \delta_\omega(s_1, \ldots, s_n)$ for the Ω-alge-
bra $S = (S, \delta)$, $V = (V, \gamma)$, $T = (T, \tau)$, $W = (W, \eta)$ we have

$$t \quad \beta\text{-models} \quad s \quad \text{with} \quad \beta \geq \prod_{i=1}^{n} \alpha_i \quad \text{(product of } \alpha_i\text{'s)}.$$

<u>Proof</u>: By definition, the confidence with which t models s is:

$$\beta = P_t[H^{-1}[A(\Phi(s))]] \; .$$

Let $\Phi(s_i) = v_i$ and $\Phi(s) = v$.

$$\therefore \; P_t[H^{-1}[A(\Phi(s))]]$$

$$= P_t[H^{-1}[A(\Phi(\delta_\omega(s_1, \ldots, s_n)))]]$$

$$\geq P_t[H^{-1}[\gamma_\omega(A(v_1), \ldots, A(v_n))]] \quad \text{(base model is permissive)}$$

$$\geq P_t[\eta_\omega(H^{-1}[A(v_1)], \ldots, H^{-1}[A(v_n)])] \quad (H \text{ is a homomorphism})$$

$$\geq \prod_{i=1}^{n} P_{t_i}[H^{-1}[A(\Phi(s_i))]] \quad \text{(by definition of induced measure)}$$

$$= \prod_{i=1}^{n} \alpha_i \; . \qquad \square$$

The following pictorial representation of the approximands in the
spaces V and W should help in the understanding of the theorem.

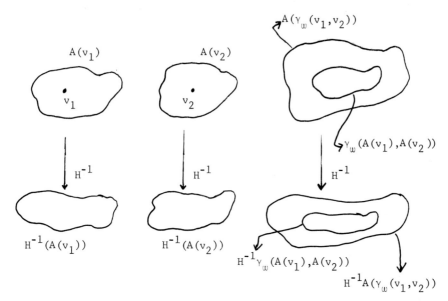

Note that since $H:W \to V$ is an algebra homomorphism, it follows that
for any $P,Q \in V$, $\eta_\omega(H^{-1}(P),H^{-1}(Q)) \subset H^{-1}(\gamma_\omega(P,Q))$. The reason
that we required H to be a homomorphism was that the base and
lumped systems had to be "connected up" in the same fashion. The
formal mechanism for this similarity of connection pattern is obtained by requiring that the two performance spaces W and V be
homomorphically related.

Our theorem has thus shown that we can indeed give an estimate of
the modeling confidence of composed elements. Obviously, as more
components are added, each with confidence less than 1 , the confidence in the entire lumped system modeling the base system is reduced. The theorem gives the precise formulation.

Consider next the following transitivity result.

<u>Proposition 4.3</u> Let $\langle S_1, \Phi_1, V_1, G_1 \rangle$ and (S_2, Φ_2, V_2, G_2) be system
4-tuples and $\langle S_3, \Phi_3, V_3, G_3 \rangle$ be a measurable system 4-tuple. Let
$H_a:V_2 \to V_1$ and $H_b:V_3 \to V_2$. Then, if s_3 α-models s_2 and s_2
strongly models s_1 , we have s_3 β-models s_1 with $\beta \geq \alpha$.
Assume $s_i \in S_i$ and $H_c:V_3 \to V_1$ is just $H_c = H_a \cdot H_b$. Let $A_i(v)$ be
the appropriate v-approximands.

<u>Proof:</u> By definition of strongly models, $H_a A_2[\Phi_2(s_2)] \subset A_1[\Phi_1(s_1)]$

$$\therefore \; \beta = P_{s_1} [H_b^{-1} \circ H_a^{-1} (A_1 (\Phi (s_1)))]$$

$$\geq P_{s_1} [H_b^{-1} (A_2 (\Phi (s_2)))]$$

$$= \alpha . \qquad \square$$

Notice that a similar statement cannot be made for strongly models being replaced by weakly models since under the weaker condition, the behavior of the entire approximand $A_2 (\Phi_2 (s_2))$ is not certain. Furthermore, we cannot consider s_3 models s_2 strongly and s_2 α-models s_1 since the systems do not match for the definition of either modeling or α-modeling.

We now investigate extending our results by considering systems that have formally defined operations of meet ($\wedge$) and join ($\vee$).

Given a system 4-tuple $\langle S, \Phi, V, G \rangle$, we notice that $\Phi : S \to V$ induces a map $f : S \to G(V)$ by associating to each $s \in S$, the corresponding approximand $A(\Phi (s))$. That is, to each s is associated a subset of V that is the approximand of $\Phi(s)$. Now, it is natural to consider the operations of $\cup$ and $\cap$ which are well defined on the subsets of V although they may not result in valid approximands. Nevertheless, if we formally define operations on S called meet and join, we get some interesting results.

Definition For $s_1, s_2 \in S$, define new formal elements $s_1 \vee s_2$ called s_1 meet s_2 and $s_1 \wedge s_2$ called s_1 join s_2. If the corresponding approximands for s_1, s_2 are $a_1 = f(s_1)$ and $a_2 = f(s_2)$, let $f(s_1 \wedge s_2) = a_1 \cap a_2$ and $f(s_1 \vee s_2) = a_1 \cup a_2$. Thus we have defined new approximands for $s_1 \vee s_2$ and $s_1 \wedge s_2$.

The interpretation that we give to the above is that for model elements s_1, s_2, the new model $s_1 \wedge s_2$ is just a model s_1 and s_2, and the conceptual model corresponding to $s_1 \vee s_2$ is just a model thought of as s_1 or s_2. In the latter case, one could consider using a component that behaves either like s_1 or behaves like s_2.

Let $\langle S, \Phi, V, G \rangle$ be a system 4-tuple with the above interpretation with $s_1, s_2 \in S$ and let $\langle T, \psi, W, \beta \rangle$ be a measurable system 4-tuple with $t \in T$.

The following propositions are easily shown to be true.

Proposition 4.4 If t α-models s_1 then t β-models $(s_1 \vee s_2)$

with confidence $\beta \geq \alpha$ for all $s_2 \in S$.

<u>Proof</u>: Follows from the fact that $A[\Phi(s_1 \vee s_2)] \supset A[\Phi(s_1)]$.   □

<u>Proposition 4.5</u> Let t α_1-model s_1 and t α_2-model s_2 . Then t β-models $(s_1 \vee s_2)$ with $\max[\alpha_1, \alpha_2] \leq \beta \leq \alpha_1 + \alpha_2$.

<u>Proof</u>: Follows from $A[\Phi(s_1 \vee s_2)] \supset A[\Phi(s_i)]$, i=1,2 and the fact that $P_t[C \cup D] \leq P_t(C) + P_t(D)$ for any $C, D \in \mathbb{B}$.   □

<u>Proposition 4.6</u> If t α-models s_1 then t α-models $(s_1 \wedge s_2)$ with confidence $\alpha \leq \beta$ for all $s_2 \in S$.

<u>Proof</u>: Similar to Proposition 4.4.   □

<u>Proposition 4.7</u> If t α-models $(s_1 \wedge s_2)$ then t α_1-model s_1 and t α_2-model s_2 with $\alpha_1 \geq \alpha$ and $\alpha_2 \geq \alpha$.

<u>Proof</u>: Again, simply notice that $A[\Phi(s_1 \wedge s_2)] \subset A[\Phi(s_i)]$, i=1,2 . □

<u>Proposition 4.8</u> Let t α_1-model s_1 and t α_2-model s_2 . Then t β-models $(s_1 \wedge s_2)$ where $\beta \geq \alpha_1 + \alpha_2 - 1$.

<u>Proof</u>: Let $Q_1 = f(s_1)$ and $Q_2 = f(s_2)$. Suppose $B = Q_1 \cap Q_2$ and $A = Q_1 - B$ and $C = Q_2 - B$. Notice that A, B, C are disjoint sets. Thus, if $A' = H^{-1}(A)$, $B' = H^{-1}(B)$ and $C' = H^{-1}(C)$, it follows that A′, B′, C′ are also disjoint. Thus, since

$$P_t[A'] + P_t[B'] = \alpha_1 ,$$

$$P_t[B'] + P_t[C'] = \alpha_2 ,$$

and

$$P_t[A'] + P_t[B'] + P_t[C'] \leq 1 ,$$

the result follows.   □

The above proposition simply generalizes the following proposition for the case of nonprobabilistic modeling that we might consider assuming that $\langle T, \psi, W, \mathbb{B} \rangle$ was a system 4-tuple.

<u>Proposition 4.9</u> If t models s_1 and t models s_2 , then t models $s_1 \wedge s_2$.

<u>Proof</u>: Straightforward using the definition of "models."   □

<u>CONCLUSION</u>

In this paper, we considered the connection between a theory of composition (or decomposition) and modeling theory. We uncovered the important characteristics that allowed us to be precise about when simplification of components of a base model allow an extrapolation to simplification of the composite model. We showed that these results held for both the case of deterministic modeling as well as for the case of probabilistic modeling. The concept of probabilistic modeling seems to be a more realistic one than the concept of weak or strong modeling since we are often faced with modeling performance uncertainties for which we have only partial information---that is, for which probabilities may in fact be known. The probabilistic theorem shows that in this situation, we can combine components into more complex models and still have some idea of the new modeling confidence, just as we did for the deterministic case.

The results obtained seem to be the best that we can hope for, since we have been working at a very general level. Our results apply to compositions of components that are connected together through the maps (δ, γ) and thus can be used in the case of true physical connections as well as more elusive connections such as information flow linkages. Further results can only come by assuming more structure on the models studied and finer characterizations of the performance evaluation maps. However, the composition space framework together with the modeling framework can form a unifying methodology in which to discuss such problems.

ACKNOWLEDGMENT

The author wishes to thank Dr. Guy C. Corynen and Dr. Stephen H. Hegner for helpful discussions. Part of the research reported here was performed at Lawrence Livermore Laboratory under the auspices of the U. S. Department of Energy under Contract No. W-7405-ENG-48. The author expresses his thanks to Mr. L. H. Fink, Division of Electrical Energy Systems, Department of Energy, for his encouragement and support.

REFERENCES

[1]   S. Aggarwal (January, 1977), Composition Spaces and Modeling, Lawrence Livermore Laboratory Tech. Report UCRL-79323, Livermore, California.

[2]   S. Aggarwal (November, 1977), Probabilistic Modeling, Lawrence Livermore Lab. Tech. Report UCRL-80413, Livermore, California.

[3]   S. Aggarwal, "Simplification via aggregation for neural net-
      works," Simulation 31(1978), no. 4, 129-137.

[4]   G. C. Corynen, "Another look at the concept of model," South-
      eastern Symposium on System Theory (April 26-27, 1976), Knox-
      ville, Tennessee.

[5]   W. Feller (1957), Probability Theory, John Wiley & Sons, New
      York.

[6]   G. Grätzer (1968), Universal Algebra, D. Van Nostrand Co., New
      York.

[7]   E. G. Manes (1976), Algebraic Theories, Springer-Verlag, New
      York.

[8]   H. L. Royden (1973), Real Analysis, MacMillan Co., New York.

[9]   B. P. Zeigler (1976), Theory of Modelling and Simulation, John
      Wiley & Sons, New York.

METHODOLOGY IN SYSTEMS MODELLING AND SIMULATION
B.P. Zeigler, M.S. Elzas, G.J. Klir, T.I. Ören (eds.)
© North-Holland Publishing Company, 1979

DETERMINISTIC SYSTEM THEORY
APPLIED TO SIMPLIFICATIONS
OF STOCHASTIC SYSTEMS

Benjamin Melamed

Industrial Engineering
and Management Sciences Department
Technological Institute
Northwestern University
Evanston, Illinois

Sample path arguments are often used in stochastic processes. When the subject matter is stochastic systems,
sample paths can be accurately and succinctly modeled
as ordinary deterministic systems.

This paper describes a methodology for proving the
"equivalence" of two stochastic sytems in the sense
that the probability law of a stochastic aspect (process)
is preserved under a simplification of the underlying
stochastic system. The methodology blends deterministic
system-theoretic tools and probabilistic reasoning in sample path arguments. More specifically, deterministic system-theoretic structures and morphisms are used to describe sample path histories and to deduce temporal relations among them. The methodology enables the analyst to
test, through sufficient conditions, the validity of stochastic system simplification. When successful, this test
provides a rigorous mathematical proof that formally jus-
tifies the intuition that customarily motivates simplifications.

Finally, the methodology is demonstrated via a working
example from the domain of queueing theory.

1. Introduction

The purpose of this paper is to argue and demonstrate that deterministic system
theory can be successfully used to rigorize sample path arguments pertaining to
the preservation of a probability law in a complex stochastic system. The context of discussion will be a simplification situation; that is, two systems (simplification pair) will be assumed given, the second being simpler than the first
vis-a-vis some complexity measure. The intent of the underlying simplification
is to map the base (first) system in a complexity reducing manner into a lumped
(simpler) one, yet preserve some aspects of interest. Since the systems of interest here are stochastic ones, they will be represented by probability spaces,
while the aspects of interest will be represented as stochastic processes over
the former. Our notion of preservation will accordingly be stochastic equivalence
of those processes, i.e., preservation of their probability law.

This paper offers a methodology for verifying such preservation of stochastic
aspects via two standard theorems from probability theory. The first one gives
sufficient conditions that establish a measure preserving transformation connecting the simplification pair and embodying the simplification process.

The second gives sufficient conditions for preserving the probability law of two
stochastic processes with common parameter set, one from each system in the simplification pair.

The crux of the methodology is to put into correspondence and compare sample histories of the two systems. Deterministic system theory enters here as a powerful
tool for describing and comparing sample histories via deterministic system structures and system morphisms respectively, thus complementing the stochastic nature
of the systems under consideration. In short, we reduce part of the problem of
stochastic equivalence to that of multiple deterministic equivalence of sample
histories using a blend of probability theory and deterministic system theory.
The combined approach then provides a rigorous proof for the desired preservation as well as intuitive insight into the simplification process.

The simplification methodology will be demonstrated by an example from queueing
theory.

2.   Probability-Theoretic Basis of the Simplification Methodology

We begin with the two coordinate probability spaces $\mathscr{A} = \langle \Omega, \mathscr{T}, P \rangle$ and
$\mathscr{A}' = \langle \Omega', \mathscr{T}', P' \rangle$ respectively representing the simplification pair. Their construction is a rather intuitive procedure that can be briefly summarized as
follows.

The analyst is given a verbal description of each system which is partly probabilistic and partly deterministic system-theoretic. For example, in a queueing
system one may specify the arrival and service distributions (probabilistic part)
as well as the queueing discipline, or more generally, the behavior of a typical
itinerant customer (deterministic system-theoretic part). It is known that under
mild regularity conditions on the prescribed distributions, one can construct a
probability space and stochastic processes having that prescribed distribution
law (see e.g. [3] pp. 49-50). These stochastic processes will be referred to as
the generating processes of the system. We mention that each sample point completely specifies a sample history of the system.

The fundamental concept that embodies the idea of a stochastic simplification is
the following variant of a measure preserving transformation (cf. [3] p. 453;
[4] Ch. VIII); we assume that the underlying $\mathscr{A}$ and $\mathscr{A}'$ have already been constructed.

Definition 2.1

Let H: $\Omega \to \Omega'$ be a **surjective** map such that

a) $\forall E' \in \mathscr{T}'$, $H^{-1}(E') \in \mathscr{T}$ (preservation of events) .

b) $\forall E' \in \mathscr{T}'$, $P'(E') = P(H^{-1}(E))$ (preservation of measure) .

Then H is called a measure preserving point morphism (m.p.p.m.) from $\mathscr{A}$ to $\mathscr{A}'$. $\square$

The reader is referred to [8], Sec. 3.3, for a discussion of the simplification
aspects of an m.p.p.m. The theoretical basis for the simplification methodology
is provided by the following two theorems.

Theorem 2.1

A surjective map H: $\Omega \to \Omega'$ is an m.p.p.m. from $\mathscr{A}$ to $\mathscr{A}'$ iff there are stochastic
processes $\mathscr{Y} = \{Y_A\}_{A \in \Theta}$ and $\mathscr{Y}' = \{Y_A'\}_{A \in \Theta}$ over $\mathscr{A}$ and $\mathscr{A}'$ respectively, such that

a) $\mathscr{Y}'$ generates $\mathscr{T}'$ up to completion

b) $\mathcal{Y}'$ and $\mathcal{Y}$ are stochastically equivalent

c) For every $\theta \in \Theta$, there is a null set $N_\theta \in \mathcal{J}$ satisfying

$\forall \omega \notin N_\theta$, $Y_\theta(\omega) = Y'_\theta(H(\omega))$.

Proof: See, e.g., [8], Theorem 3.1.1.   ☐

Theorem 2.2

Let $\mathcal{Y} = \{Y_\theta\}_{\theta \in \Theta}$ and $\mathcal{Y}' = \{Y'_\theta\}_{\theta \in \Theta}$ be stochastic processes over $\mathcal{J}$ and $\mathcal{J}'$ respectively. Suppose H: $\Omega \rightarrow \Omega'$ is an m.p.p.m. such that for each $\theta \in \Theta$ there is a null
set N_θ satisfying

a) $\forall \omega \notin N_\theta$, $Y_\theta(\omega) = Y'_\theta(H(\omega))$.

Then $\mathcal{Y}$ and $\mathcal{Y}'$ are stochastically equivalent.

Proof: See, e.g., [8] Theorem 3.2.1.   ☐

Notice that in our terminology the behavioral aspects are represented by $\mathcal{Y}$ and $\mathcal{Y}'$,
and that the preservation effect under a stochastic simplification $\mathcal{J} \xrightarrow{H} \mathcal{J}'$
(via some m.p.p.m. H) is defined as the stochastic equivalent of the behavioral
pair $(\mathcal{Y},\mathcal{Y}')$ of interest.

To sum up, the simplification methodology consists of two steps. In the first
step one attempts to identify an m.p.p.m. H: $\Omega \rightarrow \Omega'$ as per Theorem 2.1; the second step attempts to establish the stochastic equivalence of a behavioral pair
$(\mathcal{Y},\mathcal{Y}')$ of interest as per Theorem 2.2.

3. Deterministic System-Theoretic Level of the Simplification Methodology

Where and how does deterministic system theory fit into the statistical framework? The answer is that system theory is well-suited for establishing the sample path assertions of conditions c), and a) in Theorems 2.1 and 2.2 respectively, by capturing and exploiting their deterministic system-theoretic underpinning.

First, however, one needs to endow the sample spaces Ω and Ω' with a system-
theoretic representation. The following guidelines should be largely self-evi-
dent: A stochastic system can usually be thought of as one whose parameters are
nondeterministic; however, the underlying rules of operation are common to all
possible parameter settings which themselves obey probabilistic rules of chance.
In other words, a sample path corresponds to a deterministic system which is obtained from the nondeterministic prototype by choosing a parameter setting; the
sample space is merely the family of all such deterministic systems. This family could consist in any of the garden variety of formal systems developed to
date: finite or infinite automata (sequential machines), dynamic systems, iterative specifications, discrete event systems (see Zeigler [10] Ch. 9) etc. Since
the last ones constitute an important case to be used in the sequel, we shall now
give their definition which is due to Zeigler (ibid.).

Definition 3.1

A discrete event system specification (DEVS) is a structure M = $\langle X,S,Y,\mathbf{t},\delta,\lambda \rangle$
where

1. X is the external event set

2. S is the sequential state set

3. Y is the output value set

4. $t: S \rightarrow \lceil 0,\infty \rceil$ is the <u>time advance function</u>

5. $\delta: Q \times (X \cup \{\varphi\}) \rightarrow S$ is the <u>sequential state transition function</u> (where $\varphi \notin X$ is the external <u>nonevent</u> and $Q = \{(s,e): s \in S, 0 \le e < t(s)\}$ is the <u>full state</u> set) such that

$$\forall q = (s,e) \in Q, \ \forall x \in X, \ \delta(q,x) \overset{\Delta}{=} \delta_M(s,e,x)$$

$$\forall q = (s,e) \in Q, \ \delta(q,\varphi) \overset{\Delta}{=} \delta_\varphi(s)$$

6. $\lambda: Q \rightarrow Y$ is the <u>output function</u>.   □

A DEVS M evolves in leaps and bounds, its sequential state trajectories being step functions. A change in sequential state is effected either by an external event in accordance with δ_M , or due to internal dynamics in accordance with δ_φ ; the latter case is triggered when the elapsed time e in state s exceeds its allowable maximum as given by $t(s)$. Thus, M is essentially a discrete (sequential) machine operating over continuous time -- a situation common to many real-life systems (e.g. , queueing systems).

Let us define an input segment to be a pulse train arbitrarily spaced, with pulses $x \in X$ corresponding to external events. It is then easy to construct the mathematical system S_M whose state set is Q, and whose transition function gives the full state landed in after starting in full state (s,e) and receiving a pulse train input η . The derivation employs an interim system called iterative specification constructed from M on the way to obtaining S_M. The details are omitted here; the interested reader is referred to Zeigler [10] Ch.9. Each state trajectory $STRAJ_{q,\eta}: [0,\infty) \rightarrow Q$ an be well-defined under a suitable regularity condition called legitimacy. It is these state trajectories which we use as a deterministic system-theoretic representation for the sample spaces. We point out in passing that the particular representation of our (coordinate) sample space is immaterial so long as all representations are one-one.

The next step is to get hold of a deterministic system-theoretic device for matching and comparing output trajectories. The concept of system morphism is precisely the tool we seek, as it formalizes the idea of state simplification yet retains invariance under the transition function. Zeigler [10] Ch. 10 develops a hierarchy of such morphisms, from which the basic definition is reproduced in

<u>Definition 3.2</u>

A system morphism from a mathematical system $S = \langle T,X,\Gamma,Q,Y,\delta,\lambda \rangle$ to a mathematical system $S' = \langle T',X',\Gamma',Q',Y',\delta',\lambda' \rangle$ is a triple $(\widetilde{g},\widetilde{h},\widetilde{k})$ where

1. $\widetilde{g}: \Gamma' \rightarrow \Gamma$ is the <u>input segment encoding function</u>

2. $\widetilde{h}: \overline{Q} \rightarrow Q'$ (for some $\overline{Q} \subset Q$) is the <u>state decoding function</u>

3. $\widetilde{k}: Y \rightarrow Y'$ is the <u>output decoding function</u>

such that

a) $\widetilde{h},\widetilde{k}$ are surjective

b) $\forall q \in \overline{Q}, \ \forall \eta' \in \Gamma', \ \widetilde{h}(\delta(q,\widetilde{g}(\eta'))) = \delta'(\widetilde{h}(q),\eta')$

c) $\forall q \in \overline{Q}, \ \widetilde{k}(\lambda(q)) = \lambda'(\widetilde{h}(q))$ □

In the system S above, T is the time base, X the input set, Γ the input segment set, Q the state set, Y the output set, δ the transition function, and λ the output function. Likewise for S' .

For our purposes we cite, however, a brand of DEVS morphism which matches jumps in the step functions representing the respective DEVS state trajectories. These are the so-called <u>transition covering DEVS morphisms</u> (TC-DEVS morphisms) which are investigated in some detail in Melamed [8] Sec. 1.4. Following Zeigler [10] Ch. 9 let us first define the following auxiliary functions (N=set of natural numbers):

1. The <u>extended autonomous transition</u> function of M is $\overline{\delta}_\varphi : S \times (N \cup \{0\}) \to S$ where

$$\overline{\delta}_\varphi(s,n) \quad \triangleq \quad \begin{cases} s, & \text{if } n = 0 \\[2em] \delta_\varphi(\overline{\delta}_\varphi(s,n-1)), & \text{if } n > 0 \end{cases}$$

2. The <u>total time advance function</u> of M is $\sigma : S \times (N \cup \{0\}) \to [0,\infty]$ where

$$\sigma(s,n) \quad \triangleq \quad \begin{cases} 0, & \text{if } n = 0 \\[2em] \sum_{i=0}^{n-1} t(\overline{\delta}_\varphi(s,i)), & \text{if } n > 0 \end{cases}$$

3. The <u>jump counter function</u> of M is $m : Q \times [0,\infty) \to N \cup \{0\}$ where
$$m((s,e),\tau) \triangleq \sup\{n: \sigma(s,n) \leq e + \tau\}.$$

With these definitions in mind we further make

<u>Definition 3.3</u>

A <u>TC-DEVS morphism</u> from a DEVS $M = \langle X,S,Y,t,\delta,\lambda \rangle$ to a DEVS $M' = \langle X',X',Y',t',\delta',\lambda' \rangle$ is a quadruple (g,L,h,k) where

1. $g: X' \to X$ is the <u>external event encoding function</u>
2. $L: \hat{S} \to N \cup \{0\}$ is the <u>transition counting function</u> where $\hat{S} \subset S$
3. $h: \hat{S} \to S'$ is the <u>sequential state decoding function</u>
4. $k: Y \to Y'$ is the <u>output decoding function</u>

and such that the following hold:

a) h and k are surjective.

b) Let $\hat{Q} \triangleq \{(s,e) \in Q: s = \overline{\delta}_\varphi(\hat{s},m((\hat{s},0),\tau))$, $e = \tau - \sigma(\hat{s},m((\hat{s},0),\tau))$ for some $\hat{s} \in \hat{S}$, $0 \leq \tau < t'(h(\hat{s}))\}$. If $(s,e) \in \hat{Q}$ has representations via $(\hat{s}_1,\tau_1)$ and $(\hat{s}_2,\tau_2)$, then $h(\hat{s}_1) = h(\hat{s}_2)$ and $\tau_1 = \tau_2$.

c) For any $\hat{s} \in \hat{S}$,
    c.1) $t'(h(\hat{s})) = \sum_{i=0}^{L(\hat{s})} t(\overline{\delta}_\varphi(\hat{s},i))$

    c.2) $\overline{\delta}_\varphi(\hat{s},L(\hat{s})+1)) \in \hat{S}$

    c.3) $h(\overline{\delta}_\varphi(\hat{s},L(\hat{s})+1)) = \delta'_\varphi(h(\hat{s}))$

d) For every $(s,e) \in \hat{Q}$ represented via $\hat{s}$ and τ and any $x' \in X'$,
    d.1) $\delta_M((s,e),g(x')) \in \hat{S}$

    d.2) $h(\delta_M((s,e),g(x'))) = \delta'_M((h(s),\tau),x')$

    d.3) $k(\lambda(s,e)) = \lambda'(h(\hat{s}),\tau)$.

In this case we say that M is a <u>transitional covering</u> of M' (or simply that M covers M').   ☐

While the definition above appears to be complex, the basic idea here is quite simple. In fact it can be shown that it is equivalent to the existence of a system morphism $(\tilde{g},\tilde{h},\tilde{k})$ from the mathematical system induced by M to the one induced by M' such that (see Melamed [8] Sec. 1.4)

a) $\tilde{g}$ relables the pulses according to g.

b) $\tilde{h}(s,e) = (s',0) \Rightarrow e = 0$, where $s' = h(s)$.

c) $\tilde{k} = k$.

In other words ᵦ) asserts that every jump in M' has a concurrent counterpart in M but not necessarily vice versa. Since nothing happens between jumps, it suffices in many cases to match state trajectories of M and M' only at jump points of the latter and to ascertain that jumps in M without concurrent counterparts in M' do not affect the matching of the analyst's behavioral aspect of interest -- to ensure complete matching of the requisite output trajectories as required in Theorem 2.1 and 2.2. To clarify the proceedings we shall demonstrate in the next section the utility of the methodology and the foregoing observations through a typical example from the domain of queueing theory.

4. Queueing--Theoretic Examples

Consider the simplification depicted in Figure 1.

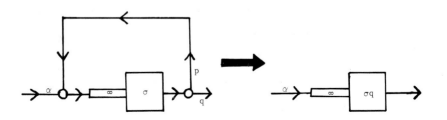

<u>Figure 1</u>:   Simplifying an M/M/1 queue with feedback to one without feedback

On the left hand side we have the base model -- an M/M/1 feedback queue (Poisson arrivals with parameter α, infinite waiting line, exponential server with parameter σ); the feedback is such that on each service completion a customer is routed back to the end of the line with probability p, or departs the system altogether with the complementary probability q. On the right hand side we have a lumped version of the base model where the feedback loop was eliminated and the service parameter was modified to σq. All arrivals, services, and routings are mutually independent processes. Using the simplification methodology described in Section 2, we shall show (in outline) the preservation of certain operating characteristics under the foregoing simplification, provided the initial conditions of both models coincide.

The first step is to construct the coordinate probability spaces representing the lumped and base models.

For the lumped model a typical history is specified by a sample point $\omega' = (\ell_0', \{a_i'\}_{i=1}^{\infty}, \{s_j'\}_{j=1}^{\infty})$ where ℓ_0' stands for the initial line length, a_i' is the i-th interarrival interval, and s_j' is the j-th service interval awarded. Recall that the family of finite dimensional distributions of the generating random variables $\mathscr{L}' = \{L_0', \{A_i'\}_{i=1}^{\infty}, \{S_j'\}_{j=1}^{\infty}\}$ is assumed specified. One can now construct $\mathscr{A}' = \langle \Omega', \mathscr{T}', P' \rangle$ as a complete probability space with sample points as above, and the stochastic processes in $\mathscr{L}'$ can be defined over it with the prescribed family of finite dimensional distributions.

The construction of the coordinate space $\mathscr{A} = \langle \Omega, \mathscr{T}, P \rangle$ for the base model is similar except that a typical sample point here is $\omega = (\ell_0, \{a_i\}_{i=1}^{\infty}, \{s_j\}_{j=1}^{\infty}, \{v_k\}_{k=1}^{\infty})$ where ℓ_0, a_i and s_j are as before, and v_k is the k-th routing decision (1 for feedback, 0 for departure). The generating random variables are $\mathscr{L} = \{L_0, \{A_i\}_{i=1}^{\infty}, \{S_j\}_{j=1}^{\infty}, \{V_k\}_{k=1}^{\infty}\}$.

The next step is to give Ω' and Ω a deterministic system-theoretic representation.

For the lumped model we associate with each $\omega' = (\ell_0', \{a_i'\}_{i=1}^{\infty}, \{s_j'\}_{j=1}^{\infty}) \in \Omega'$ a DEVS $M'(\omega') = \langle X_{\omega'}', S_{\omega'}', \bullet, t_{\omega'}', \delta_{\omega'}', \bullet \rangle$ (with unspecified output set and function) where

1') $X_{\omega'}' \triangleq \{1\}$

2') $S_{\omega'}' \triangleq \{(0,n',\infty): n' \in \mathbb{N}\} \cup \{(\ell',n',r'): \ell' > 0, n' \in \mathbb{N}, 0 \leq r' < s_n'\}$

3') $t_{\omega'}'(\ell',n',r') \triangleq r'$

4') $\delta_{\omega',\varphi}'(\ell',n',r') \triangleq \begin{cases} (0, n'+1, \infty), & \text{if } \ell' = 1 \\[2em] (\ell'-1, n'+1, s_{n+1}'), & \text{if } \ell' > 1 \end{cases}$

5') $\delta_{\omega',M}'(((\ell',n',r'),e'),1) \triangleq \begin{cases} (1,n',s_n'), & \text{if } \ell' = 0 \\[2em] (\ell'+1, n', r'-e'), & \text{if } \ell' > 0 \end{cases}$

The symbol 1 designates here the arrival event of a customer. Observe that in a sequential state $s' = (\ell',n',r')$, ℓ' is the current line length, $n'-1$ the total services completed, and r' the residual service time of the current customer. Note also that $\mathfrak{t}'$ gives this residual service time, $\delta'_{\omega',\varphi}$ describes a transition due to service completion, and $\delta'_{\omega',M'}$ describes the same due to an arrival. Let $\eta'(\omega') \triangleq \overset{\infty}{\underset{i=1}{\otimes}} 1_{a'_i}$ be the infinite concatenation of arrival events with $\{a'_i\}_{i=1}^{\infty}$ interarrival intervals, and define the initial state $q'_0(\omega') \triangleq (s'_0,0)$ where

$$s'_0 = s'_0(\omega') \quad \triangleq \quad \begin{cases} (\ell'_0,1,\infty), & \text{if} \quad \ell'_0 = 0 \\\\ (\ell'_0,1,s'_1), & \text{if} \quad \ell'_0 > 0 \end{cases}$$

The deterministic system-theoretic representation of each $\omega' \in \Omega'$ is the infinite trajectory $STRAJ_{q'_0(\omega'),\eta'(\omega')}: [0,\infty) \to Q'$.

For the base model we first define an auxiliary sequence of random variables $\{Z_j\}_{j=0}^{\infty}$ where

$$Z_j(\omega) \quad \triangleq \quad \begin{cases} 0, & \text{if } j = 0 \\\\ \min\{k: k > Z_{j-1}(\omega) \text{ and } V_k(\omega) = 0\}, & \text{if } j > 0. \end{cases}$$

$Z_j(\omega)$ is the index of the j-th 0 (departure) in the sequence $\{V_k(\omega)\}_{k=1}^{\infty}$ of Bernoulli trials. With each $\omega = (\ell_0, \{a_i\}_{i=1}^{\infty}, \{s_j\}_{j=1}^{\infty}, \{v_k\}_{k=1}^{\infty}) \in \Omega$, we associate a DEVS $M(\omega) = \langle X_\omega, S_\omega, \cdot, \mathfrak{t}_\omega, \delta_\omega, \cdot \rangle$ where

1) $X_\omega \triangleq \{1\}$

2) $S_\omega \triangleq \{(0,n,v_n,\infty): n = Z_{j-1}(\omega)+1, j \in \mathbb{N}\} \cup \{(\ell,n,v_n,r): \ell,n \in \mathbb{N}, 0 \le r < s_n\}$

3) $\mathfrak{t}_\omega(\ell,n,v_n,r) \triangleq r$

4) $\delta_{\omega,\varphi}(\ell,n,v_n,r) \triangleq \begin{cases} (0, \ n+1, \ v_{n+1}, \ \infty), & \text{if } \ell = 1 \text{ and } v_n = 0 \\\\ (\ell-1, \ n+1, \ v_{n+1}, \ s_{n+1}), & \text{if } \ell > 1 \text{ and } v_n = 0 \\\\ (\ell,n+1, \ v_{n+1}, \ s_{n+1}), & \text{if } v_n = 1 \end{cases}$

5) $\delta_{\omega,M}(((\ell,n,v_n,r),e),1) \triangleq \begin{cases} (1,n, \ v_n,s_n), & \text{if } \ell = 0 \\\\ (\ell+1,n,v_n,r-e), & \text{if } \ell > 0 \end{cases}$

The intuitive interpretation of the above is similar to that of the lumped model. Again we define an input segment $\eta(\omega) \triangleq \overset{\infty}{\underset{i=1}{\otimes}} 1_{a_i}$ and an initial state $q_0(\omega) \triangleq (s_0,0)$ where

$$s_0 = s_0(\omega) \triangleq \begin{cases} (\ell_0,1,v_1,\infty), & \text{if } \ell_0 = 0 \\\\ (\ell_0,1,v_1,s_1), & \text{if } \ell_0 > 0 \end{cases}$$

and we use $STRAJ_{q_0(\omega),\eta(\omega)}: [0,\infty) \to Q$ as the deterministic system-theoretic representation of each $\omega \in \Omega$.

We are now in a position to construct an m.p.p.m. H: $\Omega \rightarrow \Omega'$. The intuitive motivation for H may be described as follows. Consider in the base model any sequence of services that is preceded and followed by a customer departure from the system. It so happens (see Melamed [8] Sec. 5.1) that the sum of those services is exponentially distributed but with parameter σq. Why not consider each such total service time as a new service time? In other words, eliminate the feedback loop and substitute the old server parameter σ with a new parameter σq in the base model thus obtaining the lumped model of Figure 1.

We shall now proceed to formalize these intuitive ideas. Assuming that L_0 and L_0' have the same distribution, let $\omega = (\ell_0, \{a_i\}_{i=1}^{\infty}, \{s_j\}_{j=1}^{\infty}, \{v_k\}_{k=1}^{\infty})$ be any sample point in Ω.

Define $H(\omega) = \omega' \in \Omega'$ such that $\omega' = (\ell_0, \{a_i\}_{i=1}^{\infty}, \{s_j'\}_{j=1}^{\infty})$ where

$$s_j' = \sum_{n=Z_{j-1}(\omega)+1}^{Z_j(\omega)} s_i \; .$$ It should be obvious that H is surjective.

Theorem 4.1

The map H is an m.p.p.m.

Proof

Follows from the almost sure equalities

$$L_0(\omega) = \ell_0 = L_0'(H(\omega))$$

$$A_i(\omega) = a_i = A_i'(H(\omega)), \qquad i = 1,2,\ldots,$$

$$\bar{S}_j(\omega) \triangleq \sum_{i=Z_{j-1}(\omega)+1}^{Z_j(\omega)} S_i(\omega) = s_j' = S_j'(H(\omega)), \qquad j = 1,2,\ldots,$$

The remaining details may be found in Melamed [18], Theorem 5.1.1.   □

Observe that the random variables $\{\bar{S}_j\}_{j=1}^{\infty}$ are the total service times alluded to above. Now compare the operation of $M(\omega)$ and $M(H(\omega))$ as given by the respective state trajectories $STRAJ_{q_0}(\omega), \eta(\omega)$ and $STRAJ_{q_0'}(H(\omega)), \eta'(H(\omega))$ when "run" concurrently from time 0 and on. Clearly, every time a jump occurs in $M'(H(\omega))$ due to a customer departure from the system, a similar jump occurs in $M(\omega)$. However in between such jumps, nothing (except for arrivals) occurs in $M'(H(\omega))$, while $M(\omega)$ is carrying out a sequence of services coupled with feedback routings. Its arrivals are also concurrent with those of the $M'(H(\omega))$.

We again formalize these observations in

Theorem 4.2

For almost every $\omega \in \Omega$, there is a TC-DEVS state-homomorphism[+] (i,L,h) running from $M(\omega)$ to $M'(H(\omega))$.

[+]A TC-DEVS state-homomorphism (i,L,h) is a TC-DEVS morphism (g,L,h,k) with $\hat{Q} = Q$, g = i (the identity map) and an unspecified k.

Proof

Follows from Theorem 5.1.2 in Melamed [8] by defining

$$h(\ell, n, v_n, r) \triangleq \begin{cases} (\ell, j_\omega(n), \infty), & \text{if } \ell = 0 \\ (\ell, j_\omega(n), r + \sum\limits_{i=n+1}^{Z_j(\omega)} s_i), & \text{if } \ell > 0 \end{cases}$$

where $j_\omega(n)$ is the unique integer j satisfying $Z_{j-1}(\omega) < n \le Z_j(\omega)$ and

$$L(\ell, n, v_n, r) \triangleq \begin{cases} 0, & \text{if } \ell = 0 \\ Z_j(\omega) - n, & \text{if } \ell > 0 \end{cases}$$

Corollary 4.1

Since $h(s_0(\omega)) = s_0'(H(\omega))$ and $\eta(\omega) = \eta'(H(\omega))$, it follows from Theorem 4.2 and Melamed ([8], p.230) that there is a surjective map $\tilde{h}: 0 \to 0'$ such that for all $t \ge 0$

a) $\tilde{h}(STRAJ_{q_0(\omega), \eta(\omega)}(t)) = STRAJ_{q_0'(H(\omega)), \eta'(H(\omega))}(t)$. $\square$

What behavioral aspects are preserved under the simplification of Figure 1 as formalized by the m.p.p.m. H? Intuitively, we feel that customer-oriented aspects will not be preserved, since the Figure 1 simplification does not preserve the order and identity of customers. We can, however, state the following

Theorem 4.3

When the initial line sizes L_0 and L_0' are stochastically equivalent, the following processes in the base model are preserved by their counterparts in the lumped model:

a) the line size process
b) the busy/idle period process†
c) the departure process from the system

Proof

According to Theorem 2.2, it suffices to show for each case that

$$\lambda(STRAJ_{q_0(\omega), \eta(\omega)}(t)) = \lambda'(STRAJ_{q_0'(H(\omega)), \eta'(H(\omega))}(t)), \qquad \forall t \ge 0$$

through appropriately defined output functions λ and λ'. Thus for Case a) above we set

$$\lambda(q) = \lambda((\ell, n, v_n, r), e) \triangleq \ell, \qquad \lambda'(q') = \lambda'((\ell', n', r'), e') \triangleq \ell'$$

Case b) is a consequence of Case a). For Case c) we set for the departure count

$$\lambda((\ell, n, v_n, r), e) \triangleq j - 1, \qquad \lambda'((\ell', n', r'), e') \triangleq n' - 1$$

where $j = j_\omega(n)$ satisfies $Z_{j-1}(\omega) < n \le Z_j(\omega)$. The details of the proof may be found in Melamed [8], pp. 231-234. $\square$

It is known that, in equilibrium, the departure process from the lumped model is a Poisson process with the same parameter α as the arrival process, a fact first proven by Burke [2]. It immediately follows from Theorem 4.3 that this is also

† A queue is said to be busy when line size > 0, idle when line size = 0.

true for the base model -- a fact exemplifying the power of the simplification
methodology.

5. Concluding Remarks

The treatment in the previous section can be extended in essence to networks of
M/M/1 nodes with routing probabilities p_{ij} from node i to node j (the so-called
Jackson networks first treated by Jackson [5]). In fact, a feedback-removing
simplification may be effected as follows:

1. define the new routing probabilities by

$$
p'_{ij} = \begin{cases} 1, & \text{if } i = j \text{ and } p_{ii} = 1 \\ 0, & \text{if } i = j \text{ and } p_{ii} < 1 \\ \dfrac{p_{ij}}{1-p_{ii}}, & \text{if } i \neq j \end{cases}
$$

2. define the new service parameters by

$$
\sigma'_i = \begin{cases} \sigma_i, & \text{if } p_{ii} = 1 \\ \sigma_i(1 - p_{ii}), & \text{if } p_{ii} < 1 \end{cases}
$$

Again, the behavioral aspects preserved include the vector of line sizes at all
nodes, the respective idle/busy periods, and the traffic processes on any arc
(i,j), i ≠ j, including, in particular, all departure processes from the system.
(See Melamed [8] pp.234-237.)

It should be pointed out that the preservation of the vector of line size processes
can be proved by different methods. Indeed, it happens that this process in
Markovian[+]; writing down the differential equations for the absolute state probabilities reveals that they are identical for the base and lumped models, and
therefore have identical solutions since the initial conditions coincide.

The point to note, however, is that the simplification methodology is a general
method for showing stochastic equivalence, while the method just cited is ad-hoc
and depends on the Markovian nature of the underlying process. To illustrate this
point, consider two arbitrary queues whose arrivals, services and initial conditions are characterized by the same family of finite dimensional distributions, but
their queue disciplines are first-in-first-out and preemptive-resume (last-in-
first-out with immediate preemption), respectively. Their line size processes are
not, in general, stochastically equivalent. But their idle-busy period processes
are, because we can match sample points with coincident idle/busy periods in a
measure preserving manner, even though we cannot do this for every service period.
A formal proof may be supplied by the simplification methodology; a little reflection reveals that the m.p.p.m. H involved is the identity map. Of course, the
line size process is not, in general, Markovian, and thus the absolute state probability equations are not readily available. Moreover, it is difficult to arith-
metically describe realizations of the process, due to the "logical" nature of the
queueing disciplines. A deterministic system-theoretic description is more intuitive and therefore more appealing and easier to manipulate.

[+]Roughly speaking, a stochastic process is Markovian if its stochastic future is
independent of its stochastic past, given the stochastic present. See e.g. Doob[3].

Finally, we mention the reversibility and quasi-reversibility methods used by a
number of authors ([9], [6], [7], [1]) to show that in certain Markovian queueing
networks in equilibrium with independent Poisson arrival streams to each node,
the departure processes are also Poisson. The proof utilizes a technique whereby
each queueing system is associated with a dual network obtained by running time in
reverse in the original network. Both the original and dual networks have independent Poisson arrivals at each node. It is then claimed that departures from the
first system "correspond" to arrivals in the second system, so that the two are
stochastically equivalent and the claimed result follows. The argument is of the
sample path type but there are gaps in the proof, although the results are valid.
We conjecture that a formal justification by means of the simplification methodology can be attained to close those gaps and achieve a complete proof.

References

[1]   Barbour, A.D.   (1976)   "Networks of Queues and the Method of Stages," _Advances
      in Applied Probability_, Vol. 8   584-591.
[2]   Burke, P.J.   (1956)   "Output of a Queueing System," _Operations Research_, Vol.
      4   699-704.
[3]   Doob, J.L.   (1953)   _Stochastic Processes_, Wiley.
[4]   Halmos, P.R.   (1950)   _Measure Theory_, Van Nostrand, 1950.
[5]   Jackson, J.R.   (1957)   "Networks of Waiting Lines," _Operations Research_, Vol.
      5, No. 4   518-521.
[6]   Kelly, F.P.   (1975)   "Networks of Queues with Customers of Different Types,"
      J. of Applied Probability, Vol. 12   542-554.
[7]   Kelly, F.P.   (1976)   "Networks of Queues," _Advances in Applied Probability_,
      Vol. 8,   416-432.
[8]   Melamed, B.   (1976)   "Analysis and Simplifications of Discrete Event Systems
      and Jackson Queueing Networks," Doctoral Dissertation, _TR 195_, Dept. of CCS,
      University of Michigan.
[9]   Reich, E.   (1957)   "Waiting Times When Queues are in Tandem," _Annals of Mathe-
      matical Statistics_, Vol. 28   768-773.
[10] Zeigler, B.P.   (1976) _Theory of Modeling and Simulation_, Wiley.

METHODOLOGY IN SYSTEMS MODELLING AND SIMULATION
B.P. Zeigler, M.S. Elzas, G.J. Klir, T.I. Ören (eds.)
© North-Holland Publishing Company, 1979

GENERAL SYSTEMS AS A SYNTHESIS OF HOLISM AND REDUCTIONISM:
PREDOMINANCE OF LARGER SYSTEM CHARACTERISTICS

A. AVERBUKCH
Department of Health Education,
Ministry of Health, Jerusalem,
Israel

To understand the necessity and nature of the systems
approach, it is helpful to take into account the incompleteness and uncertainty principles inherent in the process
of human cognition and its results. These principles have
received some formal support from developments in science
such as the principle of uncertainty (Heisenberg) and the
theorem of incompleteness (Gödel). Uncertainty and incompleteness of our cognition on all its levels is substantial. As a result, there constantly appears the necessity
for a complex many-sided use of research processes (methods,
etc.) and their results on all levels, i.e. necessity
for systems methodology. At present we wish to stress
the complex and systematic combination of the holistic
and analytical approaches in the framework of the
systems approach. The paper is devoted also to an examination of the statement about predominance of larger
system's characteristics.

We consider the above statements to be true not only in
some specific fields, but to have a general philosophical
and methodological meaning. Therefore these statements
must be taken into account when we deal with cosmological
problems and our view of the world, as also when we deal
with more concrete theoretical and practical problems.

THE PRINCIPLES OF INCOMPLETENESS AND UNCERTAINTY

The principle of incompleteness was studied in mathematical logic by Godel in the
theorem of the incompleteness of formal systems (Gödel's First Theorem,1931).
The theorem is of major logical and epistemological significance, since it
proves the impossiblity of a complete formalization of cognition.

The principle of uncertainty was given correct expression in physics in the
correlation of uncertainty (Heisenberg), showing a specific relation between the
momentum and the coordinate of micro-objects. The principle set forth by
Heisenberg has also been examined by scientists engaged in other fields
(sociology, for instance.).

We consider the above principles to be true not only in physics and mathematical
logic, but to have general methodological and philosophical meaning.[*]

---

[*] The uncertainty and incompleteness of our cognition is also the focus of
investigation of many philosophical schools: agnostics, skeptics, phenomenologists,
positivits, etc.) However, the following analysis of the systems approach will
make sense to the reader who accepts that cognition is inherently incomplete
and uncertain on either philosphical or merely intuitive, common sense grounds.

What is very important and must be underlined is the fact, that from Heisenberg's correlation of uncertainty and Godel's First Theorem it follows that
(1) principles of uncertainty and incompleteness "work" on all levels of research, i.e. on the phenomenological, super- and subphenomenological (analytical) levels;
(2) uncertainty and incompleteness of our cognition on all these levels is not accidental and insignificant, but have substantial character.
Hence we face the problems of the substantial impossibility of giving a complete interpretation of any research based on a single one of these levels (phenomenological, etc.) or research methods. As a result, there constantly appears the necessity for a complex many-sided use of research processes (methods, etc.) and their results (facts, hypotheses, etc.) on all levels.

We should not expect, however, that such complex many-sided research can entirely overcome the uncertainty and incompleteness of our cognitive processes. Nevertheless, the complex approach may do away with many limitations inherent in each of the above levels of research.

The complex systems approach overcomes not only the limitations of analytical approach, but also those of the holistic approach, which belittles the role of the parts in the phenomena under consideration and the analytical approach in general.

In this connection we emphatically disagree with many systems theorists, who consider "holism as a methodology and even ontology" of the General Systems Theory, and who identify the systems approach with holism. (e.g. works by E. Laszlo, L Bertalanffy, P.B. Checkland etc. [1-3]

The point is that holism and the holistic approach, proposed by Smuts and developed in particular by von Meyer-Abich, represented a first, direct and simplistic reaction to the analytical, atomistic approach that prevailed in scientific research over the centuries. The holistic approach, in its essence (as a methodological aspect of the "wholeness philosophy") denies the significance and contribution of the analytical approach. At the time of its emergence (the 1920's) there existed neither the needed theoretical disciplines nor the technical possibility of combining data pertaining to both analytical and holistic approaches into one scientific framework. In its turn, the analytical approach in its various forms (methodological individualism, atomistic approach, etc.) rejects the holistic approach in principle.

The present development of many theoretical disciplines (cybernetics, mathematical logic, information theory, theory of sets, theory of games and decision-making, control and organization theory, etc.) and the resultant rapid technical progress, especially in computer techniques, has opened up qualitatively new possibilities in the field of human thinking processes.

Extremely important in this respect is the fact, that powerful technical means and new methods of carrying out logical operations have made it possible to analyse objects and processes as entities, wholes, organizations or systems, taking into account, concurrently and in unity, both the characteristics and functions on the specific level of the whole and the characteristics and functions of components (parts, etc.) inside and outside the research whole.

The complex and systematic combination of the holistic and the analytic approaches* for the solution of new epistemological problems is essentially the main characteristic of the systems approach.

---

* In our opinion, there is a substantial parallelism between the following divisions of research approaches:

The latter is a synthesis of these approaches, and hence, cannot be identified
with either of them. Owing to its synthetical character, it is qualitatively
newer and more powerful than the other approaches. Its novelty and efficiency
consist of the following:

(1) The subject of its investigation is a qualitatively new phenomenon-
organized complexity, organization, system. Bertalanffy and his followers refer
to W. Weaver [4] who distinguishes three stages in the development of research:
organized simplicity in the first stage (classical mechanics); non-organized
complexity at the second stage (classical statistical physics); and organized
complexity at the third stage, typical of science in the 20th century.

(2) The systemic stage in the development of science is characterized
not only by the emergence of a qualitatively new research subject, i.e.,the
organized complexity, the system. The statement of the new subject of research
posed a new knowledge problem [5] which presupposes the introduction and development of new means, methods and approaches for its solution. The investigators
of the General Systems Theory propose various means for the above.

At present we wish to stress the complex and systemic combination of the holistic
and analytical approaches. The prime difficulty is posed not by our initial use
of the holistic approach and the subsequent use of the analytical one, but by
the problem of finding a systematic, complex intercorrelation of the two
approaches. This should be a consistent and rigorous system combining the two
approaches as integral constituents of the new methodological approach.

The ways in which the holistic and analytical approaches can be used for the
solution of the problems can be defined by: (a) the general specifics of the
problem in question; (b) the general specifics of our Universe. The development
of the latter is to provide a broader basis for the working out of new general
methods and a research methodology, suitable for the study of the problems
related to the wider Universe as a Whole, as well as its constituents (parts);
i.e, to the study of more particular problems.

THE PREDOMINANCE OF LARGER SYSTEM CHARACTERISTICS

In our opinion, one of the most serious shortcomings inherent in the systems
approach at the present stage, is the fact that the systems principle does not
carry consistently to its logical end.

The realization of the principle presupposes the successive consideration of all
smaller objects, phenomena, as a subsystem (components) of larger systems. This
is reflected in the hierarchy principle.

In proceeding from the above principle, in order to solve any problem relevant
to a given system, one must draw a preliminary solution relating to the broader
system, of which the former constitutes a part [6] .

From this follows an extremely important theoretical and methodological conclusion, namely that in principle the solution of any problem relevant to a given
system requires a preliminary, as well as a subsequent, consideration of the

        phenomenology - sub- and superphenomenology
        holistic - subholistic (analytical)- and superholistic
        systems - sub- and supersystems.

We therefore use these terms interchangeably.

broader system, including the broadest system known to us, even if we lack comprehensive information about it. (Today, the broadest system known to us is the
Metagalaxy).

This approach to the solution of all problems (theoretical as well as practical
ones) is a methodological principle, formulated as a result of the discovery of
basic ontological relationships between the "behavior" of various large phenomena,
systems, and that of the smaller systems, which are subsystems (constituents) of
the larger ones.

This principle is widely accepted and developed in the conventional sciences.
There is an emergence of a qualitatively new tendency in these sciences (physics,
biology, sociology, psychology etc.), in the development of their theoretical
conceptions and methodology. In the past hundred years (and especially within the
last few decades) there has been a reevaluation and reorientation in the understanding of various phenomena, i.e. a transition from the explanation of the
whole as sum total of the characteristics of the parts, to regarding them as
entities and proceeding from the characteristics of the broader systems of which
the studied phenomenon is a part.

Gradually more and more data is accumulating to prove that broader systems and
their characteristics exert a determining influence on subsystems (elements) that
they include. The validity of this principle is proved not only in those
spheres of reality with which the conventional sciences deal, but also on the
frontiers of the sciences, as well as on the boundary of science and cognition in
general, on the one hand, and in the sphere of practice, on the other.

As a result of this development, there is ever more reason to regard this principle not simply as a general or abstract universal one, but as a truly universal
philosophical and methodological principle, reflecting the above stated correlation between larger systems and smaller ones, forming subsystems of the former.

We attempt to substantiate this principle as a universal philosophical and
methodological one[*]. Since all elements, parts, systems known to us are subsystems
of ever larger systems and the largest one known to us is our Universe, its
parameters must be regarded as the predominant ones, and must be taken into
account when we deal with cosmological problems and our view of the world, as
also when we deal with more concrete theoretical and practical problems. There
are grounds to consider this as a universal philosophical principle. It would be
even more correct to consider this principle and approach as a predominant
tendency within the framework of the complex systems principle and approach.[**]

------

[*] Larger systems and their characteristics predominate over smaller systems and
their characteristics, not in a rigid causal way, but in a more complicated and
indirect manner.

[**] Because of the very complexity of these ties between large sytems and smaller
ones, identifying these ties can be very difficult. Still, knowledge about the
behavior of larger systems should be taken into consideration when dealing with
smaller systems.

A similar approach is expressed by G.J. Klir[7] " A higher level system entails
all knowledge of the corresponding systems at any lower level and contains some
additional knowledge which is not available at the lower levels."

In our opinion, there are some factors which obstruct an absolutely scrupulous and consistent use of this methodological principle*. Still, the considerable difficulties encountered in the consistent realization of this principle must not make us reject the idea. Similarly, in the past, enormous difficulties faced the consistent application of the analytical principle and approach, both as in a cognitive process and in practice. But the process of coping with these difficulties had remarkable results, though gradually it became clear that the analytical method encountered obstacles of a fundamental nature in its development.

In summing up we note the wider development of: (1) the complex systems approach as a predominant principle and method in describing any phenomena; (2) the predominant significance in the framework of systems method of the approach proceeding from the larger systems (including the Universe as a system) and their most general characteristics, to smaller ones.

Both these principles and approaches are of cardinal importance to an interpretation of phenomena in the various spheres of reality, including the social realm. These principles and approaches can be effectively used in the construction of a general sociological theory as well as in the consideration of more particular social problems.

References

(1) E. Laszlo, ed: The Relevance of the General Systems Theory (Braziller, N.Y, 1972) 5-7.
(2) L. von Bertalanffy: General Systems Theory: Foundations, Developments, Applications.The Penguin Press, London,(1971).
(3) P.B. Checkland: Science and the Systems Paradigm, Int. J. General Systems, 3 (1976) 127-134.
(4) W. Weaver: Science and Complexity, American Scientist 36 (1948) 536-544.
(5) V.N. Sadovski: Osnovanija Obschej teorii sistem,Nauka, Moscow,(1974).
(6) V.N.Sadovski: ibid.,248.
(7) G.J. Klir: Identification of Generative Structures in Empirical Data, Int. J. General Systems,3(1976) 89.

---

* Among them are, for example, lack of time, lack of means (including technical ones) and of people possessing the required knowledge and ability to apply these methodological principles (and the systemic approach in general) while solving particular problems, etc.

PANEL DISCUSSION

THE FUTURE OF MODELLING METHODOLOGY:
WHICH ROADS TO TAKE?

The following is an edited record of the Symposium panel discussion. The record was developed from notes taken by the editor which he wrote up and sent to the discusants for their review. We hope that the spontaneity lost from the verbatim transcript is more than compensated for by the coherence of a second reflection.

The panel chairman was Maurice Elzas and the panelists were: Mones Berman, Ahron Nir, David Dayan, Tuncer Oren, Ghislain Vansteenkiste and D. Szekely.

Berman: As more large models are developed, the problems of portability and transferability will be increasingly important. In trans- ferring a model from developer to user, the data base with which the model was developed should be transferred as well. This will help the recipient to assess the model's domain of validity. A quantitative assessment of confidence in the model should also be transferred.

After large physiological models are validated they will have to be reduced for routine clinical use. The development of complexity reducing model transforms is thus important.

Nir: Referring to disappointment in the results of certain modelling efforts (e.g., the global CO_2 cycle), Nir felt that we may be developing more and more sophisticated tools to produce more and more doubtful models. We should recognize various modelling objec- tives. In increasing stringency of demand these are: acquiring insight, guiding research, system prediction, control, and modi- fication. Only in the latter three cases are model calibration (or subsequent validation) necessary. But for large scale problem domains such as energy, environment, etc., we are far from having reliable calibration procedures. Indeed, the more error in the calibration data, the more likely is the model to pass statistical significance tests. Perhaps we should focus on the first two objectives which do not require difficult calibration but are use- ful in themselves.

Dayan: In economic theory, the pricing mechanism is understood to bring about better resource allocation. The lack of a pricing policy in the utilization of modelling and computer resources can lead to waste of these resources. We need to be more sensitive to the utilities of the various components of systems analysis in con- tributing to better decision making. In addition, the modelling process is fraught with risk and uncertainty. So what is the value of providing the user with sophisticated modelling tools unless their expected benefits exceeded their expected cost? A clear policy of pricing computer time and modelling effort would lead to a more balanced array of software tools, an array better tailored to meet the user's needs as expressed by his or her willingness to pay.

513

Oren: Responding, Oren summarized Dayan's remarks in the form of a
         question: What is the cost of modelling? Oren's rhetorical
         response was: What is the cost of not modelling? In today's
         complex problem areas, models are a prerequisite to any form of
         rational decision making. What is needed is the development of
         advanced modelling methodologies to greatly enhance, and make
         reliable, the modelling process.

Dayan: In response to Oren, I agree that there may be a cost of not
         modelling. However, this is the same as saying that the expected
         benefits of modelling were in fact found to exceed the expected
         costs. What I would like to urge is that adequate cost/benefit
         evaluations be made before such proposed developments are under-
         taken.

Cellier: Cellier felt that this discussion was somewhat wrong in that it
         took just the economical point of view into account. It is simply
         human to raise questions as they stand, even though answering may
         not immediately pay off. We would not sit together and discuss
         modelling here at all, if the human attitude would have been dif-
         ferent throughout the centuries.

Vansteenkiste:
         Experience with environmental problems has uncovered the
         following weak points in current modelling methodology:
             - determining suitable forms of model structure for a given
         system and rejecting unsuitable forms is not easily carried out;
         methods for formulating the structure in a systematic way should
         be developed
             - Interfaces are not well established between the modeller and
         the experimenter in collecting data, and between the modeller and
         the model user in adhering to design objectives.

         If any doubt the necessity for developing better methodologies,
         they may refer to a poll of participants at a recent meeting on
         the land, air and water uses of the environment (Proceedings of the
         IFIP Working Conference 1977, Ghent, Belgium, Ed. G. C. Vansteen-
         kiste, North Holland). Among the recommendations, summarizing the
         opinions of participants:
             - the nature of the model should be better suited to the nature
         of the problem
             - physical reasoning and black box approaches should be combined
             - there is a great need for a standardized methodology in
         systems modelling.

Szekely: Szekely asserted that a revolutionary new concept was needed to
         break the current impasse. He had been working on such a concept
         called "heterocategorical logic" but time was too short to describe
         it here. He would be glad to send reprints to anyone requesting
         them.

Dayan: Referring to the discussion of integrated modelling systems (see
         papers by Oren, Jones and Zeigler in this volume), Dayan ques-
         tioned the value of such integrated systems in domains like energy
         and economics. In these domains, given the environmental and data
         uncertainties and complexities, submodels were already of doubtful

reliability so that combining submodels would lead to even less
reliability. In any case, what would be the benefit vs. cost of
having one giant model? We need to examine the tradeoffs involved.

Zeigler: Denying that the goal of integrated modelling was to build giant
models, Zeigler stressed that to the contrary, the emphasis was to
be placed on matching models to objectives. Computer assistance
is needed to help formulate objectives, to retrieve existing models
relevant to these objectives, to simplify model components to the
level required by the objectives, etc. All this to avoid giant,
everything included, models.

Berman: Sharpening Zeigler's response, Berman pointed out that large models
nevertheless have the utility of enabling the testing of assumptions made in constructing simpler models.

Kleijnen: Picking up the utility of modelling theme, Kleijnen suggested
that experiments should be done (through business games) in which
managers are exposed to making decisions with and without models.
He felt that models should be viewed as mapping raw data into
useful information.

Nir: Nir pointed out that modelling is usually a minor part of the
overall expense. On the other hand, data collection is often very
expensive and modelling, particularly sensitivity analysis, can
guide the acquisition of significant data. However, care must be
taken not to transform good data into bad models.

Odess: Odess distinguished two groups of workers: the problem havers,
who are looking for tools to help solve their particular problems,
and the problem seekers, who are developing tools and looking for
problem havers to use them. This symposium is oriented around the
latter individuals but it may be that the best approach is for the
problem havers in each field to develop the tools most appropriate
to the field. The reason is that in seeking general validity one
usually has to sacrifice efficiency and the total cost of a general
purpose tool when specialized may be much higher than a special
purpose tool equivalent to it in a restricted domain.

Oren: Replying to Odess, Oren pointed out that especially in modelling
and simulation, there are so many aspects common to the different
fields that it is a waste not to take advantage of this. Moreover,
awareness of the general concepts may open up whole new vistas not
accessible from a limited point of view.

Halfon: Ecosystem modelling is an area where traditional engineering tools
do not apply directly. This offers an opportunity for tool
generalization.

Bandler: Commenting on specificity versus generality, Bandler was reminded
          of a truism due to Eysenck: "If we make up an ad hoc hypothesis
          for every new case . . . then we shall never go beyond the
          present position where we can explain everything and predict
          nothing".

Elzas: In concluding the session, Elzas offered the following thought:
          Icarus had a (bad) model of flying birds. He was also the first
          human who ever tried to fly.

PARTICIPANTS ADDRESSES

Mones Berman
Laboratory of Theoretical Biology
DCBD, NCI
National Institute of Health
Bethesda, Maryland 20014
U. S. A.

Aharon Nir
Isotopes Department
Weizmann Institute of Science
Rehovot
Israel

David Dayan
Bell Laboratories
Room 2B 509
Holmdel, New Jersey 07747
U. S. A.

D.L. Szekely
Optarbrain Ltd.
P.O. Box 1364
Jerusalem
Israel

METHODOLOGY OF A LARGE-SCALE COMPREHENSIVE ECONOMIC MODEL

Vladimir Simunek
Department of Economics
Kent State University
Kent, Ohio 44242
U.S.A.

    descriptors: Application: Socio Economic; Model, Large Scale;

A PARADIGM FOR SIMULATION IN INTER-DISCIPLINARY SYSTEMS

D. Fairburn
Miami University
Oxford, Ohio 45056
U.S.A.

    descriptors: Methodology, Modelling;

AN INTERACTIVE MODEL-STRUCTURING PROGRAM FOR COACHING A PANEL OF EXPERTS

Judea Pearl
School of Engineering and Applied Science
University of California
Los Angeles, CA 90024
U.S.A.

    descriptors: Application: Policy making; Interactive Mode;

MODELLING OF SOCIO-ECONOMIC SYSTEMS — METHODICAL REQUIREMENTS FOR FORCASTING STUDIES

Werner Prautsch
Technical University
Berlin
W.Germany

    descriptors: Methodology, Modelling; Aggregation;

STRUCTURAL MODELS WITH DISCRETE EXOGENOUS EVENTS AS AN AID TO FORESEE THE UNFORCASTED

Lucien A. Gerardin
THOMSON
49bis Avenue Hoche
75362 Paris Cedex 08
France

    descriptors: Application: Socio Economic;

STRUCTURED SIMULATION PROGRAMMING

Jacob Palme
Swedish National Defense Research Institute
Stockholm
Sweden

    descriptors: Methodology: Software Design; SIMULA; Simulation Language, General/Special Purpose;

MULTICLASS QUEUEING NETWORKS: TWO DIVERSE APPLICATIONS

A. Kzresinski*, P. Kritzinger*, R. Reinecke** and F. Viviers**
*Department of Computer Science
**Department of Industrial Engineering
University of Stellenbosch,
7600 Stellenbosch
S.Africa

   descriptors:

MODELLING PROBLEMS IN PHYSIOLOGICAL SYSTEMS

Mones Berman
Laboratory of Theoretical Biology, DCBD, NCI
National Institute of Health
Bethesda, MD 20014
U.S.A.

   descriptors: Application: Biological; Methodology, Parameter Identification

A METHODOLOGY FOR MODELLING GENERAL DYNAMICAL SYSTEMS

G. Adomian
Center for Applied Mathematics
University of Georgia
Athens, Georgia 30602
U.S.A.

   descriptors: Model, Stochastic Differential Equation;
                  Solution Technique: Inversion of Stochastic Operator;

AN IMPLEMENTATION OF FUZZY SETS THEORY: THE L.P.L. LANGUAGE

J.M. Adamo
Universite Claude Bernard
43, Boulevard du 11 Novembre 1918
69621 Villeurbanne
France

   descriptors: Model, Fuzzy;
                  Model Description Language;

A SPACECRAFT AND TRACKING STATION SIMULATOR FOR GROUND DATA SYSTEMS CHECKOUT

Fred Lesh, Dennis Wittman
MTS Jet Propulsion Laboratory
California Institute of Technology
Pasadena, CA 91109
U.S.A.

   descriptors: Application: Spacecraft Tracking;
                  Model, Large Scale;

THE IAS-SYSTEM (INTERACTIVE-SIMULATION SYSTEM)

Klaus Plasser
Institute for Advanced Studies
Stumpergasse 56
A-1060 Vienna
Austria

   descriptors: Software Tools for Modelling and Simulation; Man Made Interface;
                  Interactive Mode; Model/Data Base;

SYSTEMS SIMULATION AND OPTIMIZATION IN THE PRESENCE OF NONLINEAR CONSTRAINTS

Joseph S. Vogel
IBM, General Guisan-Quai 26
P.O.Box 8022
Zurich
Switzerland

    descriptors: Solution Technique: Optimization;
                       Application: Nonlinear Electrical Circuits; CSMP

THE SARA SYSTEM FOR COMPUTER ARCHITECTURE DESIGN AND MODELLING

Gerald Estrin
University of California
Los Angeles, CA 90024
U.S.A.

    descriptors: Methodology, Modelling/Design;
                       Application: Computer Systems; Methodology, Software Design;
                       Model, Petri Net;

MODELLING CONCEPTS FOR UNIFYING PERFORMANCE AND RELIABILITY EVALUATION

John F. Meyer
The University of Michigan
Ann Arbor, Michigan 48109
U.S.A.

    descriptors: Application: Computer Systems; Acceptability of Design Wrt
                       Performance and Reliability;
                       Methodology, Modelling/Design; System; Theory,

STRUCTURED DESIGN IN INTERACTIVE SIMULATION SYSTEMS

Arie A. Kaufman, Michael Z. Hanani
Department of Mathematics and the Computation Center
Ben-Gurion University of the Negev
Beer Sheva
Israel

    descriptors: Methodology, Software Design;
                       Application: Traffic, Intersection;

OPTIMAL ALLOCATION OF RESOURCES USING MULTIDIMENSIONAL SEARCH TECHNIQUES

David Birnbaum, Rolan Doray
Naval Air Development Center
Warminster, PA 18974
U.S.A.

    descriptors: Solution Technique: Optimization;
                       Application: Sensor Placement;

ANALYTICAL AND SIMULATION MODELLING OF AN AUTOMATIC TELEPHONE EXCHANGE

C. Barret, M. Gougand, M. Schneider
Universite de Clermont
Aubiere
France

    descriptors: Model, Queueing:
                     Application: Traffic, Intersection;

A METHODOLOGY TO GENERATE FAMILIES OF COMPUTER SIMULATION PROGRAMS

S. Brandi, T. Pedrott
Olivetti R. & D.
Ivrea (TO)
Italy

    descriptors: Methodology: Software Design;   SIMULA;
                     Application: Computer System;

LINKING OR-MODELS IN A SYSTEMS FRAMEWORK

Christer Carlsson
Abo Swedish University School of Economics
Henriksgaten 7, 20500
Abo 50
Finland

    descriptors: System Theory; Solution Technique: Optimization;
                     Model Manipulation: Simplification;

REGULATION OF DISTURBED SELF-PRODUCING CELL SYSTEMS

Werner Duchting
Department of Electrical Engineering
University of Siegen
Siegen
W.Germany

    descriptors: Application: Biological; Model, Cellular Space;
                     Simulation Type: Discrete Event;

MISS — A USER-ORIENTED SIMULATION SYSTEM FOR BIOLOGICAL AND CHEMICAL PROCESSES
IN RESEARCH AND EDUCATION

B.A. Gottwald
Faculty of Biology
University of Freiburg
Schanzlestrasse 1
D-7800, Freiburg
W.Germany

    descriptors: Application: Biological; Simulation Language, Continuous;
                     Man-Machine Interface: Interactive Mode;

NELSON — A PACKAGE FOR COMBINED DISCRETE/CONTINUOUS SIMULATION IN A MINICOMPUTER
ENVIRONMENT

Anders Hedin
Royal Institute of Technology
Stockholm 70
Sweden

    descriptors: Simulation Language;

SYSTEM MODELLING AND FEEDBACK

Diego Bricio Hernandez
Universidad Autonoma
Iztapalapa
Mexico

   descriptors: System Theory;   Stochastic Model;

DYNAMIC PROCESS MODELS IN BEHAVIORAL SCIENCE

Rene Hirsig*, Amos S. Cohen**
*University of Zurich
**Swiss Federal Institute of Technology
Zurich
Switzerland

   descriptors: Model: Behaviorally Anticipatory; Methodology, Parameter
                  Identification;

A SOFTWARE SYSTEM FOR CHECKING THE VALIDITY OF MODEL SIMPLIFICATIONS

Arieh Hopfeld, Bernard P. Zeigler
Department of Applied Mathematics
The Weizmann Institute of Science
Rehovot, Israel

   descriptors: Model Manipulation: Comparison of models;
                  Acceptability of Model; Methodology For;

SOFTWARE ASPECTS OF AN AERIAL REFUELING SIMULATION

Samuel J. Rosengarten
ASD Computer Center
Wright Patterson AFB
Ohio 45433
U.S.A.

   descriptors: Application: Aircraft; Model, Large Scale;

WHAT ARE THE PROPERTIES OF A SYSTEM THAT ARE MORE THAN ITS CONSTITUENT
PARTS ADD UP TO?

E.C. Saxon*, H.F. Tibbals**
*Departmnet of Anthropology
**Computer Unit
University of Durham
Ireland

   descriptors: System Theory;

DIRECT SEARCH FOR OPTIMAL PARAMETERS WITHIN SIMULATION MODELS

H.P. Schwefel
P.O.Box 1913,
D-5170 Julich
W.Germany

   descriptors: Solution Technique: Optimization;

ANALYSIS OF BLOOD FLOW MEASUREMENTS BY COMARTMENTAL MODELLING

S. Shalev, S. Sasson, C. Chaimovitz
Department of Biomedical Engineering
Technion, Haifa
Israel

   descriptors: Application: Biological;
                      Methodology, Parameter Identification;

MODELLING METHODOLOGY OF PSYCHOSOCIAL SYSTEMS WITH ONE APPLICATION:
"SELF-DESTRUCTION OF THE PERFECT DEMOCRACY"

R. Starkerman
University of New Brunswick
Fredricton, New Brunswick
Canada

   descriptors: Application: Social Systems;

INTERACTIVE GRAPHICAL NETWORK SIMULATION WITH MASON GRAPHS

K. Waldschmidt, D. Tavangarian
Universitat Dortmund
Arbeitsgebeit Schaltungen der Datenverarbeitung
Postfach 50 05 00
4600 Dortmun 50
W.Germany

   descriptors: Interactive Mode; Model, Coupled Linear/Nonlinear

Philosophy of Science

The Structure of Scientific Theories, F. Suppe (Ed.) 2nd Edition University of Illinois Press, Urbana, 1977

In his afterword (page 682-728), Suppe surveys the major trends in recent philosophy of science. The undermining of both the oversimplified account of logical positivism and the irrationalistic emphasis of Kuhn and others, has left a challenging vacuum in its wake. Current work aims at characterizing coherent problem domains, rational direction of research programs and the variety of patterns of reasoning employed by scientists. As a realistic account of the methodology of theory formation and its justification, this work has implications for the development of computer assisted modelling.

Systems Theory

An Approach to General Systems Theory, George J. Klir, Van Nostrand Reinhold, New York, 1969

Perhaps the first work to suggest that systems modelling in the large could be understood, rationalized and computer assisted within a general systems framework. Klir and others have subsequently expanded on the methodological proposals suggested in the book (see papers by Klir and Broekstra in this volume). An earlier book, Cybernetic Modelling, by Klir and Valach remains a source of suggestive ideas on modelling methodology.

Applied General Systems Research: Recent Developments and Trends. G. J. Klir (Ed.) Plenum Press, (227 West 17th Street, New York, NY 10011), 1978.

The book contains review articles on the state of the art of general systems research and reports on new research findings. The three major themes developed, in separate sections, are: conceptual and methodological foundations of general systems research, advances of general systems research in the biological sciences, and the impact of general systems research on the social sciences. A fourth section is devoted to critical views of the field from an external perspective. Each of the first three sections starts with an extensive survey article, which is followed by papers relevant to the major theme. A study showing the major influences among the contributors on their view of general systems research is added as an appendix.

Systems Design

Systems Engineering Methodology for Interdisciplinary Teams, A. W. Wymore, Wiley, N.Y., 1976

Sets out a formal framework for systems design which explicitly incorporates most of its characteristic features. Design is formulated as the synthesis of a system from component systems of a given technology which exhibits both the specified behavior and is maximal in a given merit ordering. While useful as a neutral language for mediating interdisciplinary cooperation, the formalism is unlikely to be adopted without extensive computer assistance.

Software Engineering Education: Needs and Objectives, Anthony I. Wasserman and Peter Freeman (Eds.) Springer-Verlag, New York, 1976

Contains a variety of views, academic and industrial, on the nature of software engineering and the training of software engineers. Included is a list of milestones in the literature on program design methodology.

Reflections on Computer Aids to Design and Architecture, N. Negroponte (Ed.) Petrocelli/Charter, New York 1975

This book is a comprehensive anthology covering the development and practical use of computer methods to aid architectural and urban design.

Each of the contributors to the book is an outstanding personality in the field, selected for his present activity or for his historical importance.

The text is an assemblage of a broad range of positions and attitudes ranging from the rational epistemological approach to modelling and design problems all the way to holistic views on the design process.

Trends and results in research and practice, both utter failures and renowned successes, are discussed by the main protagonists in the field.

The book can be recommended as essential reading to anyone interested in the philosophy and methods that have been part of the founding work of model-based computer aided design.

Theory of Modelling

Theory of Modelling and Simulation, B. P. Zeigler, Wiley, N.Y., 1976

Provides a unifying framework, within systems theory, for the modelling and simulation enterprise. The elements (real system, experimental frame, base model, lumped model and computer) are characterized as are their interrelations (modelling, simulation). Zeigler systemizes the various model formalisms (differential equation, discrete event, etc.) and equivalence concepts into a hierarchy of system specifications and relations. Applications are made to such problems as model simplification, structure identification and a simulation program verification. The framework provides a basis for the design of advanced simulation languages and systems (see especially papers by Zeigler, Oren, and Jones in this volume).

Fundamentals of Measurement and Representation of Natural Systems, R. Rosen North-Holland (52 Vanderbilt Ave., New York, NY 10017), 1978.

This book represents an in-depth investigation of an important approach to systems complexity. It examines the relationships which exist between different descriptions (i.e., between different modes of abstraction) and the manner in which different kinds of descriptions can be combined to obtain more comprehensive ones.

The basic unit of all system description, and of all observation, is an individual observable. The author provides in this study a comprehensive theory of observables, and the description arising from them. The theory is then applicable to any situation in which objects of interest are named or labelled by definite mapping processes, measurement in physics, pattern recognition, classification or discrimination. In the process, relationships are established between the different languages which have been used to deal with such problems in specific situations: symmetry, invariance, similarity, homeostasis. The resulting formalism is in fact a general theory of modelling: two systems model each other to the extent that one can be substituted for another in such a way that some measure of system behavior is left invariant. The study of this kind of situation provides the major thrust of the book.

The basic problem of reductionism is studied in some detail. The author shows that the reductionistic hypothesis - that there is a universal mode of system analysis from which all others can be derived - is not necessarily a valid hypothesis in the natural world.

The Role of Systems Methodology in Social Science Research, R. Cavallo, Martinus Nijhoff (160 Old Derby St., Hingham, Mass., 02043), 1978.

While the conceptual significance of general systems approaches have long been recognized with respect to research in the social sciences, its transformation into working approaches has remained relatively undeveloped. This book develops and presents an integrated conceptual and operational general systems methodological framework which is directed to the special needs involved in the investigation of complex social and humanistic systems.

The framework provides specific procedures and working methods for the investigation of social systems but, more importantly, it develops an organized and well-formulated problem-solving approach which integrates these and other methods. The integration serves to effectively associate the working methods and operational tools with fundamental systems concepts such as behavior, state, structure, simplification, complexity and decomposition. These concepts are then associated with their counterparts in specific 'real-world' social systems.

The book also presents an extensive example which demonstrates the utility of the framework for social science research. In this example the ability of the framework - and thus of results in general systems research - to extend and augment the investigative capabilities of researchers in specific areas is well-demonstrated.

Modelling and Simulation - Critical Analyses

Simulation Modelling of Environmental Problems, F. N. Frenkiel and D. W. Goodall (Eds.), Wiley, England, 1978

Prepared by the subcommittee on Simulation Modelling of SCOPE (Scientific Committee on Problems of the Environment), this report surveys the objectives, methods and obstacles in model building for environmental problems. It is especially concerned with the place of modelling within the broader problem solving context and thus with the interaction between the modeller and the policy decision maker. The problems it delineates for modelling in the environmental context are quite general and stand as a challenge to the developers of multi-faceted system modelling methodology.

Critical Evaluation of Systems Analysis in Ecosystems Research and Management

G. W. Arnold and C. T. de Wit (Eds), Center for Agricultural Publishing and Documentation, Wagening, The Netherlands, 1976

Outlines the multi-level approach to modelling and simulation taken by the Wageningen Theoretical Production Ecology Group, an influential force in agroecosystem modelling. Contains a review of Grassland models developed in various countries (ca 1975) qualified by level of aggregation, scope of appli- cation, and domain of validity. A note of pessimism concerning the value of large simulation models, sounded by one author, is balanced by a range of suggested methodological approaches offered by other contribuotrs including hierarchical modelling, integration of modelling and field research, and develop- ment of model/data banks.

Global Simulation Models, John Clark and Sam Cole, John Wiley and Sons, London, 1975

Discusses the potential contribution and limitations of global simulation models of which the MIT models, including that in The Limits to Growth (see below), are a subset. These models are necessary for rational decision making on a world wide (global) scale but there are intrinsic difficulties in developing credible models at this level. The difficulties, for example, pertain to acquisition of data, level of aggregation, and opportunity for validation, which are especially troublesome in this global context. Such difficulties, well brought out in this book, stand as another challenge to methodological development. Especially of interest to the methodologist is the chapter comparing the state of the art in global modeling with that in such areas as macroeconomics, ecology, urban systems, and other socio-economic systems. Another interesting point raised concerns the extent to which the "intrusion" of the individual will (the human factor!) "disturbs" predictions of global scale responses to policy interventions.

The Limits to Growth. (A report for the Club of Rome project on the predicament on Mankind) Dennis Meadows, Universe Books, New York, 1972.

Although many scientists, coming from widely diverse disciplines, have (justly) criticized this book and the underlying modelling approach, it has opened a new era in modelling and simulation by giving wide publicity to techniques mainly discussed before in the seclusion of the ivory towers of science and technology. Although many of the assumptions in the book have been shown to be in error, the model used has been proven to be extremely coarse, and the chosen simulation vehicle is known to be numerically weak, this work can have a healthy influence on the critical reader.

Rather than clearly showing the limits to growth, it provides provocative evi- dence to the limitations of modelling and simulation.

Methods for Solving Engineering Problems (using analog computers) Leon and Arnold Levine, 1964 McGraw-Hill Book Inc., Series in Information Processing and Computers.

This book describes how a computer may be used as a simulation tool to assist the engineer and scientist in tackling the experimentation side of modelling problems.

The book emphasizes problem solving techniques in connection with (continuous) system simulation rather than specific hardware-bound approaches, and places special attention on an understanding of the inherent mathematical theories.

It was one of the first books to include in this context chapters on:
- Estimation and test of hypotheses;
- Experimental design and detection of computational errors;
- Application of statistical techniques to study simulation results.

General

Computer Power and Human Reason, J. Weizenbaum, Freeman & Co., San Francisco, 1976

Perhaps the first comprehensive work written by an outstanding computer scientist on the limitations of computers, and associated techniques, viewed as tools to assist human decision making and intelligence.

Especially the chapters devoted to "Science and the compulsive programmer", "Theories and models" and "Computer models in psychology" are worth reading and are a good basis for an exercise in modesty for the computer scientist and the modeller. As a realistic appraisal of the possible future achievements in modelling and artificial intelligence, this work clearly spells out the humane boundary conditions within which future developments in the field of computer usage, including modelling, should take place.

Systems Models for Decision Making, Sharif, N. and P. Adulbhan (Eds.), Asian Institute of Technology, Bangkok, 1978.

The book is based on a set of invited papers presented as state-of-the-art lectures during the International Conference on Systems Modelling in Developing Countries held at the Asian Institute of Technology in Bangkok, in May 1978. It contains a chapter on "The General Systems Research Movement."

Recent Publications Related to Methodology -

Basic and Applied General Systems Research: A Bibliography, G.J. Klir and G. Rogers (Ed.) in cooperation with R.G. Gesyps. Available from SAT, SUNY - Binghamton, N.Y. (see especially index terms: systems methodology, modelling, simulation; yearly addenda also available.)

Theoretical Systems Ecology, E. Halfon (Ed.), Academic Press, 1979.

Basic and Applied General Systems Research (Book series) G.J. Klir (Ed.), North-Holland, N.Y.

Frontiers in Systems Research: Implications for the Social Sciences (Book series) G.J. Klir (Ed.), Martinus Nijhoff, Holland.

Dynamic Stochastic Models From Empirical Data, R.L. Kashyap, A. Ramachandra Rao, Academic Press, 1976.

Large Scale Models for Policy Evaluation, P.W. House, J. McLeod, Wiley, 1977.

Periodicals Sometimes Offering Methodological Articles -

Simulation (especially, Frontiers in Systems Modelling, a series edited by B.P. Zeigler)

Simuletter
International Journal of General Systems
International Journal of Systems Science

International Journal of Man Machine Studies
IEEE Transactions
IMACS Journal
Applied Mathematical Modelling
General Systems Yearbook
Annals of Systems Research
Mathematical System Theory

Also of relevance:

ACM Special Interest Group on Data Bases
IEEE Transactions on Software Engineering